Molecular Principles
of Animal Development

Molecular Principles of Animal Development

Alfonso Martinez Arias

Department of Genetics, University of Cambridge

Alison Stewart

Public Health Genetics Unit, Cambridge

OXFORD

UNIVERSITY PRESS

OXFORD

UNIVERSITY PRESS

Great Clarendon Street, Oxford OX2 6DP

Oxford University Press is a department of the University of Oxford.
It furthers the University's objective of excellence in research, scholarship,
and education by publishing worldwide in

Oxford New York

Auckland Bangkok Buenos Aires Cape Town Chennai
Dar es Salaam Delhi Hong Kong Istanbul Karachi Kolkata
Kuala Lumpur Madrid Melbourne Mexico City Mumbai Nairobi
São Paulo Shanghai Singapore Taipei Tokyo Toronto

with associated companies in Berlin

Oxford is a registered trade mark of Oxford University Press
in the UK and in certain other countries

Published in the United States
by Oxford University Press Inc., New York

Library of Congress Cataloging in Publication Data
(Data applied for)

ISBN 0–19–879284–0

Typeset by SNP Best-set Typesetter Ltd., Hong Kong
Printed in Spain
by Book Print, S.L., Barcelona

Preface

The aim of developmental biology is to understand how animals and plants are put together, a process in which a cell gives rise to many different cells that become organized into functional structures and assembled into whole organisms. This involves a beautiful and reproducible choreography of cells dividing, moving and changing shape in a coordinated manner over time—a cellular ballet that is great fun both to observe and to describe. Traditionally these descriptions have fallen within the realm of embryology. However, developmental biology is not embryology. Whereas embryology describes the way a single cell gives rise to an organism, developmental biology adds experimentation and causal analysis to this description. Where embryology aims to tell us what happens during the development of an organism, developmental biology wants to know how and why it happens. Both disciplines have the cell as their unit but whereas for an embryologist the cell is the element of description, for the developmental biologist it is an agent of action.

This is a book about developmental biology. For many years, the teaching of this subject has focused on the discussion of particular embryological events that are common to all organisms, for example, gastrulation, neurulation, limb development. The analysis of these processes has been anchored firmly in an embryological framework. Although this is understandable—it is difficult to talk about what cells are doing in an embryo without referring to the spatial and temporal context—the application of genetic and molecular analyses to the old questions of developmental biology has changed the way we perceive and discuss these questions. Underlying the behaviour of cells there are molecules; moreover, it appears that these molecules and the processes they mediate are conserved across organisms. The molecular principles that govern the generation of body plans are conserved: the body of a fruit fly is organized in basically the same way as the body of a mouse, and the molecular processes that underlie this organization are the same in both organisms. This molecular conservation is even more far-reaching in that the same groups of molecules organize the building of different organs and structures in different organisms. These discoveries have brought about a revolution in the way we describe and analyse developmental processes.

In this book we explore the idea that developmental biology can be discussed from the point of view of the molecules and the processes that drive the shaping of embryos, rather than from the point of view of the organism or its constituent parts. We try to show that organisms are logical, and sometimes not so logical, products of molecular interactions within cellular contexts, and that the magic of the regulative events that are revealed in many experimental systems simply reflects responses to and activities of molecular networks. We bring in genetics as a tool to unravel and organize the molecular information that is brought into play during development. The genetic and molecular analysis of developmental events has changed fundamentally the methods, the language and often the aims of developmental biology, and we suggest that these changes should perhaps now be reflected in the way the subject is taught.

The book is organized into three broad sections that build upon each other. After an introduction that discusses the changes in perspective within a historical framework, we embark on a basic course on the molecular biology of developmental systems (Chapters 2–5). In these chapters we discuss transcriptional programs, signalling molecules and receptors, and the way these systems interact to form networks that process molecular information. We then move on, in the second section of the book, to see the information-processing networks at work in a cellular context where they carry out cellular functions, such as cell adhesion, polarity and movement (Chapter 6), cell division or cell death (Chapter 7), that are essential to al developmental processes. We also see how these molecular networks can be used to generate lineages of different cells (Chapter 8) as well as signalling systems that are used to set up differences between cells (Chapter 9). These first nine chapters form the core of the book. The last three chapters can be viewed as a selection of special topics. Here we discuss different levels of spatial organization mediated by the networks and processes discussed in the first nine chapters. We see how different cell types can be generated (Chapter 10), how different cell types can generate simple patterns in two dimensional sheets (Chapter 11) and how they organise themselves to generate complex three dimensional structures (Chapter 12).

In this book we have tried to illustrate a way of thinking rather than attempting to cover every single detail of the issues we discuss. This is particularly clear in the last three chapters, where we have selected some topics for discussion and left out many that could have served the same purpose.

The book is aimed, primarily, at higher-level undergraduate and lower-level graduate courses in developmental biology. We also hope to interest cell biologists and biochemists in the subject. There are several other excellent introductory textbooks about developmental biology; we see our book not as an alternative but as a companion to them.

At the end of every chapter we have included a few references for extra reading. The amount of information in all fields of biology is phenomenal, and growing, making the task of selecting a 'relevant' reference all the more difficult. Wherever possible we have tried to suggest papers that discuss the topic of the chapter in a particularly succinct way or that raise interesting questions or ways to look to the future. We have not included primary research papers in these reading lists, but rather in the legends to many figures where we present the experimental evidence underlying the ideas we discuss. The reader is encouraged to go to these papers for a more detailed understanding of particular experiments.

Acknowledgements

The idea for this book began to emerge several years ago in conversations with M. Bate about the way research in genetics and molecular biology was changing the fabric of an old and well established tradition. Several generations of Cambridge undergraduates taking the Part II developmental biology course during the 1990s were, unknowingly, the 'guinea pigs' for experiments in new approaches to teaching developmental biology that were eventually consolidated into this book.

Many people have contributed to this work in a variety of ways. We thank in particular M. Akam, M. Bate, M. Baylies, K. Brennan, J. Castelli, G. Craig, P. Gardner, M. Ruiz Gomez, D. Gordon, J. Gurdon, J. C. Izpisua Belmonte, S. Kerridge, T. Klein, N. Lawrence, C. Sharpe, M. Taylor and V. Zecchini for their support, discussions and ideas at various points throughout the project. The book began its journey to publication with the enthusiasm and interest of E. Browning (OUP) and has been seen through to completion with the help and patience of J. Crowe and J. Grandidge.

The chapters have been read by several colleagues who have corrected and commented upon them. Our thanks for this to P. B. Armstrong, M. Bate, D. J. R. Evans, T. P. Fleming, B. Garcia Fernandez, M. Freeman, D. Glover, L. Haynes, A. Howard, J. Gurdon, J. Kaltschmidt, E. Knust, H. Krider, T. Kouzarides, P. Martin, J. Miyan, R. Roy, C. Sackerson, C. Sharpe, F. Schweisguth, T. Tollefsbol, J. R. S. Whittle, A. W. Wiens and E. Wilder. Any remaining mistakes are our own.

Finally we want to thank our families who bore with patience and understanding our commitment to this endeavour and the toll it took upon our time and energy. As all who attempt such projects will recognize, a task that began as one we thought could be fitted in readily among our other activities inevitably took on a life of its own.

Cambridge   A. M. A.
January 2002   A. S.

Contents

Chapter 1

Introduction: Towards a molecular analysis of development

During embryogenesis a multicellular organism emerges from a single cell which divides a finite number of times, in a spatially and temporally ordered way, to generate an ensemble of different cells (Fig. 1.1). The aim of developmental biology is to understand how this occurs. What are the elements that define the different kinds of cells? Where do the instructions for the process lie? What is the language of those instructions?

Classical approaches to these questions are based on describing the embryonic development of various animals. Moving to a finer level of resolution, they emphasize the central role that cells play in development, describe the emergence of complexity in different organisms in terms of what cells are doing, and discuss the strategies used by groups of cells to guide and coordinate developmental processes. Implicit in this approach is the idea that many experimental observations can be explained by the activities of specialized groups of cells, such as 'organizers', which can instruct the development of other cells. In more recent years, as the genetic and molecular mechanisms underlying cellular behaviour have begun to be elucidated, this molecular information has been added to the analysis.

Here we shall approach development from a different standpoint, one in which the molecules and the genes that encode them, rather than the organism, take centre stage. We do this in the belief that the logic of development can only be appreciated by beginning with the genetic programs that underlie development. These programs both encode and are executed by molecular networks of proteins operating within and between cells. The protein networks create cells and direct basic cellular behaviours, or 'routines', such as cell adhesion, division or movement, which are deployed during the development of the organized cellular assemblies that make up tissues, organs and ultimately whole animals (Fig. 1.2).

This analysis does not, however, imply a simple linear transformation of genetic information into shape and size, mediated by proteins, but a progressive unfolding of information to create different levels of complexity and organization (Fig. 1.2). The unfolding is based on the activities of a small number of functional modules, organized hierarchically with increasing levels of complexity. The components of a particular module (proteins, cells, tissues) are specified by the module in the level below, but its basic organization and properties depend exclusively on its own elements and cannot be predicted from those of the module below. For example, the properties of cellular ensembles or protein networks cannot be predicted from the genetic circuits that determine their composition (Fig. 1.2).

The different modules are linked through functional relationships that enable the developmental events they mediate to function smoothly and in a regulated way (Fig. 1.3). Thus, genes and protein networks interact to create regulatory circuits that integrate the activity of these two levels. Each level lays down the molecular prerequisites for interpreting the information that instructs the next level; in turn, this level will generate a new level of information processing and will modify the one that gave rise to it. This modification may lead to a new configuration of the whole system, advancing the developmental process. For example, a new set of proteins within a cell can change its pattern of gene expression, which may determine a transformation towards a new cell type.

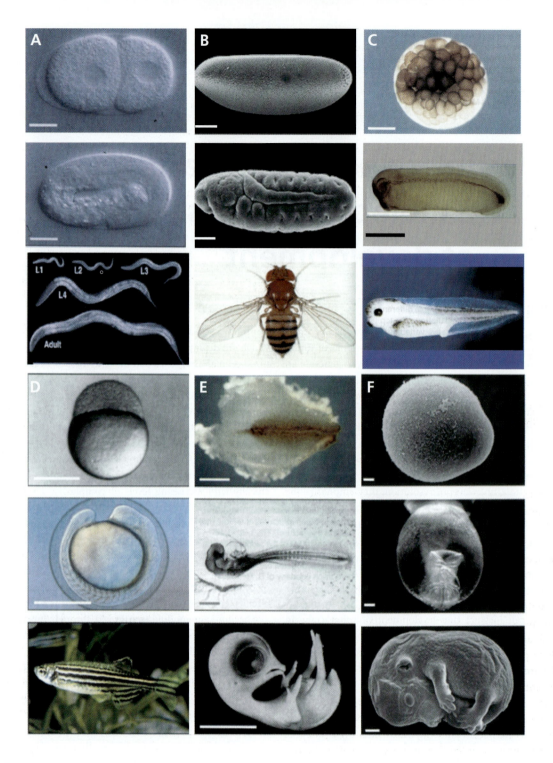

Fig. 1.1. The development of animal model systems. Three different stages of development of various animals used as developmental model systems are depicted from top to bottom in each panel. Sizes of scale bars are indicated in the legend. **(A)** Nematode *Caenorhabditis elegans*. Two-cell embryo (10 μm); larva (10 μm); various larval stages with the adult worm. **(B)** Fruit fly *Drosophila melanogaster*. Blastoderm stage (0.1 mm); extended germ band stage (0.1 mm); adult fly. **(C)** Frog *Xenopus laevis*. Blastula stage (0.5 mm); tail bud stage (1 mm); tadpole. **(D)** Zebrafish. Sphere stage with embryo on top (0.5 mm); 14 somite stage displaying the basic body plan with head to the left (0.5 mm); adult fish. **(E)** Chick. Gastrulating embryo (1 mm); 2.5 days after laying (1 mm); 8.5–9 days after laying. **(F)** Mouse. Zygote before the first cleavage (10 μm); 8 days after fertilization (0.1 mm); 14 days after fertilization. Images from Wolpert, L. (1998) *Principles of development*, Current Biology and Oxford University Press.

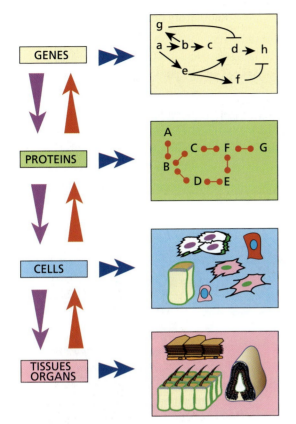

Fig. 1.2. **The hierarchical organization of biological systems.** Biological systems are organized into functional modules of different levels of complexity, which are related through the activities of their component elements. An important rule of this organization is that a particular level provides the elements and information to generate the next higher level of complexity. Genes (a, b, c . . .) encode proteins (A, B, C . . .) which organize themselves into functional networks. Some of these functional networks act back on the genes to create regulatory relationships that generate networks of gene activity. Other protein networks assemble macromolecular structures which generate cells. The different appearances or functional attributes of cells arise from their different protein composition. In the same way that proteins have properties that cannot be predicted from the structure or functioning of genes, proteins work together in complexes within the context of cells in ways that transcend the properties of individual proteins. Interactions between cells can affect the operation of the protein networks and, in this way, influence gene activity. Cells assemble themselves into tissues and organs which develop properties of their own that are not easy to predict from the properties of their individual cellular components. At the next level, tissues and organs provide the structural basis of an organism.

It follows from this view of development that it is misleading to think of the genome as embodying the 'blueprint' of an organism. This often-heard architectural analogy implies that the DNA sequence of an organism contains, in a more or less straightforward way, both the information for the materials to make an organism and the instructions for its assembly. The sequence of the genome contains some of the information to build an organism but this information is very limited. Genes encode proteins and carry information that will determine when and where these proteins are made. However, genes do not contain the information that determines how the proteins will assemble, or when, where and how will they be active. These pieces of information, essential in any 'developmental blueprint', lie in the proteins themselves, which assemble into functional and regulatory networks following instructions embodied in their three-dimensional structures (Fig. 1.2). The constituent proteins of these networks contribute to the composition and activity of cells, determining and modulating their movement, shape, information-processing ability, and patterns of division and differentiation. At the same time, the activity of the networks feeds information back to the genes that encode their elements, to create regulatory circuits of gene expression (Fig. 1.2, 1.3). In the same way that proteins can be said to be the creation of the genetic machinery, cells are the creation of the interaction between genes and proteins. Different combinations of these elements generate different cell types which will assemble into tissues and contribute to the shape and size of an organism (Fig. 1.2).

A better analogy for encapsulating the logic of development is a computer. The molecular networks can be thought of as the 'hardware' of developmental systems—a hardware made up of proteins that read and interpret different 'programs' to produce different outputs: different animals, different organs, and supracellular aggregates. The programs are written in the regulatory regions of the genome and determine the sequence of patterns of gene expression characteristic of each organism. This also means that the hardware that reads and interprets the programs is itself determined by gene expression programs.

The rationale for a different approach

The impetus for a change in approach to development has come from the explosion of information over the last twenty years as the techniques for isolating and

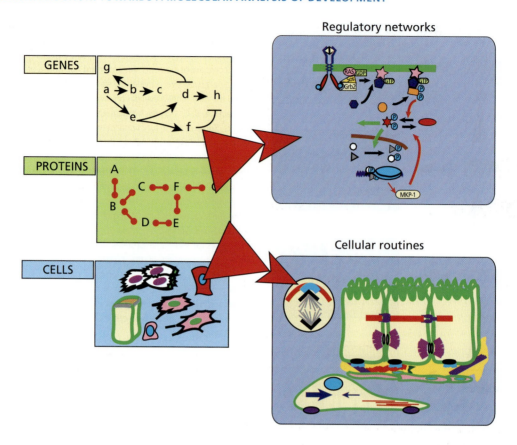

characterizing genes have been applied to the analysis of developmental processes. We now have an impressive and ever-growing list of genes and associated regulatory sequences involved in generating different cell types. We have realized that in order to understand developmental processes we need to understand the underlying genetic circuitry, that is, how the activity of one gene leads to the activity of other genes that are under its control and how these activities are modulated in space and time. We have also discovered that embryonic development and its component patterning processes rely on the activity of information-processing networks that act within and between cells. These networks are made up of proteins that function in the emission, reception, and transduction of chemical signals, and others that act as targets and effectors for these signals.

Perhaps the only justification that is really needed for moving aside from the organism-based approach to development is to point out the universal nature of these molecular mechanisms and networks. The studies of the last few years have shown that there are relatively few different

Fig. 1.3. Assembly of the components of each level into functional modules. Interactions between the different levels depicted in Fig. 1.2 generate functional modules that contribute to the building of an organism. For example, genes and proteins interact to organize regulatory networks that govern the patterns of gene activity of different cells. These patterns, in turn, determine the protein composition of those cells. The networks are usually represented as flow diagrams of interactions between different proteins and between proteins and genes. Similarly, proteins and the cells to which they contribute interact to generate cellular routines (such as adhesion, division or motility) that will participate in the construction and modulation of the large-scale cell assemblies seen in organs and tissues.

types of networks, that they are evolutionarily conserved, and that they are used both in different organisms and in different parts of the same organism. Looking at different developmental processes within the same organism, or even at the same process in different organisms, always reveals the involvement of the same signalling networks

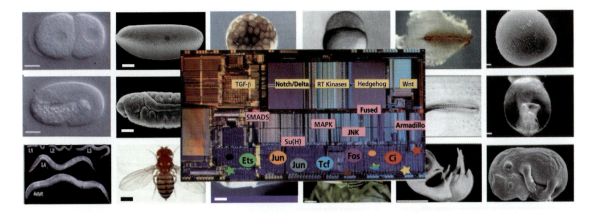

and, sometimes, related patterns of nuclear activity. What this observation suggests is that the networks act as molecular processors of cellular information (Fig. 1.4); the output from these processors is determined by the protein composition and the genetic state of the cells on which they act. The same signal input will be processed differently in, for example, a neuron or a muscle cell, because one has a 'neural' genetic and molecular environment, while the other has a 'myogenic' one.

This principle helps to explain otherwise puzzling observations such as the appearance of extra limbs when a piece of embryonic head from which the ear will develop is implanted on the flank of an amphibian embryo (Fig. 1.5). This bizarre outcome of a strange experiment was only recently explained by the discovery that the initial stages in limb development require a signal encoded by the molecule fibroblast growth factor (FGF) and that the cells that will give rise to the ear, at a certain stage of development, are a source of this growth factor. Thus, if FGF acts on cells of the head it promotes the development of anterior structures, but if released on the flank of the embryo at the appropriate time, it will trigger the development of a limb. This is because the same signal is interpreted in different ways by cells in different states.

Returning to the analogy of the computer: the hardware of the cells in the anterior head of the embryo is very similar to the hardware in the flank but they are plugged into different programs and that is why the same signal, FGF, is interpreted in such different ways as a signal to follow the developmental pathway of either an ear or a limb. The output is not specified by the signal but by the state of the cell it acts upon. This state is measured by the presence or absence of an active network capable of interpreting the signal, and its constellation of active genes. In many other

Fig. 1.4. The universal cellular microprocessor. Despite the diversity in modes of development that characterizes the animal kingdom, there is an underlying theme of molecular unity. All organisms have a similar set of proteins that can be organized into classes. These proteins are able to process information and act as elements of a microprocessor. An interesting property of this microprocessor is that it can be transferred from one organism to another and will process information according to where it is placed. If an element of one organism (e.g. in the form of a gene encoding a protein) is placed in another, the gene will be expressed, the protein made and incorporated into the routines of the host, performing the tasks and contributing to the development of the host in a host-specific manner. The background images show, from left to right, embryos of *C. elegans*, *Drosophila*, *Xenopus*, zebrafish, chick and mouse. (Images from Wolpert, L. (1998) *Principles of development*, Current Biology and Oxford University Press.)

cells of the embryo, FGF has no effect, that is, the signal cannot be interpreted or read.

A clear implication of this universal nature of the molecular networks is that, when trying to explain the mechanisms that control and execute developmental events, there might be only limited usefulness in talking in terms of model systems, whether whole organisms (e.g. the fish, the fly, the mouse) or specific organs or structures (the heart, the lung, the gut). Questions such as how to make a fly or a worm or a mouse are certainly interesting but before we can begin to set down the sequences of molecular events that underlie these processes, we need first to understand what different organisms have in common in terms of their information processing devices.

A basis for this understanding lies in appreciating the difference between mechanisms and strategies. In the context of developmental biology, mechanisms depend on the

A

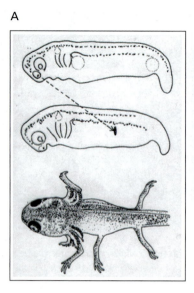

B

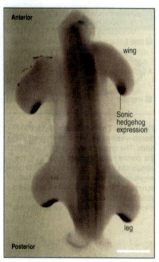

C

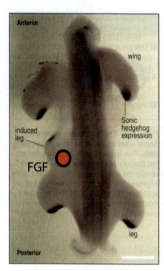

molecules that encode and execute the various functions of a cell and are, probably, finite and conserved. On the other hand, strategies are the uses that cells make of the mechanisms to generate animals, and are vastly more varied. How the activation of a cell surface receptor leads to the expression of a particular gene or to cell division, or how the transcription of sets of genes is coordinately regulated, are examples of mechanisms. The way a cell or a group of cells uses these to generate particular spatial patterns is an example of a strategy. In the example of the activities of FGF, the mechanism of action of FGF is the same in the head or in the flank of the embryo, in a chick or in a sheep. How this basic mechanism is used in the development of a particular organ or an animal, by controlling when and where the signal is released or whether the signalling system is operative, are examples of strategies.

In this book, our goal is to establish a basis for dealing with the mechanisms that lead cells to become different within groups, to organize themselves in space and time, and to give rise to the patterns that make up organisms. Because the molecular elements that govern cellular activities are conserved at some basic level it seems reasonable to assume that the problem does not require a solution for every particular case, but that each system might have something to offer that can be extrapolated to a general framework. For this reason the book will always address the issues from the point of view of the mechanism and not of the system or organism.

For the same reason, we break with the classical idea that precise descriptions of developing embryos are needed in

Fig. 1.5. Multiple functionality of single signals. (A) If a piece of tissue from the anterior region of the head of a salamander embryo is grafted onto its flank, extra limbs appear. (Modified from Balinsky, B.I. (1933). Das Extremitätenseitenfeld, seine Ausdehnung und Beschaffenheit. *Roux's Arch. EntwMech. Org.* **130**, 704–36.) **(B)** During the development of vertebrate embryos, the initiation of limb development requires fibroblast growth factor (FGF), whose activity leads to the expression of Sonic hedgehog, a protein that plays a key role in limb development. **(C)** Placing a bead impregnated with FGF on the flank of a chick embryo gives rise to an extra limb, just as transplantating the anterior head region to this position does in the case of the salamander.

order to answer the important questions. Such descriptions are measurements of the outcome of developmental processes but shed little light on the molecular elements involved or the nature of the processes they mediate. We do not think that one should dispense with such descriptions altogether. However, they only provide a framework for thinking about molecular interactions and, at best, a rudimentary language in which more mechanistic questions can be phrased.

Before embarking on an account of the molecular principles of development, it may be instructive to look briefly at the way in which the problem of embryonic development has been viewed and dealt with over the centuries. This will allow us to place the new 'molecular era' in a historical context and to introduce some classical embryological features and concepts. Although some of these classical terms are now being reinterpreted in the light of molecular information, they still form part of the language of the subject and will be referred to at many points during the book.

Beginnings

The fundamental questions of developmental biology have always fascinated human beings. How can a complex animal emerge from a simple egg? And how are the characteristics that define individuals and species transmitted from one generation to the next? In the fourth century BC, Aristotle dissected chickens' eggs at different stages after fertilization, and in describing the development of a fully formed animal from a minute speck, he recognized what later became known as the principle of 'epigenesis'—that the development of an organism is a process of growth in size combined with increasing complexity and organization. With no means of observing the microscopic sperm and ovum, however, his view of the parental contributions to the offspring was that the embryo was formed by the

shaping influence of semen acting on the blood of the mother.

It was not until the development of microscopy during the seventeenth century that it became possible to explore developmental processes at a more detailed level. By the end of that century sperm had been seen by Antonie van Leeuwenhoek. William Harvey had asserted that the maternal contribution to the fetus was an egg rather than simply blood, and the anatomist Stensen had identified the ovarian follicles as the source of mammalian ova. But the ability to see more seems also to have unleashed the full powers of the human imagination, and despite Harvey's arguments to the contrary, the notion of epigenesis was eclipsed by the concept of preformation: the idea that either the sperm or the ovum contained a fully preformed organism, and that development consisted of no more than an increase in size of this 'homunculus' in the environment of the mother's uterus (Fig. 1.6).

As the quality of microscopes improved, the claims of the preformationists became less and less tenable, and Aristotle's principle of epigenesis reasserted itself: the homunculus did not exist; instead the embryo was sketched, outlined, coloured, and fleshed out little by little. The concept of epigenesis finally gained ascendancy in the

Fig. 1.6. Preformationism. The belief developed in the seventeenth century that the sperm contained a miniature replica of a human being, the homunculus, as shown in this drawing after Nicholas Hartsoeker (1694).

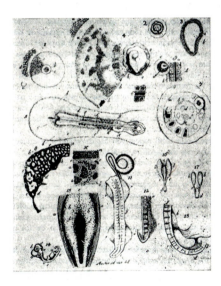

Fig. 1.7. Epigenesis. During the eighteenth century, the idea began to emerge that shape and form arise progressively during development. One of the first works to develop this concept was the 'Theoria Generationis' by Kaspar Friedrich Wolff (1759), which described different stages of chick development and the progressive emergence of form.

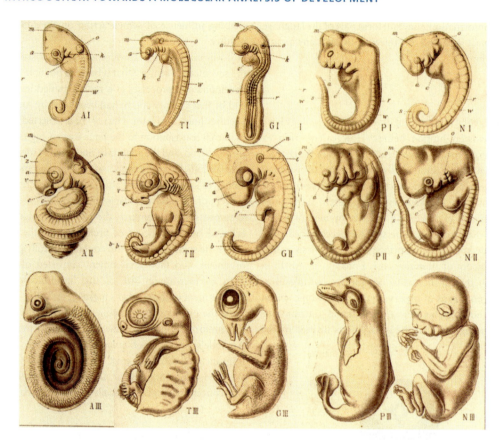

Fig. 1.8. Comparative embryology. Composite from *The evolution of man* by Ernst Haeckel (4th edn, 1903) showing the development of (from left to right) a snake, a tortoise, a chicken, a dolphin, and a monkey. In these drawings the author emphasized the similarity of all embryos during the early stages of development, something that had been documented by Karl von Baer at the beginning of the century. However, whereas Haeckel mistakenly assumed that more complex organisms recapitulate the development of the ancestral ones, von Baer correctly interpreted the progressive similarities as departures from shared stages.

closing years of the eighteenth century, with the recognition of Kaspar Friedrich Wolff's minutely observed treatise on chick development (Fig. 1.7). Embryonic development consisted of a progressive and carefully orchestrated increase in morphological complexity, synchronized with growth in size.

The heyday of descriptive embryology

By the nineteenth century, the discipline of embryology was firmly established, and embryologists were engaged in a ferment of observation and description, producing detailed accounts of the development of all manner of creatures from marine invertebrates to large, land-dwelling mammals. This century-long exercise brought with it its own set of principles, concepts, and terms, which became firmly embedded in the literature and in many cases have survived into the modern era. An important observation of this period was that, at the earliest stages of development, all animals look remarkably similar (Fig. 1.8).

Furthermore, these very early stages appeared to follow a basic, stereotyped pattern, in which the fertilized egg first divides many times to generate a mass of cells that become a hollow ball, the blastula or blastoderm (Fig. 1.9). The blastula then undergoes a series of well-defined movements and invaginations—gastrulation—that folds the ball of cells within itself and converts what was essentially one layer into three. The resulting embryonic stage, the gastrula, consists of three groups of cells, also known as germ layers, each with a characteristic position relative to the others. The term 'germ layer' refers to the observation

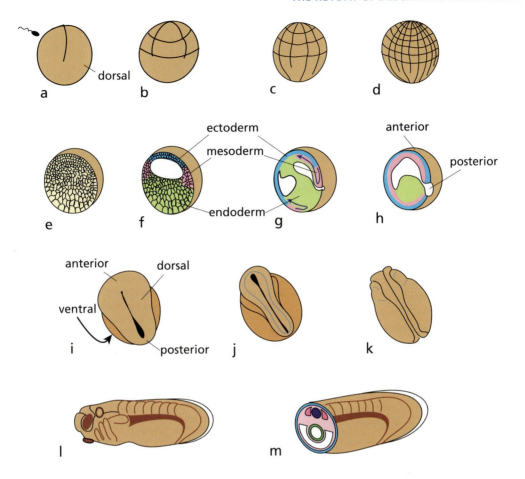

Fig. 1.9. Different stages of development of the frog *Xeno-pus laevis*. After fertilization the embryo cleaves several times (a–d) to generate a mass of cells (shown in transverse section in e–h). These cleavages are asymmetric because one side of the embryo (the vegetal side) is full of yolk, which impedes mitosis so that fewer and bigger cells are produced. After cleavage, concerted movement of the cells creates a cavity (f). Further movements of defined cell populations ensue (g) which, through the process of gastrulation, generate three layers of cells: the endoderm (green), the mesoderm (pink), and the ectoderm (blue), derived from different regions of the embryo. Gastrulation endows the embryo with a clear antero-posterior polarity (h, i). Further cell proliferation and a series of morphogenetic movements (i–k) give rise to the tadpole (l). A section through this stage of development (m) highlights the derivatives of the different germ layers.

that each of these groups of cells acts as a rudiment, or 'germ', for a specific set of elements of the different tissues of the embryo (Fig. 1.9). The external layer was named the 'ectoderm', the internal layer the 'endoderm' and the intermediate layer the 'mesoderm'.

In most organisms the process of gastrulation reveals a clear outline of the basic body plan, with defined anterior (head), posterior (tail), dorsal (back), and ventral (belly) regions. From the gastrula stage onwards, these rudimentary tissues are expanded, shaped, and decorated in a huge variety of ways to create all the rich diversity of the animal kingdom. The classical nomenclature of these early developmental stages is still used today, and serves as useful way of referring to specific times and events in development.

During the second half of the nineteenth century the vogue for comparative embryology, together with the publication of Charles Darwin's theory of evolution by natural selection, brought questions of taxonomy and phylogeny to the fore. Developmental features, such as seg-mentation, body plans, and the organization of the germ layers, became part of the criteria for describing and classifying groups of organisms and proposing evolutionary pathways.

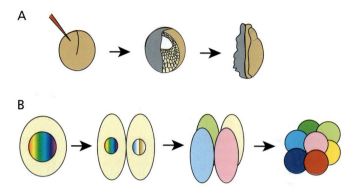

By this time, too, the description of developmental processes had reached the cellular level, and two things had become clear. The first was that one cell leads to many cells, in other words that cell proliferation is an important 'engine' of the developing system. The second was that proliferation was accompanied by a continuous but regulated increase in cell diversity. These observations prompted questions, such as 'What instructions or mechanisms make one cell give rise to several different cells?' and 'How, during development, is all the information preserved to be passed on to the next generation?'

In the last two decades of the nineteenth century, August Weismann tackled both questions with a single proposal (Fig. 1.10). With remarkable prescience, he attributed a central role to the cell nucleus in both development and inheritance. He proposed that, as cells proliferate during development, different nuclear instructions or 'determinants' were allocated to different cells, causing them to form different tissues and organs, but that a group of 'germ line' cells, segregated from other cells, retained the full complement of nuclear determinants. From very early in development, the cells of the germ line were set apart so that they could not lose their capacity. This concept became known as the soma/germ line theory. Weissmann's insight was remarkable at many levels, as he also envisaged interactions between the cytoplasm and the nucleus in this process.

Experimental embryology is born

Weismann's proposals signalled a move away from a purely descriptive approach to embryology. His idea of determinants progressively allocated to different parts of the embryo was clearly testable and, in setting out to test it,

Fig. 1.10. **Experiment of Wilhelm Roux and its derived conclusions. (A)** Roux (1888) killed one of the cells at the two-cell stage of a frog embryo, and observed the development of the remaining cell. His interpretation of what he saw was that this cell developed into a half-embryo. This led him to conclude that during normal development, each of the two cells receives a certain amount of information about its developmental fate, and that the acquisition of this information means the loss of other information. **(B)** These observations fitted well with ideas of the time. According to this view, during development determinants for specific regions of the embrge were allocated as the cells divided. These determinants allecated specific developmental characteristics to the cells that received them.

Wilhelm Roux initiated a new, empirical approach that he called 'developmental mechanics'. Taking a frog embryo at the two-blastomere stage, he killed one of the blastomeres and observed the development of the remaining one: it developed into something he interpreted as a half-embryo, thus apparently confirming Weismann's theory (Fig. 1.10).

Among those excited by the possibilities offered by the new science of experimental embryology was the young biologist Hans Driesch. Driesch was intrigued by Roux's experiments: could half-animals really exist? He decided to repeat Roux's experiment but with a different organism, the sea urchin, and under different conditions. Unlike Roux, instead of killing one blastomere he merely separated the two and left them to develop in different dishes (Fig. 1.11). In his subsequent paper, he described his thoughts as he waited for the results of the experiment: 'I awaited in excitement the picture that was to present itself in my dishes the next day. I must confess that the idea of a free-swimming hemisphere or a half gastrula . . . seemed rather extraordinary. I thought the formation would probably die. Instead, the next morning I found in their respective dishes typical, actively swimming blastula of half size.'

This was a seminal moment in the history of developmental biology. Driesch had observed that early in development the blastomeres are 'aware' of whether they are part of a whole, or are the whole itself, and will tailor their development accordingly: if they are part of the embryo they will contribute to the whole, but if they are all that is available they will adjust, regulate, and give rise to the whole embryo (Fig. 1.11). This regulative property of the sea urchin embryo was later observed in many other embryos and in some cases was not restricted to the first division of the zygote. One possible explanation for the observations of Roux might be that he left some damaged tissue behind which affected the outcome of the experiment.

The result of Driesch's experiment poses an essential question about development, and one to which we still do not have a satisfactory answer: how does a cell 'know' when it is the whole or only part of the whole? The different behaviour of the blastomeres, depending on whether they are isolated or part of an embryo, suggests that during embryogenesis there are exchanges of information between cells that affect their developmental potential. This was underlined in a second experiment by Driesch (Fig. 1.11) which reinforced the view that the components of the embryo have a regulative ability. In this experiment, a zygote was allowed to develop until the eight-cell stage and then gently perturbed in such a way as to change the relative positions of the cells. This embryo developed completely normally, an outcome that would not have been possible had the fate of each cell been 'hard-wired' from the outset as suggested by the ideas of Roux.

These experiments introduced the concept, which has since been shown to be universal, that a cell within an embryo has both a 'prospective fate' (what it will become if left undisturbed) and a 'prospective potency' (its range of possible fates if the circumstances change). In addition they show that the potency of a cell is always wider than its fate, and make it clear that an understanding of how the potency of a cell arises and is regulated lies at the heart of developmental biology.

Driesch considered the possibility that during development cells acted as machines, and he wrestled to find a mechanistic explanation for what he had observed. He developed interesting and quite modern insights into the processes he was studying, for example, he wrote that 'development starts with a few ordered manifoldnesses, but the new manifoldnesses create by interactions new ones and these are able . . . to provoke new differences and so

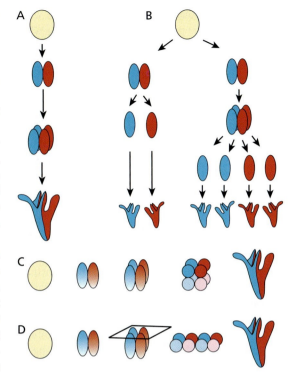

Fig. 1.11. **Experiments of Hans Driesch on the sea urchin embryo.** **(A)** The development of the larval stage of the sea urchin relies on stereotyped events early in development which allow the fates of each of the blastomeres to be traced. **(B)** In 1892, Driesch reported his experiments on sea urchin embryos. He took embryos at the two- (*left*) and four-cell (*right*) stages, separated the cells and allowed them to develop as individual embryos. The result was always the generation of whole larvae; that is, cells that would have given rise to only one-half of the larva were, in isolation, capable of generating the whole of the larva. **(C)** The third cleavage of the sea urchin embryo is perpendicular to the plane of the first two and generates two tiers of four cells each. **(D)** By applying gentle pressure with a glass slide, Driesch changed the orientation of the third cleavage, thus altering the relative positions of the eight cells, then released the pressure and left the embryo to develop. The result was a normal larva.

on. With each effect immediately a new cause is provided and the possibility of a new specific action'. However, he failed to see a way towards a mechanistic explanation of development, largely because to find such an explanation he needed to know about molecules and genes and this knowledge was still many years in the future. In the absence of more concrete options he turned to abstract concepts and summed up the essence of what he had observed by describing the developing embryo as a 'harmonious equipotential system', that is, one in which all elements had the potential to develop into the whole but this potential was restrained by the need of the system to be 'harmonious'.

Morgan's 'deviation': Genetics

As a consequence of the work of Roux and Driesch, the early twentieth century saw a steady stream of experimental manipulations that revealed the amazing potential of developing embryos for growth and regeneration. Hans Spemann has been quoted as claiming to have spent his life placing embryos in 'ever more embarrassing positions'— to which one may add that the embryo was always one step ahead. As the list of embryological manipulations and their consequences started to rival the catalogue of descriptions of embryonic development, the important questions started to become clear. Most importantly, what was the material or molecular basis of the machine that runs development? Answers were harder to come by.

The issues that defeated Driesch remained. If embryos were to be thought of as machines, where was the engine? Where were the screws, the nuts and bolts? What was the fuel; what the laws that regulated the transformation of matter into energy? As no elements had been identified that would enable developing systems to be described in mechanical terms, concepts were developed to explain the behaviour and properties of their only obvious physical components: cells. Fields and organizers are two such concepts that arose during the early days of experimental embryology.

The 'field', a concept borrowed from physics, refers to a self-organizing group of cells and became popular as a concept in development at the beginning of the twentieth century. As in physics, a field is defined by the potentials of its constituent particles, in this case, cells, and can be used to explain the regulative properties of many embryos, or groups of cells within embryos. Organizers, discovered and named by Hilde Mangold and Hans Spemann, are special groups of cells that have the ability to determine the fates and organization of cells around them. The original 'organizer' was a group of cells in the frog embryo that directs the fates and arrangements of adjacent cells as they invaginate during gastrulation (Fig. 1.12). But these concepts, although they are useful in a descriptive sense and have survived the passing of time remarkably well, brought a mechanistic understanding no closer. If the organizer emits information, for example, what is the nature of this information and how is it processed?

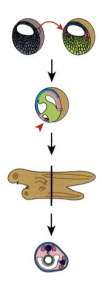

Fig. 1.12. Experimental demonstration of the activity of the organizer. In a frog embryo, transplanting the leading edge of the mesoderm during gastrulation to a different position of an embryo induces a new focus of gastrulation (red arrow head) and, associated with it, a secondary axis. Other regions of the embryo do not show this activity, indicating that the cells of the organizer have special properties.

For some time it had been suspected that the nucleus contained a key to the problem; Weismann had said as much and the work of Theodor Boveri and Edmund Beecher Wilson, in particular, provided ample evidence in support of the idea that the nucleus contained the instructions for development. At this time, too, the role of the nucleus and specifically the chromosomes in the answer to another problem, that of heredity, was also becoming clear. In a career decision that had a vital impact on the history of modern developmental biology, Thomas Hunt Morgan, although originally an embryologist, abandoned experimental embryology and turned instead to the problem of how characters were transmitted from generation to generation: Morgan's 'deviation', as the molecular geneticist Sydney Brenner has aptly called it, established the fundamentals of modern genetics, and the fruit fly *Drosophila melanogaster* as an invaluable experimental system. The concept of the gene as the unit of inheritance and the idea that genes were associated with chromosomes pre-dated Morgan's work, but it was Morgan and his group who first used linkage to map the linear arrangement of genes on chromosomes, correlated the genetic maps obtained from linkage studies with cytological maps of the chromosomes, recognized the effects of gene interactions, and systematically bred and maintained mutant stocks of *Drosophila*, thus laying the groundwork for the application of genetics to the study of basic cellular processes.

The impasse

From the 1920s to the 1960s, biologists were engaged in unravelling the molecular basis of heredity and cell biology. From the point of view of heredity, a sustained focus on the nucleus paid off and, after the seminal experiments of Oswald Avery, Colin MacLeod, and MacLyn McCarty on DNA-mediated transfer of traits between bacteria, a long onslaught on the structure of DNA ended with the elucidation of the double helix by James Watson and Francis Crick. The ensuing discovery of the genetic code led to the idea that DNA makes RNA makes protein in a directional and largely irreversible manner: the 'central dogma' of molecular biology. The implication for genetics was that the information to generate the components of a cell and an organism, which had long been known to reside in the nucleus, were shown to lie in the DNA.

In parallel with this progress at the molecular front, there were several attempts to link genetics and development. Some thought that if this could be done there would be an answer to the questions posed by Roux and Driesch. But the problem was that despite the efforts of geneticists in describing developmental mutations there was at the time no simple way of bridging the gap between the mutation and the nature and normal function of the gene associated with it. For this there needed to be a way of pinpointing the gene, decoding the information it contained, and linking specific mutations with specific functional defects.

Genetics meets molecular biology

Despite this impasse, the middle decades of the twentieth century did see an important advance in the use of genetics to study developmental processes, with work on the genetics of bacteria and their viruses, the bacteriophages. These studies sought to understand how bacteriophage lambda 'chose' between two alternative programs: integrating silently in a bacterial genome, or replicating its genome and destroying the bacterium. As a result of this work it became clear that not only phenotypic traits but also strategic developmental decisions could be traced back to regulatory interactions controlling the expression of genes. François Jacob and Jacques Monod drew on this idea in their work on the regulation of bacterial adaptation to lactose utilization. They developed the model of the *lac* operon for the control of gene expression, a model that still permeates much of our thinking on the subject. In this model, a pattern of gene expression is generated through positive and negative regulatory proteins whose balanced interactions determine the activity of the genes.

With these studies, and the subsequent identification during the 1970s of the genes and proteins involved, Genetics had met Molecular Biology. From the collecting of mutants, models began to emerge which portrayed development in terms of genetic circuits. In an approach pioneered by Ed Lewis, Eric Wieschaus, Christiane Nüsslein-Volhard and Gerd Jürgens on the fruit fly *Drosophila melanogaster*, the haphazard accumulation of developmental mutants was replaced by systematic screens to pick up mutants in specific processes (Fig. 1.13). From these, genes were identified that are involved not only in the generation but also the processing of information: the 'generation' because many of the genes encode proteins that are involved in defining cell- or tissue-specific characteristics, the 'processing' because others encode proteins involved in regulating, in time and space, the expression of those genes or the activity of the proteins they encode.

The nematode *Caenorhabditis elegans* joined *Drosophila* as a model organism with well-studied genetics and ease of experimental manipulation. With the development of molecular cloning it became obvious that the way forward was to clone, sequence, and analyse genes—an analysis that would yield not only the nature of the proteins but also of the developmental program itself. The culmination of this analysis has been the ability to sequence and analyse whole genomes (Fig. 1.14).

The reward for all this work has been the identification of the elements that form the basis of Driesch's 'harmonious equipotential system', and the ability to see how they assemble into functional modules that create diversity and order, shape and function.

Lessons from history?

This brief historical introduction shows how the central theme at the heart of developmental biology—the spatially and temporally regulated increase in cell number and diversity—has been perceived in different eras and in the light of different levels of knowledge. Broadly, it is possible

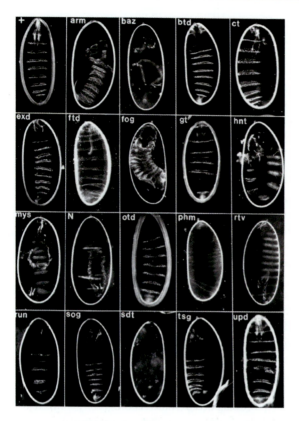

Fig. 1.13. Patterning mutants of *Drosophila melanogaster*. Gallery of mutants obtained in a saturation screen for zygotic mutations affecting the pattern of the larval cuticle in *Drosophila*. The wild type pattern is shown at the top left corner (+) and each picture represents a mutation found on the X chromosome. (From Wieschaus, E., Nüsslein-Volhard, C., and Jürgens, G. (1984) Mutations affecting the pattern of the larval cuticle in *Drosophila melanogaster*: III. Zygotic loci on the X chromosome. *W. Roux Arch. Dev. Biol.* **193**, 296–307.)

to discern three periods and one interlude. The first period was defined by the awareness that development is about change, not simply growth, while the second period was a call to arms to explain this change in mechanistic terms. These two phases were followed by an interlude or impasse while tools were developed to make this mechanistic approach possible. In the most recent period, a flood of information about the molecules that function during development has begun to make a detailed mechanistic understanding a reality.

At the end of each 'phase' there has been a brick wall because the progress made has led to questions whose answers required conceptual and technical advances. There are some interesting parallels here with the development of the physical sciences. Physics always demands a mechanical explanation for every observation but sometimes such explanations are not forthcoming in the way one would hope. For example, thermodynamics provides a conceptual understanding of certain processes like heat, but it is phenomenological: it invents quantities, such as enthalpy and entropy, that allow properties of the system to be measured, and that are needed to analyse macroscopic observations such as the formation of ice or the evaporation of water. However, thermodynamics itself does not provide a mechanistic explanation for these transitions of physical state. These explanations lie in the realm of statistical mechanics, which emerged long after thermodynamics was established. Statistical mechanics provides an account of the mechanisms that underlie the more conceptual ideas of thermodynamics (Fig. 1.15). It does this by inventing a new language and new concepts which allow us to deal with large numbers of particles and describe properties that emerge from their interactions.

In modern developmental biology, classical concepts such as 'fields' and 'organizers' are a bit like thermodynamic concepts: they are useful at a descriptive level but they do not provide an explanation of how their measurable properties are generated from their molecular components. As our knowledge of these components increases, it is very likely that a mechanistic understanding will emerge of many of the concepts that we use to represent developmental events. Along the way, it will be important to bear in mind the different levels of organization that we outlined earlier in this chapter (Fig. 1.2). Here might lie an important difference from the physical sciences, where usually only two levels have to be bridged: a macroscopic level and a microscopic one. In biology there is a scale of levels, each with emergent properties that depend on interactions with both the level above and the level below.

About this book

A change is underway in how developmental biologists view their subject, in the sorts of questions they ask and in how they address these questions experimentally. This book stems from an appreciation of this change of perspective and attempts an approach to developmental

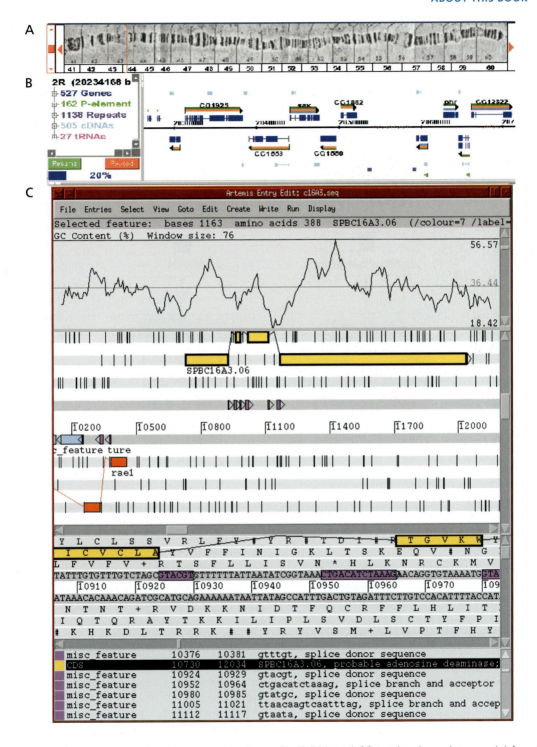

Fig. 1.14. Genome analysis. (A) Graphics from GadFly, the database for the *Drosophila* genome, displaying the right arm of the second chromosome (portrayed as a polytene chromosome). **(B)** Annotation of the predicted transcripts and genes that can be found in the region highlighted by a red bar on the polytene chromosome (in subdivision 44). **(C)** A tool used to analyse genomic information is applied to one of the regions of DNA in subdivision 44. The programs within this tool find open reading frames and putative transcripts within large stretches of DNA and are used to create the maps illustrated in (B). (Image courtesy of S. Russell.)

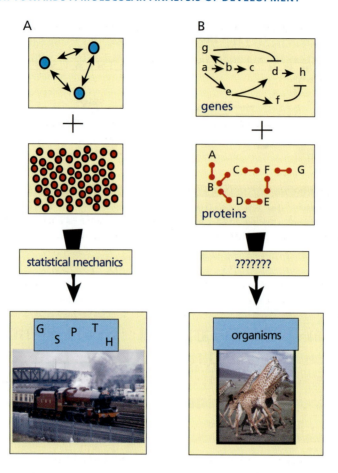

Fig. 1.15. A need for an integrated analysis of molecular events during development. (A) In physics, thermodynamics uses variables, such as Gibbs free energy (G), enthalpy (H) or entropy (S), to describe temperature (T) or pressure (P). All these variables result from interactions between the particles that make up matter but the laws of 'classical mechanics' cannot be used to infer how interactions and collisions between particles generate S, H, T or P. The reason is that the calculations become too cumbersome, and meaningless in terms of what is being explained. This problem is circumvented by 'statistical mechanics' which provides an analytical tool and a language for applying the laws of mechanics to large numbers of molecules. **(B)** Developmental biology requires a related tool that will allow the properties of cells and tissues, as represented in the shape of organisms, to be described in terms of the interactions between genes and proteins. It is unclear, at the moment, whether the description of every interaction between genes or between proteins will provide such a picture. It is hoped that modern biology will find a conceptual analogue of statistical mechanics, which will allow us to understand developmental processes in terms of 'average' interactions between genes and proteins.

biology that focuses, as far as possible, on the molecules and mechanisms that make organisms, rather than the more traditional emphasis on organisms and developmental systems.

The book is structured around the concept, introduced in this chapter (Fig. 1.2), that development involves increasing levels of complexity and organization and that in order to understand developmental processes we have to understand the organization of those levels. In Chapters 2–5, we describe the basic molecular elements of developing systems: the proteins and the DNA and how they create an information-processing unit within the cell. We begin by introducing the concept that the DNA contains a program that determines sequences of events through the proteins it encodes (Chapter 2). We then discuss how this information is decoded (Chapter 3) and move on to discuss the ways in which this decoding is regulated through cell interactions. We do this by introducing signals and receptors (Chapter 4) and seeing how these assemble into networks that can process molecular information by creating relationships between the membrane and the nucleus that allow cells to receive inputs and produce outputs, to integrate and to calculate (Chapter 5).

In Chapters 6–9, we explore how the molecular networks act within cells to create 'routines' that can be used to perform developmental operations. First (Chapter 6), we introduce basic cellular activities, such as adhesion, movement, and cell interactions, that modulate much of the behaviour of cells in developing embryos. In Chapter 7, we continue this trend by discussing cell division and cell death and the way they contribute to development. In Chapter 8, we see how large-scale regulated cell division begins to create cell diversity and, in Chapter 9, how long- and short-range cell interactions contribute to this diversity.

With these first two parts of the book we hope to lay down the molecular and cellular elements that operate in developing embryos. In the last part of the book we explore these basic mechanisms in action. In Chapter 10, we study what is basically a problem of change over the single dimension of time: how are specific cell types generated during development? In the following two chapters we gradually add further, spatial, dimensions, to see how different cell types become organized in what are essentially a series of two- (Chapter 11) or three- (Chapter 12) dimensional patterns.

Unlike most traditional developmental biology texts, the book does not contain a separate chapter or section devoted to discussing the emergence of body plans, that is the laying out early in embryogenesis of the coordinates that act as a reference system for the development of particular embryos. The reason is that we do not see this process as a special one that is different from the basic organization of a limb or the patterning of epithelial cells. We see body plans rather as examples of the outcomes of molecular and cellular interactions and, as such, aspects of the ways in which they can be set up will figure in several places throughout our discussion.

One obvious omission from this book is plant development. This is not because we think plants develop in a way that is fundamentally different or that uses completely different elements. On the contrary, the findings of the last few years have shown that plants follow essentially the same developmental principles and they have made their own contributions to the general concepts that we have discussed. Plants also have their own peculiarities, however (as do animals, from the point of view of plants). Perhaps unjustly, we follow a fairly widespread bias and base our discussion on examples from animal development. We nevertheless hope that this book will be of general use to all developmental biologists.

Although the focus of this book is on animal development, several of the examples we discuss in detail, particularly in the first half of the book, come from the yeast *Saccharomyces cerevisiae*. The reason for this is that, largely because of its small genome and its amenability to genetic approaches, the detailed molecular analysis of biological processes is in many cases more advanced in yeast than in higher eukaryotes. Yeast does undergo some simple, regulated developmental events, such as mating type switching and sporulation, that are analogous to developmental events in higher eukaryotes such as animals. This, together with the remarkable evolutionary conservation of many of the individual elements involved in biological processes in all eukaryotes, suggests that lessons learned in yeast will be relevant in higher eukaryotes as well.

SUMMARY

1. Biological systems are organized into a hierarchy of functional modules that are linked through their component elements.

2. The components of each module contain the information to generate the next higher level of complexity and to feed back to the lower one. For example, proteins, the fundamental functional components of cells, feed back information both to the genes that code for them and to other genes, creating functional interrelationships.

3. Development relies on the generation of large numbers of cells which are diversified in a regulated manner and organized in space.

4. The generation of complexity during development is a progressive process that relies on the organized unfolding of the information contained in each level of complexity. Because there is

constant feedback between the different levels, this process is an open one that carries on throughout the life of an organism.

5. The first major conceptual advance in the history of developmental biology was the realization that development is about change, not just growth. Attempts to explain development in mechanistic terms were, however, hampered by lack of the right experimental tools.

6. In the modern era, a flood of information about the molecules that function during development has begun to make a detailed mechanistic understanding of this process a reality.

7. Genetics contributes to this understanding by enabling us to interfere with the activity of proteins and their interactions and infer, from the results, how these processes normally function. Functional genomics adds to this the ability to analyse simultaneously the functions of whole batteries of genes. Molecular and cell biological techniques allow us to probe, *in situ*, the activity of proteins and their effects on the properties and behaviour of cells.

8. A major challenge for the future is to devise languages that can enable us effectively to describe and understand the dynamic molecular and cellular events that form the basis of development.

COMPLEMENTARY READING

Coen, E. (1999) *The art of genes.* Oxford University Press.

Jacob, F. (1993) *The logic of the living.* Princeton University Press.

Judson, H. F. (1996) *The eighth day of creation*: The makers of the revolution in biology. (Expanded edn). Cold Spring Harbor Laboratory Press.

Sander, K. (1992) Shaking a concept: Hans Driesch and the varied fates of sea urchin blastomeres. *W. Roux Arch. Dev. Biol.* **201**, 265–7.

Sander, K. (1991) Mosaic work and assimilating effects in embryogenesis: W. Roux conclusions after disabling frog blastomeres. *W. Roux Arch. Dev. Biol.* **200**, 237–9.

Wolpert, L. (1994) Do we understand development? *Science* **266**, 571–2.

Chapter 2

Programs and regulatory elements in DNA and RNA

The essence of embryonic development is a regulated increase in cell numbers, coordinated in space and time with an increase in cell diversity. Cell numbers increase by cell division, and diversity is achieved through differential gene expression. Cell multiplication is associated with the replication of DNA, while differential gene expression is achieved, essentially, through the regulation of transcription. Replication, transcription, translation, and the regulation of each of these processes result from the activity of proteins, which therefore create a feedback loop between the source of the information and its expression. As we have discussed in Chapter 1, if we use the analogy of a computing device then the DNA constitutes the 'software' and the proteins make up the 'hardware' that is used to scan, read, interpret and run the information encoded in the DNA. The fact that the protein 'hardware' is encoded in the DNA 'software' creates a unique relationship that is central to developmental processes and which allows the unfolding of different levels of complexity.

In this chapter we explore the nature and organization of the information that resides in the DNA and how it becomes arranged at different levels to generate programs of gene expression that underlie the generation of cell diversity in development.

The DNA during development

The DNA molecules of each cell are located in the chromosomes (Fig. 2.1). In prokaryotes there is a single, often circular, chromosome associated with a few specific types of proteins but with little higher-order organization. In eukaryotic organisms each somatic cell contains two homologous sets of chromosomes (as well as a small amount of DNA in the mitochondria, the organelles devoted to the energy metabolism of the cell). All the individuals in a given species have the same number of chromosomes. In sexually reproducing organisms, one member of each homologous pair of chromosomes is inherited from the mother, and the other from the father. Within the chromosomes, the DNA is supercoiled and packaged with proteins to form chromatin (Fig. 2.1). Most of the protein content of chromatin is accounted for by a group of basic proteins called histones and its fundamental structural unit, the nucleosome, consists of about 200 bp of DNA wound around a core of histones (Fig. 2.1). A variety of non-histone proteins participate in looping and folding the chromatin to form a complex three-dimensional structure: the chromosome.

During the lifetime of a multicellular organism, cell proliferation is associated (with some exceptions, which we shall return to later) with the flawless propagation of every nucleotide contained in the chromosome(s). The perfect replication of the software means that on the way to multicellularity, at every cell division, each daughter cell receives a full complement of the information that is required to make an organism and that, therefore, all the cells of an organism have the same basic information content. This fact has been demonstrated most clearly by experiments in which organisms have been cloned from single cells. A clone is a set of two or more genetically identical individuals. Twins represent a simple and trivial example of a clone—'trivial' because in most cases, the duplication of the embryo occurs very early, before there has been any chance for the cells to diversify or differentiate and thus before there has been much of an opportunity for information to be lost. However, experiments carried out initially with the frog *Xenopus laevis*, and more recently with other animals including sheep, mice, and cattle, demonstrate that even a fully differentiated cell can, in some circumstances, be used to clone an organism. In these experiments (Fig. 2.2), the nucleus from a

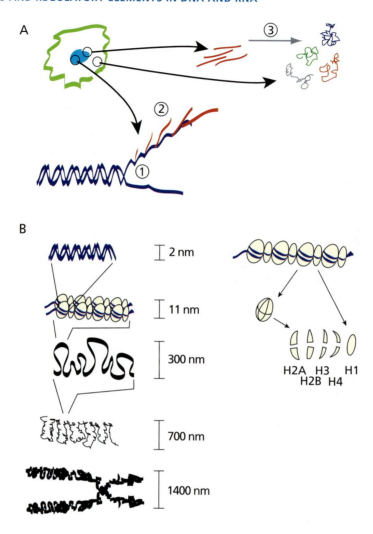

Fig. 2.1. Structure and function of the DNA in eukaryotic cells. (A) The DNA (blue ribbon) resides in the nucleus where each molecule is unwound (1) and replicated at every cell division (not shown) so that the number of copies of the genome increases as cells multiply, but the amount in each cell remains constant. Some sequences on the DNA are transcribed (2) into RNA (red lines). The process of transcription is also a process of amplification of the information, as multiple transcripts may be generated from a single gene. There are three classes of RNA molecules: ribosomal RNA (rRNA), transfer RNA (tRNA), and messenger RNA (mRNA). The RNA molecules are exported to the cytoplasm where those encoding proteins associate with ribosomes and tRNAs to be translated (3). Proteins (different colours) are involved in each of the steps of information maintenance (replication), transfer (transcription), and decoding (translation). In addition, they participate in specialized functions within the cell and the organism. **(B)** In eukaryotes the DNA is complexed with an ensemble of specific proteins to form chromatin. The fundamental structural unit of chromatin is the nucleosome, about 200 bp of DNA coiled around a core consisting of eight molecules of the basic histone proteins. A nucleosome contains two molecules each of histones 2A, 2B, 3, and 4. A stretch of about 50 bp of DNA serves as a linker between nucleosomes and provides an anchor for histone H1, which organizes nucleosomes into a higher-order fibre. The chromatin fibre acquires higher (and less well understood) orders of spatial complexity through folding and interactions with other proteins. The scale bars indicate the dimensions of the structures and give an idea of the packaging of the DNA. Chromosomes represent the highest order of packing of chromatin. During the life of a cell the chromatin condenses and decondenses in response to internal and external influences.

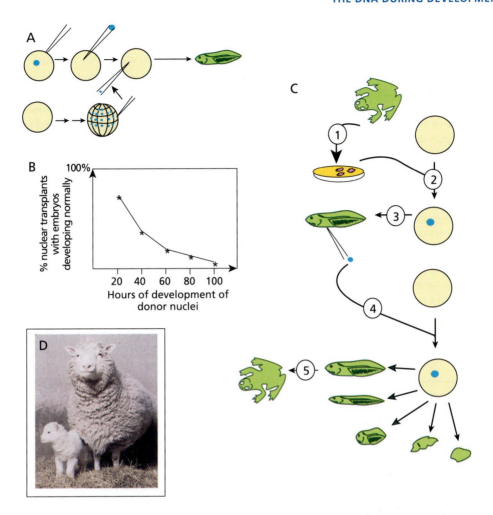

Fig. 2.2. Conservation of the genetic information during development. Experiments in animal cloning. The developmental potential of nuclei from cells that have undergone any degree of development and differentiation can be tested in cloning experiments, to see whether they are able to support the development of a complete organism genetically identical to the source of the donor nuclei. **(A)** Oocytes are enucleated, and nuclei extracted from the cells of a *Xenopus* blastula stage embryo are introduced, individually, into the enucleated oocytes. Activation of the resulting egg triggers development. In a large number of experiments with nuclei from early embryos, most nuclei support development to the tadpole stage. **(B)** As the developmental stage of the donor nuclei advances, the frequency with which tadpoles develop decreases. The 'older' the donor nuclei, the earlier development is aborted. **(C)** To test the developmental potential of fully differentiated cells, epidermal cells from an adult frog are cultured *in vitro* (1) and their nuclei placed in an enucleated oocyte (2). Subsequent activation of development leads, in a small number of cases, to the development of tadpoles (3). If the nuclei from these tadpoles are used as donors for new enucleated oocytes (4), adult frogs sometimes result (5). This experiment suggests that the reason adult nuclei did not support complete development in the first part of the experiment might have something to do with their failure to adapt readily to the early embryonic environment. (Panels A–C adapted from Gurdon, J. B. (1962) The developmental capacity of nuclei taken from intestinal epithelial cells of feeding tadpoles and embryos. *J. Embryol. Exp. Morph.* **10**, 622–640; and Gurdon, J. B., Laskey, R. A. and Reeves, O. R. (1975) The developmental capacity of nuclei transplanted from keratinized cells of adult frogs. *J. Embryol. Exp. Morph.* **34**, 93–112.) **(D)** The first mammal to be cloned from an adult cell nucleus was the sheep Dolly. In these experiments, cells from udder tissue were placed in culture and treated so that they stabilized in a quiescent stage of the cell cycle in which DNA replication does not begin. Nuclei from these cells were then placed into enucleated oocytes and development was activated by an electric shock. After a few divisions in culture, the embryo was implanted into a ewe. In one case this resulted in the birth of Dolly.

differentiated cell was inserted into an enucleated oocyte, which was activated to begin development without undergoing fertilization. In frogs, the nuclei of terminally differentiated cells were able to direct development to the tadpole stage and, under certain experimental conditions, to fertile adult frogs. The first clone from an adult differentiated cell, the sheep known as Dolly, was made using the nucleus from a mammary (udder) cell. Dolly grew normally into a fertile adult. In similar experiments, mice and cattle have been cloned by injecting nuclei from differentiated granulosa cells (a cell type that surrounds the ovum) into enucleated oocytes. In all cases, the nuclear DNA of the resulting animals was genetically identical to that of the animal from which the differentiated cell was taken (Fig. 2.2). If growth and differentiation were associated with the irreversible removal of information from the DNA, the differentiated nucleus would not be capable of sustaining the development of a new organism.

Animal cloning is not yet as easy as this description implies. Many attempts to produce live-born, viable animals have been unsuccessful. The first cloned sheep was the sole success from 277 initial clones, and this low success rate is still fairly standard. The reasons for the difficulty of cloning animals from adult nuclei are not yet fully understood but there are some possible explanations. One is that the adult cell nucleus may not readily adapt to an early embryonic cytoplasmic environment. This would explain why, compared to nuclei from adult cells, nuclei from embryonic cells give rise to clones with greater ease (Fig. 2.2). An associated factor may be that during the development of an animal, modifications to the chromatin make it difficult for certain genes to be reactivated once their participation in a particular developmental process is over. This implies that, as an animal develops, information that will not be used is parcelled away in structural forms that are difficult to access.

The observation that a differentiated cell contains all the information to make an organism highlights an important question in developmental biology: how is it that although every somatic cell of an organism contains the same DNA, each displays a unique identity? At the simplest conceptual level, the answer is that by the end of a developmental process, cells have different appearances because they have decoded different bits of their DNA and have done so in a particular order. In order to understand development at the molecular level, we need to build on this concept to explain how the differential expression of the information is achieved and organized in space and time, and how the interactions between the software (DNA) and the hardware (proteins) are orchestrated and regulated.

Types of DNA sequences

The organization of the linear sequence of nucleotides in the DNA forms the basis for understanding the developmental program. Each eukaryotic chromosome contains

Fig. 2.3. Types of DNA sequences. (A) Chromosomes contain three types of sequences with specialized functions involved in maintaining the functional integrity of the chromosomes as cells multiply. Telomeres are located at the end of the chromosomes, and are composed of long stretches of short repeated DNA sequences with a special structure that shields nearby single-copy DNA from possible losses during replication. Centromeres are specialized sequences, at chromosome-specific locations, through which the two DNA molecules are held together after replication. The role of the centromere is to ensure that during mitosis each daughter cell receives a complete complement of information. Origins of replication are scattered throughout the chromosome and provide sites at which DNA replication starts. The frequency of these sites allows rapid replication of the information after mitosis. **(B)** At the origins of replication, the strands are separated so that they can be replicated individually. At the sites of transcription, specific enzymes make RNA from the DNA template. (Note that, for ease of representation, the diagrams show some features that would not be simultaneously evident on the same chromosome. For example, the origins of replication would not normally be active after the chromosome has already replicated to form two chromatids held together at the centromere, as shown.) **(C)** Genes have sequences that are transcribed (coding region). The coding region is bounded by special sequences which determine the beginning (blue box) and the end (red box) of transcription, and nontranscribed sequences (regulatory or control sequences) which determine the time, place, and duration of transcription. The regulatory sequences are scattered throughout the gene: they may be upstream from the start site, downstream from the gene or even in noncoding intervening regions within the gene. **(D)** The product of transcription is an RNA molecule that contains a start site (5′) characterized by a chemical modification and an end (3′) with a characteristic stretch of multiple adenine nucleotides known as the polyA tail. The original transcript is much longer than the one that is translated into protein. The process of maturation requires the removal of noncoding sequences, or introns, which break the coding region into segments called exons (green boxes). The removal of the introns, a process called splicing, is very precise and is catalysed by special proteins. **(E)** The mature mRNA contains an untranslated 'leader' sequence, and a translatable portion that starts with an AUG codon and is terminated at one of three termination codons (UAG, UAA, UGA).

some basic structural features which ensure that the information in the DNA can be reproducibly propagated (Fig. 2.3). Replication origins act as initiation sites for the ordered copying of the entire sequence of nucleotides at every cell division. The centromere ensures that the replicated chromosomes segregate from each other when the cell divides, so that each daughter cell receives a full complement of information. Telomeres, the physical ends of chromosomes, help buffer the genetic content of the chromosome against losses resulting from the difficulty of replicating the ends of a DNA molecule. The job of these

sequences is to carry other sequences, including those containing the information to make cells and organisms, along through the lifetime of an organism and, through the germ line, to the next generation.

Broadly speaking, there are two kinds of 'informational' sequences: those that can be expressed, that is transcribed, and those that cannot. Sequences that can be expressed are known as genes (Fig. 2.3). Gene sequences can be thought of as the instructions both for building the 'hardware' that executes the developmental program, and for the execution of the program itself. The RNAs transcribed from

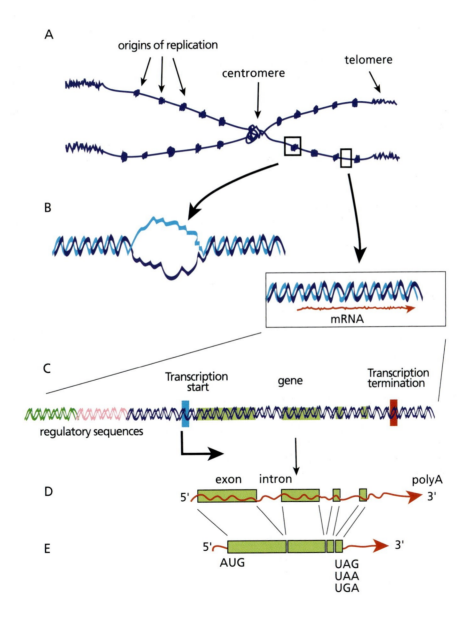

genes may be functional molecules in their own right (e.g. ribosomal RNAs, or transfer RNAs which play a part in protein synthesis), or messenger RNAs, which can be translated into proteins to execute tasks that include catalysing metabolic reactions, contributing to the structure of cells and organelles, and interacting with the DNA 'software' to direct the expression of further genes. The second type of informational sequences are special non-transcribed sequences on the DNA that are concerned with determining when, where and for how long genes are expressed. Usually known as 'control' or 'regulatory' sequences, they tend to lie on either side of or sometimes within the expressed sequences, usually fairly close to the gene but in some cases many thousands of nucleotides away (Fig. 2.3).

In higher eukaryotes, informational sequences can account for as little as 10% of the genome, or even less. Of the remaining 90–95% of sequences, some are introns: noncoding sequences that interrupt genes and are spliced out of RNA before it is translated into protein (Fig. 2.3). Others are members of various classes of repetitive DNA that are scattered through the genomes of most eukaryotic organisms. It is not clear what the noncoding DNA, sometimes referred to as 'junk DNA', is for; one of the challenges of the genome projects for humans and other organisms is to sequence successfully through the vast tracts of 'meaningless' DNA and to identify functionally meaningful information within them.

Genes and genomes

Organisms contain variable amounts of non-informational sequences. This means that, in general, there is no correlation between the DNA content of an organism and its complexity, either morphological or developmental (Fig. 2.4). There is a general tendency for prokaryotes and 'lower' eukaryotes, such as fungi, to have small, compact genomes with largely intron-less genes. However, among higher eukaryotes there is a huge variation in the ratio of noncoding to coding DNA. The fruit fly *Drosophila melanogaster* has a genome of 120 Megabases (Mb) and about 13 600 genes, whereas an apparently more 'complicated' organism, the puffer fish *Fugu*, appears to accommodate about four times the number of genes (30 000–40 000 according to current estimates) in about twice as much DNA (400 Mb). The genomes of mice and humans, with almost ten times as much DNA (3300 Mb), contain roughly the same number of genes as *Fugu*. It has been suggested that all animals contain approximately the same

number of 'core' genes—around 12 000—encoding the components of basic biochemical pathways and cellular functions. The difference between this number and the total number of genes in most vertebrates may be accounted for by polyploidization of the genome during evolution, followed by adoption of various specialist functions by different members of the duplicated gene families. It has been suggested that two wholesale duplications of the genome have occurred during the evolution of vertebrates (Fig. 2.4); these duplications might account for the numbers of copies of corresponding genes in vertebrates relative to those of flies.

A telling comparison at a finer level of detail gives an idea of the enormous range in coding capacity of a given amount of DNA. Chromosome III from the baker's yeast *Saccharomyces cerevisiae* (whose complete genome sequence was published in 1996) spans 320 kilobases (kb) and houses 182 open reading frames (ORFs, sequences framed by start and stop codons and therefore potentially capable of being transcribed and translated). Most of these ORFs have been correlated with genes. Roughly the same amount of DNA in *Drosophila* contains the seven genes (only five of them coding for proteins) of the Bithorax complex (BX-C), a set of genes that contributes to determining the identity of cells along the anterior–posterior axis of the animal (Fig. 2.5). Many theories have been put forward to explain why some genomes are more compact than others but they remain hypothetical—the truth is that we simply do not know.

The decoding of the information contained in the DNA is one of the challenges of modern biology. An important benefit of acquiring genome sequence information from a variety of organisms will be to use the power of comparative sequence analysis to identify both coding and noncoding sequences that are widely conserved and thus likely to be of functional significance. The rate of progress in genome sequencing is staggering. The sequencing of complete prokaryote genomes has almost become routine, and work on eukaryote genomes is gathering pace. The landmark sequencing of the *S. cerevisiae* genome in 1996 was followed at the end of 1998 by the reporting of the first complete genome sequence for a multicellular organism, the nematode worm *Caenorhabditis elegans* (Fig. 2.5). The *Drosophila* genome and the first draft of the human genome sequence followed in 2000. Within the next few years the genomes of several other important 'model' eukaryotic organisms will have been deciphered, including our own, which is now scheduled for completion to a higher degree of accuracy by 2002. The acquisition of this

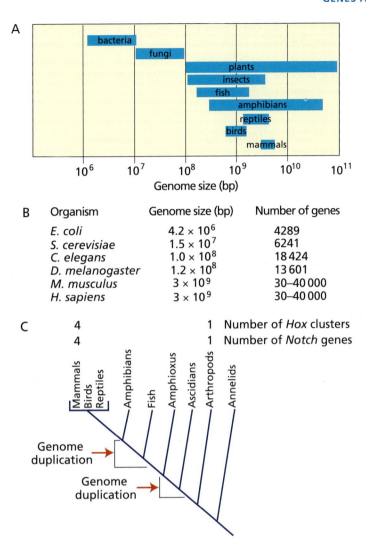

Fig. 2.4. DNA content and number of genes. (A) There is no relationship between the DNA content of an organism and its complexity in terms of cell number, diversity, and arrangement. The genome of humans is smaller than those of some other organisms that are developmentally simpler. **(B)** There is also no linear relationship between the size of a genome and the number of different genes it harbours. Although the genomes of mice and humans are one order of magnitude larger than those of flies or worms, the numbers of genes they contain are not in proportion to this size difference. **(C)** Phylogenetic analysis indicates that there have been two whole-genome duplications since the divergence of the arthropod and vertebrate lineages. This means that for every gene of the fruit fly, there are four homologues in humans and mice, as shown here for genes encoding the receptor Notch, or the *Hox* clusters of transcription factor genes. This type of duplication allows, over evolutionary time, functional diversification of the duplicated genes.

huge amount of information promises eventually to lay the foundations of a new approach to developmental biology, though its interpretation is likely to be a much slower process than the sequencing itself.

The first step in the genome projects for most organisms has been to 'map' the locations of landmark genes or other DNA sequences, known as markers, along the chromosomes. Genetic mapping of genes has been a fundamental tool of geneticists for more than seventy years and has underpinned the use of genetics to unravel gene action during development. It has been gratifying, in the era of genomic sequencing, to discover the generally good correlation between genetic maps, in which the order and spacing of markers are calculated on the basis of how frequently the markers are inherited together, and molecular or 'physical' maps, which are constructed by breaking the DNA down into smaller pieces and determining the content and order of markers on each fragment. Thus one important message from the genome projects has been to validate the role of genetics in developmental analysis. The relative organization of genes on the chromosome provides useful information about regulatory interactions and, when compared across species, about the evolution of genomes.

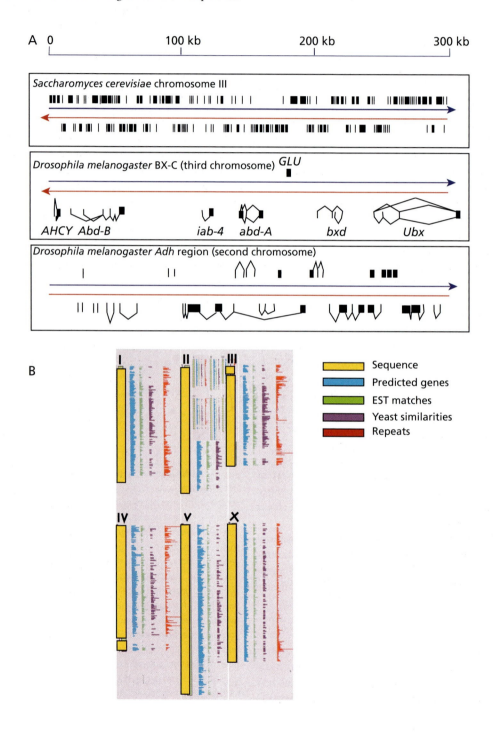

Regulatory sequences in DNA

The regulatory sequences associated with genes, also called *cis*-acting or *cis*-regulatory sequences because they act only on genes that are encoded in the same DNA molecule, contain the molecular instructions that govern when, where and for how long the genes are expressed. Because these regulatory sequences are not transcribed, it is by no means a straightforward task to find them among the huge morass of noncoding DNA. Regulatory sequences are often identified by testing regions of DNA surrounding a gene for their effect on the expression either of that gene (two methods are shown in Fig 2.6) or of a more easily assayed 'reporter gene', *in vitro* or in a specialized expression system (Fig. 2.6). Sequences can also be assessed for a possible function as regulatory sequences by comparing the noncoding DNA sequence surrounding a gene in different species. The rationale for this approach is the assumption that conserved sequences are likely to be functionally important. Once a regulatory region has been broadly defined, deletion and mutation experiments can be used to perform a fine-structure analysis of the important elements within it, down to the level of single nucleotides.

There are two types of regulatory sequences (Fig. 2.6). Promoters, common to all genes and essential for gene expression, are located close to and around the site of transcriptional initiation, usually within about 50–100 bp. Comparison of these sequences in many genes reveals highly conserved sequences such as the 'TATA box', which in most genes directs the assembly of the transcriptional machinery near the start site for transcription, and sequences surrounding the start site itself. These sequences have prokaryotic counterparts and contain little qualitative information, being concerned with the initiation of transcription at times and places that are defined by other factors. In addition, each gene is associated with a regulatory region(s) containing segments of DNA that are also required for proper expression but that vary in sequence, number, and structure from gene to gene. In contrast with the gene-proximal promoter around the transcription start site, these distal regulatory elements, also known as enhancers, upstream activating sequences (UASs) and (in yeast) upstream regulatory sequences (URSs), are located at variable and sometimes very large distances from the site of transcription initiation, and contain qualitative information for gene expression; that is, they play a role in the decision of when and where a gene is transcribed. There are also distal regulatory sequences whose role is to turn off the transcription of a gene. Such sequences are relatively rare in eukaryotes, which generally use less direct mechanisms to repress gene activity, but they are common in prokaryotes.

A large number of 'cutting and pasting' experiments, in which different regulatory sequence elements are combined with reporter genes, have highlighted an important feature of regulatory regions: their modular nature. An enhancer is made up of short sequences, each of which confers some information about expression. The regulatory region of a eukaryotic gene is a mosaic of sequences that

---

Fig. 2.5. Informational content of large stretches of DNA. (A) Examples of the coding capacity of 300 kb of DNA in various organisms highlight the enormous range that is observed. In the yeast *Saccharomyces cerevisiae*, the third chromosome spans 320 kb of DNA and contains a total of 182 ORFs (black bars of varying thickness). These are shown arrayed along the two strands of the DNA molecule (represented by a blue and a red arrow). In contrast, in *Drosophila*, a similar amount of DNA contains a variable number of genes. For example, 300 kb of the Bithorax complex (BX-C) contains just seven genes, of which only five encode proteins: three transcriptional regulators (*Ubx*, *abd-A*, and *Abd-B*) a glucose transporter (*GLU*) and S-adenosyl homocysteine hydrolase (*AHCY*). Exons are represented by black bars of varying thickness, and splicing patterns by zig-zag lines linking the exons. Note that six of the transcripts are encoded in one strand, and only one on the other. In other regions of the genome the same amount of DNA contains a large number of genes, as seen in the region from the the second chromosome that contains the *Adh* gene. (From Lewis, E.B. *et al.* (1995). Sequence analysis of the cis-regulatory regions of the bitho-rax complex of *Drosophila. Proc. Natl Acad. Sci.* **92**, 8403–7; and Ashburner, M. *et al.* (1999) An exploration of the sequence of a 2.9-Mb region of the genome of *Drosophila melanogaster*: the *Adh* region. *Genetics* **153**, 179–219.). **(B)** Analysis of potential coding regions in the genome of *Caenorhabditis elegans*. The yellow bars on the left represent the DNA sequence of the six chromosomes. The sequence has been analysed in a number of ways and the results are shown alongside the chromosomes: potential coding sequences i.e. predicted genes (blue); sequences called expressed requence tags (ESTs) (green). ESTs are identified from copy DNAs (cDNAs), which are produced by reverse transcription of mRNAs. The same programs identify sequences that can also be identified in yeast (purple). The analysis also identifies tracts of repeated sequence (red), which are more frequent in gene-poor regions towards the ends of the chromosome arms. (From fig. 3 in: The *C. elegans* sequencing consortium (1998) Genome sequence of the nematode *C. elegans*: a platform for investigating biology. *Science* **282**, 2012–17; 2016.)

together determine if, when, where and for how long the gene is transcribed. Regulatory sequences can lie upstream or downstream of the transcribed region, and the pattern of expression of a particular gene is the integrated outcome of all the information contained in these sequences. The *even skipped* (*eve*) gene of *Drosophila* provides a good example of the basic structure of these regulatory regions. During development, *eve* is expressed at various times in several different tissues, including an early phase of expres-sion in seven bands of cells during the blastoderm stage, and later expression in particular cells of the mesoderm, nervous system, and epidermis. Deletion and reporter gene experiments have defined discrete distal regulatory elements that are responsible for each facet of this complex expression pattern (Fig. 2.7). These regulatory elements are scattered throughout the noncoding region of the gene without any apparent order or logic to their arrangement (Fig. 2.7).

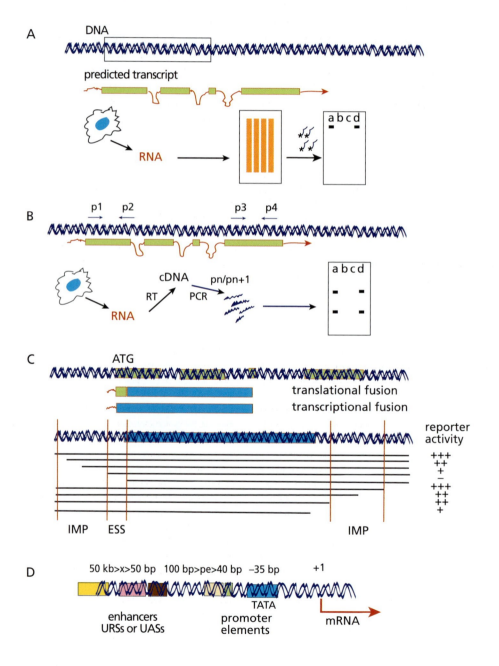

A striking example of spatial control of gene expression is provided by the regulatory regions of the genes known as *Hox* genes (Fig. 2.8). These genes are expressed in defined domains along the anteroposterior axis of an embryo and their function is to contribute to the development of the different morphological and physiological characteristics of each level. Both in vertebrates and some invertebrates, these genes are grouped together in one or more clusters. In each cluster, the order of the genes reflects the spatial order in which they are expressed during development along the anteroposterior axis of the body. This ordered expression is directed by an intricate array of regulatory sequences surrounding the genes within the cluster. The full details of how this integrated control system works are not yet known and are likely to vary to some extent in different types of organisms. Experiments in mice and observations in some mutants of *Drosophila*, in which individual regulatory sequences have been transposed to different parts of

the cluster, suggest that the overall pattern of expression depends both on local interactions between neighbouring genes (e.g. some promoter and enhancer sequences control more than one gene) and on higher-order control mechanisms involving larger groups of genes or the whole cluster (Fig. 2.8).

As mentioned above, comparative approaches are also very useful in pinpointing regulatory sequences. In some experiments, regulatory sequences transferred into a different organism have been found to be able to direct expression of a reporter gene in a manner reminiscent of the action of the homologous control sequence in the host organism. For example, a regulatory element of the human *Hoxb4* gene directed expression of a reporter gene in a posterior head segment in *Drosophila*, mimicking part of the expression pattern of the *Drosophila* homologue of *Hoxb4*, the *Deformed* (*Dfd*) gene. Interestingly, the corresponding regulatory region of the *Dfd* gene directs expression in the

---

Fig. 2.6. Mapping regulatory regions on the DNA. The activity of a gene (i.e. its level of transcription), can be followed in a number of different ways. Two common methods are to assay activity directly, by Northern analysis or RT-PCR (reverse transcriptase polymerase chain reaction), or indirectly by reading the activity of a reporter gene. **(A)** In a Northern analysis, whole RNA is extracted from different cells (a, b, c, d), size-separated by gel electrophoresis, and transferred to a membrane that then is probed with a labelled sequence (asterisks) complementary to part of the region of interest (boxed). If the region of interest is being transcribed, the labelled probe will hybridize to RNA fragments of defined sizes. In the example shown, the region is transcribed into the predicted transcript in samples a and d, but not in b or c. This method also provides a measure of the length of the transcript, which is related to its motility in the gel. **(B)** The polymerase chain reaction (PCR) allows the specific amplification of small amounts of DNA by using primers for specific regions of DNA and repeated rounds of DNA replication *in vitro*. This technique can be adapted to the analysis of RNA sequences and thus of transcripts: the RNA is first copied 'backwards' into cDNA with reverse transcriptase (RT) and then pairs of primers (p1/p2, p3/p4 . . . pn/pn+1) specific to the region of interest are used to amplify it by PCR. The products of the PCR are then run on a gel. A band of the appropriate size will indicate that the gene has been expressed in the sample. Usually more than one pair of primers is used to avoid artefacts. The experiment gives the same result as in (A) (i.e. the gene is expressed in samples a and d). **(C)** In the reporter gene method, the coding sequence of the gene is replaced by the coding region of a gene encoding an enzyme with easily assayable activity. Commonly used enzymes are chloramphenicol acetyl transferase (CAT), luciferase, and *Escherichia coli* β-galactosidase (lacZ). For each of these there is a simple quantitative enzymatic assay that al-

lows the rate of expression to be measured. There are two types of fusions: translational fusions, in which the ATG used by the reporter is that of the gene under study, and transcriptional fusions, in which the ATG is that of the reporter. Expression differences between transcriptional and translational fusions for the same gene suggest the existence of translational controls (not shown). Sequences that regulate the expression of a gene can be identified by following expression of the reporter. Sections of the DNA surrounding the expressed sequence are deleted, and the ability of the remaining DNA to drive expression of the gene is measured. This analysis usually reveals multiple regions that influence the levels of expression. In the example shown there are sequences that are essential (ESS) for expression, and others that are important (IMP). Transcription is only abolished completely when deletions very near the transcription start site are made. At a much finer level of analysis, very short sequences in a suspected regulatory region can be replaced by a known sequence and the process repeated systematically along the whole sequence under study. This method, called 'linker scanning', can define requirements for expression to the single base pair, depending on the length of the linker. **(D)** Deletion analyses of the type described in (C) have revealed a basic set of *cis*-acting sequences close to the transcription start site that are essential for expression of most eukaryotic genes: the TATA box, usually located about 35 bp from the transcriptional start site (the +1 position) and other sequences between about 40 bp and 100 bp upstream from the start site (promoter elements, pe), which regulate the efficiency of transcriptional initiation. Upstream (and sometimes also downstream) of the promoter there are elements involved in the modulation and spatial regulation of transcription; these are called enhancers, upstream regulatory sequences (URS) or upstream activating sequences (UAS).

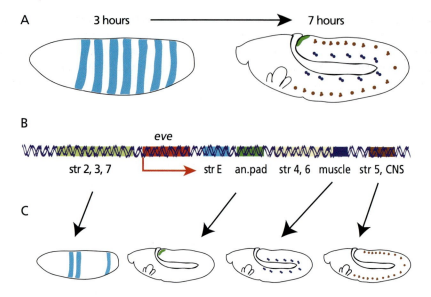

Fig. 2.7. Regulatory regions of the *Drosophila* gene *even skipped*. (A) Expression of the *even skipped* (*eve*) gene at two different stages of *Drosophila* embryogenesis: in the blastoderm and halfway through the proliferating phase of embryogenesis. At the blastoderm stage (left), *eve* (blue) is expressed in seven stripes that span the whole circumference of the embryo. Later (right) the stripes disappear and expression becomes prominent in elements of the mesoderm and the nervous system. The lateral view of an embryo shows *eve* expression in muscle precursors in the mesoderm (dark blue), in elements of the nervous system (brown), and also in an epidermal structure: the anal pad (green). **(B)** Modular nature of the *cis*-regulatory sequences of the *eve* gene. Regulatory sequences are located upstream and downstream of the transcription unit (indicated in red), and span about 20 kb. There are discrete elements regulating the expression of each of the subpatterns revealed in the normal pattern of expression: each of the different stripes (str), a general enhancer for all the stripes (strE), the muscle, the nervous system (CNS), and the anal pads (an.pad). **(C)** These elements can be identified by isolating them and using them to drive specific patterns of expression of reporter genes from heterologous promoters. (Modified from Sackerson, C., Fujioka, M., and Goto, T. (1999) The *even-skipped* locus is contained in a 16-kb chromatin domain. *Dev. Biol.* **211**, 39–52; and Fujioka, M. *et al.* (1999) Development analysis of an *even-skipped* rescue transgene reveals both composite and discrete neuronal and early blastoderm enhancers, and multi-stripe positioning by gap gene repressor gradients. *Development* **126**, 2527–38.)

Fig. 2.8. The regulation of *Hox* gene expression in *Drosophila* and mouse. (A) The *Hox* genes are a conserved set of genes that control cellular and morphological differences along the anterior–posterior (AP) axis in all animals. The *Hox* genes are organized in clusters; a given species has one or more *Hox* gene clusters, each comprising a set of genes that are expressed along the body axis in the same order as their arrangement along the chromosome. The gene expressed at the most anterior position in the embryo is located at the 5′ end of the cluster. In *Drosophila* there is just one cluster separated into two halves. Mice and other vertebrates have four clusters, some of which contain an incomplete complement of genes relative to the canonical set in *Drosophila*. Genes that are homologous between *Drosophila* and mouse are indicated by the same colours. Corresponding genes from different clusters within the same organism are known as paralogues. Genes expressed at a comparable AP level in different species are very similar in their primary structure. *Top*: a *Drosophila* embryo at a late stage of development, showing the spatial pattern of expression of the different *Hox* genes. *Bottom*: the *Hox* gene expression pattern in a mouse embryo. **(B)** The *Dfd* gene of *Drosophila* corresponds to the *Hoxb4* genes of mice and humans (see A), which are expressed within a small domain in the anterior region of the embryos of these animals. If the control region of the *Drosophila Dfd* gene is placed in front of a reporter and inserted in the mouse genome, it will drive expression of the reporter in a pattern that corresponds to a subset of the expression pattern of the *Hoxb4* gene. Conversely, the control region of *Hoxb4* (in this experiment the gene used was from humans) will drive expression of a reporter in *Drosophila* within a domain of expression of the *Dfd* gene. (Modified from Awgulewitsch, A. and Jacobs, D. (1992) *Deformed* autoregulatory element from *Drosophila* functions in a conserved manner in transgenic mice. *Nature* **358**, 341–4; and Maliccki, J. *et al.* (1992) A human *Hox4b* regulatory element provides head specific expression in *Drosophila* embryos. *Nature* **358**, 345–7.)

mouse in a pattern that is a subset of the normal expression pattern of the *Hoxb4* gene (Fig. 2.8). This remarkable and unexpected reciprocal interchangeability between insect and mammals suggests that both the basic software and the machinery to read it are to a large extent conserved across the animal kingdom.

Cell differentiation can involve more than simply switching specific genes on or off. The differentiated stage of a cell type may require that the expression of some genes is stably maintained, while others are kept stably 'silenced'. There are regulatory sequences in eukaryotes whose role is to maintain genes in an active or inactive state; these

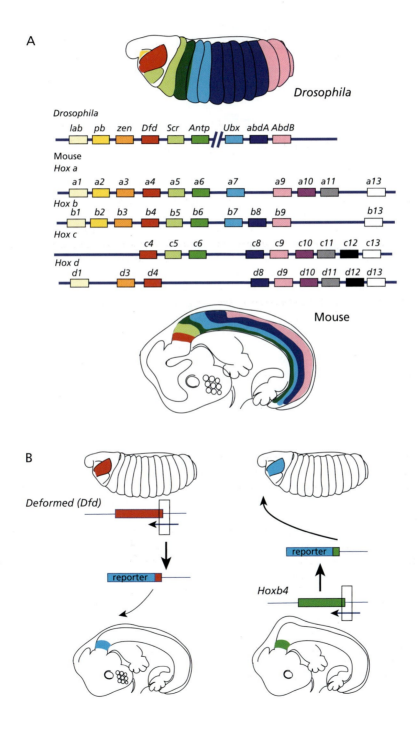

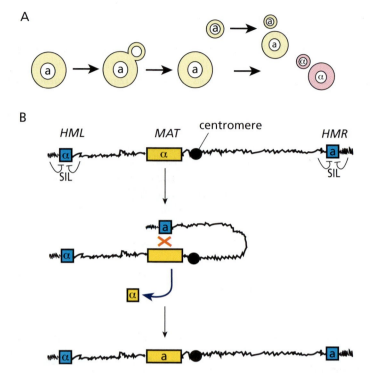

Fig. 2.9. **Regulatory events governing cell mating type in the yeast *Saccharomyces cerevisiae*. (A)** There are two haploid cell types in the yeast *S. cerevisiae*: **a** and α, each determined by a specific gene. The cell types are also known as mating types. The cells of this yeast divide by budding. A large cell (mother) buds a smaller daughter cell. Mother cells (i.e. cells that have previously budded) are able to change mating type: that is, an **a** mother cell can switch to the α mating type, such that after division its two daughters are both α. The larger (mother) cell resulting from this division is able to switch mating type again, but the smaller (daughter) cell cannot. The switch always occurs after division of the cell as shown. **(B)** The mating type is determined by the information present at a locus on the third chromosome, the *MAT* locus. If the DNA at the *MAT* locus codes for the **a** transcripts, the cell type will be **a**; if α it will be α. Near each end of the third chromosome, at locations known as *HMR* and *HML*, there is information to code for the **a** and α cell types. However, at these locations, the information is not expressed because of surrounding sequences known as silencers (SIL). The switch in mating type occurs because the information at *MAT* is substituted by information from that present at *HMR* or *HML*, through a special recombination process known as gene conversion. In this process, the DNA present at *MAT* is lost, and new information is copied from one of the silent loci. Because *MAT* is devoid of silencer sequences, the information is now expressed.

sequences often act via effects on the conformation of chromatin (Fig. 2.9). In haploid cells of the yeast *Saccharomyces cerevisiae*, for example, the sequence of the DNA at the *MAT* locus determines the cell's mating type (**a** or α alleles), and thereby which of two different cell differentiation pathways it will follow. On either side of the *MAT* locus, and a considerable distance away from it, there are extra copies of the **a** and α alleles at locations called *HMR* and *HML*. These copies are kept in a transcriptionally repressed state by specialized DNA sequences called 'silencers'. The silent alleles are only activated if the silencer sequences are experimentally removed or if the cell switches mating type by copying the appropriate allele into the expressed *MAT* site. If other genes are experimentally inserted into the genome at the silencer sites, they are also transcriptionally inactivated. The silencer sequences also have some ability to repress transcription of nearby genes if they are moved to other positions in the genome. It is thought that silencers act by binding specific proteins that package the chromatin in their vicinity into a transcriptionally silent or 'closed' conformation (Fig. 2.9).

Sequences that act to keep chromatin 'open' are less well defined, but there is some evidence that certain enhancer sequences associated with the *Drosophila Hox* genes become active as enhancers only after initial expression has been established. It has been suggested that these enhancers have an expression maintenance function that may be exerted via effects on chromatin conformation. The effects of chromatin on gene expression will be discussed in more detail in Chapter 3.

Irreversible modifications of the DNA sequence

Occasionally, the requirement to express a particular set of genes during differentiation is associated not with reversible control mechanisms such as gene activation and inactivation, but with changes to DNA that are irreversible within the lifetime of the cell. As mentioned above, changes in the sequence at the yeast *MAT* locus switch the cell's mating phenotype; this change is irreversible in that cell, and the new mating type is inherited by the cell's daughters after cell division. Some parasites, such as African trypanosomes, are able to change the proteins expressed on their surface by transferring genes from transcriptionally silent sites to 'expression sites'; those cells that have switched their surface proteins in this way are better able to escape the host's immune system, and survive to pass their genome on to their progeny. Many pathogenic bacteria are also able to vary their surface proteins as a result of DNA rearrangements.

In vertebrates, the differentiation of each antibody-producing B cell of the immune system involves the assembly and transcriptional activation of a unique pair of light- and heavy-chain immunoglobulin genes. Each immunoglobulin gene is created by the joining together of a unique set of three or four DNA segments that are 'chosen' by the cell from pools or sets of gene segments that are present in germ line cells (see Fig. 2.10 for details). In the process, many of the unused gene segments are irreversibly lost from the B cell genome. As far as we know at present, it appears that the irreversible loss of germ-line DNA associated with the development of the immune system is very much the exception among cell differentiation mechanisms in eukaryotes. However, there are hints that it may also operate in other systems that have a need for large diversity at the cellular and molecular levels, such as the nervous system. It is interesting that mice mutant for the specialized recombination system that mediates DNA rearrangements in the immune system also display striking defects in the development of the nervous system.

Regulatory regions in RNA

Regulatory sequences and instructions are not entirely restricted to DNA. RNA also contains instructions, though these are, of course, 'inherited' from the DNA. Instructions in RNA molecules can determine, for example, how the initial transcript is spliced. In some cases, alternative splicing can give rise to cell-type-specific protein variants, as in the case of the proteins fibronectin or tropomyosin (Fig. 2.11). Alternative splicing can also determine the function of the protein product of an mRNA. For example in *Drosophila*, alternative splicing of the *dsx* gene produces male- and female-specific proteins that contribute to sex determination (Fig. 2.11).

Other instructive RNA sequences can be found in the 5′ and 3′ untranslated regions (UTRs) of mRNA molecules—the sequences upstream of the translational start codon and downstream from the stop codon (Fig. 2.3). Both 5′ and 3′ UTR sequences can influence the rate of translational initiation of a mRNA: 5′ UTRs by forming secondary structures that affect ribosome binding and scanning, and 3′ UTRs in a variety of ways including modulation of the polyadenylation level of the mRNA, or binding of regulatory protein factors specific for a particular mRNA. The effects of regulatory UTR sequences on translation can be either positive or negative.

Regulatory mechanisms involving UTRs are particularly effective in localizing mRNAs to specific regions of the egg or early embryo, and in spatially or temporally restricting the translation of mRNAs that have been deposited by the mother into the egg. There are striking examples of these processes in *Drosophila* and *C. elegans* early embryos. In *C. elegans*, the *glp-1* mRNA is deposited by the mother into the egg and is present in all the cells of the early embryo. However, it is only translated during a specific time window, and only in some of the blastomeres (Fig. 2.12). The sequences controlling this selective spatial and temporal translation are experimentally separable and lie within the 3′ UTR of the *glp-1* mRNA, as demonstrated in elegant experiments in which they were used to direct the translation of reporter *lacZ* genes (Fig. 2.12). In *Drosophila*, the 3′ UTR sequences in the mRNAs for genes such as *bicoid*, *oskar*, *nanos*, and *hunchback*, are responsible for localizing the mRNAs to specific regions of the *Drosophila* oocyte or embryo and/or for controlling their translation. As we shall discuss in more detail in Chapter 3, these regulatory mechanisms acting at the level of RNA set up asymmetries in the distribution of the protein products of the mRNAs

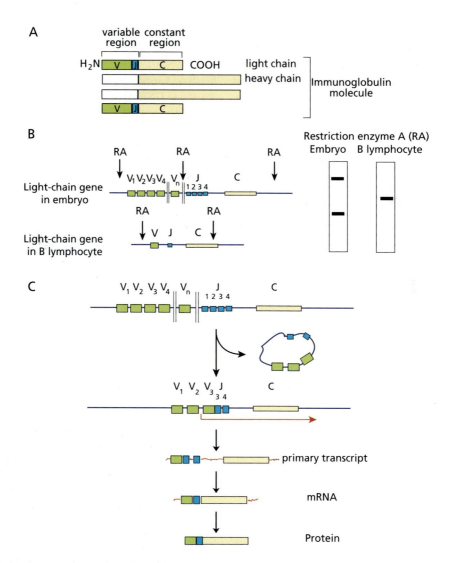

Fig. 2.10. The light chain of an immunoglobulin molecule is generated by DNA rearrangements in an immunoglobulin-secreting cell in the mouse. (A) An immunoglobulin G molecule consists of four chains of amino acids: two of the light type and two of the heavy type. Each of these chains has a constant (C) region at the carboxy-terminal end, and a variable (V) region, which is different from one immunoglobulin molecule to another, at the amino terminus. The variable region also contains sequences called joining (J) sequences. An adult B lymphocyte makes a particular type of immunoglobulin molecule with specific sequences of amino acids at the variable ends of the light and the heavy chains. **(B)** In the adult B lymphocyte, the DNA coding region for the light chain lies in a contiguous piece of DNA; the same applies to the heavy chain. However, in the embryo, the same piece of DNA is split. This can be shown by digesting the DNA in this region with a restriction enzyme: in the example shown, restriction enzyme A (RA) produces two fragments from embryonic DNA, but only one fragment from adult B lymphocytes. In the embryo, or in cells that are not from the immune system, there is a collection of related sequences each coding for a V region, illustrated here for the light chain only (though the same is true for

the heavy chain). This number can be as high as 200. Downstream from this there are four or five additional J variable regions, and further downstream again there is the constant (C) region. **(C)** During the development of the immune system a specific recombination event takes places that links a particular V region to a J region in a lymphocyte. This process is random: any V exon can be recombined with any J exon. As a result, individual B lymphocytes acquire a particular sequence of VJ and thus a particular variable region that enables them to mediate part of the function of the immune system. Transcription, splicing, and translation of the gene in each adult B lymphocyte produces the light chain characteristic of that cell. The information intervening between the V and J regions is lost from the B cell's genome. The arrangement of the exons for the heavy chain (not shown) is very similar, but with one variation. The variable region is made up from three different exons: a V and a J similar to those of the light chain, and a region called D, which joins the V and the J. There are many V and D exons for the heavy gene, and several J exons. Two recombination events make a VDJ exon which will be transcribed with and spliced to a heavy-chain-specific C exon to make the mature message for the heavy chain.

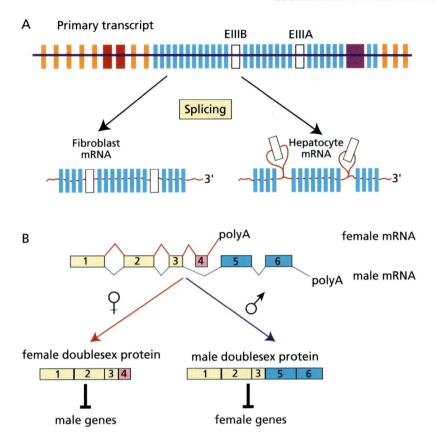

Fig. 2.11. Alternative splicing. (A) The gene encoding fibronectin is spread over 70 kb of DNA and contains about 50 different exons (white and coloured bars). The gene is transcribed as a single unit in all cells, but then tissue- and species-specific splicing produces mRNAs that are translated to produce proteins with particular characteristics. For example, the presence or absence of the EIIIA and EIIIB exons (white bars) is regulated in a tissue-specific manner. In hepatocytes they are spliced out whereas in fibroblasts they are not. **(B)** Sex determination in *Drosophila* provides an example of the use of alternative splicing in development (see also Chapter 3). The *doublesex* gene (*dsx*) has six exons and encodes two different regulatory proteins that result from alternative splicing. In females the third exon is spliced on to the fourth, which contains a polyA addition site and generates a regulatory protein that promotes female development and represses male-specific genes. In males the third exon is spliced on to the fifth, making an mRNA for a different protein which implements the development of male characteristics and represses female-specific genes.

that in turn participate in bringing about regional differences in cell fates (see Fig. 3.24).

Coordination of gene expression

In the preceding sections we have discussed how sequences within the regulatory regions of a gene or its mRNA contribute to determining the pattern of expression of that gene. We can now begin to build up the more complex picture of how the complete ensemble of regulatory sequences associated with a gene directs the correct expression of the gene in time and space.

As we have already discussed, the detailed characterization of the regulatory regions of many genes reveals that they are made up of a combination of short modules, each with a specific set of instructions for the expression of that particular gene. The *Endo16* gene from the sea urchin *Strongylocentrotus purpuratus* encodes a secreted protein that shows a dynamic expression pattern as development proceeds. A 2300 bp segment of DNA contains all the information needed to generate this pattern faithfully

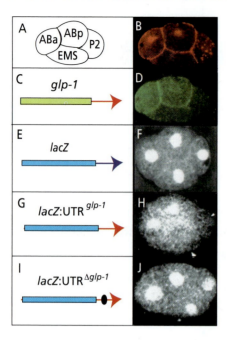

Fig. 2.12. Sequences that regulate location and translation of the *glp-1* mRNA in *Caenorhabditis elegans*. (A) *C. elegans* embryo at the four-cell stage. **(B)** A four-cell embryo stained for actin to highlight the surface of the cells and with a DNA dye to highlight the nucleus. **(C)** The *glp-1* mRNA has a coding region and a 3′ untranslated region (UTR, red arrow) and is present in the four cells of the embryo. **(D)** Four-cell embryo stained for glp-1 protein. Although the *glp-1* mRNA is expressed in all four cells, the glp-1 protein is produced only in the ABa and ABp cells. **(E)** An mRNA is engineered that contains a β-galactosidase gene (*lacZ*) with its own 3′ UTR (blue arrow). **(F)** If the *lacZ* mRNA is expressed in all four cells, the *lacZ* UTR directs expression of β-galactosidase in the four cells in which the mRNA is present. **(G)** In a different construct, the 3′ UTR of the *lacZ* reporter is substituted for that of *glp-1*. **(H)** Expression of the hybrid mRNA in all four cells of the embryo results in the expression of β-galactosidase only in the ABa and ABp cells (i.e. the pattern of expression of glp-1). **(I)** A small region of the *glp-1* 3′ UTR is deleted. **(J)** Expression of the hybrid RNA with the small deletion in the *glp-1* UTR abolishes the translational specificity of the *lacZ* gene. This experiment confirms the role of the *glp-1* 3′ UTR in the translational control of *glp-1* and identifies sequences important for this process. (From Evans, T. *et al.* (1994) Translational control of maternal *glp-1* mRNA establishes an asymmetry in the *C. elegans* embryo. *Cell* **77**, 183–94.)

(Fig. 2.13). Finer analysis of the 2300 bp segment reveals that it consists of a set of shorter sequences that each contains information for some facet of the pattern. Some sequences are associated with regulating levels of expression

and others with its spatial domains; the regulatory functions of the different sequence modules can interact, and both positive and negative interactions are observed (Fig. 2.13). This type of analysis reveals that regulatory sequences act as integrators of information and that the pattern of expression of a particular gene results from the processing of information at its promoter (Fig. 2.13).

Developmental programs

The ensemble of instructions in the regulatory regions of a gene determines many of its parameters of expression. The output of each set of gene activities in a cell includes the tools (i.e. the proteins) that read the instructions for the next set of operations. This, in molecular terms, is what we mean by saying that in developmental programs, the software encodes the hardware. In a given cell, the sequential activities at gene-regulatory sequences generate a program of gene expression characteristic of that cell type. Different cell types have different programs.

Because the output from each step of the program is required to decode the next step, it follows that, at least to some degree, 'timing' in development means 'order'. The precise timing of developmental events might suggest the existence of some sort of clock within the embryo. A closer look at what actually happens, however, shows that some DNA sequences that appear to work as 'timers' do so because they are triggered only when a specific set of precursor steps in the program have been completed and have thereby created the correct molecular environment. These sequences do not directly 'measure' elapsed time, like a clock, but respond to a set of molecular conditions that have taken a specific period of time to be generated.

The life cycles of some bacteriophages illustrate the concept of the developmental program and its relationship to timing in a simple and elegant way (Fig. 2.14) and provide a basis for thinking about more complicated events such as those involved in the development of eukaryotes. Bacteriophages are viruses that predate on bacteria and use them to reproduce. A typical bacteriophage consists of a DNA molecule and a proteinaceous coat, or capsid. In some bacteriophages the capsid has a 'head' region that surrounds the DNA molecule and a 'tail' through which the DNA is injected into a bacterial cell.

There are two main classes of bacteriophage: lytic and temperate (or lysogenic). In the lytic phages, such as bacteriophage T4 (Fig. 2.14), the virus finds its host and injects the DNA, which then subverts the replication machinery of the bacterium to its own ends, replicating its genome

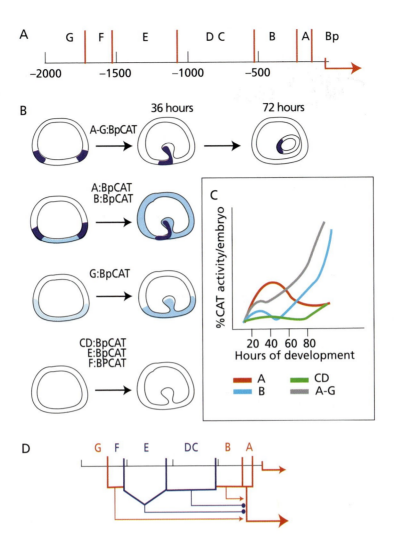

Fig. 2.13. Integration of information from the modular elements of a *cis*-acting control region. The sea urchin *Endo16* gene. The *Endo16* gene encodes an enzyme that is expressed and functions in the midgut of the sea urchin *Strongylocentrotus purpuratus*. **(A)** A 2300 bp sequence of DNA in the regulatory region of *Endo16* can direct the expression of a reporter gene in a pattern identical to that of the wild-type gene. Within this fragment, there are promoter elements essential for expression (basal promoter, Bp), and six distal regulatory elements (A, B, CD, E, F, G). **(B)** The different transcriptional regulatory elements were identified by testing their ability to direct transcription *in vivo*. In these experiments, the regulatory elements were linked to the *Endo16* Bp elements and a CAT reporter gene. The complete *cis*-regulatory region (elements A–G) recapitulated the full expression pattern (dark blue) of the *Endo16* gene: the gene is first expressed at the blastoderm stage in a group of cells on one side of the embryo (the vegetal plate). By 36 hours, during gastrulation, it is expressed in some of the cells that are moving inside the embryo; by 72 hours it is restricted to the midgut. Elements A and B were able to direct much of the wild-type expression pattern, to varying degrees. However, in both cases there was ectopic expression (light blue). This suggests that some of the other elements of the regulatory region are involved in the suppression of

the activation provided by A and B in these cells. The G elements direct an expression pattern similar to A and B, but at lower levels. A composite regulatory region containing A, B, and G behaves similarly to any one of these elements on its own (not shown). CD, E, and F are not active, either alone or in combination, but when combined with either A or B (not shown) they suppress the activity of the reporter in the cells in which it is not expressed, suggesting that they provide a negative element to the ABG combination. **(C)** Kinetic assays, in which CAT activity was measured quantitatively over time, show how the different *cis*-regulatory regions combine to produce the wild-type levels of expression. **(D)** These experiments suggest the following function for the *Endo16 cis*-regulatory region at the blastula and early gastrula stages. The AB and G elements provide a positive input for expression over all cells of the vegetal plate. The three elements have to work in concert to achieve the correct and properly regulated levels of activity of the basal promoter. The CDEF elements provide a repressive function for the activity of the A element in some of the surrounding regions of the embryo. Additional experiments show that each of the elements has a different role in this repression. (Modified from Yuh, C-H. and Davidson, E.H. (1996) Modular *cis*-regulatory organization of *Endo16*, a gut-specific gene of the sea urchin embryo. *Development* **122**, 1069–82.)

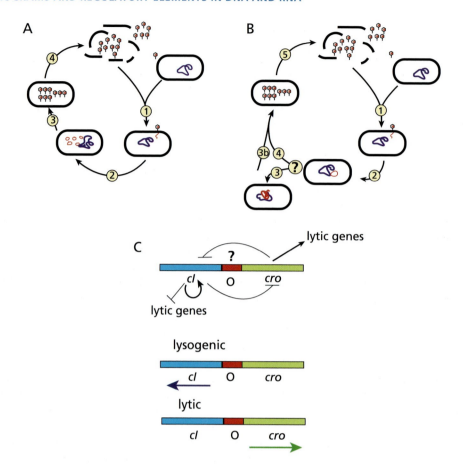

Fig. 2.14. Life cycle and genetic programs of bacteriophages.
(A) Lytic phages infect bacteria (1) and initiate an ordered program of gene expression that first arrests the activities of the infected bacteria, then replicates the DNA of the bacteriophage (2) and finally triggers the synthesis of the proteins (3) that will coat the DNA and generate new phage particles (4) which destroy the cell and propagate themselves. **(B)** Lysogenic phages, like bacteriophage lambda, upon infecting bacteria (1,2) have a choice: they can follow a lytic program (4), or they can activate a series of genes whose function is to inhibit the lytic program and to accomplish the silent insertion of the bacteriophage DNA into the bacterial chromosome (3). This second option, known as lysogeny, is essentially a parasitic state in which the bacteriophage DNA is replicated along with the bacterial chromosome. Under certain environmental conditions, the lytic program can be activated from the inserted DNA (3b), leading to the generation of new phage in a similar manner to the lytic phages (5). **(C)** Regulatory region that controls the lysis/lysogeny decision in bacteriophage lambda. Two genes, one encoding a repressor (*cl*), and the other an activator (*cro*) share a regulatory region (O). The product of *cl* performs three functions: it suppresses the expression of *cro* by occupying the O sequences which direct gene expression, it suppresses the expression of genes involved in the lytic phase, and

it reinforces its own expression. Together, these three activities suppress the ability of lambda to destroy the bacteria and lead to the insertion of the lambda genome into the bacterial chromosome. The product of the *cro* gene activates the lytic program of gene expression and suppresses the expression of *cl* from O. The cl and cro gene products compete for occupancy of O; which program prevails depends on which of the two proteins prevails at O. This depends on a mixture of chance and environmental conditions inside the bacteria, for example nutrition, or DNA damage.

many times, making its coat proteins in appropriate proportions, packaging the DNA within the coat and emerging from the lysed host cell as a burst of progeny virus particles that can go on to infect other bacterial cells. Once the phage DNA has entered the host cell, its constituent genes are expressed in a precise and orderly manner which ensures that each step of the life cycle is completed before the next begins. The first genes to be expressed are those devoted to DNA synthesis and gene expression, followed by genes involved in the synthesis and assembly of the head and the tail. The program is a simple one and the timing of individual events simply reflects their place in the sequence: the products of some of the 'early' genes are required to switch on the 'intermediate' genes, and synthesis of certain intermediate gene products is a prerequisite for activation of the 'late' genes. Although there is a large number of genes overall, they are constrained to be expressed in a specific order because each step of the program decodes the next.

The life cycles of temperate phages, such as bacteriophage lambda, illustrate another very important feature of developmental programs: the binary switch (Fig. 2.14). Upon infection, bacteriophage lambda can 'choose' between lysing the bacterial host cell in a similar way to bacteriophage T4, or integrating its DNA within the host cell genome to become a dormant 'prophage'. The 'choice' that is made initiates one of two complete and mutually exclusive developmental programs. The initial switch between the two states is activated by environmental conditions and is mediated by a region of DNA in which regulatory sequences that control the two programs partially overlap each other (Fig. 2.14).

The life cycles of the bacteriophages show how a sequence in the DNA can determine what happens next, either by means of the protein it encodes or by regulating the expression of other proteins. The principle turns out to be the same in multicellular animals. Development depends on differential gene expression and cells constantly face 'choices' as to what genes they should express; these choices are fundamentally binary choices—just like those that bacteriophage lambda makes between the lytic and lysogenic cycles—though the simplicity of the underlying switch can often be obscured if the 'decision' leads to the expression of different regulatory proteins that themselves control large batteries of genes. The process of sex determination in mammals makes this point clearly. Here, the presence of a single gene (in this case the sex-determining *Sry* gene on the mammalian Y chromosome) leads to a profound difference in the developmental program that is followed. The decoding of the *Sry* gene initiates a cascade of steps that leads irreversibly to the generation of the male phenotype; in the absence of *Sry*, a different series of molecular operations results in the female phenotype. Simply switching the *Sry* gene on at a later developmental stage, however, does not convert a female mammal to a male. *Sry* can only exert its function at a specific developmental stage, when the molecular context will allow the developmental program to be set in motion.

Higher-order programs

Ultimately, the development of a multicellular organism is the integrated output of a coordinated set of molecular programs operating at the cellular level. This higher-order program is what we observe in the precise and reproducible manner in which embryos of a given species develop.

A striking example of a developmental program operating at the level of the whole organism is provided by *C. elegans*. Each adult worm contains a fixed number of cells—959 in the hermaphrodite—which are produced through an invariant pattern of cell divisions that allows the descent of every constituent cell to be followed. In *C. elegans* development, everything is predetermined: how many cells are born; how many die; when and where these cell births and deaths occur; which cells differentiate to contribute to which tissues and organs. Considerably less plastic and variable than more complex eukaryotes, *C. elegans* may prove to be the first multicellular organism for which the molecular underpinnings of the developmental program are elucidated. The availability of its complete genome sequence, its determinate development, and the ease of genetic analysis may combine to make understanding the connection between the DNA program and its output a feasible achievement in the near future. As one moves to higher levels of complexity in other types of eukaryotes, the numbers of cells and their patterns of division and behaviour are not so precisely determined; this means that there are other ingredients coming into play, as we shall see in subsequent chapters.

Underlying the patterns of cell division and diversification in multicellular animals are complex patterns of gene expression—programs of a complexity many times greater than that seen in phages. Developmental biology consists in large part of attempting to unravel these programs. The realization that the molecular machinery is conserved has made it clear that the differences between organisms must lie in the programs: in the sequences of gene expression characteristic of each organism and how these are organized in space and time. The decoding of complete genomes is beginning to open up avenues that could allow glimpses of these programs in action. These new approaches, which have become known as 'functional genomics', rely on the analysis of whole-genome responses to particular perturbations or under certain conditions.

A study of the program of gene expression during sporulation in the yeast *S. cerevisiae* gives an example of what can be achieved. Sporulation, a response of yeast cells

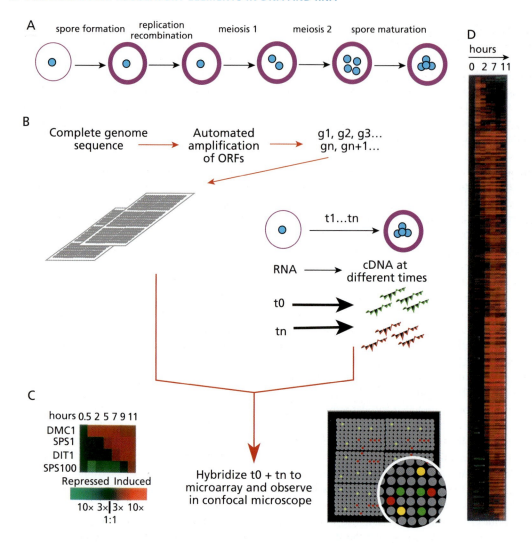

Fig. 2.15. Whole genome analysis of a simple developmental process: Sporulation in the yeast *Saccharomyces cerevisiae*. (A) When yeast cells are starved of nutrients, they stop dividing, enter into a quiescent phase, and undergo meiosis. Subsequently, they undergo a morphogenetic process in which a spore forms that contains the products of meiosis. **(B)** The patterns of transcriptional activity of the whole genome can be observed by the use of DNA microarrays that display sequences representative of the complete set of yeast genes. The complete set of ORFs of the yeast genome (g1, g2 . . . gn, gn + 1) is synthesized *in vitro* and individual products are placed on a grid (the size of a microscope cover slip) in an order that permits their identification. cDNA is prepared from a yeast culture at time zero (t0) and labelled with a fluorochrome (Cy3 green in this example). The culture is then induced to undergo sporulation. RNA is prepared from cells at different stages of the process (t1 . . . tn) and used to synthesize stage-specific cDNA, which is labelled with a different fluorochrome (Cy5 red). The cDNA from each stage is then hybridized to the arrayed ORF library, together with the t0 cDNA, and the results are observed using a confocal microscope. If an ORF fluoresces green, this means that the corresponding gene has been repressed. If it appears red, it has been induced, and if yellow, its expression has not altered. **(C)** Example showing the representation of four genes in this experiment. The genes are arranged from the top to the bottom of the panel according to the order in which they are induced (red) during the course of the experiment. **(D)** A time series carried out in this way produces a dynamic picture of the activity of the whole genome during the process of sporulation. Each thin horizontal line represents one gene. Because the identity of each ORF is known, it is possible to begin to build up a picture of the sets of gene functions that are deployed (or down-regulated) during the developmental program. Ultimately, the aim is to decipher the program itself. (Adapted from Chu, S. *et al.* (1998) The transcriptional program of sporulation in budding yeast. *Science* **282**, 699–705.)

to various environmental conditions, can be viewed as a simple process of cell differentiation (Fig. 2.15). Upon starvation, the yeast cell slows down its metabolism and undergoes a series of physiological and cellular changes that lead to meiosis and to the production of a cyst containing gametes for the next generation. The availability of the whole yeast genome sequence has allowed the development of microarrays or 'chips' of expressed sequences that encompass the whole genome (Fig. 2.15). These arrays can be probed with RNA specific to different stages of the sporulation process. Because the identity of genes at particular locations of the arrays is known, this procedure makes it possible to follow the time course of changes in concentrations of transcripts from each gene. A time series after a yeast culture is transferred from a rich medium to one that induces sporulation provides a complete picture of the activity of the genome throughout this period. In this way, more than 1000 genes were identified that are either induced or repressed during sporulation (Fig. 2.15), compared with 50 known from classical genetic studies, and the timing of their expression changes could be correlated with visible events during sporulation.

Similar approaches in other organisms have the potential not just to uncover all the genes involved in specific developmental programs, but also to provide a basis for ordering their action and predicting their functions.

SUMMARY

1. The DNA of an organism contains the information to generate all the proteins of that organism. There are two classes of DNA sequences: coding sequences, which determine the structure of the proteins, and noncoding regulatory regions, which determine where and when the coding regions are transcribed into mRNA.

2. There is no simple relationship between the DNA content of an organism and its coding capacity: the same amount of DNA can encompass different numbers of genes both within one organism and in different organisms.

3. The regulatory regions of genes are composites of short, discrete DNA sequences. The combination and arrangement of regulatory sequences determines when and where a particular gene is expressed.

4. Functional relationships between the regulatory regions of different genes create programs of gene expression.

5. In viruses and bacteria, programs of gene expression are usually linear, with few decision points. External influences on the program tend to be mainly from the environment, in the form of fluctuations in levels of nutrients or other substances. Thus cells generally behave as individual units in the running of their developmental programs.

6. The developmental programs of multicellular organisms are subject to multiple decision points in individual cells. The decisions made at these points are regulated by communication between cells, which exchange information about their individual states.

7. The development of a multicellular organism results from the unfolding, in every cell, of a specific set of information contained in the DNA. The instructions for the unfolding of this information are encoded in the regulatory regions of the DNA.

COMPLEMENTARY READING

Brent, R. (2000) Genomic biology. *Cell* **100**, 169–83.

Chu, S., DeRisi, J., Eisen, M., Mulholland, J., Botstein, D., Brown, P. O., and Herskowitz, I. (1998) The transcriptional program of sporulation in budding yeast. *Science* **282**, 699–705.

Davidson, E. (2001) *Genome regulatory systems.* Chapters 1 and 2. Academic Press.

Ferea, T. L. and Brown, P. O. (1999) Observing the living genome. *Curr. Opin. Genet. Dev.* **9**, 715–22.

Furlong, E. E., Andersen, E., Null, B., White, K., and Scott, M. P. (2001) Patterns of gene expression during *Drosophila* mesoderm development. *Science* **293**, 1629–33.

Gurdon, J. B. and Colman, A. (1999) The future of cloning. *Nature* **402**, 743–6.

International Human Geneme Sequencing Consortium (2001) Initial sequencing and analysis of the human genome. *Nature* **409**, 860–921.

Rubin, G. M. *et al.* (2000) Comparative genomics of the eukaryotes. *Science* **287**, 2204–15.

Ruvkun, G. and Hobert, O. (1998) The taxonomy of developmental control in *Caenorhabditis elegans*. *Science* **282**, 2033–41.

Venter, J. C. *et al.* (2001) The sequence of the human genome. *Science* **291**, 1304–51.

Chapter 3

Decoding the program: Transcription

As we have seen in Chapter 2, the information for initiating and regulating gene expression during development is located in the noncoding sequences of the DNA. Accessing this information requires the activity of proteins that can 'see', read, and execute the instructions in the regulatory sequences. Consequently, the pattern of activity of any cell at a given point in development, although encrypted in the DNA, is determined by its complement of proteins and, in particular, by those that can read the DNA.

This has been demonstrated in experiments in which cells from different mammalian species are fused to form hybrid cells with more than one nucleus: heterokaryons (Fig. 3.1). In such interspecific heterokaryons, the proteins produced by each type of nucleus can be distinguished. When an undifferentiated human cell is fused with a differentiated mouse muscle cell, human muscle proteins can be detected in the hybrid, indicating that *trans*-acting factors (proteins) in the differentiated mouse cell have activated expression of muscle genes in the human cell nucleus (Fig. 3.1). Fusions between mouse muscle cells and some (but not all) types of differentiated human cells gave similar results, again showing that at least in some circumstances the protein milieu of a cell can reprogram gene expression, even if a different program of differentiation has already begun. In effect, this is the basis of the cloning experiments discussed in Chapter 2.

Proteins that are involved in interpreting the instructions in the regulatory regions of the DNA are called transcription factors. Many transcription factors bind DNA; these factors represent the direct molecular interface between the cell and the program encoded in its DNA: regulatory (control) sequences in the DNA define molecular facets that are recognized and bound by cognate molecular facets in the transcription factor protein. In recent years many other proteins have been identified that also play a role in transcriptional regulation, although they may not directly bind DNA. We shall also refer to these proteins as transcription factors though they are sometimes given alternative names such as coactivators or coregulators.

The key common player in the expression of every gene is the enzyme RNA polymerase, which catalyses the synthesis of an RNA molecule complementary to the DNA sequence of the gene. Prokaryotes have only one type of RNA polymerase, while eukaryotes have three: RNA polymerase I is devoted to the transcription of rRNAs, RNA polymerase II mostly to the expression of protein-coding genes, and RNA polymerase III to the transcription of 5S rRNAs, tRNAs, and snRNAs.

In this chapter, we shall discuss in some detail the mechanisms that lead to the initiation of transcription by RNA polymerase II and thus to the synthesis of proteins. The reason for this emphasis is that it is the differential activity of this enzyme that is the ultimate molecular basis for the developmental process. The outcomes of the activities of polymerases I and III are, for the most part, common to all cells and, while there is some regulation of these activities, their contribution to developmental events is less clear.

Transcription in prokaryotes

Prokaryotes perform the same fundamental biological processes as eukaryotes but with considerably less complexity. Transcription is no exception and so a review of the basic mechanics of this process in bacteria serves as a useful way of introducing some of the molecular principles that also apply to eukaryotic cells.

The RNA polymerase of *Escherichia coli* is made up of five subunits (Fig. 3.2). Four of these—two identical α subunits and two larger ones called β and β′—form the

A

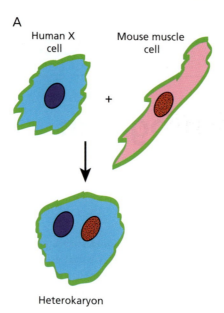

B

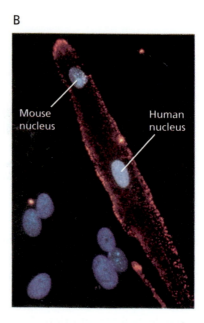

basic holoenzyme. A smaller subunit, σ, acts in the initiation of transcription and dissociates from the enzyme after the first few nucleotides have been polymerized. The role of the σ subunit, by contacting both the DNA and the other subunits, is to ensure that these interactions occur at the right place, the TATA box. This basic RNA polymerase enzyme can direct transcription *in vivo* and *in vitro* from naked DNA and the efficiency of the process is determined, in most cases, simply by the sequence of the promoter.

RNA polymerase exists in two isomeric states: a 'closed' complex that is unable to initiate transcription, and an 'open' complex that is able to loosen the strands of the DNA double helix so that transcription can begin (Fig. 3.2). In many cases, the polymerase alone can initiate transcription of the gene: the holoenzyme tracks along the DNA looking for specific initiation sequences and, when it finds them, begins the polymerization of nucleotides and the process of transcription. In other cases, specific protein factors, transcription factors, help the activity of the enzyme either by recruiting it to the relevant sites, or by making it active after it is bound to the DNA but unable to effect isomerization from the closed to the open state. In some instances, different types of σ factors function in the initiation of transcription of these different types of genes. For example, transcription of the *lac* operon (Fig. 3.3), which requires recruitment of RNA polymerase to the promoter by activator proteins, uses the initiation factor σ^{70}, while transcription of the *glnAp2* gene, which requires

Fig. 3.1. Regulation of nuclear activity by cytoplasmic factors in heterokaryons. (A) Heterokaryons result from the fusion of two different cells types. In a particularly informative set of experiments, cells from a muscle cell line derived from mouse were combined with a variety of human cell types derived from different germ layers (keratinocytes from the ectoderm, chondrocytes and fibroblasts from the mesoderm, and hepatocytes from the endoderm). The heterokaryons can be easily identified by the presence of multiple nuclei, and the nuclei from each cell type and species can sometimes also be identified: in this case, upon staining with a DNA dye, such as Hoechst, the mouse nucleus displays a granular appearance while the nucleus from the human cell looks smooth and complete. The effect of the mouse cell on the human nucleus is then investigated by testing for the appearance of muscle-specific proteins derived from the activity of the human nucleus. **(B)** Binucleate heterokaryon of a human hepatocyte and a mouse muscle cell showing the activation of a human-specific muscle gene encoding the protein 5.1H11 (pink stain). This gene is is not expressed by the hepatocyte cell under normal conditions. The mouse nucleus can be easily identified by its punctate appearance. Note that other heterokaryons around this one have not activated 5.1H11. (From Blau, H. *et al.* (1985) Plasticity of the differentiated state. *Science* **230**, 758–66.)

binding of an additional protein, NTRC, to activate a prebound polymerase, uses the σ^{54} initiation factor. Interestingly, the binding site for NTRC is a considerable distance upstream from the initiation site, a situation usually more characteristic of eukaryotic promoters.

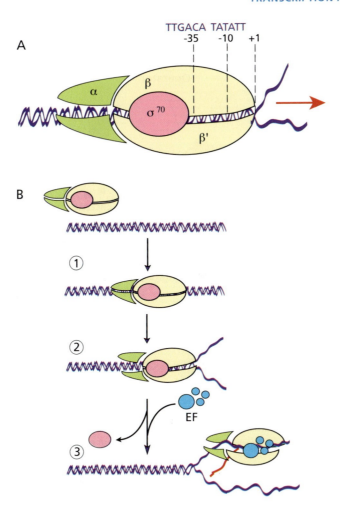

Fig. 3.2. Transcription in a prokaryote, *Escherichia coli*. (A) Cartoon of the RNA polymerase molecule from *E. coli* bound to DNA. The complete enzyme is composed of two identical α subunits of 37 kDa each, two closely related β subunits—β (151 kDa) and β′ (156 kDa)—and (for the gene illustrated) a 70 kDa σ subunit called σ^{70}. The σ^{70} subunit binds the DNA at specific sequences located 35 and 10 bp upstream from the transcription initiation site and is responsible for the specificity of the initiation. **(B)** The polymerase complex binds to the DNA and the selection of the site of transcription initiation is mediated by the σ^{70} subunit. This 'closed' complex (1) has to isomerize to an 'open' form (2) in order for transcription to start; this can be achieved upon interaction with a specific DNA sequence or with the aid of other proteins. As transcription proceeds (3), σ^{70} is released and other proteins, called elongation factors (EF), join the complex. The complex now proceeds to polymerize the RNA molecule in the 5′ to 3′ direction, until it encounters termination signals on the DNA and recruits termination factors that facilitate the termination reaction.

Not all specific transcription factors are activators. The *lac* operon (Fig. 3.3) also provides an example of a transcription factor that functions by inhibiting transcription: the lac repressor which prevents binding of the polymerase by occupying a region of DNA that physically overlaps the binding site of the RNA polymerase. This can be observed by the technique of footprinting, which delineates the binding sites of different proteins on the DNA (Fig. 3.4). Transcription proceeds when specific inducer molecules, such as lactose derivatives, bind the repressor and provoke its dissociation from the DNA through a conformational change (Fig. 3.3). This balance between positive and negative controls, achieved by relatively simple molecular mechanisms, allows the bacterium to make the metabolic adjustments necessary for survival and growth in a changing external environment. The efficiency of transcription of the *lac* operon is increased by the activity of the

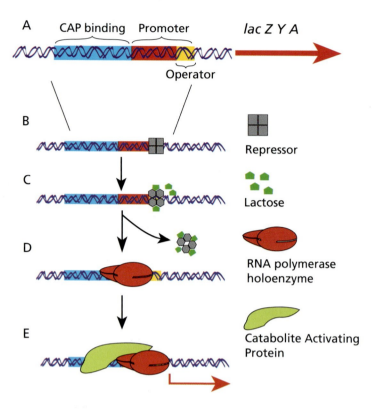

catabolite activating protein (CAP), which interacts with the RNA polymerase to aid the initiation process (Fig. 3.3).

The transcriptional machinery in eukaryotes

The transcription of a eukaryotic gene is a more complex affair than gene expression in prokaryotes for a variety of reasons. First, in contrast with prokaryotes, the DNA of eukaryotes is wrapped around histones to form nucleosomes, which in turn are packaged with other proteins to form the higher-order structures of chromatin (see Fig. 2.1). This means that neither regulatory nor coding sequences are readily available for interaction with the proteins that initiate, execute, and terminate transcription. Second, unlike, say *E. coli*, which has evolved to run programs poised to adjust to a relatively limited number of environmental circumstances, eukaryotes—particularly multicellular ones—have to find ways of expressing many different combinations of genes to generate a wide variety of cell types, and correspondingly of inactivating all those genes that are not required. As a consequence, transcription in eukaryotes employs much larger numbers of both specific and

Fig. 3.3. **Regulatory interactions at the *lac* operon.** The regulation of the expression of the *lac* operon of *Escherichia coli* provides a simple model to illustrate the basic features of the control of transcriptional initiation. **(A)** When *E. coli* grows on lactose, it transcribes and translates three genes that are part of the same primary transcript: the *lac* operon. The first gene (*lacZ*) encodes the enzyme β-galactosidase, the second one (*lacY*) a permease, and the third one (*lacA*) a transacetylase. Genetic analysis identified three regions upstream of the protein-coding DNA that determine whether the gene is expressed. These regions are the CAP (catabolite activating protein)-binding region, required for maximal expression; a promoter region, which represents the site for RNA polymerase binding; and a small region within the promoter that determines whether or not RNA polymerase can operate. This region is called the Operator. **(B)** When *E. coli* is not growing on lactose the genes are not expressed because a repressor protein is bound to DNA in the operator region, obliterating part of the binding site for the RNA polymerase holoenzyme. **(C)** When there is lactose in the medium, it binds the repressor and induces a conformational change that results in the repressor dissociating from the DNA. **(D)** RNA polymerase can now bind to the promoter and initiate transcription. **(E)** Transcription of the *lac* operon genes is more effective when a protein, catabolite activating protein (CAP), which responds to the presence or absence of glucose in the medium, is bound upstream of the promoter site and contacts the RNA polymerase to increase its efficiency at initiation.

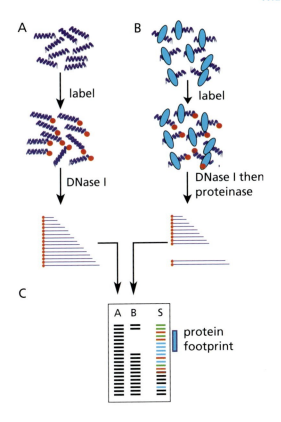

Fig. 3.4. Use of DNA footprinting to delineate protein-binding regions on the DNA. (A) A piece of DNA, radioactively labelled at one end (red), is treated with the enzyme DNase I, which cuts DNA non-specifically, at a concentration such that every DNA molecule in the solution will be cut once. The site of the cut in each molecule is random, and when a sample of the reaction is run on a polyacrylamide gel a series of bands is revealed that differ 1 bp in length. **(B)** If the experiment in A is repeated with a protein bound to the DNA, the protein–DNA interaction will shield certain fragments of the DNA from DNase I and, after termination of the reaction, elimination of the protein and running on a gel, a gap or 'footprint' will appear in the pattern of bands. The gap lies exactly where the protein was bound. **(C)** Comparison of the two patterns A and B reveals the precise place where the protein is bound in that sequence. If a sequencing reaction of the region (S) is run in parallel, the sequence to which the protein binds can be determined.

general transcription factors, to ensure that the necessary discrimination between expression patterns in different cells can be achieved.

A variety of experiments have shown that in eukaryotes, transcription is initiated at specific sites on the DNA. It is possible to isolate an RNA polymerase II (pol II) complex containing 10–12 subunits that is capable of synthesizing RNA *in vitro* from a DNA template. However, in contrast to the complex from *E. coli*, the initiation specificity of this preparation is very low, even on an optimal DNA sequence, unless it is supplemented with a cellular extract containing other proteins. Six 'general initiation factors' (or general or basal transcription factors) named TFII A, B, D, E, F, and H, have been isolated from such extracts (Fig. 3.5 and Table 3.1). Some of the general initiation factors are themselves composed of more than one subunit, for example, TFIID consists of the TATA-binding protein (TBP) which, as its name implies, recognizes the TATA box, and several non-DNA-binding proteins known as TBP-associated factors (TAFs). The amino acid sequences of the general initiation factors show a remarkably high degree of evolutionary conservation among eukaryotes and have been shown to be functionally interchangeable between organisms. The complex of pol II and general initiation factors is sufficient to initiate transcription *in vitro*; in this respect, the general initiation factors can be considered the functional equivalents of the σ factor of prokaryotes.

Genetic and biochemical experiments in yeast have revealed that *in vivo* the transcriptionally competent pol II holoenzyme contains, in addition to general initiation factors, a number of other proteins. These include 11 proteins that are part of a complex called mediator (MED) and a set of nine proteins in a second complex called SRB (for suppressors of RNA polymerase B), which interact with the carboxy-terminal domain (CTD) of the polymerase and with the MED complex (Fig. 3.5). For this reason the complex is called SRB/MED. The SRBs were identified first from genetic experiments as proteins that when mutated could compensate for functional deficiencies in transcription assays caused by truncations of the carboxy terminus of pol II. This suggests that some of them play an inhibitory role in the process of initiation. Biochemical experiments have shown that members of the SRB/MED complex interact directly with the pol II complex. Although the components of SRB/MED are not required for the activity of the enzyme *in vitro*, mutations in these cofactors do have effects on transcription *in vivo*.

The characterization of SRB/MED complexes in other eukaryotes, and the finding that several of their components are structurally similar to components of the yeast complex, suggest that the picture drawn from the studies in yeast can be extrapolated to more complex organisms. However, in contrast to yeast which has one, or perhaps two different SRB/MED complexes, higher eukaryotes appear to have a range of cell-type-specific complexes with

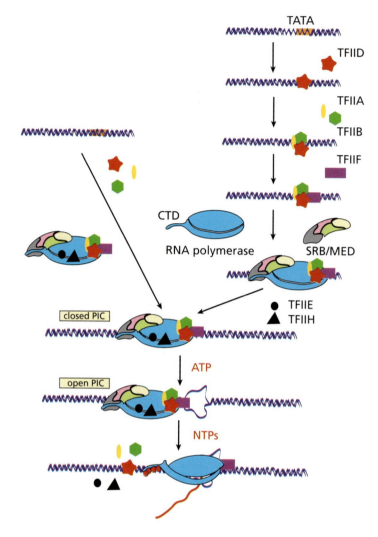

Fig. 3.5. Outline of the initiation of transcription in eukaryotes. (A) The process of transcription, both *in vivo* and *in vitro*, requires the assembly, at the promoter, of an RNA polymerase holoenzyme and a complex of proteins termed SRB/MED. How this complex assembles is not well understood, but a picture is emerging from analysis of the nature and order of the protein–protein interactions required to elicit transcriptional activity. There are two extreme views of the process. In one (right), the assembly is ordered and begins with the binding of TFIID to the TATA box. TFIID is a complex of the TBP (TATA-binding protein) and several TBP-associated factors (TAFs). The binding of this complex to the DNA acts as a nucleation event for the binding of other general initiation factors, such as TFIIA and TFIIB, followed by TFIIF and the 12 subunit complex of RNA polymerase. The carboxy-terminal domain (CTD) of polymerase, which is characterized by a 26 tandem repeats of a hep-

tapeptide, interacts with the SRB members of the SRB/MED complex, which are required for efficient transcriptional initiation. In an alternative model (left), a largely pre-assembled holoenzyme finds the promoter region and binds to the TATA region through TBP. This association generates a closed pre-initiation complex (PIC). In both models, once formed, the PIC is activated in an ATP-dependent manner to form an unstable open PIC. In the presence of nucleotide triphosphates (NTPs), dissociation of most of the general initiation factors and TFIIH-dependent phosphorylation of the CTD of RNA polymerase triggers the synthesis of mRNA. (After Orphanides, G. *et al.* (1996) The general transcription factors of RNA polymerase II. *Genes Dev.* **10**, 2657–82; and Roeder, R. (1996) The role of general transcription factors in transcription by RNA polymerase II. *Trends Biochem. Sci.* **21**, 326–37.)

Table 3.1 General transcription initiation factors

Factor	Nr of subunits	MW (kDa)	Function
TFIIA	3	12, 19, 35	Stabilization of initiation complex, anti-repression
TFIIB	1	35	Recruits pol II and selects start site
TFIID/TBP	1	38	Binds TATA
TFIID/TAFs	12	15–250	Recognition of non TATA promoter elements and regulatory functions
TFIIE	2	34, 56	TFIIH recruitment; promoter melting
TFIIF	2	30, 74	σ-like function stimulates elongation
TFIIH	9	35–89	Promoter clearance
RNApol II	12	10–220	Catalytic functions

partially overlapping sets of constituents. These complexes affect the efficiency of transcription and might therefore contribute to the variation in the activity of promoters in different cell types.

Genetic and biochemical analysis of transcription initiation in yeast is providing an increasingly detailed picture of the mechanics of the process but the large number of components and the multiple possible interactions mean that there is undoubtedly still much to be learned.

Despite the different scale of the machinery and the different template on which it acts, initiation of transcription in eukaryotes (Fig. 3.5) has many similarities to the process in *E. coli*. For example, collectively the general initiation factors play a role analogous to that of the σ factors in bacteria, in selecting the correct sites for assembly of the pre-initiation complex of pol II and associated factors. Once the complex has assembled, the double-stranded DNA molecule melts to expose the template strand, and, like the bacterial polymerase, the pre-initiation complex is converted from a 'closed' to an 'open' conformation competent to begin transcription (Fig. 3.5).

After this point, further enzymatic activity is required for the elongation and termination steps. Elongation and clearance from the promoter is associated with the phosphorylation of the CTD of pol II, probably by a subunit of the TFIIH factor. At this stage, too, some of the general initiation factors (TFIIB, E, and H) appear to dissociate from the polymerase complex, like the σ factors of bacteria. A further set of general factors, not part of the initiation complex, has been found that are specific for the elongation and termination steps.

The complicated architecture of the transcriptional initiation machinery in eukaryotes might be related to the very stringent control that is required to execute the types of programs we have discussed in Chapter 2. A relaxed or non-restrictive transcriptional state, as in *E. coli*, could easily result in indiscriminate and/or inadequate patterns of transcription.

Enhancers and gene-specific transcription factors

As we have discussed in Chapter 2, efficient transcription in cells or within developing organisms requires, in addition to the coding region of the gene and its neighbouring sequences, regulatory sequences [known as enhancers, upstream activating sequences (UASs) or upstream regulatory sequences (URSs)] that can lie many hundreds of bases away from the transcription start site. Footprinting of these sequences indicates that they are the binding sites for proteins that cannot mediate transcription on their own but act together with the transcriptional machinery at the promoter to determine the rate and frequency of transcription initiation (Fig. 3.6). The molecular interaction between these transcription factors and the transcriptional machinery may be a direct one, or may be mediated by other, non-DNA-binding factors.

A DNA-binding transcription factor has two basic elements: a DNA binding domain and an interaction or activation domain through which it interacts with other proteins involved in the process of transcription. Although there are many different DNA-binding transcription factors, their DNA-binding domains are characterized by a relatively small number of different amino acid sequence motifs that have been selected and conserved during evolution. Examples of these motifs, which include the helix–turn–helix, zinc finger, and leucine zipper motifs, are shown in Table 3.2 and Fig. 3.7. The motifs are often basic (as might be expected for a protein binding to the negatively charged DNA molecule) and are characterized by their ability to form structures that can bind sequence-specifically to DNA. For many of these motifs, notably the helix–turn–helix, X-ray and NMR structural studies of the complexes formed by the binding of the motif to DNA

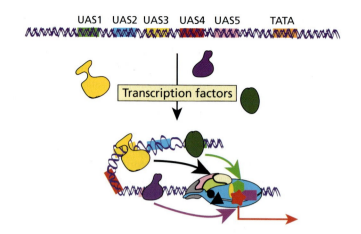

Fig. 3.6. Interactions between factors bound at enhancers and the transcriptional machinery at the promoter. Mapping of sequences required for efficient transcription *in vivo*, usually through DNA footprinting (see Fig.3.4) highlights short sequences [called variously enhancers, upstream activating sequences (UASs) or upstream regulatory sequences (URSs) and designated as UASs in the figure] that are sometimes located many base pairs from the TATA box. Transcription factors bind to these enhancers and, in many cases, interact with components of the initiation complex (the SRB/MED complex and/or the TAFs). There is no evidence that transcription factors interact directly with RNA polymerase. The function of these interactions is to enable the effective activation of the initiation complex.

Table 3.2 DNA binding motifs

DNA binding motif	Structure and properties	Examples
Hox homeodomain	Helix–turn–helix motif. Related to prokaryotic regulators	Hox proteins
POU homeodomain	Helix–turn–helix with an adjacent helical region that modulates the interactions of the helices with the DNA	Oct1, Oct2, Pit1
PRD homeodomain	Helix–turn–helix related to the Hox homeodomain and often coupled to a homeodomain	Pax factors
PBC homeodomain	Helix–turn–helix related to the Hox homeodomain	Pbx, exd
LIM homeodomain	Cysteine-rich domain; mediates protein-protein interactions, coupled with homeodomain	Isl1, apterous
C_2H_2 zinc finger	Multiple fingers each coordinating a Zn atom. β-sheet and adjacent α helix. Monomer	Kr, Sp1, TFIIIA
C_4 zinc finger	Single pair of fingers, each with a Zn atom. Dimer	Steroid receptors
bZip	Amino-terminal basic DNA-binding domain and carboxy-terminal leu-rich helix. Forms dimers through the leu-rich domain	Jun, Fos, Myc
bHLH	Basic DNA-binding domain adjacent to helix–loop–helix motif involved in dimerization	MyoD, E12, twist
Winged helix	Helix–turn–helix which binds DNA as a monomer	forkhead, HNF3A
T domain	Two perpendicular helices with an amino-terminal β-sheet that mediates dimerization	Brachyury, Tbx
MADS domain	Helix–loop–helix with a carboxy-terminal dimerization domain	SRF, MEF2, Mcm1

have revealed which parts of the motif recognize the bases of the cognate DNA sequence, and how the motif binds with respect to the major and minor grooves of the double helix.

Transcription factors bound at an enhancer must interact with other proteins, particularly with those of the basal transcriptional machinery, as well as with DNA, in order to carry out their regulatory role. In some transcription factors, domains have been identified that are required for the protein to stimulate transcription. Some of these 'activation domains' display common features, such as stretches rich in a particular amino acid (e.g. glutamine, proline, isoleucine), or highly charged regions (e.g. 'acid blobs'). Conversely, some transcription factors have amino acid domains that have been found to have a transcriptional repressor function: alanine-rich, proline-rich, and highly charged domains have been identified among these. There is no obvious way of discerning from its amino acid sequence what function the protein–protein interaction domain of a transcription factor will have; these domains tend to be much more varied than DNA-binding domains.

Many transcription factors bind to DNA as dimers and in some cases the DNA-binding domain also acts as a dimerization domain. An example is the dimer between the leucine zipper transcription factors Jun and Fos. Jun can dimerize with itself and promote transcription activation, but it is much more efficient if it forms a heterodimer with Fos which, by itself, cannot activate transcription. The ability of transcription factors to interact with other members of the same or related families increases the versatility of their regulatory functions (Fig. 3.7), for example, a ubiquitously distributed factor can acquire tissue-specificity if its activity requires interaction with a partner that has a specific distribution.

DNA-binding specificity of transcription factors

Each DNA-binding transcription factor recognizes a specific DNA sequence and this specificity is conferred by the sequence of amino acids in the DNA-binding domain of the protein. Generally, there is a finely tuned interaction between the amino acid sequence in the transcription factor and the DNA sequence of the binding site. Mutation and substitution experiments have shown that in some cases changing just one or a few amino acids in the part of the motif that recognizes the DNA-binding site can alter the sequence-specificity of binding *in vitro* and *in vivo*

(Fig. 3.8). For example, the homeodomain transcriptional regulators fushi tarazu and bicoid from *Drosophila*, although having similar homeodomains, each recognize a specific DNA sequence. Changing a single amino acid in the homeodomain of fushi tarazu allows it to change its DNA binding-specificity and recognize bicoid-binding sites *in vitro* (Fig. 3.8). However, *in vivo*, the mutant fushi tarazu retains some wild-type activity and does not activate the targets of bicoid efficiently. This is because the function and specificity of a DNA-binding protein is not determined simply by the DNA sequence that it can recognize optimally, but by the context of that sequence in a particular promoter. The interaction of a DNA-binding domain with DNA can be modified by intramolecular interactions with surrounding amino acid sequences, or by interactions of the transcription factor with other proteins.

In some instances, different DNA-binding domains of a particular family have the same DNA-binding specificity *in vitro* but proteins containing them recognize different sequences *in vivo*. What makes each protein specific for its target gene(s) differs from one case to another. Some transcription factors contain more than one DNA-binding domain, thus refining the specificity beyond that achieved with either domain alone, for example, the POU family of transcription factors contain both a homeodomain and a second DNA-binding motif known as the POU-specific domain. The homeodomain alone interacts with the DNA with much lower affinity and specificity than when it is linked to the POU domain.

In other cases, specificity in the initiation of transcription is achieved through interactions of the transcription factors with other proteins (Fig. 3.8). As already mentioned, many transcription factors bind to DNA either as homodimers or as heterodimers with other proteins of the same or different families; the different combinatorial possibilities can be specific for different target genes. An example of this is provided by the effect of the interactions of Hox proteins with members of the PBC family of homeodomain proteins. For example, *in vitro* the Hox factors labial (lab) and Ultrabithorax (Ubx) bind very similar DNA sequences, but *in vivo* they show fine discrimination among related sequences. The specificity that underlies this effect is conferred by interaction of the lab and Ubx proteins with extradenticle (exd), a member of the PBC family (Fig. 3.8).

There are many possible twists on the theme of regulatory protein–protein interactions. To take another example, two sets of proteins may share protein–protein interaction domains that allow the formation of homo- or heterodimers (e.g. Jun and Fos), whose combinations recognize targets with differing affinities (Fig. 3.8). In these situations, if one set lacks a DNA-binding domain its members will act to titrate the transcriptional regulation function of the other set. This is characteristic of some members of the bHLH family of transcription factors which have the HLH domain but lack the basic domain. They can form non-functional heterodimers with other members of the family and thus regulate their DNA binding ability.

Assembly of complexes at enhancers: Enhanceosomes

The number of genes to be transcribed in an organism is much greater than the number of transcription factors encoded in the genome. Gene transcription also has to be tightly regulated, both in time and in specific cell types. This regulation is achieved by establishing relationships between particular genes and specific sets of transcription factors. Although the repertoire of transcription factors is limited, the number of different combinatorial possibili-

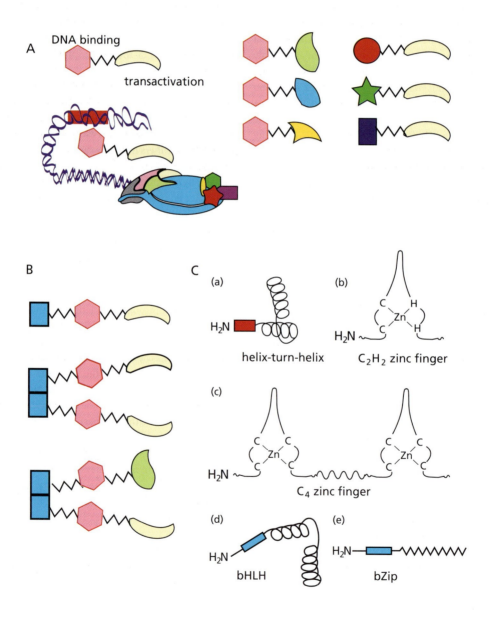

ties is immense, with the result that every gene can be regulated with exquisite precision. The mosaic of cell types that makes up an organism is a reflection of the mosaic of transcription factors that assemble at particular genes over time. Under different cellular conditions, different combinations of transcription factors may assemble at the same regulatory region, with a different regulatory effect on transcription. Conversely, different genes can be brought under the same regulatory control through the sharing of binding sites for the same regulatory proteins. This creates batteries of genes whose activity is coordinated to achieve expression of a particular cell phenotype. For example, many genes that are required to make functional muscle contain the same constellation of binding sites for regulatory proteins so that, when muscle cell differentiation is activated, these genes are expressed simultaneously (Fig. 3.9). The conditions for activity imposed by the binding sites can also dictate a particular temporal order of gene expression, as well as temporal relationships between different genes.

The presence of a particular set of regulatory sequences is not the only important factor for enhancer function. Equally important is the context of these sequences, in other words, the nature and spatial relationship of the surrounding sequences. This affects the accessibility of regula-

tory sequences to the proteins that bind to them, and the types of interactions (e.g. cooperative, anticooperative, allosteric) that occur between bound proteins. Analysis of complex regulatory regions indicates that they are built from relatively small regulatory modules that interact in space and time to produce specialized patterns of transcriptional activity (Fig. 3.9).

The requirements for transcription of the human β-interferon gene provide a good illustration of how spatial interactions between different binding sites contribute to specific, high-level expression of a particular gene (Fig. 3.10). Transcription of this gene in response to viral infection requires a 60 bp enhancer region upstream of the TATA box. This segment of DNA contains binding sites for several transcription factors as well as molecules of the protein HMG I(Y). Experiments involving the deletion or addition of sequences within the enhancer have shown that the presence of all the transcription factor-binding sites, and their precise spacing and organization, are essential for the optimal and specific activation of transcription of the β-interferon gene by viral infection (Fig. 3.10). In particular, sequence modifications that disrupt the helical phasing of the elements of the enhancer—their spatial relationship with respect to the turns of the DNA double helix—drastically lower the rate of transcription in re-

Fig. 3.7. DNA-binding transcription factors. (A) DNA-binding transcription factors have a DNA-binding domain, which allows them to interact with specific sequences on the DNA, and a transactivation domain through which they interact with elements of the transcriptional machinery. This modular structure enables the same DNA-binding domain to be linked to different transactivation domains and the same transactivation domain to be linked to different DNA-binding domains, as shown on the right. It also allows DNA-binding transcription factors to bridge large distances on the DNA, linking sequences that are far apart. **(B)** In addition to the DNA binding and transactivation domains, some transcription factors also have structural motifs that allow them to interact with each other. This, together with the modular structure described in A, creates the possibility of a large combinatorial repertoire for the reading and interpretation of DNA sequences. **(C)** There is only a limited number of DNA-binding structural motifs (see Table 3.2). Some of them are shown here in cartoon form. (a) The helix–turn–helix motif consists of two α helices joined by a short amino acid chain. The two helices are held at an angle that allows the more carboxy-terminal helix (termed the recognition helix) to interact with specific bases of the DNA. The amino acid residues in the recognition helix can vary from protein to protein and thus create a variety of interfaces specific for different regulatory sequences on the DNA. The helix–turn–helix motif is very common in bacterial regulatory pro-

teins and homeodomain proteins, which tend to bind as dimers to two half-sites on the DNA; work with these proteins has provided much of the understanding of how the motif works. The amino terminus of the motif is often associated with other motifs (box) that confer variety and specificity on the DNA-binding motif. (b) A C_2H_2 zinc finger is a binding domain created by the coordination of a zinc atom by strategically placed cys (C) and his (H) residues. The resulting three dimensional structure resembles a finger, hence the name. Some proteins have multiple C_2H_2 zinc fingers on the same molecule, enabling multiple interactions with the DNA. (c) A second type of zinc finger results from the coordination of a zinc atom with four cys residues. This class of zinc finger is called a C_4 zinc finger. Transcription factors containing this motif always have two of them, and work as dimers. (d) A basic helix–loop–helix (also known as bHLH) motif is characterized by two α helices separated by a hinge and with an adjacent basic sequence (blue). The basic sequence is involved in the recognition and interaction with DNA, whereas the helices are involved in interactions with similar proteins. (e) The leu zipper (or bZip) motif is very similar to the bHLH. In this case a basic DNA-binding domain is adjacent to a stretch of leu residues within a helical region, organized in such a way that all leu residues are on the same side of the helix. This structure creates a platform that allows these proteins to form dimers with themselves and with other members of the family.

sponse to virus, presumably because they prevent the correct steric interactions among the various bound factors. HMG I(Y), which, unlike the other factors, is unable itself to activate transcription *in vitro*, appears to interact both with DNA at its binding sites and with all three of the other β-interferon transcription factors, helping to bring about the correct assembly of the complex and the synergistic interactions between its components that mediate an

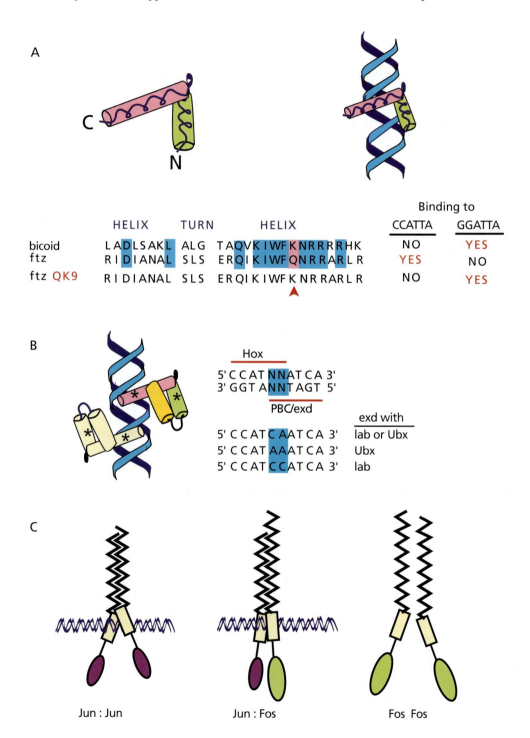

A

	HELIX	TURN	HELIX	Binding to	
				CCATTA	GGATTA
bicoid	L A D L S A K L	A L G	T A Q V K I W F K N R R R R H K	NO	YES
ftz	R I D I A N A L	S L S	E R Q I K I W F Q N R R A R L R	YES	NO
ftz QK9	R I D I A N A L	S L S	E R Q I K I W F K N R R A R L R	NO	YES

B

Hox

5' C C A T N N A T C A 3'
3' G G T A N N T A G T 5'

PBC/exd

	exd with
5' C C A T C A A T C A 3'	lab or Ubx
5' C C A T A A A T C A 3'	Ubx
5' C C A T C C A T C A 3'	lab

C

Jun : Jun   Jun : Fos   Fos Fos

effective transcriptional response. It carries out this function in part by bending the DNA in the enhancer region. The complete functional complex of transcription factors and enhancer has been called an enhanceosome.

The physical constraints set by the organization of the binding sites in the β-interferon enhancer ensure that, although the individual transcription factors that bind to the enhancer also participate in the activation of other genes, the β-interferon gene is only activated by viral infection when all four factors bind in the correct arrangement to the β-interferon enhancer. Thus, the same transcription factor proteins can, in different combinations and contexts, contribute to the activation of different genes.

Another example of this combinatorial specificity is provided by the yeast transcription factor Mcm1, a MADS box protein that participates in the regulation of genes that specify the cell's mating type. The yeast *Sacharomyces cerevisiae* has three cell types determined by the genotype at the *MAT* locus (also see Chapter 2): two haploid types **a** and α, and the diploid **a**/α. The products of these genes are transcription factors. MAT**a** cells make the **a**1 protein, whereas α cells make two proteins, α1 and α2. The **a** and α

factors activate the expression of sets of genes characteristic of the specific haploid cell type. Diploid cells express **a**1 and α2 but do not express the sets of genes characteristic of either of the two haploid cell types. Mcm1, which is expressed in all three yeast cell types, interacts with the **a** and α gene products and plays a central role in the regulation of cell-type-specific gene expression. Depending on the context, it functions either as an activator or a repressor (see Fig. 3.11 for details). How Mcm1 acts is determined by the availability of cell-type-specific transcription factors (**a** or α proteins) and the spatial context of its binding sites with respect to those for other factors.

Like the component proteins of the β-interferon enhanceosome, Mcm1 binds cooperatively with other transcription factors and acts synergistically with them to regulate transcription of target genes. The spatial constraints on the binding sites for Mcm1 with respect to those for other factors are extremely precisely specified: for example, inserting only 3 bp between the binding sites of Mcm1 and α2 in the promoter of an **a**-specific gene abolishes cooperative binding and de-represses the promoter in α cells (Fig. 3.11).

Fig. 3.8. Regulation of binding specificity of transcription factors. (A) The helix–turn–helix motif and the configuration of its binding to DNA are shown at the top in cartoon form. Single amino acid changes in a DNA-binding domain can alter the binding specificity of a transcription factor. The amino acid sequences are shown for the helix–turn–helix motifs in two homeodomain-containing proteins from *Drosophila*: bicoid (bcd) and fushi tarazu (ftz). Each of these proteins recognizes a specific DNA sequence that is a variation of the consensus sequence NNATTA, where N is any nucleotide. In the case of ftz, NN is CC and in the case of bcd it is GG. Mutational and structural studies indicate that the residues that mediate binding specificity are located in the carboxy-terminal recognition helix and that mutations in position 9, but not others, can abolish the recognition of a specific DNA sequence. This position is occupied by a lysine (K) in bcd and a glutamine (Q) in ftz. Ftz will recognize the bcd sequence albeit with very reduced affinity, and the same is true for bcd with regard to the ftz-binding sequence. However, a change from Q to K in position 9 of the recognition helix in ftz (ftz QK9) will change its binding specificity to that of bcd. **(B)** Protein–protein interactions increase the binding specificity between proteins and DNA. The figure shows the example of interactions between Hox proteins and members of the PBC family of related DNA-binding proteins. The diagram on the left shows the configuration of the DNA-binding domains of the Hox (pink green yellow) and PBC (pale yellow) proteins bound to the DNA helix. (The asterisks indicate the helices that form the helix–turn–helix motif. Both proteins also contain an additional amino-terminal helix.) The binding sites

for the two DNA-binding domains overlap in the composite Hox/PBC-binding site (*centre, top*), with the Hox protein bound to the left side of the site and the PBC protein to the right. The nucleotides, N, in the centre of the site are specific for a particular Hox protein. The interaction between the two DNA-bound proteins determines the specificity of binding. Extradenticle (exd) is a *Drosophila* member of the PBC family of homeodomain proteins, and binds DNA in partnership with many Hox proteins, determining their specificity. In the example shown, exd interacts with the Hox proteins labial (lab) and Ultrabithorax (Ubx) to define their specificity. Exd–lab and exd–Ubx will both bind the sequence CCAT-CAATCA but, in addition, each heterodimer will bind to a particular version of the consensus and thus exd–lab and exd–Ubx each recognize additional specific sites. The interactions between the PBC protein and the Hox proteins which are mediated by a specific amino acid sequence at the amino terminus of the homeodomain, are essential for modulating the binding specificity of the Hox proteins. **(C)** The interactions between Jun and Fos to form the AP1 transcription factor provide a further example of protein–protein interactions that define the function of a transcription factor. Jun and Fos combine to form a high-affinity heterodimeric transcription factor that will bind to the sequence TGACTCA. Jun forms homodimers that can bind to the site, but with low affinity. Fos does not dimerize and does not bind DNA. However, the heterodimer between Jun and Fos binds the consensus site with very high affinity and promotes transcription efficiently.

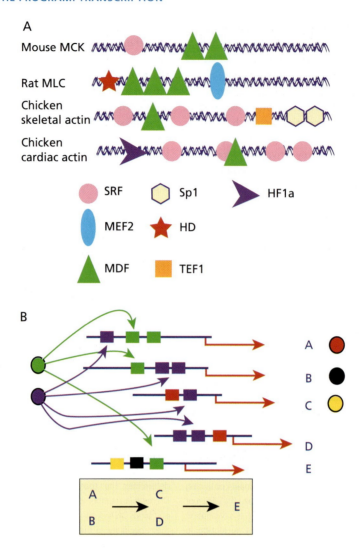

Fig. 3.9. Regulatory sequences serve to recruit genes to particular functions through the action of specific transcription factors. (A) The promoters of four genes expressed in muscle, showing binding of the shared regulatory proteins which ensure that they are all coordinately expressed. All the promoters contain binding sites for the serum response factor (SRF) and myogenic determination factors of the bHLH family such as MEF2 and MDF. There are other binding sites that are not found in all the promoters, including sites for a C_2H_2 zinc-finger protein (Sp1), transcription enhancer factor 1 (TEF1), and cardiac myocyte factor 1a (HF1a). (After Arnone M. and Davidson, E. (1997) The hardwiring of development: organization and function of genomic regulatory systems. *Development* **124**, 1851–64.) **(B)** The distribution of regulatory sequences in the promoters of various genes can determine their profiles and patterns of expression. Boxes represent regulatory sequences which are recognized by transcription factors of the same colour. The red arrows represent gene expression directed by the different promoters. In the example shown, A and B are transcribed first because the factors required for their expression (green and purple) exist already. C and D will follow once A is expressed because the A transcription factor (red) is required for transcription of the C and D genes. As time passes, E is also turned on, once the genes required for its activation have been expressed. In this way, the regulatory requirements of the different promoters create a sequence of expression in time.

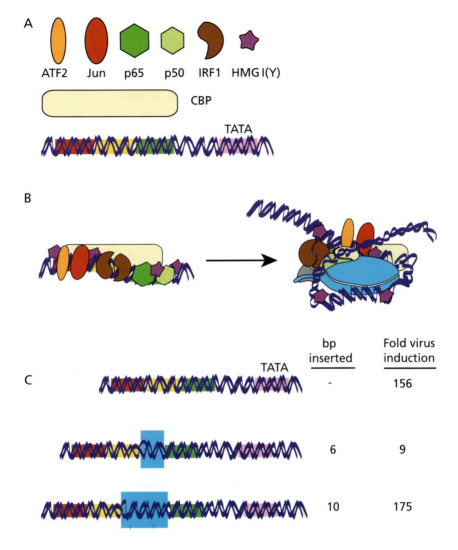

Fig. 3.10. An enhanceosome. (A) The activity of the 60 bp enhancer of the human β-interferon gene requires three regulatory elements: a binding site (red) for a heterodimer of the activating transcription factor 2 (ATF2) and c-Jun; a binding site (yellow) for interferon regulatory factor 1 (IRF1); and a binding site (green) for the two subunits (p65 and p50) of the transcription factor NFκB. In addition, the high mobility group protein HMG I(Y) binds to two sites flanking the ATF2/c-Jun site and to a site within the NFκB binding site. These proteins interact with the basal transcriptional machinery through a number of protein–protein interactions, some of which are mediated by the CBP/p300 protein. **(B)** The response to viral infection requires the precise assembly of these interacting proteins over a 60 bp segment of DNA that contains the binding sites; the resulting protein–DNA complex has been called an enhanceosome. As a consequence of this binding the DNA is bent, and precise spatial interactions, both among the different factors and between the factors and the transcriptional machinery, ensure effective and stable initiation of transcription. **(C)** The relative spatial arrangement of the transcription factors on the DNA is essential for the activity of the enhancer. This can be shown in an experiment in which human HeLa cells are transfected with different constructs containing the β-interferon regulatory region linked to the CAT reporter. The transfected cells are infected with virus and the activity of the different enhancers is measured by CAT activity. The three regulatory elements are colour-coded as in in A. When the binding sites are placed out of phase by about half a helical turn (6 bp or 4 bp) by insertion of extra DNA (light blue) the activity of the enhancer is dramatically reduced, but inserting the right amount of DNA to introduce one complete extra helical turn (10 bp) keeps all protein–protein interactions in the same phase and restores the activity of the enhancer. (After Thanos, D. and Maniatis, T. (1995) Virus induction of human IFNβ gene expression requires the assembly of an enhanceosome. *Cell* **83**, 1091–100.)

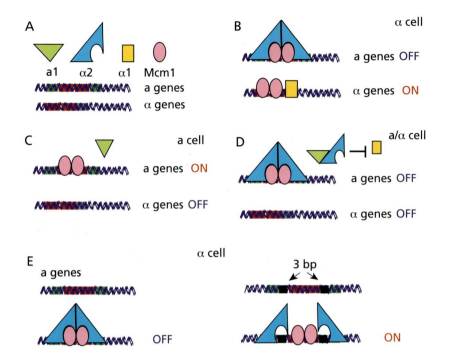

Fig. 3.11. Function of the Mcm1 transcription factor in cell type specification in *Saccharomyces cerevisiae*. (A) The Mcm1 transcription factor binds as a dimer to the promoters of **a** and α-specific genes and, together with the **a** and α transcriptional activators, plays a central role in the development and expression of the **a** and α cell types. Binding sites are colour-coded on the DNA: Mcm1 (red), α1 (grey) and α2 (green). **(B)** In α cells, Mcm1 and the α2 homeodomain protein form a heterotetramer that binds to the promoters of **a**-specific genes and represses their transcription. In addition, complexes of Mcm1 and the α1 protein stimulate transcription of α-specific genes. **(C)** In **a** cells, on the other hand, Mcm1 binds to **a**-specific gene promoters and, in the absence of α2 activates transcription because α1 protein is absent, Mcm1 does not bind to α-specific promoters. Thus, in **a** cells Mcm1 activates **a**-specific genes and in α cells it activates α-specific genes. **(D)** Diploid cells express both **a**1 and α2 genes, but do not express the haploid-specific genes characteristic of the **a** and α cell types. The **a**1 and α2 products form a complex that represses the expression of α1. In the absence of α1 there is no transcription of α-specific genes, and the interaction of α2 and Mcm1 represses the expression of **a**-specific genes, as it does in α cells. As a result, both **a** and α genes are off. **(E)** In an α cell, the functional binding and interactions of Mcm1 and α2 at the promoter of **a**-specific genes require very precise spacing of the binding sites. This is demonstrated in an experiment in which 3 extra bp are introduced between the Mcm1 and α2-binding sites. This mutation does not alter the binding parameters of either of the transcription factors to the promoter, but it does prevent their proper interaction with each other and abolishes repression of the **a**-specific gene. (After Smith, D. and Johnson, A. (1992) A molecular mechanism for combinatorial control in yeast. MCM1 protein sets the spacing and orientation of the homeodomain of an α2 dimer. *Cell* **68**, 133–42.)

Interactions between promoters and enhancers

Transcription factors bind to the DNA and enhance the enzymatic activity of the RNA polymerase holoenzyme. But how do transcription factor complexes assembled at enhancers, sometimes many hundreds of bases from the promoter where the transcriptional machinery assembles, contribute to the activation of transcription?

Virtually all models for enhanceosome function involve molecular interactions between the enhanceosome and components of the transcriptional machinery (Fig. 3.10). These interactions may be accomplished by the activation domains of DNA-binding transcription factors, and/or more indirectly by non-DNA-binding transcription factors, sometimes called coactivators, acting as molecular bridges between an enhanceosome and the pre-initiation complex. Bending and looping of the intervening DNA plays a part in these interactions. One important group of coactivator proteins are the members of the p300/CBP family. These are very large proteins which can interact both with a variety of activators and with the basal transcriptional machinery, thus acting as a platform that transmits the effects of the activators to the pol II holoenzyme.

What are the molecular targets of activators within the transcriptional machinery? *In vitro* studies on the β-interferon enhanceosome have suggested that it acts first to recruit TFIIB to a pre-formed complex containing TFIID, TFIIA, and other proteins, and then, via an interaction with CBP, recruits the pol II holoenzyme (Fig. 3.10). Other enhanceosomes may have different targets. In general, it appears that activators establish contact with pol II indirectly through other components of the initiation complex. Some activators may be able to interact with more than one target, the one that is used in a particular context depending on the cell type and the array of other protein factors present.

The SRB/MED complexes are important elements in the interaction between enhancers and the basal transcriptional machinery at the promoter. Individual components of some of these complexes show specific interactions with some enhancer proteins which might be involved in the fine-tuning of the activity of specific promoters. This might also explain the existence of different SRB/MED complexes in higher eukaryotes: these complexes contain some shared components and some that appear to be specific to certain promoters.

The role of chromatin in transcription

Eukaryotic DNA is wound around nucleosomes into a higher-order structure, chromatin (Chapter 2). This organization complicates the interaction between the transcriptional machinery and the sequences it recognizes. For example, nucleosomes prevent TBP from binding to the TATA box region and, at least in yeast, TBP will not associate with most promoters in the absence of activators. This means that the assembly of enhanceosomes, binding of the transcriptional machinery to the initiation site, and movement of the transcriptional machinery along the DNA require the remodelling of chromatin from a closely packed nucleosomal conformation to one that will accommodate these processes. More importantly, the experiments in yeast suggest that part of the function of complexes at enhancers is to open the chromatin for the transcriptional machinery. This point was made clear in an experiment in yeast, in which reducing the overall concentration of

histones (by reducing the number of histone coding genes) led to the spontaneous activation of many genes that are usually expressed only under specific conditions.

The enzyme DNase I can be used to probe the state of chromatin (Fig. 3.12); the reasoning is that segments of DNA that are particularly sensitive to small amounts of the enzyme *in vivo* are also likely to be accessible to other proteins such as transcription factors. This type of experiment, called a DNase I hypersensitivity test, has been used to show that different regions of the genome are differentially accessible both in different cell types and in the same cell type at different stages of development. For example, the human β-globin gene has two variants, β1 and β2, that are reciprocally expressed in embryonic and adult cells. DNase I hypersensitivity tests show that the DNA in the region of the β1 variant is more resistant to digestion than β2 in adult cells whereas the reverse is true in the embryo (Fig. 3.12).

The regulated temporal expression of the globin genes provides a good example of very long-range effects on transcription. The globin genes are arranged in a cluster in the order in which they are expressed during the embryonic, fetal and adult stages of life. Between 5 kb and 25 kb upstream of the cluster, there are five erythrocyte-specific DNase I hypersensitive sites that control the transcriptional activity of the cluster. This far-upstream region is called the locus control region (LCR) and although only some of its elements can act as enhancers in standard tests *in vitro*, all of them can direct high-level and erythrocyte-specific expression of linked genes in transgenic animals, regardless of where the construct is integrated into the genome. The LCR defines a domain of open chromatin, but in addition it appears to play a role in determining the order in which the different genes of the cluster are expressed: those closer to the LCR earlier than the others.

Changes in the structure of the chromatin rely on multisubunit 'chromatin remodelling machines' that have been revealed by genetic and biochemical experiments. These complexes (including, for example, four different complexes in *Drosophila* and two in yeast) are able to perturb chromatin structure. All of these complexes contain a subunit that is a member of a family of DNA-dependent ATPases but there are no other obvious similarities among them and it seems likely that they are specialized for different chromatin-remodelling functions during transcription and chromatin assembly.

One of the best studied, the yeast SWI/SNF complex, contains about 11 polypeptides. Similar complexes have been identified in *Drosophila* (the BRM complex, which

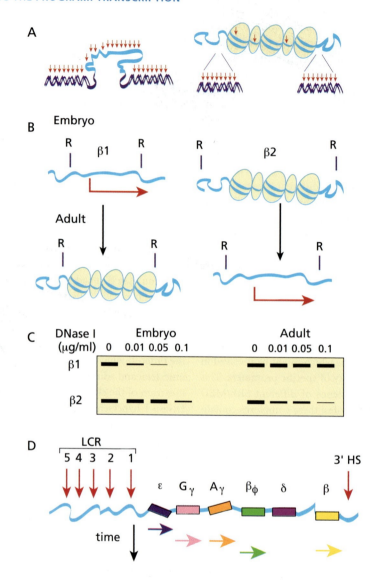

Fig. 3.12. Large-scale regulation of chromatin configuration.
(A) Sensitivity to DNase I digestion (red arrows) provides a test for the state of chromatin. DNA tightly packaged into chromatin (right) contains fewer DNase I sensitive sites, because the chromatin-associated proteins impede the access of the enzyme to the DNA. For this reason, a higher concentration of DNase I is needed to digest chromatin-packaged DNA than DNA in a transcriptionally active state (left) that is not packaged into chromatin. **(B)** In chickens, two globin genes (β1 and β2) are differentially expressed during development: β1 is expressed during embryogenesis whereas β2 is not, and the situation is reversed in the adult. At each developmental stage, the unexpressed gene is packaged into a nucleosomal configuration. R indicates a restriction site that defines fragments of specific lengths. **(C)** The differential DNase I sensitivity of the globin genes in embryonic and adult cells. Nuclei are isolated from embryonic and adult cells and treated with varying concentrations of DNase I. After treatment, DNA is extracted and cut with a restriction

enzyme that yields DNA fragments diagnostic for each gene. The DNA is run on a gel and probed. The DNA from the β2 locus is very sensitive to DNase I in the adult, where it is expressed, while it is resistant in the embryo. The opposite is true for the β1 gene. (Based on Stalder, J. *et al.* (1980) Hb switching in chickens. *Cell* **19**, 973–80.) **(D)** The locus control region (LCR) of the human globin gene cluster, which appears to be responsible for the large-scale order of the chromatin in this region. There are five hypersensitive sites within a range of several kb upstream of the globin gene cluster. There is also another site (3' HS) located 3' to the cluster. Mutations indicate that the 5' sites are essential for expression of the globin genes. The three switches that occur during development are indicated: first the cells express ε, then Gγ and Aγ and finally in the adult, the two β genes. The expression of each of these genes is associated with a large domain of open chromatin, and switches between them are associated with reorganization of the chromatin. At each of these switches, the LCR directs the expression of particular genes.

contains the brahma protein) and in human cells. It is not yet known how chromatin remodelling machines, such as the SWI/SNF complex work, but recent studies suggest that they should be thought of as both positively and negatively acting agents that can either stabilize or destabilize chromatin depending on the gene and the cellular context.

There is considerable evidence that, in addition to the activities of complexes like SWI/SNF, the acetylation and deacetylation of the lysine residues in histones has a strong effect on the stability of histone–DNA interactions. Deacetylation favours chromatin compaction, while acetylation by histone acetyltransferase proteins has the reverse effect. Several different protein complexes that have histone acetyltransferase (HAT) activity have been identified both in yeast and in animal cells, and there are indications that these complexes may play important roles in transcriptional initiation and/or elongation. In yeast, the SAGA complex contains specific proteins with HAT activity. Similar complexes, dubbed PCAF and hGCN5, have been identified in mammalian cells.

Intriguingly, all of these complexes also contain a subset of TAFs, and some of these have HAT activity themselves: TFIID has been shown to destabilize nucleosomes, and to be able to exclude nucleosomes if pre-bound to a promoter. The TAFs found in HAT complexes are those known as the 'histone-like' TAFs, but the significance of this similarity to histones is not yet clear. The p300/CBP coactivator proteins have also been shown to have intrinsic HAT activity and can form complexes with other HATs at some promoters, suggesting that their coactivator function may depend at least partly on their ability to destabilize nucleosomal chromatin. There is evidence that specific HATs may be required at specific promoters.

The functional relationship between the HAT complexes and the chromatin remodelling complexes is not clear, but it is intriguing that some of their components share common features. For example, the p300/CBP protein, some components of the yeast SWI/SNF and SAGA complexes, and *Drosophila* brahma all share not only HAT activity but also an amino acid sequence motif known as the 'bromodomain'.

As in the case of SRB/MED and the TAFs associated with RNA polymerase, the emerging picture seems to be one of a range of chromatin remodelling complexes that contain different but overlapping subsets of proteins and can assemble, perhaps in a promoter- and cell-specific manner, to remodel chromatin and effect functional interactions between enhanceosomes and the transcriptional machinery.

The dynamics of transcriptional activation and chromatin remodelling

The analysis of protein complexes that participate in the remodelling of chromatin during transcription initiation does not by itself tell us how these complexes act, together with other protein complexes bound at enhancers and promoters, to convert chromatin from a closed to an open conformation and to propagate this change along the gene as transcription proceeds. Various models for the dynamics of chromatin remodelling during transcription have been proposed (Fig. 3.13). In the case of the enhanceosome described above, a series of protein–protein interactions might act in two ways to promote transcription initiation. They might act to generate a series of molecular interfaces that can interact with the SRB/MED complexes, TFIID and general initiation factors and thus nucleate the initiation of transcription, and they might also create a local destabilization of the chromatin which would allow the transcription initiation complex to assemble (Fig. 3.13). Both effects, though, act close to the initiation of transcription. Effects at a greater distance from the initiation site can be accounted for by a variation of this model, in which chromatin remodelling complexes are recruited to enhancers by specific activators, resulting in the local generation of active chromatin. This state could be propagated along the DNA by successive rounds of recruitment of complexes whose function is to open the promoter for the assembly of the transcription initiation complex. In this situation, the open configuration remains in that state for a long time, as long as the different complexes are stably bound to the DNA (Fig. 3.13).

Transcriptional control at the *HO* promoter of *Saccharomyces cerevisiae* provides an example of the importance of the ordered modulation of chromatin structure in a developmental context (Fig. 3.14). Haploid *S. cerevisiae* cells divide by budding. Mother cells (i.e. cells that have previously budded), but not daughter cells, are able to switch mating type (see Chapter 2 and Fig. 2.9). Mating type switching depends on the activity of the *HO* gene, which is transcribed in mothers but not in daughters. Correct *HO* transcription involves remodelling of the chromatin in a 1400 bp region upstream of the *HO* gene. This remodelling only occurs if the region is first bound by the transcription factor Swi5, which then attracts chromatin remodelling complexes to the regulatory region. Swi5 is only needed for the first step, however, and indeed

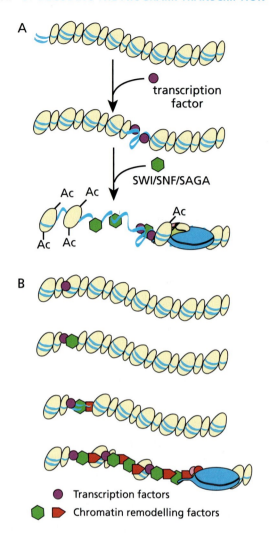

Fig. 3.13. The dynamics of transcription-associated remodelling and modification of chromatin. (A) Before the binding of transcription factors, the DNA surrounding a promoter is tightly bound in a chromatin configuration. Some transcription factors can recognize some sequences on this DNA and the resulting protein–DNA interactions might locally disturb the interactions between the DNA and histones, providing access for ATP-dependent chromatin remodelling machines (SWI/SNF) that in turn make the DNA more accessible to further proteins. These remodelling machines might simultaneously recruit proteins or complexes (such as SAGA) with histone acetyltransferase (Ac) activity; by covalently modifying histones, this further destabilizes histone–DNA interactions. This will result in the region of DNA becoming accessible to other proteins, such as those of the transcription initiation complex, so that an enhanceosome can now assemble near the initiation site. **(B)** The same process might be initiated far upstream from the initiation site and the 'open chromatin' state then propagate itself along the DNA. A transcription factor can bind to a site within a region of unmodified chromatin far upstream of the transcription-initiation region. This results in the recruitment of chromatin-modifying activities that in turn can lead to the opening of the surrounding chromatin and the recruitment of further transcription factor molecules. This effect can be propagated for some distance, without the need for the initial transcription factors to remain bound to the DNA, as long as the chromatin remodelling complexes are stable when bound to DNA. Finally, a transcription factor will establish contact with the transcription machinery and lead to the initiation of transcription. The various molecular events have to happen in the order shown. (After Struhl, K. (1999) Fundamentally different logic of gene regulation in eukaryotes and prokaryotes. *Cell* **98**, 1–4.)

it is degraded shortly afterwards. Although both mother and daughter cells inherit Swi5, only mothers switch because daughters inherit a DNA-binding protein, Ash1, which blocks the ability of Swi5 to bind DNA and recruit the remodelling complex.

Transcriptional repression in eukaryotes

In prokaryotes, the general lack of chromatin-mediated restraints on the activity of RNA polymerase means that the expression of a gene can only be repressed by blocking access of the polymerase to the DNA. In the case of the *lac* operon this is achieved by a repressor whose binding site

overlaps that of the RNA polymerase (Fig. 3.3). In eukaryotes, on the other hand, the properties of the RNA polymerase complex mean that gene expression requires an activation step which, as we have seen, is mediated by the stepwise assembly of large transcription complexes at specific promoters. Nevertheless, genetic and biochemical analyses indicate that repression also plays a role in the regulation of transcription in eukaryotes, even though there are not many situations where steric hindrance is used as it is in prokaryotes.

There are many molecular strategies for regulatory proteins to act as transcriptional repressors (Fig. 3.15). We have seen already one example, the Mcm1 gene product, which in different molecular contexts acts as a repressor or as an activator (Fig. 3.11). The different molecular contexts are provided by its association with other proteins. This versatility is common to many transcription factors and simply reflects the fact that the complex functional architecture of enhanceosomes and promoter regions can be

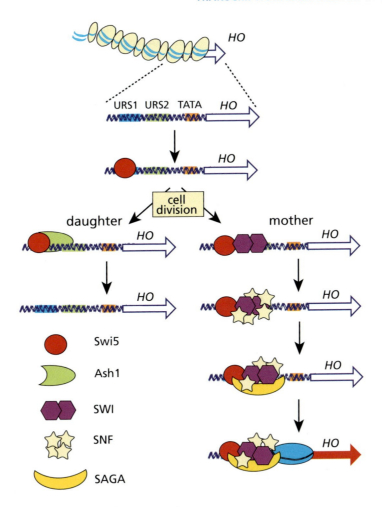

Fig. 3.14. Chromatin remodelling and modification during cell type specification in *Saccharomyces cerevisiae*. In budding yeast, the *HO* promoter, whose activity is crucial for the ability to switch mating type, is active in mother cells (i.e. cells that have previously budded) but inactive in daughters (i.e. cells that arise as a bud from a mother cell). This differential activity of *HO* is set up by events during the division that gives rise to each of these cells. The *HO* promoter contains two upstream regulatory sequences (URS1 and URS2). URS1, which binds the transcriptional activator Swi5, is required for mother/daughter differences and URS2 for other aspects of *HO* regulation. At the time of cell division, Swi5 is produced and binds to the *HO* URS1 in both the cells arising from the division: the new mother cell and the bud (daughter). In the mother cell, it recruits the SWI/SNF chromatin remodelling complex to URS1 and URS2. SWI/SNF in turn recruits the chromatin-modifying complex SAGA. SAGA binding acetylates histones and stabilizes the remod-elled state so that the mother cell arising from the division has an active *HO* promoter and is able to switch mating type. The proteins at URS1 and URS2 can now recruit specific elements of the transcriptional machinery to the TATA box and allow transcription of the *HO* gene to proceed. It is important to note that, after their recruitment, binding of SWI/SNF and SAGA does not require Swi5, highlighting the fact that the main role of Swi5 is to initiate the nucleation of the transcription complex. The bud that gives rise to the daughter cell also expresses Swi5 transiently while the division that separates it from the mother cell is in progress, but the presence of the protein Ash1, which is specifically localized to the bud, prevents Swi5 from recruiting SWI/SNF to the *HO* promoter. (After Cosma, M. P., Tanaka, T., and Nasmyth, K. (1999) Ordered recruitment of transcription and chromatin remodeling factors to a cell cycle- and developmentally regulated promoter. *Cell* **97**, 299–311.)

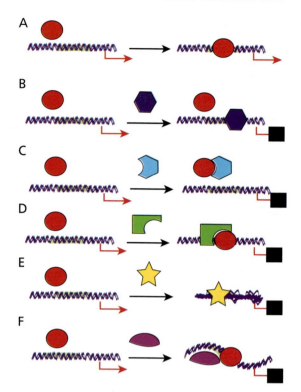

Fig. 3.15. **Strategies for transcriptional repression in eukaryotes. (A)** An activator (red) binds to DNA and this leads to the activation of transcription as shown here in a very schematic and oversimplified form. **(B)** In the presence of a repressor (blue) the activator cannot bind to the DNA because its binding is sterically inhibited by the binding of the repressor to an adjacent or overlapping site. **(C)** The activator is sequestered away from the DNA by an interacting protein which, in this case, does not bind DNA itself. This is often seen with transcription factors of the bZip and bHLH classes (see Fig. 3.7), some members of which have the dimerization but not the DNA binding motif and therefore can titrate active members away from the DNA. **(D)** An interaction between an activator and another protein can turn the activator into a repressor. The interaction of yeast Mcm1 with α2 (see Fig. 3.11) is an example of this type of molecular interaction. **(E)** A transcription factor (star) can repress transcription either by interfering with the transcriptional machinery directly and blocking its activity (not shown) or by recruiting chromatin remodelling complexes that close rather than open the chromatin (e.g. by promoting histone deacetylation). The latter is represented here diagrammatically as a disruption of the structure of DNA. **(F)** Some transcription factors, such as members of the HMG box family of transcription factors, modulate DNA bending. This can result in transcriptional repression by altering the phasing of various protein–protein interactions within the promoter region of a gene.

easily disrupted by slight alterations of the molecular interactions.

We have discussed the role played by chromatin remodelling in the activation of gene expression, but it must be kept in mind that the reverse of this process is an important facet of transcriptional regulation. At many promoters of genes that are not expressed in a particular cell type, proteins recruit chromatin-modifying activities that will tend to 'close' the chromatin. Some repressor proteins contain or recruit histone deacetylase activities that make nucleosomes inaccessible to the elements of the transcriptional machinery. Members of the groucho and SMART families of regulatory proteins are examples. These proteins have many functional facets that, through interactions with other proteins, enable them to be recruited to specific promoters to promote repression.

The genome-wide analysis of transcriptional regulation in yeast (see Fig. 2.15) has also emphasized the importance of negative effects in gene-specific regulation: disrupting the SWI/SNF chromatin remodelling complex, for example, actually increased the expression of a large number of genes, suggesting that an important function of this complex may be either to facilitate repressor binding or to 'close' chromatin domains at non-expressed genes.

The transcriptional machinery and regulated transcription

Overall, the picture that emerges of the transcriptionally active complex that assembles at eukaryotic promoters is one of an enormous molecular machine that consists of dozens of proteins and is probably about the size of a ribosome. Recent experiments in yeast, using the powerful approach of high-density whole genome oligonucleotide arrays described in Chapter 2 (Fig. 2.15), suggest that, rather than being an essentially invariant complex whose activity is modulated by gene-specific transcriptional activators bound to UASs, the transcriptional machinery itself is also a flexible complex containing certain core components that are required for transcription of most protein-coding genes, augmented by a huge and variable array of associated proteins required for specific subsets of genes (Fig. 3.16).

Interestingly, common features can sometimes be discerned among sets of genes requiring a particular protein factor. For example, the set of genes dependent on the TFIID component TAF145 contains many genes required for cell-cycle progression, suggesting coordinated regulation of these genes.

A

Complex	Subunit	Function	% genes
RNApol II	Rpb1	Largest subunit mRNA catalysis	100
SRB/MED (core)	Srb4	Target of activators	93
	Med6	Activation of some genes	10
SWI/SNF	Swi2	ATP dependent chromatin remodelling	6
General TFs	TAF 145/TFIID	TBP associated factor	16
	TAF 17/TFIID	Component of TFIID/SAGA	67
	TFIIE	Promoter opening	54
SAGA	Gcn5	Histone acetyltransferase	5

B

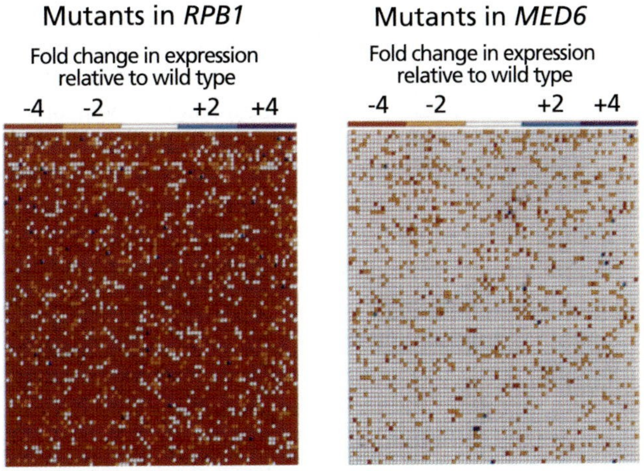

Mutants in *RPB1*

Fold change in expression relative to wild type

-4   -2   +2   +4

Mutants in *MED6*

Fold change in expression relative to wild type

-4   -2   +2   +4

Fig. 3.16. Genetic analysis of the requirement for elements of the basic transcription machinery in yeast. The number and identity of genes that require a particular gene product for their expression can be revealed by the use of whole genome microarrays (see Chapter 2, Fig. 2.15). This approach has been used to study the requirements for the different components of the transcriptional machinery in the yeast *Saccharomyces cerevisiae*. **(A)** RNA populations are compared in wild-type yeast and yeast carrying a null mutation for one of the subunits of the basic transcriptional machinery. In this way, the percentage of genes whose transcription depends on the function of each subunit can be estimated. The figure also lists the known functions of the subunits. **(B)** Grids showing whole genome transcriptional activity of mutants for *RPB1* and for *MED6*. The genes are arrayed on the grids in order, with the first gene of chromosome I at the top left, and the last gene of chromosome XVI at the bottom right. The bar at the top provides a guide for interpreting the colours of the fluorescence and the relative changes of transcriptional activity between the wild type and the mutants. The results show that almost all genes are strongly repressed by mutations in *RPB1*, encoding the catalytic subunit of RNA polymerase. In contrast, only a few genes are repressed by mutations in *MED6*, encoding a component of the mediator complex. (After Holstege, F. C. *et al.* (1998) Dissecting the regulatory circuitry of a eukaryotic genome. *Cell* **95**, 717–28.)

Transcriptional regulation during development: From domains to stripes in the *Drosophila* blastoderm

We have already discussed how the combinatorial action of sets of protein transcription factors and chromatin remodelling complexes regulates the activity of individual genes by controlling the accessibility and initiation rate of RNA polymerase. It is the integrated sum of these activities, over time, that specifies the identity of cells. In this section we illustrate these principles with two well-characterized examples.

The blastoderm stage of *Drosophila* embryogenesis

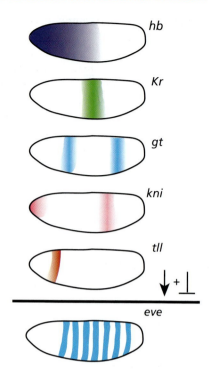

Fig. 3.17. Transcriptional regulation of the *even skipped* gene in the blastoderm stage embryo of *Drosophila*. The illustrations show the broad distribution of the regulators of *eve* expression in the blastoderm stage embryo of *Drosophila*, and the resulting expression pattern of *eve*. Anterior is to the left and posterior to the right in all cases. Expression patterns are illustrated for *hunchback* (*hb*), *Krüppel* (*Kr*), *giant* (*gt*), *knirps* (*kni*) and *tailless* (*tll*). The barbed and blunt-ended arrows indicate that there are both activating and repressive effects on *eve* expression.

around 3 hours after fertilization, cytoplasmic membranes begin to form but the cells remain connected by cytoplasmic channels that allow the free diffusion of large molecules such as proteins. Within the central region of the blastoderm, a cascade of transcriptional regulation acting on a group of 60–70 nuclei leads to the expression of specific sets of genes in a pattern of transverse stripes that prefigure the formation of the head, thoracic, and abdominal segments. The stripes of gene expression can be visualized by the use of reporter genes or antibody staining (Fig. 3.17).

One of these genes, *even skipped* (*eve*), is initially expressed throughout the embryo, but by the onset of the 14th nuclear division its expression is restricted to the central domain of the embryo and in the ensuing 20–30 minutes it is further restricted to an increasingly sharply focused pattern of seven stripes, each 5–6 nuclei in width and separated by 3–4 nuclei (Fig. 3.17). This pattern results from the combinatorial action of a set of transcription factors that act on the promoter of the *eve* gene (see Fig. 2.7).

provides an excellent experimental system for demonstrating how transcriptional control can direct very precise spatial domains of gene expression (Fig. 3.17). The first 13 nuclear divisions of *Drosophila* embryogenesis are not accompanied by cytokinesis; during the blastoderm stage,

---

Fig. 3.18. Stripe-specific enhancers of the *Drosophila even skipped* gene. At the time when the blastoderm-stage embryo becomes cellularized, the *even skipped* (*eve*) gene is expressed in seven evenly spaced transverse stripes along the embryo. The expression of *eve* in each stripe is driven by a specific enhancer region (see Fig. 2.7). Each of these regions, if placed upstream of a reporter, can drive expression in a subset of the original pattern. **(A)** The enhancers that regulate the expression of stripes 2 and 3 of *eve* are located within a 4 kb piece of DNA 5′ to the transcription start site. The stripe 2 element is about 500 bp in length and contains sites for Krüppel (Kr), giant (gt), bicoid (bcd), and hunchback (hb). Kr and gt act as repressors and bcd and hb as activators. The stripe 3 enhancer (which also drives expression in stripe 7) is about 500 bp long and contains binding sites for knirps (kni) and hunchback (hb), which act as repressors, as well as a putative activator (DStat). Each site has a distinct affinity for the cognate factor; the enhancer is activated ubiquitously through the DStat binding sites and is repressed in the central domain by kni. **(B)** Expression of a reporter under the control of the stripe 3 enhancer in a *kni⁻* mutant results in ectopic activity of the enhancer so that there is expression in the region between the two stripes. **(C)** Elimination of the gt-binding sites in the stripe 2 enhancer results in ectopic expression of a reporter under the control of this enhancer. (A and B adapted from Arnosti, D. *et al.* (1996) The *eve* stripe 2 enhancer employs multiple modes of transcriptional synergy. *Development* **122**, 205–14; and Small, S. *et al.* (1996) Regulation of two pair rule stripes by a single enhancer in the *Drosophila* embryo. *Dev. Biol.* **175**, 314–24.) **(D)** The sequences that separate the enhancers for stripe 2 from 3, and stripe 2 from other elements, are called spacers (sp) and are important for the generation of the precise pattern, perhaps because they prevent interference between the activities of the factors bound to the different enhancers. Removing both spacers (a) alters the pattern of the stripes but inserting either spacer 1 (b) or spacer 2 (not shown) between the enhancers restores a pattern that is similar to the wild type. Interestingly, a wild-type-like pattern is also observed even if the relative locations of the stripe 2 and stripe 3 enhancers are inverted (c). (After Small, S. *et al.* (1993) Spacing ensures autonomous expression of different stripe enhancers in the *even skipped* promoter. *Development* **119**, 767–72.)

The distribution of these factors, which include both activators and repressors, is neither uniform nor constant: this dynamic distribution arises as a result of a cascade of events that is set in motion by the graded distribution of certain maternal mRNAs in the egg. The genes transcribed at each level of the cascade produce proteins that themselves act as transcription factors regulating the expression of genes at the next level. In this way, the information contained in the broad initial gradients is transformed into precise patterns of expression of the genes involved in establishing segments (see Chapter 9, Fig. 9.19, 9.20).

Perhaps surprisingly, it appears that each *eve* stripe has an independent transcriptional control system (Fig. 2.7). Most is known about stripes 2 and 3, whose regulation provides examples of many of the aspects of transcription factor action that have been discussed in this chapter. Deletion analysis of the *eve* promoter identifies a region of about 4 kb that can drive expression of a reporter gene in the pattern of stripes 2 and 3. Within this region, there are separate regulatory elements for the two stripes, each around 500–600 bp and separated by a 1.5 kb 'spacer' region (Fig. 3.18).

The stripe 2 promoter (Fig. 3.18) contains an array of binding sites for transcription factor proteins: three for the

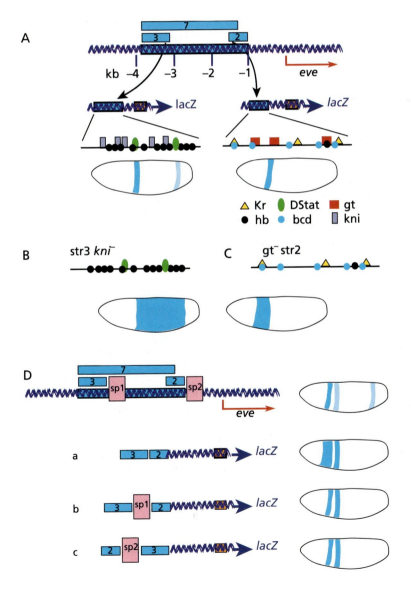

zinc-finger protein Krüppel and three for the leucine zipper protein giant, which both act as repressors; and five for bicoid and one for hunchback, which act as activators. The arrangement of these sites on the DNA, and their affinities for their cognate factors, result in the expression of *eve* stripe 2 at a reproducible position in the *Drosophila* embryo. Altering the number and arrangement of these binding sites has revealed how fine control is achieved by the interplay between repressors and activators. Expression is initially activated by the cooperative action of bicoid and hunchback in a broad band towards the anterior of the embryo. The borders of *eve* expression then retract and sharpen as the repressors giant and Krüppel are expressed: giant represses *eve* expression in the anterior part of the embryo and Krüppel in the posterior domain.

Thus, the *eve* stripe 2 control element integrates the changing and overlapping concentration gradients of these four transcription factors to generate an expression pattern that is finely controlled both in space and time. As in the β-interferon enhancer, the close spacing of the transcription factor binding sites within the control region acts to promote cooperative and synergistic interactions between adjacent bound factors. However, there is evidence that the organization of the stripe 2 enhancer has more inherent flexibility than the β-interferon enhancer: for example, giant can effectively mediate repression even if its binding sites are moved relative to the binding sites of the bicoid and hunchback activators.

Stripe 3 (Fig. 3.18) arises by the same basic logic as stripe 2 (i.e. an interplay between activators and repressors), but in this case, repression is all-important. Two ubiquitous activators (one known as DStat92E and the other so far unidentified) activate *eve* expression throughout the embryo, but expression is restricted to the region of stripe 3 by the hunchback and knirps proteins acting as repressors to set the anterior and posterior boundaries of expression, respectively. The stripe 3 enhancer contains two binding sites for DStat92E, five for knirps and 11 for hunchback. Initially, it seems remarkable that the same protein—hunchback—can activate *eve* expression in one part of the embryo and repress it in another. Once again, the explanation lies in the combinatorial nature of transcriptional control: the transcriptional outcome for each nucleus depends on the local concentration of hunchback protein, the relative concentrations of other transcription factors, the spatial arrangement of binding sites in each *eve* enhancer, and the binding affinities of the various factors at those sites.

An interesting feature of the *eve* enhancers is their ability to act independently of one another; this is ensured by the spacer region, which acts as an insulator to prevent interference between the enhancers. Removing the spacer region leads to abnormal expression patterns, presumably by creating new combinatorial interactions between factors bound at the now-adjacent enhancers (Fig. 3.18). Similar insulator or boundary elements have been observed in many gene-regulatory regions and are an important element of their modular structure.

The transcriptional control of *eve* expression, complex though it seems, is a relatively simple case. For genes regulated by a much greater number of transcription factors, the task of unravelling what is going on at the molecular level becomes dauntingly large.

However, it is important to keep in mind that the principles of transcriptional control are the same however complex the system: in each case the transcriptional activity of the gene represents the integrated output of the constellation of factors bound at its regulatory regions, effected via their interactions with each other and with core and gene-specific factors of the transcriptional machinery.

Transcriptional regulation during development: Regulation of the immunoglobulin μ heavy-chain gene

The transcriptional regulation of the mammalian immunoglobulin μ (Igμ) heavy-chain gene (the heavy chain specific to the immunoglobulin M class of antibodies) provides additional examples of the sorts of strategies that can be employed to effect tissue-specific regulation of gene expression. The immunoglobulin genes are expressed only in the B cell lineage. Mature transcripts begin to accumulate after gene rearrangement (see Chapter 2, Fig. 2.10). Fig. 3.19 outlines the composition of the regulatory regions of the rearranged murine Igμ gene. Both the promoter region located just upstream of the transcriptional start site, and the enhancer region located within an intron of the gene, contain an array of binding sites for both ubiquitous and cell-type-specific DNA-binding transcription factors. In addition, the intronic enhancer is flanked by sequences known as matrix attachment regions that appear to be involved in creating an open chromatin domain that allows continuous expression of the gene in B cells.

Mapping of *cis*-acting regulatory regions has revealed a number of sites for the binding of transcription factors whose interactions are required for the B-cell-specific

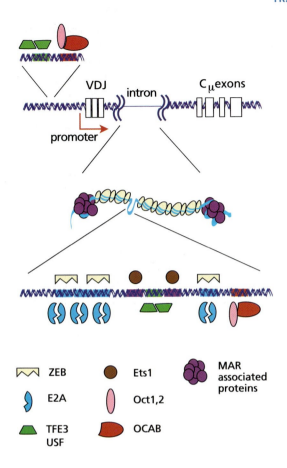

Fig. 3.19. Configuration of the promoter and enhancers of the immunoglobulin heavy-chain gene Igμ. The Igμ gene has two major regulatory regions: a promoter upstream of the transcription start site and an enhancer within an intron of the gene. The intronic enhancer lies in a special chromatin domain flanked by MAR (matrix attachment region) sites. Some of the regulatory elements and cognate transcription factors that are involved in the cell-type-specific transcription of the heavy-chain Igμ gene are shown. Two octamer sequences (red boxes), one in the promoter and one in the intron enhancer, are essential for B cell transcription of Igμ. These sites bind two POU domain proteins: Oct1, which is ubiquitous, and Oct2, which is largely restricted to B cells. A lymphoid-specific coactivator protein, OCAB, does not bind DNA but interacts with Oct1/Oct2 to promote transcription. The Oct proteins combine with a large array of bHLH and bZip transcription factors to ensure the efficient transcription of the Igμ gene in B cells. bHLH family members such as E2A that are particularly abundant in B cells form homo- or heterodimers that can bind specifically to regulatory elements (blue) in the Igμ enhancer. These dimers disrupt binding of repressor proteins, dubbed ZEB proteins, which help to prevent transcriptional activation of the Igμ gene in non-lymphoid cells. Members of the bZip family (TFE3 and USF) also function as dimers, with binding sites (green) in both the promoter and the intron enhancer; these proteins may form a molecular bridge that contributes to activation of the Igμ gene. In the intron enhancer the bZip binding site is flanked by two sites (purple) that bind members of the Ets family of transcription factors. Transfection experiments suggest that there are cooperative interactions between transcription factors bound at neighbouring sites and that all sites are needed for strong activation of the gene. Although the expression of most of the factors is not restricted to B cells, only in B cells does their expression overlap. (Adapted from Ernst, P. and Smale, S. (1995) Combinatorial regulation of transcription: II. The immunoglobulin μ heavy chain gene. *Immunity* **5**, 127–38.)

expression of the Igμ gene (Fig. 3.19). Although most of these proteins are not expressed exclusively in B cells, only in B cells do they act together to direct regulated, high-level expression of the Igμ gene. The requirements for the various transcription factors to enable expression of the Igμ gene have been worked out by experiments in which different factors or combinations of factors were transfected into non-lymphoid cells or added to non-lymphoid cell extracts and their effects on the transcription of the Igμ gene tested.

It is the regulated interactions among different proteins bound to the three sets of regulatory elements that ensure that the Igμ gene is expressed only in B cells. At the Oct-binding sites, a B-cell-specific protein that does not bind DNA (OCAB) interacts with two DNA-binding proteins (Oct1, 2) to make them B-cell-specific and functional. At the other two sets of elements, B-cell-specific combinations of more widely expressed proteins are formed: their specific interactions, determined by the spe-

cific arrangement of binding sites, help to ensure a B-cell-specific output. The situation is similar in some ways to the β-interferon gene discussed previously in that cell type specific gene expression arises from a unique cell-type-specific combination of transcription factors which are expressed in other cells at other times. In a way, the regulation of Igμ expression can be viewed as the output of the coordinated action of several enhanceosomes.

Transcriptional elongation and termination

Work in recent years, although not diminishing the importance of the pre-initiation and initiation steps as the major control point for transcription in eukaryotes, has uncovered evidence for further control exerted at the level of elongation or termination of the transcript. Purified pol II preparations from higher eukaryotes synthesize mRNA *in vitro* at only a fraction of the rates measured *in vivo* (100–300 nucleotides per minute compared with 1200–2000); moreover, transcription *in vitro* is interrupted by frequent pauses and in some cases by arrest before the transcript is complete. Reconstitution experiments have identified a number of 'general elongation factors' that suppress polymerase pausing or arrest, helping to ensure that large genes, such as the 70 kb *Drosophila Antennapedia* gene and the 2 Mb mammalian dystrophin gene, can be transcribed at rates compatible with the timing of the developmental program.

There is also evidence that some specific transcriptional activators are able to increase the efficiency of elongation, as well as participating in activating initiation. These factors appear to release the polymerase from pausing at nucleosomes downstream from the initiation site. Polymerase release also involves both ATP-dependent chromatin remodelling and phosphorylation of the polymerase CTD. It has been suggested that the function of phosphorylation may be to disrupt interactions between SRB/MED complex and the CTD, clearing the CTD for interaction with other proteins required for elongation and RNA processing.

Although a candidate eukaryotic transcriptional termination factor (known as Factor 2) has recently been isolated, little is known about the mechanism of termination in eukaryotes. Like the prokaryotic transcriptional termination factors, Factor 2 disrupts transcription in an ATP-dependent manner. It seems likely that future research will uncover control mechanisms operating at the level of transcription termination, at least for some genes. The observation that in many genes there are alternative sites of polyadenylation suggests that, like transcription initiation, termination is subject to spatial and temporal control.

The stability of the transcriptionally active or inactive state

As mentioned in Chapter 2, the execution of a developmental program may require not only that specific genes are activated or repressed, but that their activity is maintained in that state as development proceeds. Examples include the sex-determining genes, and those that relay identity along the anterior–posterior axis of the body. The proteins that are responsible for the initial activation or repression of transcription may be only transiently present, so there is a need for a mechanism to take over from them and 'lock' transcriptional activity into the correct state in a way that is heritable by the cell's descendants. This locking mechanism appears to involve chromatin modulation.

In yeast, the Sir proteins are involved in the mechanism that silences the *MAT* genes at the *HML* and *HMR* loci (see Chapter 2, Fig. 2.9). The Sir proteins are recruited to chromatin by other proteins (Rap1, Abf1, and the origin recognition complex ORC) that bind to specific sequences in the silencer regions. Silencing begins at these nucleation centres and spreads from nucleosome to nucleosome as the Sir proteins interact with the amino-terminal tails of the histones. Mutations in the *SIR* genes lead to the expression of otherwise silent genes and to the unwinding of chromatin.

In higher eukaryotes, the Polycomb group (PcG) of proteins are involved in a related process of locking chromatin into the inactive state (Fig. 3.20). These proteins were first discovered because of the role of the Polycomb protein in maintaining *Hox* genes in a silent state in specific regions of the *Drosophila* embryo (see Fig. 2.8). The *Hox* genes are activated early in development in a region-specific manner and are maintained in a silent state where they are not activated. This appears to be due to the activity of proteins of the PcG which ensure that genes that have not been activated early, are not activated later. Mutations in PcG genes can cause the development of abnormal structures by derepressing *Hox* gene activity in the wrong part of the embryo at any time in development (Fig. 3.20). This is analogous to the effects of mutations in the *SIR* genes in yeast.

The PcG proteins are a very heterogeneous group whose only common feature is the ability to bind to chromosomes (though they do not bind to DNA *in vitro*). Some members of the group share domains such as the zinc finger or the 'chromodomain', but there is no overall sequence similarity. Silencing appears to involve the formation of large complexes of PcG proteins and chromatin, accompanied by looping and folding of the chromatin into complicated structures that silence regions of up to tens of kilobases. The mechanism of silencing is not yet understood but it has been suggested that weak but cooperative binding of several PcG proteins to an array of Polycomb response elements in DNA nucleates the recruitment of further proteins and the formation of stable complexes. It

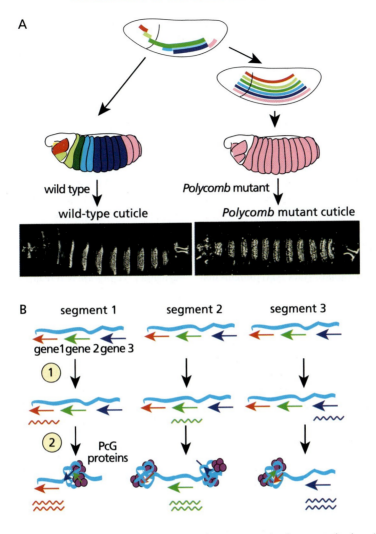

Fig. 3.20. **Repression of the active chromatin state by members of the Polycomb group.** **(A)** In *Drosophila*, *Hox* genes are activated along the anteroposterior axis in a particular order, and are responsible for directing morphological differences between segments (see also Chapter 2, Fig. 2.8). In wild-type animals (left), expression within these initial domains of activity in the embryo is maintained and propagated during development and is the basis for the segmental diversity that is observed later in the larval cuticle. In mutants for *Polycomb* (*Pc*) and related genes (right), the initial activation of *Hox* genes is normal, but soon afterwards they become deregulated and are expressed throughout the embryo. Later, the gene product that is normally active in the most posterior region down-regulates all the others and establishes a posterior segmental identity. As a result, all segments develop with similar morphological characteristics. The lower panels show dark-field photographs of the cuticle of a wild-type animal (*left*) and a mutant of the *Pc* group (*right*); the bright spots identify denticles secreted by epidermal cells. Note that the wild type has a segment-specific pattern of denticles, while in the mutant all segments look alike and their pattern resembles that of the last segment. **(B)** Diagrammatic illustration of the effects of the members of the PcG on the DNA after the initial activation of expression (1) of three genes of a *Hox* cluster in three different segments. Those genes whose expression has not been established and stabilized by the initial mechanism become incorporated in a heritable chromatin configuration in which they are not expressed (2).

seems likely that, as in yeast, histones are also involved in chromatin silencing by PcG proteins, but this has not yet been demonstrated directly.

The differential activity of *Hox* genes during development requires that a particular gene may need to be stably repressed in one region of the animal but stably activated in another. A heterogeneous group of proteins, the Tritho-rax group (TrxG) has been shown to be involved in the maintenance of the correct levels of expression, and once again, establishing a stable heritable state involves modifications of chromatin structure. Interestingly, several TrxG proteins share domains with members of the PcG group; the shared domains seem likely to be involved in the protein–protein interactions required for complex formation. Unlike chromatin silencing, the distinction between initial transcriptional activation and the maintenance of the active state is less clear, and it appears that TrxG proteins may be involved in both functions.

In vertebrates, in addition to the function of PcG and TrxG proteins, methylation of CpG dinucleotides in DNA also plays a role in maintaining large tracts of chromatin in a transcriptionally repressed state. A methyl CpG DNA-binding protein, MeCP2, isolated from mammalian cells, has been shown to be essential for embryonic development. The role of methylation may be to guide chromatin-mediated repression: recent studies have shown that MeCP2 exists as a complex with histone deacetylase, suggesting that methylation may template specific chromatin domains for condensation to a transcriptionally silenced state. In female mammals, the dosage of X-linked genes is adjusted to the same level as in males (which have only a single X chromosome) by inactivation of one entire X chromosome in all somatic cells. The highly methylated inactive X chromosome is heritably maintained in a compacted, heterochromatic state.

The chromatin architecture created by these complexes and the processes they mediate are very different from the chromatin remodelling and modification associated with transcriptional regulation. An important distinction is that whereas the latter create structures that are dynamic, non-heritable, and easily reversible, those created by the Polycomb group and related proteins turn transient patterns of gene activity into stable and heritable ones (Fig. 3.20). Chromatin silencing acts both on genes that have been transcriptionally repressed, and on genes that have simply not been activated, ensuring that they cannot be inappropriately activated later in development. It is possible that some of the problems associated with the efficient and reproducible cloning of animals from somatic cells discussed in Chapter 2 (see Fig. 2.2) are related to the state of the chromatin in differentiated cells and to the difficulty of reactivating repressed domains created by these proteins.

RNA-binding proteins

Transcription factors decode the information contained in the promoters and enhancers of genes by affecting the efficiency of the various steps of mRNA synthesis, in particular transcriptional initiation. Further tiers of regulation of gene expression can be achieved at post-transcriptional steps (for example, 3′ end formation, polyadenylation, mRNA stability, translational initiation) via regulatory regions that exert their effects as RNA rather than DNA sequences (see Chapter 2). The 'instructions' in these RNA regulatory regions are also read, interpreted, and executed by the binding of proteins.

Splice site recognition and commitment to splicing involve both the U1 small nuclear ribonucleoprotein (snRNP) and a factor known as U2AF, as well as members of the SR family of ser-arg-rich proteins. In addition to these constitutive splicing factors, a range of both positive and negative factors may participate in the control of developmentally regulated splicing by enhancing or inhibiting splice site recognition. These factors typically contain one or more of a set of RNA-binding domains. The cascade of alternative splicing events that controls sex determination in *Drosophila* illustrates how the regulation of splicing by *trans*-acting protein factors can set in motion an entire developmental pathway (Fig. 3.21). The Sxl protein is produced only in females, as a result of regulated splicing of the *Sxl* gene. Sxl itself regulates splicing of the *transformer* (*tra*) gene by blocking a splice site that would lead to the production of a truncated *tra* mRNA—thus only females produce tra protein. Tra in turn, acting together with other proteins at a 'splice enhancer' site in exon 4 of the *doublesex* (*dsx*) RNA, allows recognition of a female-specific 3′ splice site and thus production of a female-specific dsx protein that blocks expression of male-specific genes.

The length of the polyA tail that is added to mRNAs after transcriptional termination can markedly affect mRNA stability. For some genes, changes in polyA length during development have been shown to regulate gene expression. There is evidence that polyadenylation control is achieved by the binding of both nuclear polyadenylation factors and specific regulatory proteins to elements within the 3′ UTR of the mRNA.

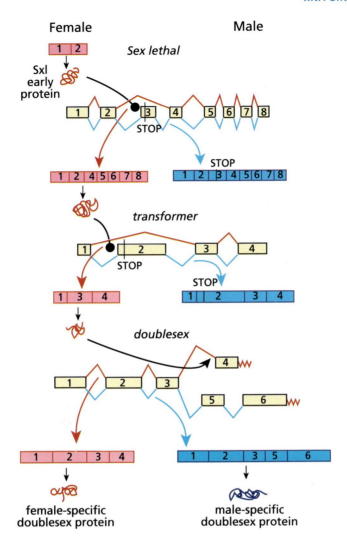

Fig. 3.21. RNA splicing and sex determination in *Drosophila*.
Sex determination in *Drosophila* relies on a cascade of sex-specific splicing events that result in the production of male- and female-specific forms of the transcription factor doublesex. The cascade is initiated with the expression, in females only, of a form of the *Sex lethal* (*Sxl*) gene product that encodes a sequence-specific RNA binding protein. Transcription of the mRNA for this product is driven by a female-specific promoter and the transcript contains two exons of the gene. Shortly afterwards, the *Sxl* gene is transcribed in both males and females, from a different promoter, producing an RNA that is related to but distinct from the early one. The third exon of this RNA contains a stop codon and the early female-specific Sex lethal product ensures that the splice from the second to the third exon does not occur. As a consequence of this, males, which lack the early Sex lethal product, do not make functional Sex lethal, but females do. A function of the late Sex lethal RNA-binding protein is to control the splicing of the *transformer* (*tra*) gene, which encodes another RNA binding protein. The *tra* gene is transcribed both in males and females and its second exon contains a stop codon. The Sex lethal protein ensures that this exon is not spliced onto the first one; as a result, females make transformer but males do not. The function of transformer is not to inhibit splicing but rather, in conjunction with another RNA-binding protein, transformer 2 (not shown), to promote a specific splicing event from an unusual splice site in the *doublesex* gene, which is expressed in both males and females. As a result, the *doublesex* mRNA is spliced differentially in males and females and generates two related transcription factors that have different properties. In males, doublesex represses female-specific genes, while in females doublesex represses male-specific genes.

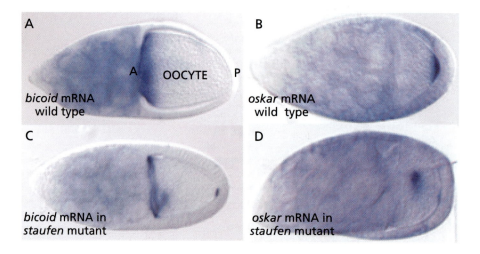

A

A   OOCYTE   P

bicoid mRNA
  wild type

B

oskar mRNA
  wild type

C

bicoid mRNA in
staufen mutant

D

oskar mRNA in
staufen mutant

Oogenesis and the early stages of embryonic development in *Drosophila*, *Xenopus* and *C. elegans* provide several examples of how translational control exerted by sequences in 3′ UTRs can be used to achieve exquisitely fine regulation of when and where a gene is expressed. The control of localization and selective translation of *Drosophila* mRNAs involved in establishing anteroposterior polarity in the egg and embryo has been mentioned in Chapter 2 (Fig. 2.12). Genetic experiments have identified several proteins required for this translational control; these proteins are assumed to include *trans*-acting factors that work on RNA, but RNA-binding ability has been shown so far only for the staufen protein, which is required for localization of the mRNA of the *oskar* gene to the posterior pole of the oocyte (Fig. 3.22).

Some of the proteins involved in iron metabolism and homeostasis in mammals provide further examples of translational control mediated by 3′ and 5′ UTRs and illustrate details of these mechanisms. Transferrin is a protein that transports iron in the serum. The transferrin receptor on the cell surface binds transferrin and brings iron into the cell where it binds to proteins and enzymes, such as haemoglobin, that require iron as cofactors. The 3′ UTR of the mammalian transferrin receptor mRNA contains a series of iron-responsive element (IRE) motifs that enable the expression of the transferrin receptor to respond to the concentration of iron (Fig. 3.23). Under conditions of abundant iron, the mRNA for the transferrin receptor is efficiently degraded through signals contained in its 3′ UTR. However, when iron is scarce, an IRE-binding protein, the

Fig. 3.22. **RNA-binding proteins.** RNA localization (shown by *in situ* hybridization) in developing oocytes of *Drosophila*. In each case the oocyte is the large cell on the right; the nuclei on the left belong to nurse cells which act as sources of RNA and nutrients for the oocyte. Anterior to the left and posterior to the right. **(A)** *bicoid* RNA is tightly localized to the anterior region of the oocyte. **(B)** *oskar* RNA is localized to the posterior end of the wild-type developing oocyte. **(C)** In the absence of the RNA-binding protein staufen, the *bicoid* RNA is more diffuse and now appears also at the posterior end of the oocyte. **(D)** In the absence of staufen, the *oskar* RNA is not localized and now appears in the middle of the oocyte. (Images of *in situ* hybridization for *bicoid* and *oskar* mRNAs courtesy of Hernan Lopez Schier.)

enzyme aconitase, is converted to a form that can bind to the IREs of the transferrin receptor mRNA and prevent its degradation, enabling the cell to import more iron.

Aconitase also binds to the 5′ UTRs of mRNAs encoding proteins involved in iron metabolism; in this case, however, it prevents their translation. When iron is low, in addition to stabilizing the mRNA for the transferrin receptor, aconitase also binds the 5′ UTR of the mRNAs for ferritin and of an enzyme involved in the synthesis of the haem group of haemoglobin, impeding their translation and lowering the concentration of iron-binding and storage proteins inside the cell (Fig. 3.23). When iron is high, the RNA binding properties of aconitase are inactivated and homeostatic regulation is achieved.

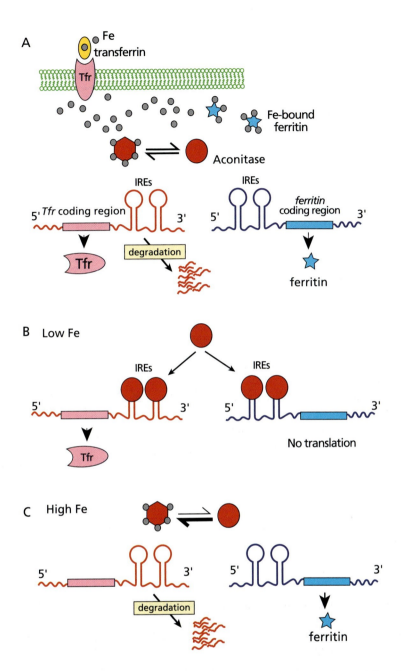

Fig. 3.23. Translational control of proteins involved in iron metabolism in mammalian cells. (A) Transferrin is an iron-binding protein that carries iron (Fe) in the serum to all cells of the body. A cell-surface receptor for transferrin (Tfr) binds transferrin and brings Fe into cells, where it is bound by enzymes or other proteins, including the storage protein ferritin (stars) and, in erythrocytes, the haem group of haemoglobin. The translation of the mRNAs for some of these proteins is under homeostatic control mediated by iron. The translational control mechanism is targeted at the 5′ and 3′ untranslated regions (UTRs) of these RNAs, which have a special three-dimensional structure and are called iron response elements (IRE). The IREs are located at the 5′ end of the ferritin mRNA and at the 3′ end of the transferrin mRNA. An IRE-binding protein (red), which turns out to be the enzyme aconitase, is the central regulator of the translation of these RNAs. The activity of aconitase is modulated by the concentration of iron. **(B)** Under conditions of low iron, the aconitase binds to the 3′ UTR of the transferrin receptor mRNA and stabilizes it so it is efficiently translated and generates high levels of transferrin. Aconitase also binds to a UTR located at the 5′ end of the ferritin mRNA and inhibits its translation, thus ensuring that the available iron is not kept in high-affinity stores. **(C)** When the concentration of iron is high, it inactivates aconitase, which therefore does not bind to the UTRs. The result is the degradation of the mRNA for the transferrin receptor and the translation of the mRNA for ferritin. These interactions provide a self-sustaining homeostatic loop for the control of iron concentration.

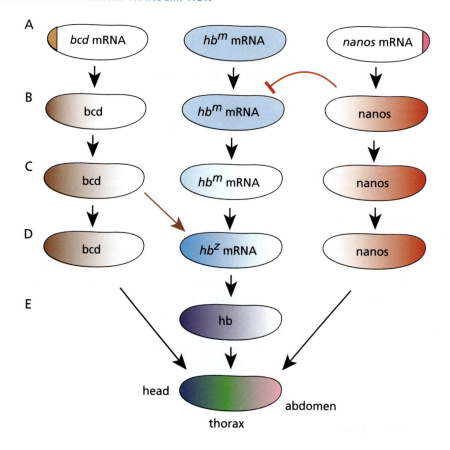

RNA localization and the establishment of the body plan in *Drosophila*

Spatial localization of specific mRNAs is an important mechanism in the definition of the axis of many embryos. For example, the basic body plan of *Drosophila* is established during the early stages of development through the translation of maternally deposited mRNAs that have been strategically localized in the egg during oogenesis. The setting up of the anteroposterior axis relies on the interactions of a homeodomain-containing transcription factor, bicoid, and of an RNA binding protein, nanos, with the *hunchback* gene and its product. The mRNAs for bicoid and nanos are synthesized in the nurse cells and are transferred to the oocyte where they are distributed differentially. The *bicoid* mRNA is retained at the anterior end through its interactions with a number of RNA-binding proteins. The *nanos* mRNA is concentrated, together with other proteins which make up the pole or germ plasm, at the posterior end as part of a protein complex centred around the oskar protein (see Fig. 3.24). These RNAs are translated upon fertilization.

Fig. 3.24. Translational regulation and the establishment of the body plan in the *Drosophila* embryo. The establishment of the basic coordinates that direct the organization of the body plan along the anteroposterior axis relies on the activities of three proteins: the PRD homeodomain transcription factor bicoid (bcd), the zinc-finger transcription factor hunchback (hb), and the translational regulator nanos. **(A)** During oogenesis, the mRNA for bicoid is deposited at the anterior end of the oocyte, that for nanos at the posterior end and that for hunchback ubiquitously (hb^m). This arrangement is maintained after fertilization. **(B)** During the early stages of embryogenesis, the *bicoid* and *nanos* RNAs are translated and the proteins diffuse away from their sources, creating a gradient of bicoid from anterior to posterior, and of nanos from posterior to anterior. **(C)** Nanos represses the translation of the mRNA for hunchback and this generates a very shallow gradient for the maternal hunchback product from anterior to posterior. **(D)** Bicoid directs the expression of *hunchback* in the zygote (hb^z) and generates a gradient of zygotic hunchback that overlaps with that of bicoid. **(E)** As a result, hunchback defines the middle region of the embryo. Thus, early in development the embryo is subdivided into an anterior region (the head and the thorax), directed by bicoid and hunchback, and a posterior (abdomen) specified by the nanos-mediated repression of hunchback activity.

The bicoid protein is not anchored to the cell cortex and diffuses towards the posterior pole of the embryo, creating a gradient through the syncytium. A number of experiments have shown that these different concentrations of bicoid are instructive for anterorposterior patterning (also see Chapter 9 (Fig. 9.19)). For example, mutants for *bicoid* lack the anterior half of the embryo and, in many cases, display a double posterior. This indicates that, in addition to specifying anterior structures, an important function of bicoid is to suppress posterior development. In the posterior region of the embryo, nanos protein also forms a concentration gradient after fertilization, but its role there is different from the role of bicoid in the anterior region. During oogenesis the mRNA for nanos is tightly localized at the posterior pole of the egg. Like that for *bicoid*, the *nanos* mRNA is translated after fertilization and, through a combination of diffusion and short half life, generates a protein gradient. The function of nanos is not to modulate the transcription of genes, however, but to prevent the translation of the maternal mRNA for hunchback (Fig. 3.24). Hunchback is very important in the patterning of the anterior region of the embryo. Its zygotic transcription

is under the control of bicoid, but there is also a store of maternal *hunchback* mRNA that is distributed throughout the egg during oogenesis. Hunchback will promote anterior development and therefore its expression in the posterior region has to be suppressed; this is done by nanos. In the same way that bicoid can suppress posterior development, nanos suppresses anterior development. That the function of nanos is to suppress the activity of hunchback is demonstrated by the fact that double maternal mutants for *hunchback* and *nanos* develop normally: that is, if no maternal hunchback is around, nanos is not necessary (Fig. 3.24).

Thus, the *Drosophila* embryo outlines its anteroposterior axis early in development by two strategies centred around the handling of mRNA: differential spatial localization of mRNAs for regulators of transcription and translation, and as a consequence, tightly localized transcription and translation of proteins that will begin to create differences (in terms of the batteries of genes that they activate or repress) that will be reflected in the anteroposterior organization of the embryo.

SUMMARY

1. The initial step in the decoding of the information that resides in the DNA is its transcription into mRNA.

2. Transcription is mediated by the enzyme RNA polymerase, a complex of several proteins that binds to the DNA at specific sequences and catalyses the synthesis of RNA.

3. In eukaryotes, RNA polymerase has a number of associated proteins that contribute to the specificity and stability of transcriptional initiation. The ensemble of RNA polymerase and associated proteins is often referred to as the 'basal transcriptional machinery'.

4. The selective activity of RNA polymerase in different cells is regulated by a number of specialized proteins, transcription factors, many of which bind to specific regulatory sequences on the DNA to generate ensembles of proteins that determine the time, place, and efficiency of transcription. These protein complexes work by promoting or inhibiting the activitiy of the basal transcriptional machinery.

5. Cell-type-specific transcription emerges from the assembly, at regulatory regions, of unique combinations of transcription factors, each one of which need not be expressed exclusively in that cell type. Genes that are coordinately regulated (i.e. expressed in the same cell at the same time) often share arrangements of regulatory sequences.

6. In eukaryotes, the DNA is complexed with a number of basic proteins to form chromatin, which imposes constraints on the process of transcription by limiting the access of the basal transcriptional machinery and transcription factors to the regulatory regions of genes. The regulation of chromatin structure is a crucial element in the regulation of gene expression during development.

7. The process of translation refers to the transformation of the information contained in some RNAs (mRNAs) into proteins. In some cases, specific sequences in mRNAs can act as regulators of the translational process. This mechanism, which can be used to transform a homogeneous distribution of mRNA into an inhomogeneous distribution of the protein, is important in several settings during early development.

COMPLEMENTARY READING

Arnone, M. and Davidson, E. (1997) The hardwiring of development: organization and function of genomic regulatory elements. *Development* **124**, 1851–64.

Berk, A. (1999) Activation of RNA polymerase II transcription. *Curr. Opin. Genet. Dev.* **11**, 330–5.

Carey, M. (1998) The enhanceosome and transcriptional synergy. *Cell* **92**, 5–8.

Hampsey, M. and Reinberg, D. (1999) RNA polymerase II as a control panel for multiple coactivator complexes. *Curr. Opin. Genet. Dev.* **9**, 132–9.

Latchman, D. S. (1998). *Eukaryotic transcription factors.* (3rd edn). Academic Press.

Li, Q., Hargu, S., and Peterson, K. (1999) Locus control regions coming of age at a decade plus. *Trends Genet.* **15**, 403–8.

Parvin, J. and Young, R. (1998) Regulatory targets in the RNA polymerase holoenzyme. *Curr. Opin. Genet. Dev.* **8**, 565–70.

Ptashne, M. (1992) *A genetic switch.* (2nd edn). Cell Press and Blackwell Press.

Ptashne, M. and Gann, A. (1997) Transcriptional activation by recruitment. *Nature* **386**, 569–77.

Peterson, C. and Workman, J. (2000) Promoter targetting and chromatin remodelling by the SWI/SNF complex. *Curr. Opin. Genet. Dev.* **10**, 187–92.

Struhl, K. (1999) Fundamentally different logic of gene regulation in eukaryotes and prokaryotes. *Cell* **98**, 1–4.

Chapter 4

Cell surface proteins: Receptors, ligands, and their environment

The last two chapters have focused on the nucleus as the centre of operations during development, and as the place where the developmental software is stored. At the level of the DNA, bacteria, their pathogens and eukaryotes use very similar basic strategies to run programs of transcriptional activity, in which the products of each step of the program include the proteins that decode the next step. However, once the focus moves away from the basic mechanics of transcription and the decoding of the program, the differences between prokaryotes and eukaryotes become more significant. Eukaryotic cells, with their multiple membranes and organelles, are structurally much more complex than prokaryotes. Furthermore, while prokaryotes are generally unicellular and for the most part need only to receive and integrate information about their environment, the multicellular organization of many eukaryotes relies on intricate systems of communication and coordination between cells. The coordinated operation of this system of information processing is particularly important as the developmental program unfolds during embryogenesis. The unfolding, as we have seen, begins in the nucleus where the primary instructions lie.

In a developing embryo, each cell is 'aware' of what other cells, in particular its neighbours, are doing. This is demonstrated by the experiments of Hans Driesch we discussed in Chapter 1 (Fig. 1.11): because each cell of a two- or four-cell sea urchin embryo is capable of generating a whole embryo if separated from the other cells, mechanisms must exist by which each cell detects whether it is surrounded by other cells, and through which it can communicate information about its state to other cells. This implies that each cell has a prospective fate (the fate it will

adopt if left undisturbed) and a prospective potency (the range of developmental possibilities it is capable of fulfilling). The principle underlying this observation can be generalized: during development, cells receive signals from their neighbours that do two things: restrict their potential and lead them to adopt specific fates.

In this chapter we introduce the molecular devices that cells use to communicate with each other, to transmit information about their state or to convey instructions that will contribute to development and pattern formation. The transduction and processing of this information will be the subject of Chapter 5.

Intercellular communication: Cell signalling

The idea that decisions about the fates of cells during development depend on cell interactions is illustrated in a simple manner by the way in which the oocyte is selected from other germ-line cells during oogenesis in the higher Diptera (Fig. 4.1). In the ovaries of *Drosophila*, the oocyte is selected from a group of 16 cells that are clonally derived from a single precursor through a series of four incomplete and stereotyped cell divisions. As a result of the pattern of divisions, not all cells end up with the same number of contacts. The cell that will become the oocyte is selected from the two that have the highest number of connections, probably by a stochastic process. As we have seen, cell fates are closely associated with programs of gene expression, so the adoption of the oocyte fate means the activation of a particular program of gene expression that is different

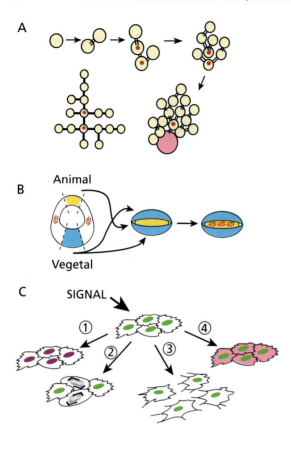

A

B
Animal

Vegetal

C
SIGNAL

Fig. 4.1. Cell fates are determined by cell interactions. (A) The oocyte of *Drosophila* is selected from among 16 sister cells. These cells are derived from a single cell that undergoes 4 mitoses with incomplete cytokinesis, so that the cells remain in communication after mitosis. The pattern of mitoses (shown in the figure) is stereotyped and results in two of the 16 sister cells (the pink cell and the cell with a central red spot) having four cytoplasmic bridges each with surrounding cells. This can be seen more clearly in the lower left of the figure, in which the group of cells has been flattened out and the connections indicated. One of these two cells stochastically becomes the oocyte and undergoes meiosis. The other cells, which have fewer connections, become 'nurse cells' whose function is to feed the oocyte with RNA and protein. **(B)** The early *Xenopus* embryo has two visibly different regions: the 'animal' region, from which the epidermis and the neural tissue will largely develop, and a 'vegetal' region, characterized by large yolky cells, that gives rise to most of the endoderm. During the early development of *Xenopus*, the *brachyury* gene becomes activated (red) at the boundary between animal and vegetal cells, suggesting that its activation is the result of an exchange of information between these two cell types. This can be shown by taking cells with high yolk content (vegetal; blue) and cells with low yolk content (animal; yellow) and placing them together in a 'sandwich'. After some time, *brachyury* expression can be detected in the animal cells, suggesting that it is the result of an interaction between the two cell types. **(C)** As a result of a signal, a cell can (1) change its state of nuclear activity and—associated with this—its fate; (2) divide; (3) change its morphology and adhesive properties; or (4) change its physiological state in terms of the biochemical activity in its cytoplasm (e.g. the intracellular concentration of Ca^{2+} ions or of a regulatory substance such as cyclic AMP). Sometimes the outcome can consist of more than one of these responses simultaneously.

from that of the other cells. This suggests that an important outcome of the oocyte's interaction with its neighbours is the modulation of its transcriptional activity.

The relationship between cell signalling and gene expression has been demonstrated in many experimental systems. In one such system, cells from early *Xenopus* embryos that can express a signal are placed next to responsive cells and the effects of the release of the signal are monitored by measuring gene expression in the responding cell population (Fig. 4.1). These 'sandwiches' demonstrate that cells can influence the patterns of gene expression in their neighbours. Although different types and quantities of signals can modulate the response of a cell, its repertoire of possible responses is limited by its own molecular constitution at a given time in development, in other words, by its complement of cell-surface receptors, signalling pathway components, transcription factors, and other molecular machinery. A given signal will have no effect if, for example, a cell that encounters it has not expressed the

cognate receptor. The ability of a cell to respond to a specific influence from its environment is known as its 'competence' (see Chapter 9, Fig. 9.12, for further discussion).

Varying patterns of gene expression are not the only reflection of cell signalling (Fig. 4.1). Cells in all embryos undergo dramatic changes of shape, arrangement, and location that are also modulated by signalling. Here, the cellular variables that are modulated are not nuclear programs of gene expression, but rather motility, adhesion, patterns of cell division, and other properties that sometimes bypass the regulation of transcriptional activity. An example of this is the development of connection in the nervous system, where extracellular cues guide the tips of axons along growth pathways.

In other situations, molecular interactions between

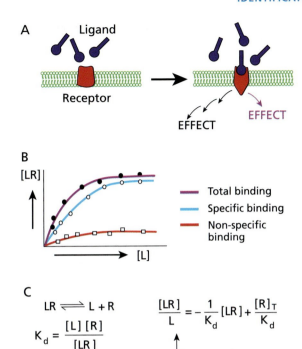

A

Ligand

Receptor

EFFECT

EFFECT

B

[LR]

— Total binding
— Specific binding
— Non-specific binding

[L]

C

$$LR \rightleftharpoons L + R$$

$$K_d = \frac{[L][R]}{[LR]}$$

$$[R] = [R]_T - [LR]$$

$$K_d = \frac{[L]([R]_T - [LR])}{[LR]}$$

$$\frac{[LR]}{L} = -\frac{1}{K_d}[LR] + \frac{[R]_T}{K_d}$$

$$\frac{[LR]}{L}$$

$$-\frac{1}{K_d}$$

[LR]

Fig. 4.2. Parameters of ligand–receptor interactions. (A) A signalling event can be triggered by an interaction between a ligand (L) and a receptor (R), usually accompanied by conformational changes in the receptor. The interaction can cause an effect directly, or can trigger a cascade of molecular interactions leading to an effect. The interaction between a ligand and a receptor has two important parameters: its specificity and its affinity. **(B)** These parameters can be demonstrated and calculated in experiments in which cells expressing the receptor are first challenged with ligand that is labelled (e.g. radioactively or with an enzymatic activity such as alkaline phosphatase) in such a way as to allow binding to be measured quantitatively. Then the experiment is repeated but in the presence of an excess of unlabelled ligand. If the binding of the ligand to the receptor is specific, the unlabelled ligand should effectively compete with the labelled ligand for binding to the receptor. Any remaining binding of labelled ligand is non-specific binding. **(C)** The affinity of an interaction can be calculated from the ratio of free ligand [L] to bound ligand [LR] and the total concentration of receptor $[R]_T$. The equations show how the dissociation constant, K_d, which provides a measure of the affinity of the interaction, can be calculated from a plot of the ratio of bound to free ligand [LR]/[L] against the concentration of bound ligand [LR]. This type of plot is known as a Scatchard plot. In the simplest case, the result is a straight line with a slope of $-1/K_d$. In reality the Scatchard plot often produces complex curves with more than one slope which reflect the presence of more than one binding site for the ligand in the receptor. Typical values for K_d range from 10^{-6} to 10^{-12}.

ligands and receptors trigger electrochemical changes that can then be propagated. This is how electrical signals are transmitted from neuron to neuron in the nervous system. The frequencies and amplitudes of the electrochemical signals constitute the information of the system.

Identification of signalling molecules and receptors

Cells exchange information by emitting, receiving, and responding to signals. A signal is transmitted by a specific chemical interaction between a signalling molecule (often referred to as a ligand) and a receptor (Fig. 4.2). The terms 'signal' and 'ligand' are sometimes used interchangeably. In this book, we will use 'signal' to refer to the transmission of information, and 'ligand' or 'signalling molecule' for the entity that delivers that information by binding to a cognate receptor. Binding of a ligand often causes a conformational change in the receptor. This may lead directly to a cellular response, or may trigger a series of molecular in-

teractions that propagate and process the information, eventually producing an output. This series of signalling interactions is known as signal transduction. Important parameters of the interaction between ligand and receptor are its affinity and specificity, which can be calculated experimentally (Fig. 4.2) and are usually very high. As a result, many signalling molecules can exert profound effects even at very low concentrations, in the range of nano- to picomolar.

The identification and characterization of signals and receptors has been of immense importance to developmental biology. A classical way of identifying a molecule with a particular function is to devise an assay for the function and use it to screen cell extracts. Extracts that contain the activity can be fractionated and the assay is then repeated to track and purify the protein responsible for the particular effect. This approach has been used to identify many signalling molecules. For example, fibroblasts in culture will proliferate if provided with serum but not in the presence of plasma alone. Extracts of platelets (present in whole serum but not in plasma) were fractionated (as described in general in Fig. 4.3), and a signalling protein named platelet-derived growth factor (PDGF) was purified on the basis of its ability to support fibroblast proliferation. A similar principle can be applied to other situations, for example, proteins involved in the attraction or repulsion of axons during neuronal development were identified by screening for molecules that could attract or repel the growth cones of cultured axons. Once the sig-

nalling proteins have been identified, working backwards from the proteins to the corresponding DNA (sometimes known as reverse genetics, see Fig. 4.3) allows the corresponding genes to be cloned. Molecular biology can also be used to create systems, such as phage display (Fig. 4.3), which can be used to isolate proteins that interact with cell-surface receptors. In this case, however, one of the partners of the interaction must be known.

For every signalling molecule, there is at least one

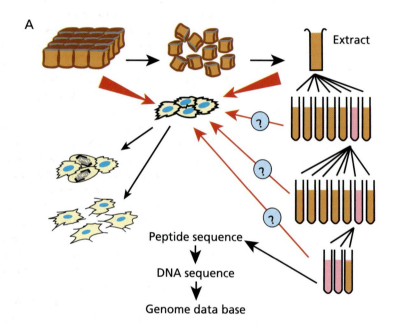

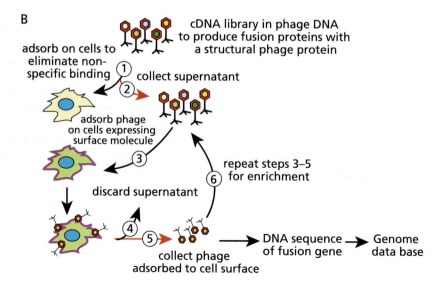

receptor, that is a molecule (or complex of molecules) that can receive a signal transmitted by the binding of a ligand, process the information it contains, and generate a molecular response to it. One way to identify the receptor corresponding to a signalling molecule is to label the particular ligand (e.g. radioactively, or with a chemical substituent that will link it covalently to any molecule it binds to) and use it as a baited hook to 'fish' for the receptor. The high affinity of the ligand for its receptor means that the receptor can sometimes be purified in a single step from a reasonably concentrated source (Fig. 4.4). A variation on this technique is called expression cloning (Fig. 4.4). This involves creating a cell culture assay for the activity of the signal and making use of a cell line that lacks the response to the signal. cDNA libraries are screened for their ability to restore the ability of the defective cell line to respond to the signal. A cDNA that can restore the function will encode a component of the signalling system (Fig. 4.4).

These types of techniques have been used to identify dozens of signalling molecules and their cognate receptors. In addition, the ability to use gene sequences as probes to detect similar sequences has led to the realization that many signalling proteins and receptors are members of large families of proteins that may be specialized for distinct but related functions in cell–cell communication.

Characterization of signalling molecules and receptors

As the various genome sequencing projects progress, scores of additional proteins are being recognized as potential signals or receptors on the basis of specific amino acid motifs characteristic of known signalling proteins. Many receptors, for example, have transmembrane domains, and ligands often carry a sequence that indicates that they are secreted molecules (Fig. 4.5). However, a classification of such proteins as ligands or receptors must be provisional until they are actually shown to have this function.

One way to distinguish the individual elements of a ligand–receptor pair is to use the genetic approach of

---

Fig. 4.3. **Identification of molecules with signalling activity or that interact with cell surface receptors. (A)** Activity fractionation. A particular cell population is capable of inducing a response in a second cell population (e.g. it might cause the cells to divide or to move). This implies that the inducing population contains a substance that exerts these effects. A simple way to isolate this substance is to make an extract from the inducing cell population and show that the extract mimics the inducing activity of the cells. The extract is then fractionated biochemically into individual samples with different components. Each sample is tested for its effects upon the responsive cell population. The sample(s) that display(s) the activity is subject to further rounds of purification until a homogeneous fraction is obtained that contains a single polypeptide. A peptide map and sequence of this protein is then obtained which leads to a DNA sequence and, through searches in genomic data bases, to a specific gene. This identifies the inducer. **(B)** Biopannning or phage display. A second way to identify ligands relies on previous knowledge of either the receptor for a putative ligand, or of proteins that the ligand interacts with. The basic premise of the method is that a protein or a protein domain incorporated on the external surface of a bacteriophage will allow the phage to adhere to the surface of cells that express a cognate protein. If the phage, for example, displays a cell-surface receptor or an associated molecule, other proteins that interact with the receptor can be identified. A cDNA library is made from a particular organism or cell type under study and the cDNAs are fused to the gene encoding a particular protein of the bacteriophage capsid or tail. This leads to the production of a large collection of fusion proteins that should, ideally, represent the complete protein complement of the organism or cell type. Phages bearing the fusion proteins are assembled and then adsorbed on cells that do not express the protein for which partners are being sought (step 1). The purpose of this step is to eliminate phages carrying fusion proteins that might interact non-specifically with the surface of the cell and so interfere with the outcome of the experiment. Any phage that have not been adsorbed under these circumstances are collected in the supernatant (step 2). This collection of bacteriophages is then adsorbed on to cells that express on their surface the protein for which partners are being sought (step 3). Some phages will adsorb to the cell surface and some will not. The supernatant, containing phages that do not interact with the experimental cells, is then discarded (step 4) and the phage particles that have adsorbed to the cell surface are eluted from it (step 5). The solution of phage recovered from step 5 is used to repeat steps 3–5, to enrich for phage particles containing fusion proteins that interact specifically with the cell-surface proteins (step 6). After a suitable number of enrichment rounds, individual phages are purified and their DNA bearing the fusion protein is sequenced. This yields partial sequences of unique proteins which can be used to screen available data bases to identify the genes encoding the interacting proteins.

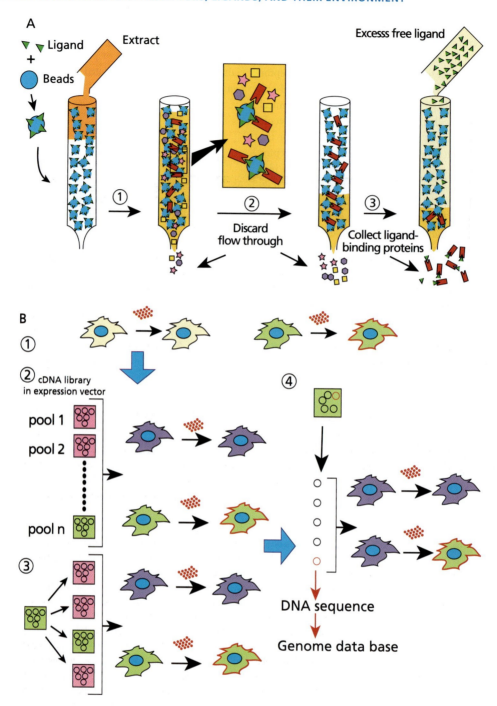

EGF receptor

Lysozyme

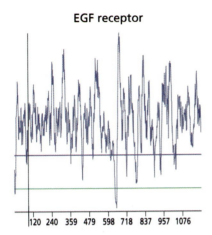

120 240 359 479 598 718 837 957 1076

14 27 40 54 67 80 94 107 120

Fig. 4.5. Signal and receptor 'signatures' in protein sequences. In a hydropathy plot, successive overlapping segments of a protein are assigned a 'hydropathy' value according to their average degree of hydrophobicity or hydrophilicity. Strongly hydrophobic amino acids confer negative values and strongly hydrophilic amino acids have positive value. A 'moving average' hydropathy value for each segment (vertical axis with zero at the position of the horizontal dark blue line) is plotted against the location of the segment in the protein (horizontal axis with the amino terminus on the left and the carboxy terminus on the right). Segments of the protein that have hydropathy values sufficiently negative to cross the horizontal green line are likely to be membrane-spanning. The figure shows hydropathy plots for a transmembrane protein (the EGF receptor) and a secreted protein (lysozyme). The EGF receptor has a single membrane-spanning segment in the centre of the protein. Lysozyme has a strongly hydrophobic segment at its amino terminus; this identifies the signal peptide, which allows the protein to enter the secretory pathway and is later cleaved.

mosaic analysis (Fig. 4.6). If a mutation behaves in a cell-autonomous fashion in a mosaic of mutant and wild-type cells (i.e. cells carrying the mutation always show a mutant phenotype, regardless of the nearby presence of wild-type cells), this means that both the protein in question and its function are restricted to the cell in which the protein is encoded and produced. If the protein is thought to be a component of a ligand–receptor pair, cell autonomy indicates that it is likely to be the receptor. If, on the other hand, mutations in the protein are not cell-autonomous, but can be 'rescued' by the presence of wild-type cells, the protein may act as a ligand. Experiments of this type in the developing

Fig. 4.4. Identification of receptor molecules. (A) If the molecular identity of a particular ligand is known, affinity chromatography can be used to identify a receptor or other proteins that can associate physically with the ligand. The ligand is coupled to special beads which are then packed in a column in a semi-aqueous phase. An extract from cells thought to contain interacting proteins is then passed through the column (step 1). Any molecule in the extract that can interact with the ligand will do so and will be trapped on the beads whereas other molecules will flow through the column and be discarded (step 2). This phase of the experiment is carried out under biochemical conditions known to favour protein–protein interactions. The column is washed to eliminate any molecules non-specifically attached to the beads, then a solution containing an excess of free ligand is poured through the column. The free ligand will bind to the interacting molecule and displace it from the beads so that it is eluted from the column (3). Suitable biochemical manipulations allow its separation from the ligand. The interacting protein is then subject to a peptide mapping and sequence analysis which provides a basis for the identification of the corresponding gene. **(B)** Expression cloning. For this method, a cell population is needed which is known to express a cell-surface receptor that can interact with a specific ligand of known physical and biological properties. The interaction can be visualized by labelling the ligand radioactively or with a fluorescent label (either directly or via a labelled antibody against the ligand). In the figure, the green cell can bind the ligand and this is reflected in a change in the colour of its cell surface (1). A cDNA library is made from the cells expressing the cell-surface receptor. The library is constructed in such a way that the cDNAs will be expressed efficiently in tissue culture cells. The library is divided into pools and the individual pools are used to transfect cells which are then challenged with the labelled ligand (2). The pool that produces a positive result is divided into smaller pools which are tested separately (3). Eventually, a pool is identified that is small enough to allow the screening of individual cDNAs and the identification of a gene encoding the receptor (4). The cDNA is then sequenced and the gene identified from sequence data bases.

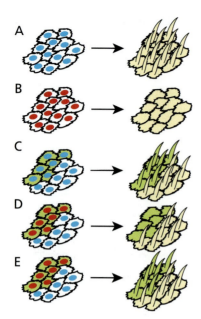

Fig. 4.6. Mosaic analysis for putative signalling proteins and receptors. A mosaic is a mixture of different kinds of cells; a genetic mosaic is a mixture of cells of different genotypes. If the structure of a substance X suggests that it can be either a signalling protein or a receptor, for example because it contains a transmembrane domain, mosaic analysis can help resolve which of the two functional classes it belongs to, by revealing whether mutations in the gene encoding X are cell-autonomous or non-autonomous. **(A)** A patch of wild-type cells (blue nuclei), each of which differentiates a hair. **(B)** Mutations in the gene X (represented by red nuclei) result in the cells failing to differentiate hairs. However, this observation does not indicate whether gene X is required in every cell that displays the mutant phenotype, as might be the case for a receptor or a transcription factor acting only in that cell, or whether it is a signalling molecule that moves across the field and instructs cells to develop hairs; such a molecule would only need to be expressed in a subset of the cells of the population. **(C)** To help resolve this using mosaic analysis, it is necessary to be able to identify the cells that are mutant for the gene in question, independently of the effects of the mutation. A common way of marking cells is to use a phenotypic marker, such as a particular colour, which does not interfere with their development or differentiation. A mosaic patch is shown here in which some cells have been labelled with a phenotypic marker (green cytoplasm) and some are wild type. Upon differentiation it is possible to identify these cells from their appearance (green). **(D)** Mosaic in which all the labelled cells are also mutant for gene X (hence the red nuclei associated with the green cytoplasm). In this case, all of the labelled cells show the loss-of-hair phenotype characteristic of mutations in gene X (i.e. the phenotype of a cell always corresponds to its genotype). This indicates that mutations in gene X behave cell autonomously and that therefore the protein encoded by gene X is likely to act as a receptor. **(E)** Mosaic in which all labelled cells are also mutant for gene X (as in D). In this case, however, the mutant cells do produce hairs; that is, the phenotype of a cell does not necessarily correspond to its genotype. This indicates that the mutation behaves non-autonomously and that therefore the protein encoded by gene X is likely to act as or through a signalling molecule.

Drosophila eye were used to distinguish the sevenless receptor from its membrane-bound ligand, boss, encoded by the *bride of sevenless* gene (Fig. 4.7). The use of mosaics and chimaeras (whole organisms containing mixtures of cells of different genotypes) has proven very useful in studying cell interactions in *Drosophila* and *Caenorhabditis elegans* embryos, and also in vertebrates such as the zebrafish.

Types of signalling molecules

In order to rationalize the vast amount of available information about signalling molecules, it is useful to be able to group or classify them in some manner. There are many possible ways of doing this, for example, they could be classified on the basis of their size, their molecular nature, their mode of synthesis, or whether they can penetrate cell membranes. None of these classifications is entirely satisfactory, in that none of them succeeds in ordering the vast variety of molecules into sets whose members share common structural and functional features.

One classification, which has generated some commonly used terminology, groups signalling molecules according to the distance over which they act. Long-range signals can act many cell diameters away from the cell in which they are synthesized, in some cases as far away as another part of the body. Long-range signals are also called endocrine signals. Hormones, secreted by specific organs and reaching receptive cells via the bloodstream, are the most familiar example of this type. Short-range signals, whose action is restricted to the immediate neighbourhood of the cells from which they originate, can be either paracrine or autocrine. Paracrine signals act upon cells surrounding the signalling cell, and this effect extends only a few cell diameters. An autocrine signal, acting upon the same cell that synthesizes it, represents a form of feedback of information from the cell to itself.

In the following discussion, we shall divide signalling molecules into two simple categories: low-molecular-mass signalling molecules (smaller than 20 kDa), and peptides and proteins. The peptide and protein signalling molecules are further subdivided into those that are freely diffusible and those that are associated with membranes.

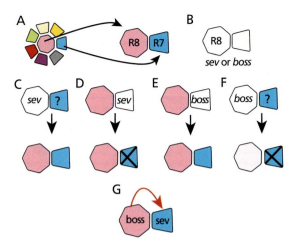

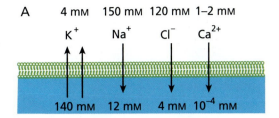

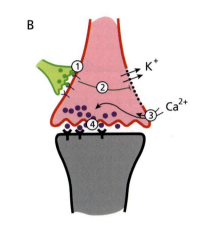

Fig. 4.7. Mosaic analysis of sevenless and bride of sevenless in the eye of *Drosophila*. (A) The *Drosophila* eye is made up of a number of units, ommatidia. One ommatidium contains eight photoreceptors (R1–R8 each labelled in a different colour), each of which can be identified by their physiological and molecular properties. Two cells (R7, blue and R8, pink) of a wild-type ommatidium are indicated. **(B)** Two mutations, *sevenless* (*sev*) and *bride of sevenless* (*boss*), affect the identity of R7. The products of both genes are transmembrane proteins so there is no clear indication as to which of these molecules is the ligand and which the receptor. Mosaic experiments of the kind shown in Fig. 4.6 can be used to determine in which cells the products of *sev* and *boss* are required for the correct development of R7. In this case, the mutant cells are marked by a mutation in the gene *white* (*w*) which is involved in making pigment and enables mutant cells to be identified because they lack pigment against a background of pigmented cells. **(C)** When R8 is mutant for *sev*, R7 develops as R7; **(D)** when R7 is mutant for *sev*, R7 does not develop its normal fate. **(E)** When R7 is mutant for *boss*, R7 develops normally; **(F)** When R8 is mutant for *boss*, R7 does not develop its identity. **(G)** These experiments indicate that boss is required in R8 to signal R7 its fate, and that sev is required in R7 for R7 to develop its fate. A reasonable conclusion is that boss acts as a signal and sev as a receptor for boss. This has been corroborated by molecular experiments.

Low-molecular-mass signalling molecules

The smallest possible biological signalling molecules are ions and gases. Na^+, K^+, and Cl^- ions cannot pass passively across membranes. Active cellular processes ensure that these ions are differentially distributed across the plasma membrane and provide a physicochemical basis for the generation of various kinds of information. In normal conditions, Na^+ and Cl^- are more abundant outside the cell (about 140 mM) than they are inside (Na^+ 12 mM and Cl^- 4 mM). The opposite is true for K^+ (140 mM inside and 4 mM outside). The concentration gradients created by

Fig. 4.8. Low-molecular-mass water-soluble signals. (A) Differential distributions of ions between the inside and the outside of the cell create potential differences across the plasma membrane. Fluctuations in these potential differences can be used as information. **(B)** The nervous system provides multiple examples of biological activities that make use of electrochemical potential differences. In the example shown, an interaction between two neurons, a motor neuron (grey) and a sensory neuron (pink), is modulated by a third neuron in the sea slug *Aplysia*. This third neuron (green) secretes serotonin (1) which triggers a signalling system in the sensory neuron (2), leading to the closing of potassium channels and, as a result, to the entry of calcium (3). As a consequence of these events, there is increased release of synaptic vesicles by the sensory neuron at the synapse with the motor neuron (4), facilitating the interaction between these nerve cells.

these differentials help maintain the shape of cells and create an electrochemical potential difference between the outside and the inside. Transient changes in this potential difference provide information that is used for the function and the activity of the nervous system (Fig. 4.8).

Ca^{2+} ions also play important roles, both in cell physiology and development. The concentration of Ca^{2+} is kept very low in the cytoplasm (100 nM or less, compared with about 2 mM outside the cell), and small changes have profound effects within the cell. Fertilization, for example, is associated with the propagation of a wave of calcium influx into the cytoplasm from stores beneath the plasma

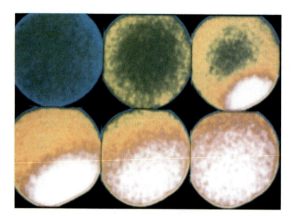

Fig. 4.9. Calcium waves during fertilization. Images show, from top left, a calcium wave spreading through a sea urchin egg after fertilization. The concentration of calcium is monitored with a calcium-sensitive dye which emits fluorescence that can be seen under a confocal microscope. The highest calcium concentration is indicated by white with successively lower concentrations by red, yellow, green then blue. The point of sperm entry is in the lower right quadrant of the egg. (From Whitaker, M. and Swann, K. (1993) Lighting the fuse at fertilization. *Development* **117**, 1–12).

membrane; the influx begins at the point of sperm entry, spreads across the egg and triggers a number of biochemical reactions that initiate embryogenesis (Fig. 4.9).

In contrast to ions, gases such as nitric oxide and carbon monoxide, which have been shown to have signalling functions, can readily permeate cells. Nitric oxide, synthesized within cells by the enzyme nitric oxide synthase, diffuses rapidly to surrounding cells. NO has a rapidly growing list of functions in development, including a role in neurogenesis, and sperm-derived NO has recently been shown to be the primary activator of the calcium waves in the egg at fertilization.

At a slightly higher level of complexity, organic compounds synthesized by cells can convey signals. Water-soluble signalling molecules of this type tend to have a polar organic backbone that is sometimes derived from amino acids. Members of this class, which have a wide variety of functions, include neurotransmitters, such as glycine, acetylcholine, and gamma-aminobutyric acid (GABA), hormones, such as epinephrine, and nucleotide derivatives, such as cyclic AMP (cAMP). Many of these molecules are involved in regulating the physiology of cells. For example, acetylcholine, serotonin, and GABA function in the transformation of chemical information into electrical information in the nervous system by triggering the opening and closing of ion channels (Fig. 4.8).

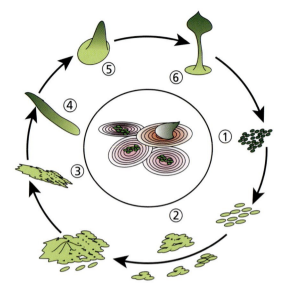

Fig. 4.10. Cyclic AMP as a signalling molecule during *Dictyostelium* development. (1) Under normal conditions, the slime mould *Dictyostelium discoideum* lives as independent amoeboidal cells feeding on bacteria and yeast. (2)–(4) When food is scarce, these cells stop dividing and begin to aggregate together to form multicellular 'slugs' which crawl through the soil attracting ever more cells. The slug has a leading edge with specialized cells that become different from cells at the rear of the slug. (4)–(5) At a certain point the slug develops into a reproductive structure made up of a stalk and a fruiting body containing spores (6). *Centre*: the aggregation of individual amoebae into a slug is a dramatic process in which groups of cells attract single cells, and the strength of the attraction seems to increase with the size of the aggregate. The attractant signal is conveyed by cAMP, which is synthesized and released by the cells (indicated as red concentric rings) and can be mimicked by an artificial source of this chemical. Individuals move towards high concentrations of cAMP. When they receive the signal, they begin to emit the signal themselves and also become temporarily refractory to it. These properties of the response to cAMP make the process directional and also allow the information to be conveyed by the frequency of the pulses of cAMP rather than by a concentration gradient.

cAMP is an extremely versatile signalling molecule. As we shall see in Chapter 5, cAMP is an intermediary in the relay of information through many different signal transduction pathways. However, it is also used as an extracellular signalling molecule during the differentiation of the cellular slime mould *Dictyostelium*, where it acts as the signal causing individual amoeboid cells to change shape and aggregate to form a multicellular slug-like structure specialized for spore formation (Fig. 4.10).

Most water-soluble low-molecular-mass signalling

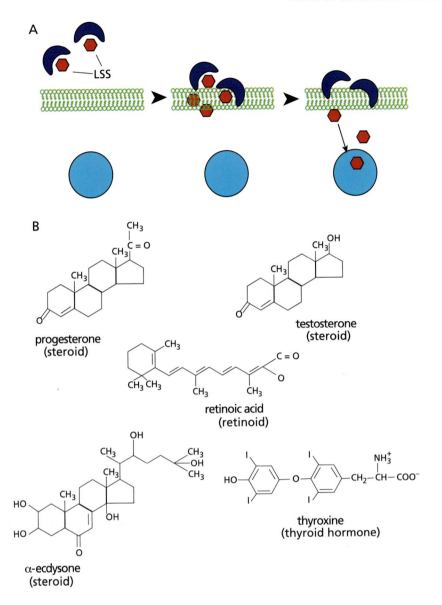

Fig. 4.11. Low-molecular-mass lipid-soluble signalling molecules. (A) Lipid-soluble signalling (LSS) molecules do not diffuse readily through the aqueous cellular environment. For this reason they are usually bound to specialized water-soluble molecules that can deliver them to responding cells. On reaching the membrane they can cross it on their own and interact directly with their targets either in the cytoplasm or the nucleus (blue). **(B)** Chemical structures of some lipid-soluble molecules that participate in developmental events.

molecules, including ions, cannot penetrate cells. Their functions are mediated by binding to receptors in the cell membrane and triggering a signal transduction chain that leads ultimately to their effectors. There is, however, a group of small lipophilic signalling molecules which, because they can pass through the plasma membrane and in many cases through the nuclear membrane as well, are often able to interact directly with their effectors inside the cell. Signalling molecules of this type, which include the steroid hormones, thyroid hormones, and retinoids, usually circulate in the bloodstream complexed with special carrier proteins (Fig. 4.11).

Figure 4.11 shows the chemical structures of some lipid-soluble signalling molecules. The steroid hormones, all derived from cholesterol, include the vertebrate sex hormones (e.g. oestrogen, progesterone, and androgens), which play a role in sexual differentiation; the insect ecdysteroid hormones, which regulate metamorphosis, and vitamin D, which regulates calcium metabolism in the gut and kidneys (Table 4.1).

Thyroid hormones, derived from the amino acid tyrosine, also have multiple effects on cell metabolism. During amphibian development, they function in a tissue-specific way to stimulate key events of metamorphosis, promoting tail regression, limb growth, and remodelling of the head and liver.

Retinoids, synthesized from vitamin A, have been found to have profound effects on several developmental processes in vertebrates, such as the development and (in some animals) regeneration of limbs, and the basic organization of the body plan.

Peptide and protein signalling molecules

At the other end of the size spectrum for signalling molecules lie peptides and proteins. All known signalling molecules of this type are membrane-impermeable, and therefore interact with their effectors indirectly, via cell-surface receptors and signal transduction systems. Peptide and protein signalling molecules are sometimes referred to as hormones and cytokines. Hormones are long-range or endocrine signals, whereas 'cytokine' is a rather loosely defined term for a group of signals that act in a cell different from the one in which they are manufactured (paracrine) or in the same cell (autocrine) to control the survival, growth, and differentiation of specific cell and tissue types.

For the purposes of discussion we shall divide signalling peptides and proteins into two large groups, one encompassing those that act as freely diffusible agents and

another including molecules that, because they are closely associated with the cell membrane or surface, are constrained to act on neighbouring cells. We will not attempt to cover every type of signalling protein, but will highlight some families of molecules that have been found to have particularly prominent roles during development.

Despite the diversity of signalling proteins, the brief survey that follows does reveal some common themes. One theme is the importance of post-translational modification in the generation of signalling proteins. Signalling peptides and proteins are synthesized by ribosomes attached to the endoplasmic reticulum (Fig. 4.12). They are then incorporated into the Golgi apparatus, via which they reach the cell surface. During this transport, many extracellular signalling proteins are post-translationally processed by addition of substituents, such as sugar groups or lipophilic moieties, that are essential for their function. Another important modification is proteolytic processing, which can take place either during or after delivery of the protein to the cell membrane. This processing is often essential for function. Insulin, an important regulator of carbohydrate and lipid metabolism, provides an example. Insulin is synthesized in the pancreas as a larger precursor molecule known as preproinsulin, which is proteolytically cleaved to proinsulin in the endoplasmic reticulum. After passage through the Golgi apparatus, proinsulin is enclosed in secretory granules, where it is endoproteolytically cleaved to the mature hormone (Fig. 4.12) and stored until it is required.

Most signalling proteins tend to be globular and compact in conformation and form oligomers, either with themselves or with other proteins, that can modulate their function. In some cases, pairs of family members are able to form heterodimers as well as homodimers, thus increasing the number of different signals that can be generated using a limited number of components.

Like DNA-binding transcription factors, signalling mol-

Table 4.1 Membrane permeable signals

Molecule	Biological properties
progesterone	Produced in ovaries and placenta. Assistance for tissues during reproduction
testosterone	Produced in testis. Promotes development of male characters
α-ecdysone	Produced by endocrine glands of various arthropods. Involved in differentiation and maturation of larvae
vitamin D	Regulation of Ca^{2+} metabolism
retinoic acid	Derived from vitamin A. Wide range of developmental effects
thyroxine	Produced in thyroid. Maintenance of metabolism

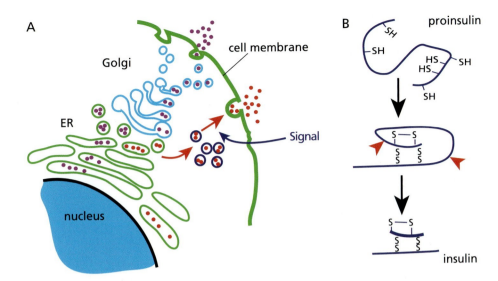

ecules are often made up of combinations of characteristic structural motifs, including sequences that mark them for export from the cell, for addition of substituents such as sugars, for oligomerization, or for interaction with specific receptors. Many signalling proteins can be grouped into large families of structurally—and often functionally—related proteins on the basis of shared sets of such structural motifs. Structural and functional characteristics are very often conserved across species, allowing the easy identification of corresponding proteins. Why are there such large sets of signalling proteins? Is each family member specific for a particular signalling function, or is there functional overlap between them? Both statements appear to be true: in some cases, different members of the same family are specialized to function at particular times or places, whereas in other cases, the members of a family seem to be functionally interchangeable.

Diffusible high-molecular-mass signalling molecules

Many signalling proteins are secreted by the cells to the extracellular space (which in some cases can mean the bloodstream), where their presence or activity can be detected several cell diameters away from their source. These proteins are freely diffusible and can also be synthesized *in vitro* in soluble form. Many diffusible signalling proteins have been discovered from fairly nonspecific functional assays in cell cultures, and have acquired generic names from these assays. For example, some have been found as molecules that affect cell growth or differentiation (growth factors), immune responses or cell survival (interferons and interleukins), or cell survival in the nervous system (neurotrophins). The members of these broad groups of

Fig. 4.12. Post-translational modification of signalling molecules. (A) Proteins are synthesized in the endoplasmic reticulum (ER) from where they are translocated to their functional destinations. In the case of signalling proteins and receptors a common route from the ER is to enter the Golgi apparatus, through which they are allocated to the cell-surface compartment. In their passage through the Golgi, many of these proteins are modified in a variety of ways: processed, assembled into complexes, modified by sugar or lipid additions. Some signalling proteins (red) are stored in specialized vesicles until they need to be released after the cell receives appropriate signals. **(B)** Proteolytic cleavages can activate the signalling function of some proteins. Insulin is an example of a signalling protein that is synthesized in the ER as a large precursor (preproinsulin, not shown) which is cleaved to a shorter precursor form, proinsulin. Cleavage of proinsulin allows the protein to fold into its functional configuration.

signalling proteins are diverse in both structure and mechanism of action.

Growth factors were first identified as molecules capable of sustaining the growth of cells in culture, for example, fibroblast growth factors (FGF), platelet-derived growth factors (PDGF), nerve growth factors (NGF) and associated neurotrophins, epidermal growth factors (EGF), and insulin-like growth factors (IGF) (Table 4.2). Another large group of molecules isolated in cell-culture assays are the transforming growth factor proteins, which include the alpha (TGFα) and beta (TGFβ) families. These proteins are capable of altering the morphology of certain cells in culture and, in the case of TGFβ, of inhibiting cell proliferation (Table 4.2).

Table 4.2 Soluble signalling proteins

Growth factor	Mouse	*Drosophila*	*C. elegans*
fibroblast growth factor (FGF)	FGF1–17	FGF1,2	FGF
epidermal growth factor (EGF)	EGF, TGFα	vein, spitz	lin-3
	herregulin	gurken	
	neuregulin	argos	
TGFβ/BMPs	nodal	dpp, gbb	dbl-1
	BMP1–9	screw	daf-7
	TGFβ1–3	Dactivin	
	activins		
neurotrophins	NGF, BDNF	(1)	—
platelet derived growth factor (PDGF)	PDGFA, B	—	—
interleukins	IL1–17	(1)	—
insulin and insulin-like growth factors	insulin, IGF1,2	(1)	(1)
interferon	IFN1–30	—	—

(1) Several homologues found.

The molecular cloning and sequencing of growth factors has enabled them to be grouped into families of molecules with shared structural features (Table 4.2). The family groupings suggested by these features are not always identical to those that arose from broadly based functional assays. For example, members of the EGF family all contain as an active domain a motif, dubbed the EGF repeat, that is also present in a number of unrelated ligands and receptors. Members of the EGF family of signalling molecules are synthesized as single transmembrane-pass proteins that are processed at the cell surface to generate the active ligand. In most cases, however, growth factors are secreted freely into the extracellular space.

Genetic analysis of the functions of many growth factors, and in particular of members of these families from lower invertebrates, has shown that they do not simply regulate proliferation but that they play specific roles during development and differentiation. For example, FGFs function in many developmental processes including, in vertebrates, the patterning of neurons in the nervous system, the induction of mesoderm (discussed in Chapter 9), and the shaping of the developing limb (see Chapter 12). Similarly, EGF signalling is a key component in basic processes of cell fate specification in many different tissues.

The TGFβ/BMP family of signalling molecules (Table 4.2) encompasses a very large number of proteins which, on the basis of structural criteria, can be grouped into three classes: activins, bone morphogenetic proteins (BMPs), and TGFβs. Members of this family are synthesized as precursors approximately 400 amino acids in

length, which are cleaved to yield a carboxy-terminal region of about 110 amino acids. The active signalling protein consists of disulphide-bonded dimers of this unit. TGFβ/BMP family members provide a striking example of the involvement of growth factors in a wide range of developmental processes. For example, *Xenopus* Vg1 (a BMP-related protein) and activin were identified as signalling molecules involved in inductive interactions in the early amphibian embryo. The *Drosophila* TGFβ protein decapentaplegic (dpp) has several roles in development, including dorsoventral patterning in the blastoderm and imaginal discs (structures in the larva that give rise, during metamorphosis, to adult organs including the wings, legs, and eyes). In mammals, members of the TGFβ family can either stimulate or inhibit cell proliferation, depending on their concentration and what other growth factors are present. The BMPs were first identified as proteins that could induce ectopic bone formation in rats, but have subsequently been found to function in many situations, for example, during neural induction and in patterning of a variety of tissues and organs.

Interferons and interleukins make up a large group of proteins that were initially recognized for their involvement in the development and function of the mammalian immune system, but many of which have subsequently been found to play a role in the development of diverse cell types, particularly in the nervous system. Multiple different interferons have been identified, and over a dozen different interleukins. The proteins do not share any obvious

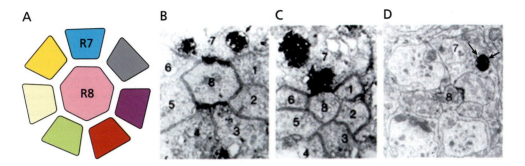

Fig. 4.13. Signalling by the bride of sevenless protein. (A) Arrangement of the photoreceptors in an ommatidium of *Drosophila*; R7 and R8 are indicated. **(B–D)** Electron microscope images of antibody stains for the sevenless and bride of sevenless proteins which are involved in the specification of the R7 fate (see Fig. 4.7). The staining can be observed as black spots. **(B)** Sevenless protein accumulates at the interfaces between a few different photoreceptors. Note some vesicles containing sevenless in the R7 cell. **(C)** In R7 in particular, sevenless becomes engulfed in large intracellular vesicles in the later stages of development. **(D)** Expression of bride of sevenless (boss) protein late in development at the interface between R7 and R8. Note that R7 has a large vesicle containing boss (arrows). (From Rubin, G.M. (1991) Signal transduction and the fate of the R7 photoreceptor in *Drosophila*. *Trends Genet.* **7**, 372–7.)

common structural features but motifs present in other extracellular proteins are common. This group of signalling molecules presents a particularly remarkable example of functional redundancy: in the absence of one protein, others will often take over its role and thus mask the effects of its loss.

Protein signalling molecules that are not freely diffusible

The common feature of these molecules is that they exert their effects either as intrinsic membrane proteins or in close association with the surface of the cell that produces them. Anchorage within the membrane may be required for specifying a very precise spatial relationship between signal and receptor. The *Drosophila* bride of sevenless (boss) protein, for example, is an intrinsic membrane protein with seven membrane-spanning domains. Although the cell expressing boss is surrounded by several other cells containing the cognate sevenless receptor, it only signals to one of them: the boss protein has been found to be localized to one face of the cell that expresses it, and the only cell that responds is the one exactly abutting this face (Fig. 4.13).

Signalling molecules that are intrinsic membrane proteins include the DSL family and members of the ephrin and semaphorin families. Other signalling molecules tightly associated with the membrane are members of the hedgehog and Wnt families of signalling molecules (Fig. 4.14).

The DSL family, first identified in *Drosophila* and *C. elegans*, is named after three proteins of this class: Delta, Serrate, and lag-2. Members of the DSL family are proteins with a single transmembrane domain that function in the assignment of cells to specific developmental pathways. They have extracellular domains of variable size with three prominent motifs: a series of tandemly arranged EGF-like repeats and two cys-rich regions, one (the DSL domain) that is located at the amino terminus and is common to all members of the family, and another near the transmembrane region that is characteristic of the Serrate but not the Delta family members (Fig. 4.14). Experiments in invertebrates have indicated that these molecules signal to neighbouring cells and are involved in short-range signalling events that regulate cell fate assignments and cell behaviour (details in Chapters 5 and 8).

Members of the ephrin (for 'eph receptor interacting proteins') and semaphorin families (Fig. 4.14) function during the development of the nervous system, in attraction and repulsion processes associated with cell migration and pathfinding of neurons during the laying down of neural networks, and in the formation of tissue boundaries. Some ephrins and semaphorins contain a single transmembrane domain that tethers the extracellular domain to the membrane and allows the delivery of information over a very short range. Semaphorins have a modular extracellular region that includes a semaphorin domain characterized by 15 conserved cysteine residues and, in some cases, domains with homology to the immunoglobulin superfamily. The ephrins have two characteristic motifs: an extracellular cysteine-rich domain and a domain with sequence similarity to fibronectin proteins (which are constituents of the extracellular matrix).

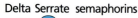

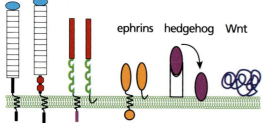

Fig. 4.14. Molecular properties of some families of membrane-associated high-molecular-mass signalling molecules. Examples of membrane-associated proteins that signal in association with the cell surface. Each belongs to a large family and all have been found in all animals. Delta and Serrate are members of the DSL (Delta, Serrate, and lag-2) family of signalling molecules. They contain a short intracellular domain and an extracellular domain with two dominant features: a number of tandemly arranged EGF-like repeats (boxes) and an amino-terminal domain, the DSL domain (blue), which is a cysteine-rich region conserved among all elements of the family. Serrate and Delta may represent two different types of members, with the distinguishing feature being some cysteine-rich repeats adjacent to the transmembrane domain (red hexagons) that are present in Serrate and Serrate-like molecules but not in Delta. Semaphorins are signalling molecules characterized by an extracellular domain with a conserved structural motif, the semaphorin domain (red block), and often a number of immunoglobulin-like repeats (green) close to the membrane. There are two types of members of this family, those associated with an integral transmembrane domain, and those associated with the membrane through a GPI link (represented as a hook). Ephrins are small membrane-associated molecules that may also be either integral membrane proteins or linked to the membrane through GPI. Their extracellular domain contains a cysteine-rich domain and a domain with similarity to fibronectin (not shown). Hedgehog is made as a precursor with proteolytic activity that frees an amino-terminal fragment (purple). The carboxy terminus of this fragment is modified by the addition of cholesterol and remains tightly associated with the cell surface. Wnt proteins are cysteine-rich proteins that are tightly associated with the cell surface.

Some other cell-surface-associated signalling proteins are synthesized as soluble proteins, but are post-translationally processed by the addition of a lipophilic tail that anchors them to the membrane. These proteins are only active as signalling molecules in this form. For example, some members of the ephrin and semaphorin families, instead of being anchored to the membrane through a transmembrane domain, are tethered to it by the covalent addition of a glycophosphatidylinositol (GPI) substituent (Fig. 4.14). This attachment to the membrane has been shown to be essential for eliciting a response.

Members of the hedgehog (hh) family of proteins (Fig. 4.14) are also indirectly anchored to the membrane, in this case by a covalently attached lipid moiety. Hh proteins are synthesized as large precursors whose carboxy-terminal domain contains a protease activity that cleaves the molecule to generate an active amino-terminal half and inactive carboxy-terminal half. The amino-terminal half is further modified by the addition of cholesterol, which associates the protein to the membrane. Hh proteins are involved in a wide variety of processes from the generation of different cell types to the assignment of positional information during limb and neuronal development.

The Wnt family of signalling molecules (Fig. 4.14) is another large family of proteins with prominent roles during development. The prototype members of this family—the *Drosophila* wingless protein and the product of the mammalian proto-oncogene *int1*—emerged from very different lines of investigation in different organisms, but were revealed after cloning and sequencing to be related proteins. The cloned genes were then used as probes to detect similar proteins in other organisms. Although Wnt family proteins are secreted glycoproteins, they are not freely diffusible and are tightly associated with the extracellular matrix.

Receptors

As well as launching signals to communicate with its neighbours, each cell in a multicellular organism is also equipped with the molecular machinery to allow it to detect and respond to the signals it receives from nearby cells and distant organs. The first tier of this machinery is the cell's complement of receptors, which act as sensors, decoders, and transducers of information.

Most receptors are cell-surface molecules anchored within the cell membrane by a domain consisting of one or more membrane-spanning chains of amino acids. A domain exposed on the outer surface of the cell allows the receptor to monitor the external environment and bind cognate ligands, while an intracellular domain connects the receptor into the cell's signal transduction machinery (Fig. 4.15). 'Mix and match' experiments, in which chimaeric receptors have been made from various combinations of extracellular, transmembrane, and intracellular domains, have shown that receptor structure is often modular, with each domain capable of functioning more or less independently. It has been possible to alter completely the response of a cell to a given signal by combining

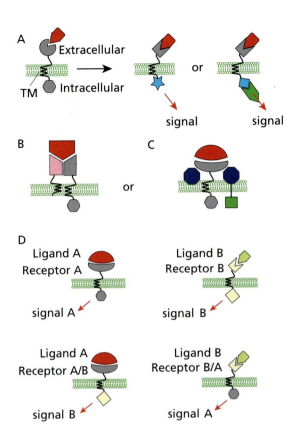

Fig. 4.15. **Receptors.** **(A)** A typical receptor (*left*) consists of an extracellular domain that can sense the environment and be recognized molecularly by signalling ligands, a hydrophobic section (usually called the transmembrane domain, TM) that anchors the protein to a membrane and an intracellular domain that, when the receptor is activated, effects a biochemical event resulting in transmission of a signal. The intracellular domain can itself have enzymatic activity (*centre*), or ligand binding may cause it to be linked to another intracellular protein that has such activity (*right*). **(B)** Receptors sometimes operate in complexes. Individual subunits in these complexes may have a variety of roles. For example, a subunit (pink) might bind ligand without directly eliciting a signal, its role instead being to make the ligand recognizable to another subunit that has signalling activity. **(C)** In other cases, signal transmission might involve the combined activity of more than one subunit. **(D)** The extra- and intracellular domains of receptors are very specific both in the molecules they recognize and in their effects. This specificity can be tested by constructing chimaeric molecules in which the intra- or extracellular domains of two receptors are exchanged. As shown in the figure, replacing the extracellular domain of receptor B by the extracellular domain of receptor A will cause the cell to transmit signal B in response to the binding of ligand A and vice versa.

the extracellular domain of the cognate receptor with the transmembrane and intracellular domains of a different receptor molecule (Fig. 4.15). The specificity for the ligand is provided by the structure of the extracellular domain.

As we have already seen, some signalling molecules are able to cross membranes; these ligands interact with receptors located in the cytoplasm of the cell or within its internal membrane system and tend to have a direct effect on their targets without relying on the mediation of a signal-transduction pathway. In the following sections, we shall first discuss cell-surface receptors, and then intracellular receptors.

Cell-surface receptors

A common classification of cell-surface receptors groups them on the basis of the nature of their interaction with the downstream components of the signal transduction pathway: two major groups that are important during development are the enzyme-linked receptors and the seven transmembrane domain family of receptors. In addition, there are some receptors with unique structures that make

use of unique signalling mechanisms that are not, as yet, well characterized. The downstream pathways associated with these receptor types will be discussed in detail in Chapter 5; here we shall restrict ourselves to a discussion of the events associated with the interaction between the ligand and the receptor.

Enzyme-linked receptors may themselves have intrinsic enzymatic activity, or the activated receptor may be closely linked to protein(s) with such activity. Receptors with intrinsic enzyme activity almost invariably have a single transmembrane segment and are characterized by an intracellular domain that can catalyse either the addition of an ATP-derived phosphate group to a substrate molecule (kinases), or its removal (phophatases). The activity of this important group of receptors highlights the role of phosphate bonds as a major 'currency' of biological signalling (also see Chapter 5). The substrate for the kinase or phosphatase activity is usually another protein, and frequently the receptor also phosphorylates itself. The function of autophosphorylation can be both to activate the kinase activity of the receptor, and to help create a 'docking site' for binding downstream targets in the signal transduction pathway.

Receptor tyrosine kinases

The enzymatic receptors can been grouped into families on the basis of the amino acid residue that is the target for their activity. The receptor tyrosine kinase (RTK) family encompasses a large number of receptors for a variety of ligands that include EGF, insulin, NGF, PDGF, FGF, and ephrins (Fig. 4.16). The EGF receptor is the prototype of this class. It acts as a dimer and the function of the ligand, EGF, is to bring two receptor monomers together such that they phosphorylate each other and act in concert to phosphorylate appropriate residues in target proteins (Fig. 4.16). The phosphorylation events result in the generation of a docking site for transducer proteins that determine

which downstream pathway is activated and thus the cell's response to the signal (see Chapter 5, Fig. 5.12).

There are some variations on this basic theme (Fig. 4.16). For example, some RTKs can form heterodimers or oligomers in response to the binding of specific ligands. The PDGF receptor, for instance, can form heterodimers or homodimers depending on whether the binding ligand is a heterodimeric or homodimeric form of PDGF (PDGF has two different chains, α and β, which can dimerize

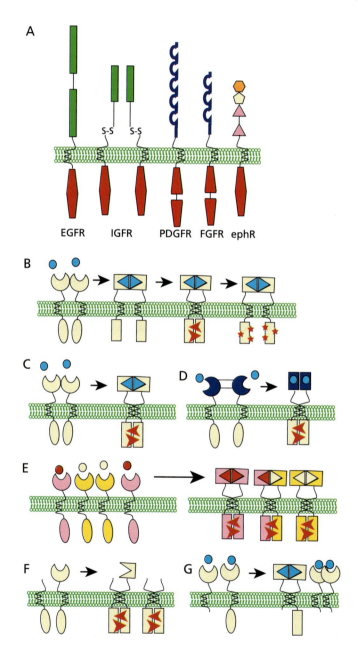

in any combination). This versatility enables the cell to achieve subtle modulation of its response to different signals (Fig. 4.16). Dimerization, oligomerization or clustering of RTKs may be a universal characteristic of signalling through these receptors.

Truncation of RTKs by removal of the extracellular domain results in constitutive signalling that is independent of ligand (Fig. 4.16). This observation suggests that the role of the ligand is to induce the dimerization of the receptor which is otherwise prevented by the configuration of the extracellular domain. Removal of the extracellular domain or ligand binding must release steric hindrances and allow the receptor to dimerize or oligomerize. The existence of point mutations in the extracellular domain of RTKs that lead to spontaneous oligomerization of the receptor also suggests that receptor activation is associated with conformational changes. These mutations generate receptors that are constitutively active and that, in particular cell types, can lead to alterations in the proliferative activity of the cell. In fact, several tumour-causing oncogenes turn out to be RTKs that have been rendered constitutively active by mutation.

Truncation of the intracellular domain can create an RTK that can still dimerize in response to ligand binding, but is unable to transduce a signal. Such mutant receptors can titrate both ligand and endogenous (unmutated) receptor molecules, and thus reduce the activity of the receptor in a dominant manner (Fig. 4.16). This type of mutation, which can be constructed artificially, has proven very useful in the analysis of some developmental processes.

Serine/threonine kinase receptors

This family is exemplified by the receptors for members of the TGFβ/BMP group of signalling molecules (Table 4.2). These receptors contain two different types of subunits, known as Type I and Type II (Fig. 4.17), both of which have kinase activity but have different properties and substrates. Binding of ligand to the Type II subunit, which contains a constitutively active kinase, triggers recruitment of the Type I subunit to the complex and enables it also to bind ligand. Within the complex, the Type II subunit phosphorylates and activates the Type I subunit, which in turn phosphorylates an intracellular signal-transducing target and activates the signal transduction pathway (Fig. 4.17). Analysis of receptor complexes cross-linked to labelled ligand suggests that on the cell surface, both Type I and Type II subunits are present as dimers, and that therefore the active signalling complex is a heterotetramer consisting of two subunits of each type.

Fig. 4.16. Receptor tyrosine kinases. (A) The prototype of the receptor tyrosine kinase (RTK) family is the EGF receptor (EGFR), which is a monomer with a single kinase domain (red) in its intracellular domain. The insulin receptor and insulin-like growth factor receptor (IGFR) proteins are covalently linked tetrameric versions of the EGFR. Members of the PDGF receptor (PDGFR) and FGF receptor (FGFR) families have an insert in their kinase domains. In all cases, the intracellular domain contains effector as well as regulatory sites. The extracellular domains of the receptors are characterized by multiple copies of motifs characteristic of each receptor family: in the case of EGFR and IGFR, they are special cysteine-rich regions (green boxes); the PDGF and FGF receptors contain immunoglobulin-like domains (blue loops). The ephrin receptors (ephR) contain fibronectin III (FNIII) repeats (pink boxes) and ephrin-characteristic cysteine-rich repeats (yellow and orange shapes). **(B)** Mechanism of RTK signalling. Upon binding ligand, the receptor dimerizes and undergoes a conformational change. This brings the two kinase subunits into close contact and allows the reciprocal phosphorylation of their intracellular domains (red arrows indicate the phosphorylation event, and the stars indicate the phosphoresidues) followed by the recruitment and phosphorylation of specific residues in target intracellular proteins. It is the phosphorylation of these residues that triggers the intracellular events (see Chapter 5). Depending on the structure of the receptor, the ligand may have one of three functions. **(C)** The ligand might bring about a conformational change that allows the formation of dimers and so triggers the auto- and cross-phosphorylation events (see also B). **(D)** In other instances, as in the case of the insulin receptor, two receptor subunits are covalently linked and the main function of ligand binding is to induce a conformational change in the receptor that triggers its kinase activity. **(E)** The function of the ligand may be to select different combinations of receptor subunits, thereby bringing different qualities to the signalling event. For example, the PDGF receptor has two different monomers that can dimerize in any combination to form an active receptor. Depending on what dimeric form of the ligand is present, different receptor dimers with different signalling properties will be formed. **(F)** Removal of the extracellular domain of an RTK bypasses the requirement for ligand, and results in the constitutive activation of the receptor. Such dominant active receptors are locked in a configuration of active kinase. **(G)** On the other hand, receptors lacking the intracellular domain will bind ligand but will not transduce a signal and therefore can compete with the normal receptor for ligand. These truncated receptors are called 'dominant negative' and achieve their effects either by titrating ligand away from productive dimers, or because they become incorporated into inactive heterodimeric complexes that dilute signalling monomers.

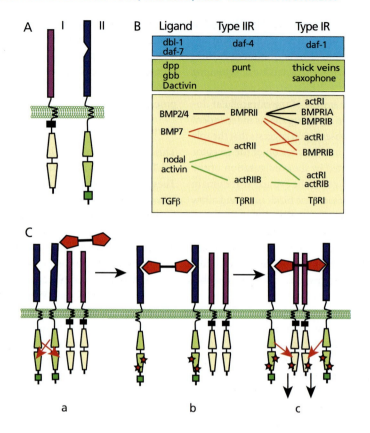

Fig. 4.17. Receptor serine/threonine kinases. Structure and mechanism of action. (A) Receptor ser/thr kinases are receptors for members of the TGFβ/BMP family of growth factors. There are two related but different receptors for these growth factors, Type I and Type II, which combine in a ligand-dependent manner to receive and transduce the signal. Both contain a short cysteine-rich extracellular domain (purple in I and blue in II) and a characteristic intracellular domain with an interrupted kinase domain (yellow in I, pale green in II). In the intracellular domain there are two short sequences that are characteristic of each type: Type I receptors have a short glycine/serine-rich domain (black box) and Type II receptors have a short sequence rich in serine and threonine located at the carboxy-terminal end of the kinase domain (green box). **(B)** For a given organism there are different Type I and Type II receptors, which interact with specific members of the ligand family. Shown here are some examples from *Caenorhabditis elegans* (blue), *Drosophila* (green), and vertebrates (yellow) which illustrate the possible relationships between ligands and receptors. Note that, particularly in vertebrates, some ligands can interact with more than one Type II receptor and some Type II receptors can combine with more than one Type I receptor. This creates a large number of combinatorial possibilities enabling versatile modulation of signalling. **(C)** Signalling through a typical TGFβ receptor complex. (a) Type II receptors have a constitutive kinase activity (indicated by red arrows) that does not

on its own seem to have any effect on the physiology of the cell. Type I receptors have no enzymatic activity of their own. (b) The ligand (red), which is a dimer, binds first to a dimer of the Type II receptor and induces an interaction with the Type I receptor. (c) As a result of the ligand-driven interaction between Type I and Type II receptors, the Type II receptor phosphorylates the Type I receptor (red arrows) and triggers signalling activity.

There are several different Type I and Type II receptors, as might be expected from the large number of different TGFβ/BMP family ligands. Each ligand can bind to more than one receptor and can drive the formation of different Type I–Type II heterodimers (Fig. 4.17). There are, however, some general rules that govern these combinatorial possibilities. The Type II receptors for TGFβs and activins can bind their ligands in solution, but the Type I receptors cannot and will only recognize ligands when they are bound to particular Type II receptors. This suggests two levels of constraint in the formation of heterodimers: one imposed by the ligand and what receptors it recognizes, and another by the compatibility of different Type I and

Type II receptors. On the other hand, Type I and Type II receptors for BMPs will each bind BMPs in solution with a low affinity, but together they bind very efficiently. Subject to these constraints, the existence of different subtypes within the Type I and Type II receptor classes makes possible a wide variety of different combinatorial interactions and thus of different responses to a given ligand.

Receptor phosphatases

In order to prevent constitutive signalling, there must be a way for tyrosine or serine/threonine kinase receptors that are activated by phosphorylation to be returned to their ground state. In the majority of cases, little is known about how this occurs, but at least for some the reaction has been shown to be carried out by protein phosphatases (Fig. 4.18). Some of these phosphatases are transmembrane receptors but others are not.

Receptor protein tyrosine phosphatases (RPTPs) down-regulate the activity of RTKs and are members of a larger group of protein tyrosine phosphatases (PTPs), which also includes many cytoplasmic members that participate in intracellular signalling pathways. PTPs act on a wide variety of phosphorylated targets, most of which have still to be defined. Many of the known RPTPs are expressed in the nervous system and have been implicated in the regulation of the processes of cellular recognition associated with the development of this tissue (see Chapter 6).

RPTPs are characterized by an intracellular region that includes one or two phosphatase domains, a single transmembrane segment and a variable extracellular domain that often contains motifs characteristic of adhesion molecules, such as immunoglobulin-like domains and fibronectin repeats (Fig. 4.18). Both soluble and membrane-anchored ligands have been identified for different RPTPs, and some RPTPs can participate in homophilic interactions that link neighbouring cells, suggesting that they may be able to function both as receptors and as ligands.

Receptors associated with cytoplasmic kinases

Some receptors do not have an intrinsic biochemical activity and instead recruit cytoplasmic proteins to aid in transmission of the signal. Many cytokine receptors, for example, lack intrinsic enzyme activity but are closely associated with cytoplasmic kinases: activation of the receptor by ligand binding serves to recruit or activate cytoplasmic or membrane-bound tyrosine kinases that convey the signal to downstream signal-transduction

pathway components. Another feature of many of these receptors—one that might turn out to be a general property of this receptor class—is that ligand binding triggers the formation of complexes. In some cases, different receptor complexes share some subunits that are differentially recruited by different ligands.

Members of the cytokine receptor group (Fig. 4.19) share a structurally homologous extracellular domain of about 200 amino acids. This domain is embedded within various combinations of other domains typical of receptors. The ligands for the cytokine receptors include interleukins, growth factors, and some signalling factors involved in development and differentiation of the nervous system. Ligand binding to the receptors stimulates formation of homo- or heterotypic multimers of receptor components. Different subunits of these complexes may have different functions. For example, the gp130 protein, a common subunit of many different heterodimeric

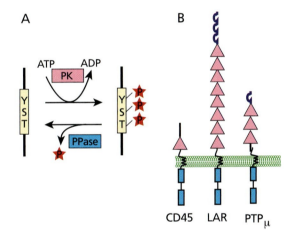

Fig. 4.18. Receptor protein phosphatases. (A) Protein phosphatases (PPase) reverse the effects of protein kinases (PK) and catalyse the removal of phosphate residues from serine (S), threonine (T), or tyrosine (Y) residues. **(B)** Basic structure and representative examples of receptor protein tyrosine phosphatases. The extracellular domain is composed of combinations of multiple copies of Ig-like domains (blue), FNIII repeats (pink triangles) and other less well-defined cysteine-rich repeats. The intracellular domain usually contains two phosphatase domains (blue boxes), of which the membrane-proximal one appears to have activity. The other phosphatase domain does not show much activity *in vitro* and in some cases is mutated in key residues. It is not clear how these proteins work. In experiments with chimaeras it appears that forced dimerization inhibits activity of the phosphatase domain and that therefore they act as monomers.

interleukin receptor complexes, participates in the signalling reaction but does not itself bind ligand (Fig. 4.19). Conversely, the interleukin-binding components of these receptors recruit ligand to the membrane but do not themselves have any signalling properties (see Fig. 4.19 for details). Mechanisms such as these enable different ligands to share receptor complex components, contributing to the redundant functions observed for many of these proteins. The signalling event is triggered by the

recruitment and activation of cytoplasmic proteins to the complexes.

Although oligomerization of receptor components is a frequent theme for enzyme-linked receptors, not all receptors share this feature. The IL1 receptor, and its *Drosophila* counterpart Toll, for example, function as monomers. These receptors are involved in the development of immunity and, in the case of Toll, some early patterning events of the embryo. Through a conformational change, ligand

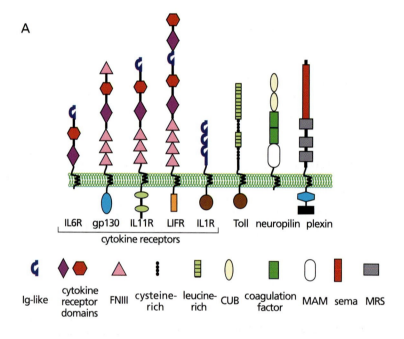

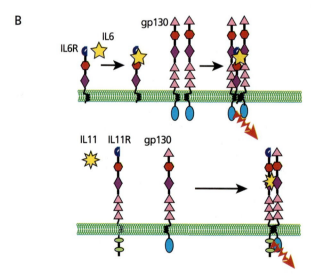

binding induces the association of the receptor with a cytoplasmic complex that is responsible for the signalling event. Neuropilins and plexins form a complex that can bind semaphorins and is involved in transmitting the semaphorin signal. It is clear that the signalling event involves the plexin subunit of the complex, but it is not yet clear what the mechanism is.

There are some receptors that are assumed to be linked to an intracellular kinase, though the evidence is as yet incomplete. In some cases, cytoplasmic protein kinases have been shown to interact with the intracellular domain of the receptor; in others, the intracellular domain of the receptor has been shown to be phosphorylated during the signalling event but a specific kinase has not been found.

Seven transmembrane domain receptors

This large family of receptors, which includes over 100 different proteins, is characterized by the presence of seven membrane-spanning segments within the same polypeptide, forming a compact barrel-like three-dimensional structure for the transmembrane domain (Fig. 4.20).

Most of the known members of this family of receptors convey signals to the interior of the cell via a group of small signal-transducing proteins called G proteins, whose role will be described in more detail in Chapter 5. G-protein-coupled receptors are found across the entire animal kingdom, from yeast, where they are used to detect mating-type pheromones, and *Dictyostelium*, where they mediate the chemotactic response to cAMP (see Fig. 4.10) to vertebrates, where they carry out many different roles from regulation of cellular metabolism to nervous system function and cell fate assignment. G-protein-coupled receptors have a huge variety of ligands, including hormones, neurotransmitters, peptides, and glycoproteins. In some receptors, such as the NMDA receptor which functions in the nervous system, the extracellular domain is very small and ligand binding occurs within the transmembrane domain.

Another group of seven transmembrane domain receptors is the frizzled class. The proteins of this group were identified as receptors for members of the Wnt family of proteins, but also include the smoothened protein which is involved in signalling by hedgehog. These proteins are characterized by a very large cysteine-rich extracellular domain, which also contains some heparin-binding motifs. The intracellular domain is very variable in size. In some receptors of this class it is very small, and mutational experiments suggest that it has little function; signalling seems to rely on domains formed by intracellular loops between the membrane-spanning segments. In contrast, the smoothened protein has a large intracellular domain containing a variety of motifs that suggest possible links to several different intracellular signalling systems.

Fig. 4.19. Receptors associated with cytoplasmic kinases. Members of this class of receptors lack a signalling ability of their own but upon ligand binding they manage to convey a signal to the cell, usually through an association with intracellular signalling elements. **(A)** Many receptors for interleukins (IL), *Drosophila* homologues of these proteins and receptors for neural signalling molecules belong to this class. Some examples are shown here. Their intracellular regions share no obvious structural features, as might be expected from the fact that they link in to different intracellular transduction systems. In some cases they contain tyrosine residues that can be phosphorylated. Their extracellular regions, however, contain characteristic motifs (colour code and key in figure) such as the FNIII repeat and Ig-like motifs which appear in some other extracellular proteins. In addition, their extracellular regions sometimes contain domains that are shared among receptors of a particular class, for example, the interleukin receptors share a specific domain composed of two conserved motifs (red and purple polygons) one of which is cysteine-rich. Toll is a *Drosophila* receptor with an intracellular region similar to that of IL1R, however, in its extracellular region it has, instead of Ig-like motifs, two blocks of leucine-rich repeats (green rectangles), followed by a domain with four cysteine residues (black dots). Plexins and neuropilins, which are both receptors for semaphorins, do not have much in common and contain a variety of different domains, including a semaphorin domain in the case of plexins. **(B)** Two examples of signalling through receptors linked to cytoplasmic kinases. Ligand binding usually triggers the assembly of a complex, in which there may be a subunit that is common to the receptors for several different ligands. For example, the gp130 protein is a common subunit of many interleukin receptor complexes. Two such complexes, specific for the IL6 and IL11 ligands, are shown here. In the case of IL6 (top panel), the ligand (yellow star) binds first to its receptor, which does not have an intracellular domain, and the complex is then recognized by gp130 which dimerizes, interacts with intracellular signalling elements and triggers a signal. In this case, the active ligand is a combination of IL6 and its receptor, IL6R. The case of IL11 (bottom panel) is more conventional: binding of the ligand (yellow star) to its cognate receptor, IL11R leads to association with gp130 and an intracellular signalling molecule.

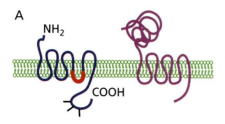

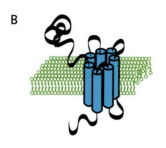

Fig. 4.20. **Seven transmembrane domain receptors. (A)** Basic structure of members of the seven transmembrane domain family of cell-surface receptors. The protein contains an amino-terminal domain of variable length and structure, seven closely apposed hydrophobic segments, and an intracellular domain of variable length. Two examples are shown. *Left*: a G-protein-coupled receptor with a short extracellular domain and carboxy-terminal domain with some phosphorylation sites (black) which are involved in the downregulation of receptor activity. The red segment between transmembrane domains 5 and 6 is involved in interactions with intracellular transducers. In some cases, the ligand binds to a region created by the extracellular loops of the transmembrane regions. *Right*: an example of a receptor of the frizzled family, which has a large cysteine-rich extracellular domain whose amino acid sequence identifies members of this family. **(B)** The folding of the receptor in the membrane creates a barrel shape in the three-dimensional structure of the protein.

Other cell-surface receptors

There are several families of cell-surface receptors whose mechanism of action does not depend on intrinsic enzymatic activity or links with enzymatic proteins. Many of these receptors do not fall readily into any of the classes mentioned above.

Some receptors are linked to ion channels. Members of this family of receptors are proteins with multiple membrane-spanning segments. When the receptor is activated, either electrically or by ligand binding, the membrane-spanning domain undergoes a conformational change to form a pore or channel that allows the passage of ions across the membrane (Fig. 4.21). Voltage-gated ion channels allow the transient fluxes of Na^+ and K^+ ions that propagate the action potential of a nerve impulse along axons in the nervous system. Ligand-gated ion channels in the post-synaptic membrane, as mentioned previously in this chapter, open and close transiently in response to neurotransmitter binding, conveying the electrochemical signal of a nerve impulse from one neuron to the next (Fig. 4.21).

One unusual type of receptor is the receptor patched, which participates in signalling by the hedgehog protein. Patched has multiple transmembrane domains and its function appears to be to affect the stability of the seven transmembrane domain protein smoothened and prevent its localization to the cell surface. Binding of hedgehog to patched inactivates this activity and allows smoothened to be localized at the cell surface (Fig. 4.20).

The Notch family of receptors for ligands of the DSL family (Fig. 4.22) has a unique mode of activation. These receptors play a central role in the assignment of cell fates during development (see Chapter 8). Notch proteins contain a single transmembrane segment, and extracellular and intracellular domains that are characterized by a number of different types of motifs (Fig. 4.22). The most striking of these are a set of EGF-like repeats, in the extracellular domain, that are largely conserved between *Drosophila* and humans. Mutational experiments have succeeded in associating some motifs with particular functions, for example, the EGF repeats are ligand-binding domains, while the extracellular LN repeats and the intracellular cdc10/ankyrin repeats are necessary to transduce the signal.

Notch exists in two forms at the cell surface, one as a single amino acid chain and the other in the form two polypeptide chains joined by disulphide bonding. Binding of the DSL ligand Delta to the heterodimeric form of Notch triggers a sequence of cleavages: first in the extracellular domain near the membrane, and then in its intracellular domain. This process releases a soluble intracellular fragment. The intracellular fragment then goes into the nucleus where it acts as a transcription factor. The extracellular domain of Notch can bind other molecules, such as some Wnt family members, that are not members of the DSL family of signalling molecules and this binding is likely to have functional consequences.

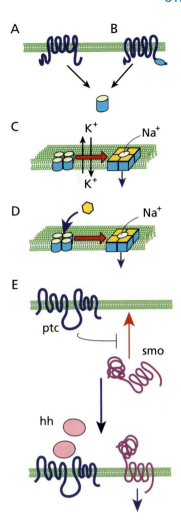

Fig. 4.21. Receptors with multiple transmembrane domains.
(A) Diagrammatic representation of the basic structure of one subunit of a voltage-gated potassium channel. **(B)** Basic structure of a subunit of a cAMP-gated channel protein. This type of channel responds to changes in the intracellular concentration of cAMP which binds to a specific region of the intracellular domain (blue), inducing a conformational change. Receptors of this kind fold tightly into structures which then assemble with different subunits. They are usually multisubunit complexes which form pores or channels that open and close depending on specific conditions (C–D). In neurons they are used to regulate neural activity. **(C)** Voltage-gated channels open and close in response to changes in the membrane potential (represented as fluxes of K^+ ions), allowing the influx or outflux of ions. **(D)** Ligand-gated channels respond to the binding of a specific ligand (polygon) to one or each of the subunits, enabling opening or closing of the pore. **(E)** Patched (ptc) is a membrane protein with several transmembrane domains, giving it a similarity to some channels. In the membrane, patched destabilizes the association of the seven transmembrane domain receptor, smoothened (smo), with the cell surface. Binding of the amino-terminal moiety of hedgehog (hh) to patched inhibits this activity and allows the accumulation of smoothened at the cell surface (see Chapter 5, Figs. 5.22, 5.23 for further details).

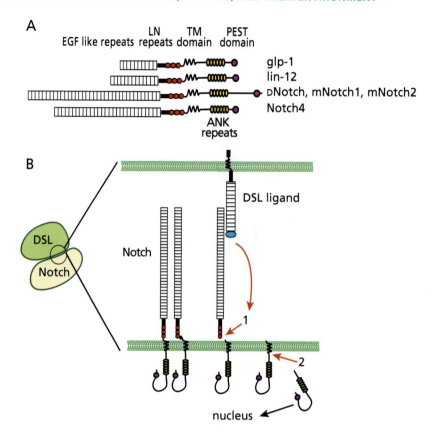

Signalling proteins as receptors

As we have seen, some signalling proteins are membrane-tethered, a property that is more usually associated with receptors. In fact, there is evidence that at least some membrane-tethered signalling proteins are also able to act as receptors. For example, it has been shown that the cytoplasmic domain of the ephrin B ligands is phosphorylated upon receptor binding, suggesting that the signalling event may transfer information back into the signalling cell as well as to the receptor-bearing cell (Fig. 4.23 and 4.24). The ephrins and their receptors are involved in axonal pathfinding and in some cell–cell recognition processes that determine whether cell populations are able to mix with one another. In the vertebrate hindbrain, for example, there is a pattern of ephrin expression that is consistent with a role in the segregation of distinct cell populations (Fig. 4.23). Mutational analysis supports this interpretation and suggests that sorting out of ephrin-expressing cells requires reciprocal signalling between ephrins and ephrin receptors (Fig. 4.24).

Fig. 4.22. Notch receptors. (A) Members of the Notch receptor family are characterized by an extracellular domain with a series of tandemly arranged EGF-like repeats (boxes) and a membrane-proximal set of three cysteine-rich repeats peculiar to this family, called LN repeats (red). The intracellular domain contains six cdc10/ankyrin repeats (yellow ovals). Lin-12 and glp-1 are *Caenorhabditis elegans* members of the family. *Drosophila* Notch (DNotch), as well as vertebrate Notch1 and Notch2, have 36 EGF-like repeats and a long tail carboxy-terminal to the cdc10/ankyrin repeats. Vertebrates have two more Notch proteins, Notch3 and Notch4, that have lost some EGF repeats and some regions of the intracellular domain. These deletions are more extreme in Notch4. **(B)** Mechanism of signalling by Notch (see Chapter 5, Fig. 5.9 for details). The Notch protein at the cell surface may be in either of two forms: a single polypeptide chain, or a form with two polypeptide chains joined by disulphide bonding; one of the chains contains the transmembrane and intracellular domains while the other is entirely extracellular but is tethered to the membrane by disulphide bonding with the transmembrane chain. It is possible that the receptor has more than one activity (see Chapter 5), but only one has been characterized in detail: binding of a ligand of the DSL family triggers two sequential cleavages (labelled 1 and 2) in the disulphide-bonded form of Notch that lead to the intracellular domain being released from the membrane and translocating into the nucleus.

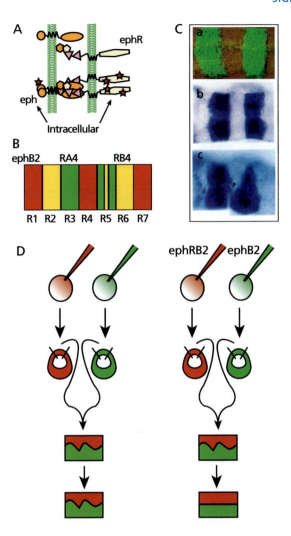

Fig. 4.23. Signalling between ephrins and their receptors.
(A) Interactions at the cell surface between ephrins (eph) and their receptors (ephR). The intracellular domain of the ephrin is also phosphorylated (stars) upon interaction with the eph receptor, suggesting that the interaction may be two-way. **(B)** Diagram indicating the distribution of ephrinB2 (ephB2, red) in the segmental structures, called rhombomeres, at the anterior end of the developing nervous system of the zebrafish, and of the ephrin receptors A4 (RA4, green) and B4 (RB4, yellow). Rhombomeres are indicated below (R1–R7). Note the sharp boundaries of expression that coincide with the rhombomere boundaries and which suggest that the boundaries might be a consequence of signalling triggered by an interaction between the two molecules. **(C)** Expression of elements of ephrin signalling in the hindrain of the zebrafish. (a) Three-dimensional reconstruction of confocal images of ephA4 receptor protein expression in rhombomeres 3 and 5 (Courtesy of Qiling Xu and D. Wilkinson.) (b) Expression of ephA4 receptor RNA. (c) Effects of ex-

pressing a dominant negative form of the ephA4 receptor on the segmentation of the hindbrain revealed through the expression of ephA4 receptor RNA. Note the disruption of the sharp boundaries and the tendency of the segments to fuse (After Holder, N. and Klein, R. (1999) Eph receptors and ephrins: effectors of morphogenesis. *Development* **126**, 2033–44). **(D)** An assay in zebrafish embryos that shows that ephrins and ephrin receptors participate in sorting out different cell populations. *Left*: zebrafish embryos are injected at the one-cell stage with either a red or a green fluorescent dye. After gastrulation, pieces of ectoderm tissue from two differently labelled embryos are placed in contact. The pieces adhere and show mingling of cells at the boundary between them. *Right*: if, in addition to the dye, RNA for an eph receptor (e.g. ephRB2) is injected into one single-cell embryo and RNA for its cognate ephrin (ephB2) is injected into the other, and then the experiment is repeated, the two cell populations segregate with a sharp boundary.

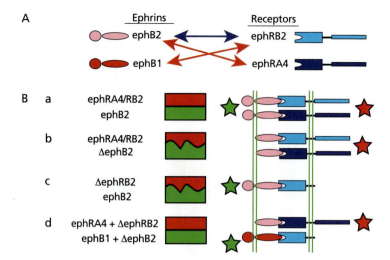

Intracellular receptors

The detection of signals is not restricted to the cell surface: signalling molecules that can cross cell membranes interact with receptors inside the target cell. These receptors may be cytoplasmic, confined to the nucleus, or associated with the internal membrane system of the cell.

The nuclear receptor family of cytoplasmic and nuclear proteins includes receptors for the lipophilic steroid and thyroid hormones as well as retinoids (Fig. 4.25). This large group of structurally related proteins all contain a ligand-binding domain, a C_4 zinc-finger-type DNA-binding domain specific for response elements in the regulatory regions of target genes, and transcription-regulatory domains through which they interact with the transcriptional machinery. In addition, nuclear receptors may have domains for such functions as translocation to the nucleus, dimerization, and interaction with other regulatory proteins (Fig. 4.25). These functions are generally activated by conformational changes that occur upon ligand binding.

There are two main classes of nuclear receptors. Members of one class, which includes most of the steroid receptors, are found in both the cytoplasm and the nucleus. In the cytoplasm they are bound to specific proteins that prevent their translocation to the nucleus. Binding of ligand causes a conformational change that releases the receptor from the binding protein and enables it to enter the nucleus, where in a dimeric form it activates transcription of target genes (Fig. 4.25).

The second class of nuclear receptors, which includes the retinoid receptors as well as the receptors for vitamins A and D and thyroid hormone, are mainly nuclear and

Fig. 4.24. Experimental demonstration of bi-directional signalling between ephrins and their receptors. (A) Some ephrins can interact with more than one receptor while some ephrin receptors can interact with more than one member of the family of ligands. **(B)** The functional interactions outlined in A can be used to test whether ephrin signalling is uni- or bi-directional. The experiments rely in the assay described in Fig. 4.23 but make use of full-length or truncated (Δ) forms of a receptor (ephRB2) that is recognized by two ephrins (ephB1 and ephB2), and of an ephrin (ephB1) that only recognizes one receptor (ephRB2). The truncated forms of the ligand and receptor lack the intracellular domain and are not expected to function in interactions with intracellular signalling components. The basis of the assay is whether or not two cell populations, one containing ligand and the other receptor, sort out as a result of the interactions between the two molecules. The combinations of eph and ephR used in four experiments are shown on the left, the results of the assays (cell mingling or sorting) in the middle, and the deduced molecular interactions on the right. The stars indicate activation of the molecule and a signalling event. Sorting of the two cell populations occurs when both full-length ephrin and its receptor are supplied (a), but is abolished if either is truncated in its intracellular domain (b,c). A combination of two different mutant ligand–receptor pairs can restore the ability of the populations to sort out (d), provided that one of the pairs has a receptor with an intact intracellular domain, and the other has a ligand with an intact intracellular domain. This result shows that bi-directional signalling has occurred, enabling a functional complementation. Although no single pair of interacting molecules signals both ways, each pair provides signalling in one direction, and combining them provides the required bi-directional signalling. (After Mellitzer, G. *et al.* (1999) Eph receptors and Ephrins restrict cell intermingling and communication. *Nature* **400**, 77–81.)

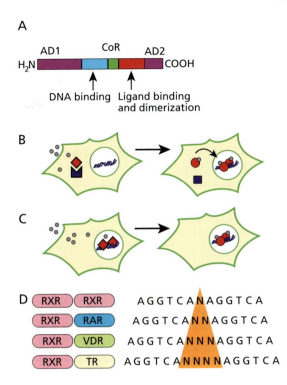

A

B

C

D

Fig. 4.25. **Nuclear receptors. (A)** Linear representation of the sequence domains in a typical member of the nuclear receptor family. The amino terminus is characterized by an activator domain (AD1), adjacent to the C₄ zinc-finger DNA-binding domain. Next to this, a short region of about 45 amino acids appears to be involved in the recruitment of corepressor proteins (CoR). At the carboxy terminus is the ligand binding and dimerization domain and a second activation domain (AD2). **(B)** The steroid receptor class of nuclear receptors, such the glucocorticoid receptor and receptors for steroid hormones, are both nuclear and cytoplasmic proteins. In the absence of ligand, the glucocorticoid receptor (red diamond) is kept in the cytoplasm by association with the Hsp90 protein (dark blue). Binding of the ligand (small grey circle) induces a conformational change in the receptor that dissociates it from Hsp90 and induces its translocation to the nucleus where it activates transcription. **(C)** The retinoic acid receptors and receptors for vitamin A, D, and thyroid hormone reside in the nucleus, where they act as dimers (red diamonds). In the absence of ligand, some of these receptors recruit corepressors (not shown), forming complexes that repress gene expression. Ligand binding induces a conformational change that releases the repressor complex and induces the binding of activators. **(D)** Nuclear receptors recognize a well-defined palindromic sequence. In the case of the retinoic acid receptors the spacing between the two halves of this sequence is different in different enhancers and so determines if the activating molecule is a homodimer or a heterodimer, as only the optimal combination will recognize the sequence. As shown, only the retinoid X receptor (RXR) forms homodimers and can bind to a repeat with one intervening nucleotide. In all the other cases, the sequence is recognized by a heterodimer between RXR and another receptor. Different combinations respond to different retinoids: RXR with the retinoic acid receptor (RAR) responds to all *trans* retinoic acid, RXR and the vitamin D receptor (VDR) responds to vitamin D, and RXR with thyroid hormone receptor (TR) responds to thyroid hormone.

many of them play central roles during development. The retinoid receptors are a complex class of molecules and fall into two groups, the retinoic acid receptors (RARs) and the retinoid X receptors (RXRs), which differ in the form of retinoic acid that acts as a high-affinity ligand. Within these groups, alternative promoter and splice-site usage generates different subtypes (α, β, and γ). Retinoid receptors bind to DNA as dimers: they can form either homodimers, or heterodimers with other retinoid receptors or with other nuclear hormone receptors. Different dimeric combinations recognize subtly different binding sites in DNA (Fig. 4.25), allowing them to regulate specific subsets of genes and to act in concert with various combinations of other transcription factors to achieve a vast number of different regulatory effects. Retinoid receptor subtypes are expressed in specific patterns during development and in adult animals. The particular set of receptors expressed by a cell fine tunes its response to circulating or diffusing hormone levels.

In contrast to the steroid receptors, retinoid receptors bind DNA both in the presence and the absence of ligand. In the absence of ligand, they bind together with other proteins to repress transcription of target genes. Ligand binding induces a conformational change that releases

the receptor from the corepressor protein and enables it instead to bind a different cofactor and act as a transcriptional activator (Fig. 4.25).

Cloning and sequencing efforts have revealed the existence of a large number of other proteins that appear to be nuclear receptors. In many cases, however, the structures of the ligand-binding domains of these receptors do not allow them to be classified with any of the known families of nuclear receptors. These proteins are known as orphan receptors.

Regulation of ligand–receptor interactions

An important feature of signalling systems is the ability to switch the signalling process on and off in a regulated manner. When signalling is used during development to pattern groups of cells, the strength and extent of signalling has to be controlled over precise spatial ranges. Unregulated signalling can have disastrous consequences, such as uncontrolled cell proliferation and tumour formation, as we have already seen in the example of receptor-derived oncogenes.

There are several strategies for achieving regulation of signalling. In principle, regulation can take place either within the signal transduction network or at the level of the interactions between signalling molecules and receptors. Here, we shall discuss the regulation of ligand–receptor interactions; regulation of signal transduction networks will be discussed in Chapter 5.

The activity of the signalling molecule can theoretically be controlled in a variety of ways and for most of these there seem to be molecular mechanisms that can implement them. Intrinsic molecular properties of the ligand, such as its half life or its binding specificity for the receptor, are commonly used. A signalling molecule with a short half life, for example, will be rapidly removed from the signalling environment so that signalling is limited to a brief interval of time. In other instances, the affinity or availability of the ligand is used to determine the timing and intensity of the signal. Another mechanism for controlling the activity of a signalling molecule is by sequestration in an internal cellular compartment until it is required for use; insulin, as we have mentioned earlier in this chapter, is an example of a signalling molecule whose availability is regulated in this way.

A further way of regulating signalling activity—and one that has been found to be particularly important in a number of developmental contexts—is by proteolytic activation or inactivation of the signalling molecule. TGFβ and hh, for example, are synthesized as inactive precursors whose signalling function is activated by proteolysis. The asymmetric activation of the receptor Toll during early embryogenesis in *Drosophila* provides an example of a situation in which proteolytic activation of a signalling molecule is precisely spatially regulated, enabling it to play a key role in a developmental patterning process (Fig. 4.26). Toll is uniformly distributed throughout the membrane of the oocyte and embryo, but its signalling function is specifically required on the ventral side of the embryo, in order to initiate the formation of ventral structures. The ligand for Toll is the product of the *spätzle* gene, which is also present all around the embryo. However, the active form of spätzle is concentrated in the ventral region where it activates Toll. The activation of spätzle is a consequence of a proteolytic cascade involving three serine/threonine

proteases encoded by the *gastrulation defective*, *snake* and *easter* genes which reside in the perivitelline space between the embryo and the surrounding vitelline membrane. The snake and gastrulation defective proteases are activated only on the ventral side of the embryo, by a complex of the nudel, windbeutel, and pipe proteins that is deposited before fertilization by the ventral follicle cells (somatic cells that surround the oocyte) and triggered after fertilization. Snake and gastrulation defective in turn activate the easter protease, which cleaves the inactive spätzle precursor, releasing its active form which binds to the Toll receptor (Fig. 4.26).

The ligand–receptor interaction can also be regulated by other proteins. Signalling by members of the TGFβ/BMP family, for example, is inhibited by the binding of proteins with the colourful names of chordin, short gastrulation, noggin, and follistatin, which prevent interaction of the signalling protein with the appropriate receptor (Fig. 4.27). In some cases, these interactions are counteracted by the proteins of the tolloid family of metalloproteases, which bind the inhibitor and release the signalling protein (Fig. 4.27). These interactions have been demonstrated both *in vitro* and *in vivo*, and there appears to be some specificity of particular inhibitors for specific subgroups of TGFβ/BMP family signalling proteins: chordin and short gastrulation have no effect on activins or TGFβ, but prefer BMPs, while follistatin binds both BMPs and activin. *In vivo*, ectopic expression of chordin in *Xenopus* embryos disrupts the normal development of the dorsal–ventral axis of the animal, causing the embryos to develop only ventral structures; this effect can be counteracted by excess BMP4. In all of these types of control systems, the result of the interaction between the inhibitor and its target is to raise the threshold for the interaction between ligand and receptor.

The formation of multiprotein complexes also offers opportunities for regulation of the signalling event at the cell surface, and can be used to help ensure that signalling occurs only when and where it is required. We have already seen examples of this in the ligand-driven formation of receptor complexes for TGFβ/BMP proteins, as well as in the case of the semaphorin receptors, where the formation of the complex precedes the presence of the ligand. In both cases the specificity of the signal is not simply determined by the ligand–receptor pair, but by the cell-type-specific receptor subunit constitution.

The activity of the receptor can also be regulated by proteolytic cleavage. The activity of the seven transmembrane domain thrombin receptor provides a clear example. This

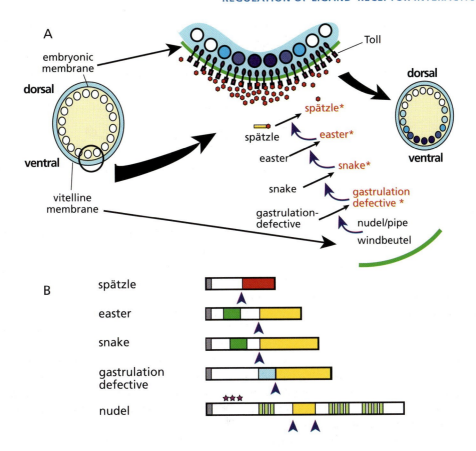

Fig. 4.26. **Mechanisms that regulate the signalling ability of ligands: Activation of the receptor Toll by its ligand spätzle in *Drosophila*. (A)** Activation of Toll is a crucial event in the specification of dorsal and ventral in *Drosophila*. In the early embryo a gradient of activation of Toll is set up from one side of the embryo towards the other. In the region where the concentration of active Toll is high, the dorsal transcription factor is translocated into the nucleus (blue) and those cells become specified as ventral. Toll is activated by its ligand, a proteolytically activated form of spätzle (small red hexagons show the proteolytically active form). Both Toll and spätzle can be found all round the embryo: Toll in the embryonic membrane and spätzle in the space between the embryo and an extraembryonic envelope, the vitelline membrane. However, a spatially localized proteolytic event leads to the activation of spätzle over one side of the embryo which becomes its ventral side. The protease that achieves this is encoded by *easter* which is the final element of a protease cascade involving various components (coloured names followed by asterisks represent the proteolytically active forms of these proteins). **(B)** Representation of the structures of the different members of the protease cascade that leads to the activation of spätzle as shown in A. Arrows indicate processing sites for each of the proteins and yellow indicates active serine protease domains. The active domain of spätzle is at its carboxy terminus (red). Easter, snake and gastrulation defective contain a carboxy-terminal protease domain (yellow). The activity of this domain is regulated by domains present in the amino-terminal half of the protein: a cysteine-rich domain (green) in easter and snake and, in gastrulation defective, a domain (blue) that is also found in complement proteins. Nudel is a very large protein with several glycosaminoglycan (GAG) addition sites (purple stars) and 11 copies of a motif (pale green bars) found in the low density lipoprotein receptor ligand-binding site. Pipe (not shown), a sulphotransferase, is thought to be involved in the modification of a GAG that triggers the activation of the protease cascade on a 'solid' phase attached to either the membrane of the embryo or the vitelline membrane. (After LeMosy, E. *et al.* (1999) Signal transduction by a protease cascade. *Trends Cell Biol.* **9**, 102–6.)

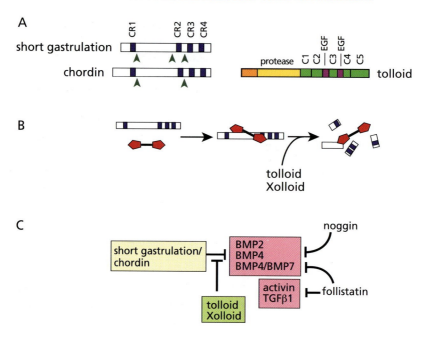

receptor has a large extracellular domain; the receptor is activated by an intramolecular reaction mediated by a sequence within the extracellular domain that is exposed by an intramolecular cleavage triggered by ligand binding (Fig. 4.28).

The ligand–receptor interaction may also be regulated by mechanisms that affect the ability of the receptor to recognize its ligand. For example, glycosylation of specific residues in the extracellular domain of the Notch receptor differentially affects its interactions with its ligands, reducing its ability to signal with Serrate and enhancing its ability to recognize and signal with Delta. 'Mock ligands', which bind to the receptor but do not activate it, can also regulate signalling by competing with the normal ligand. This kind of competitive interaction is common in the immune system, for example 'mock' IL1 ligands compete with IL1 for binding to the IL1 receptor. In *Drosophila*, the product of the *argos* gene competes with spitz, a TGFα homologue, for binding to the EGF receptor (see Chapter 9, Fig. 9.15). In an analogous type of interaction, 'false receptors' may bind ligand but do not transduce a signal and thus reduce the effective concentration of the signal. The frzB proteins are examples of such receptors. These proteins consist of the soluble part of members of the frizzled family of receptors; they appear able to bind Wnts and in-

Fig. 4.27. Regulation of the activity of members of the TGFβ/BMP family of signalling molecules. (A) Basic structure of proteins involved in the activity of TGFβ/BMP. *Drosophila* short gastrulation (sog) and vertebrate chordin contain four highly conserved cysteine-rich repeats (CR1–4) with a degree of homology to procollagens. The green arrows indicate sites of cleavage by members of the tolloid family of proteases. Tolloid-related proteins (right) have a signal sequence followed by a protease domain. This domain is related to that of the astacin family of proteases. The rest of the protein contains two kinds of motifs probably involved in protein–protein interactions: two EGF-like repeats (purple) and five CUB domains (C1–C5; green). **(B)** Members of the TGFβ/BMP family of signalling molecules are dimers (red) whose interaction with receptors is blocked by the binding of large soluble proteins, *Drosophila* short gastrulation and vertebrate chordin. This interaction is inactivated by the cleavage of these TGFβ/BMP binding proteins by members of the tolloid/Xolloid family of proteases. **(C)** Diagram of the interactions between members of the TGFβ/BMP family and regulators of their activity. Short gastrulation/chordin interact with BMP ligands as described in B. Noggin and follistatin are involved in inhibiting the same ligands and bind them directly, but have no sequence similarities with short gastrulation/chordin. Xolloid and tolloid interfere with the interactions between short gastrulation/chordin and BMPs.

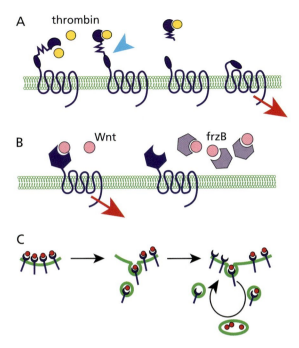

Fig. 4.28. Modes of control of receptor activity. (A) Proteolytic cleavage can regulate the activity of receptors. Binding of thrombin (yellow) to its receptor, a member of the seven transmembrane domain family, leads to the proteolytic cleavage (blue arrow) of the amino terminus of the extracellular domain of the receptor, exposing a segment of the receptor that acts as a membrane-tethered ligand and triggers the signalling function of the receptor (red arrow). **(B)** Mock receptors can act as decoys that lower the concentration of active ligands. For example, frzBs (purple) are proteins related to the frizzled family of receptors for Wnt signalling molecules. FrzBs show strong homology to the extracellular domain of frizzleds (blue) and act as antagonists of the effects of Wnt signalling by binding and neutralizing Wnt. In this way they can abolish or reduce the effect of ectopic expression of Wnt (pink). **(C)** Regulation of the concentration of receptors at the cell surface also provides an effective way of modulating signalling. Many ligand–receptor pairs are internalized by the cell after signal transduction. In some cases, this results in a down-regulation of the signalling event by reducing the effective receptor concentration at the cell surface. Once inside the cell, ligand is released into the endosome where it is degraded, while the receptor may either recycle to the cell surface or be degraded in the endosomes.

activate their signalling capacity *in vitro*, but their precise regulatory role *in vivo* remains to be clarified (Fig. 4.28).

Finally, receptor activity may be regulated through control of the number of receptor molecules at the cell surface.

In some cases (e.g. the EGF receptor), binding of ligand triggers endocytosis of the receptor with consequent down-regulation of signalling (Fig. 4.28). The free receptor may also be down-regulated, or may be recycled to the cell surface. The balance between receptor down-regulation and recycling enables the cell to adjust the sensitivity of its response. Slow recycling will reduce the number of receptor molecules at the cell surface, affecting the response to a signal. More rapid adaptation of the target cell is also sometimes achieved not by receptor endocytosis but by interaction of the receptor with regulatory molecules that affect the affinity of binding to ligand.

The vertebrate organizer as a source of signal modulators

Mechanisms that modulate the ligand–receptor interaction play a vitally important role in the strategies used to generate cell diversity and pattern in developing embryos. The molecular activity of the vertebrate organizer provides a striking example. In vertebrate embryos, this piece of mesoderm leads the process of gastrulation. Transplantation experiments show that the organizer has the ability to induce adjacent cells to generate a new axis of the embryo with new pattern elements. It achieves this by acting as a source of signal modulators that antagonize, and thus modulate, the activities of members of the BMP and Wnt families of signalling molecules, which are distributed from local sources over the developing embryo (Fig. 4.29).

These antagonists (noggin, follistatin, and chordin for TGFβ/BMPs and frzBs for Wnt) are diffusible and therefore generate gradients that induce graded distributions of BMPs and Wnts. Different concentrations of these signalling molecules lead to the activation of different genes and so to the patterning of different tissues (Fig. 4.29 and Chapter 9, Fig. 9.29, for further details).

The extracellular environment and cell signalling

The interactions between ligands and cell-surface receptors do not take place in empty space, but in the context of a complex meshwork of molecules that make up the extracellular matrix (ECM). In Chapter 6 we shall discuss in detail the composition of the ECM and its role in modulating cell–cell interactions and cell behaviour. Here, we briefly introduce just one particular ECM component that has

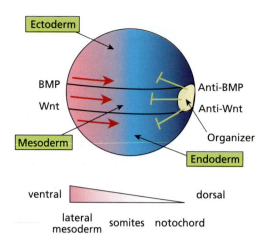

ventral ◄▬▬▬► dorsal

lateral mesoderm somites notochord

Fig. 4.29. **Use of signal antagonists during the patterning of the mesoderm in *Xenopus*.** Early in development, the *Xenopus* embryo is divided into three territories each of which will develop into a particular cell type: ectoderm, mesoderm, and endoderm. The specification and patterning of these cell types relies on the integration of signals from the major signalling pathways, in particular, Wnt and TGFβ/BMP, with cell-autonomous programs of gene expression. The activity of a group of cells, the organizer, located at the leading edge of the mesoderm during gastrulation, plays a central role in many of these pattern formation events. The organizer is a source of antagonists of Wnt and BMP signalling. In the mesoderm, the activity of these antagonists generates gradients of Wnt and BMP signalling (pink) that result in the different mesodermal derivatives. Low-level signalling induces cells to develop as notochord, medium-level signalling induces development as somites, and cells exposed to high levels of Wnt and BMP, furthest away from the organizer and thus from the antagonists, develop as lateral mesoderm.

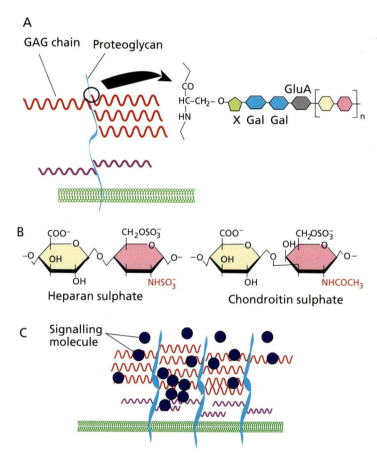

Fig. 4.30. **The extracellular matrix. (A)** The space between cells is filled by a complex matrix made up mostly of proteins and proteoglycans—proteins covalently associated with large chains of sulphated disaccharides called glycosaminoglycans (GAGs). X, xylose; Gal, galactose; GluA, glucuronic acid. The sulphated disaccharide is indicated by the yellow and pink hexagons. A proteoglycan can have different GAG chains (red and purple) attached to different parts of the protein core. **(B)** There are several classes of GAGs, each characterized by a different fundamental repeating unit. Two very common classes are derived from chondroitin sulphate and heparan sulphate. **(C)** Proteoglycans are large complexes of proteins and GAGs, often with molecular weights of the order of 10^6. Some proteoglycans are extracellular and others are associated with the cell surface. In the latter case, the protein moiety can be a transmembrane protein, which serves as an anchor for the GAGs and thus creates a high-concentration matrix for proteins that associate with GAGs. Many growth factors (circles), for example, FGFs, Wnts, and TGFβ/BMPs, use proteoglycans as signalling partners. Sometimes this creates a high local concentration of ligand; in other cases it allows the signalling molecules to adopt the correct configuration for recognition by the receptor.

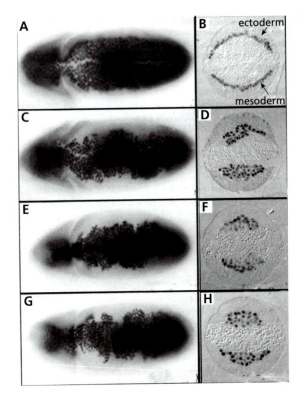

Fig. 4.31. Effects of mutations in enzymes synthesizing proteoglycans on the activity of the FGF receptor in *Drosophila*. The genes *sugarless* (*sgl*) and *sulfateless* (*sfl*) encode enzymes involved in the biosynthesis of glycosaminoglycans. In mutants for *sgl* and *sfl* the mesoderm fails to spread properly over the ectoderm after gastrulation, a phenotype that is similar to that of mutations in the FGF receptor *heartless*. **(A)** Ventral view of wild-type *Drosophila* embryo shortly after gastrulation, stained for a mesodermal marker. Note the even spread of the mesodermal cells on either side of the midline. **(B)** Section through a similar embryo; the mesoderm and the ectoderm are indicated. **(C–H)** Ventral views and sections through different mutants in which the mesoderm has failed to migrate properly. **(C, D)** FGF receptor mutant *heartless* (*htl*); **(E, F)** *sfl* mutant; **(G, H)** *sgl* mutant. (From Lin, X. *et al.* (1999) Heparan sulfate proteoglycans are essential for FGF receptor signalling during *Drosophila* development. *Development* **126**, 3715–23.)

been found to have specific functions in signalling: the proteoglycans.

Proteoglycans (Fig. 4.30) are large basic proteins decorated with long chains of highly hydrated glycosaminoglycan substituents. Some proteoglycans are integral membrane proteins, while others are attached to the membrane by a GPI anchor, although this link is often broken to yield a free extracellular protein. Signalling by a number of growth factors has been shown to be regulated by specific proteoglycans, and it has been suggested that most if not all growth factors may be sequestered at the cell surface by these ECM molecules. For example, binding of FGF to a heparan sulphated proteoglycan called syndecan is required for signalling by FGF, and the proteoglycan β-glycan can bind TGFβ and present it in high concentration to low-affinity receptors or increase its local concentration in regions where there is little of it. In *Drosophila*, the *dally* gene, which encodes a member of the glypican family of GPI-linked proteoglycans, is required for signalling by the BMP family member dpp, and by the Wnt family member wingless.

In addition to the roles of specific proteoglycans, general components of the ECM have been shown to play a role in signalling. Biochemical studies show that heparan sulphate proteoglycans play a role in Wnt and FGF signalling, and genetic analysis of enzymes involved in the synthesis of glycosaminoglycans reveals a stringent requirement for these signalling events. For example, in *Drosophila*, loss of function of the enzymes involved in sulphation of the sugars, for example, mimics loss of function of wingless. If wingless is overexpressed in these mutant backgrounds, the mutant phenotype is rescued, suggesting that the function of the glycosaminoglycans is to increase the effective local concentration of the signalling molecule. Loss of glycosaminoglycans also has been a dramatic effect on FGF receptor signalling. In *Drosophila*, this has several consequences, a dramatic one being a failure of mesodermal cells to migrate properly (Fig. 4.31).

SUMMARY

1. The coordination of cellular activities is an essential element during development because it allows a coordinated and regulated unfolding of the information contained in each cell's DNA.

2. This coordination is achieved through a molecular system of signals and receptors which act as information processors.

3. Signalling molecules are either secreted or displayed by cells. They convey information about the state of the cells that secrete them as well as instructions for the behaviour of other cells.

4. Receptors have a basic structure that allows them to recognize signals and relay the information contained in them to the cell. This is usually achieved through conformational changes, which are sometimes associated with an enzymatic activity.

5. The interactions between signals and receptors can be modulated by other molecules that interfere in a positive or negative way with the informational exchange.

6. The programs of gene expression determine dynamic distributions of signals and receptors during development that contribute to the generation of pattern.

COMPLEMENTARY READING

The identification and characterization of signalling molecules and receptors is in constant flux and it would not be helpful to single out individual references that might soon be out of date. The reader is referred instead to some journals in which there are periodic updates on these molecules and their functions. See, for example, *Current Opinion in Genetics and Development*, *Current Opinion in Cell Biology*, *Nature* reviews or *Trends in Genetics*.

Chapter 5

Intracellular effectors of cell interactions: Signalling pathways and networks

In Chapter 4 we explored the nature of, and relationships between, ligands and their receptors. For these interactions to have biological meaning, the information they convey must reach physiological targets and effect some change in them that results in a cellular response. The targets of signals can be very diverse: enzymes are common targets (e.g. metabolic enzymes, or polymerases that work on DNA), but other possible targets include membrane channels whose activity alters the electrical properties of the cell, or cytoskeletal proteins that modulate cellular behaviour such as movement or adhesion. Some receptors, after interacting with a ligand, can in turn interact directly with a target; this is the case for the steroid receptors. More frequently, however, the initial stereochemical interaction between ligand and receptor is transmitted to the target by a relay system of molecular interactions known as a signal transduction pathway.

The backbone of the simplest signal transduction pathway consists of a ligand and its receptor, a transducer, and an effector that acts on the targets of the pathway (Fig. 5.1). There are many possible variations on this theme; in one common variation, the transducer section of the pathway involves several components rather than a single one.

Signal transduction pathways have a number of important features (Fig. 5.1). They are *directional*, that is, information flows in one direction, from source to target (though sometimes individual elements in a pathway may be capable of transferring information in opposite directions, as we shall see later in this chapter). The transfer of information via a relay system involving several steps allows the potential for *amplification* of the signal, particularly if the different steps operate with different effi-

ciencies. This is an important property because it enables a small number of signalling molecules to activate a larger number of effectors, thus maximizing the efficiency of small inputs that can be easily regulated. For example, activation of one RTK can lead to the phosphorylation of many molecules of a downstream target. In signalling pathways that have multiple steps, some elements can act as nodal points that connect with other pathways, allowing *diversification* of the response to a single signal or, conversely, *integration* of information from more than one signal to produce a single response. Finally, an additional, and vital, property of signal transduction systems is that they have inbuilt *control* mechanisms that allow the signal flow to be turned on and off according to need, thus preventing futile cycles and coordinating the activity of intersecting transduction pathways. A common way to effect this control is through feedback loops to crucial steps in the pathway.

Signal transduction systems rarely represent exclusive routes for the flow of biological molecular information. Rather than isolated pathways that converge on particular targets, they are networks. Just as in a road system, there may be a hierarchy of routes, with subsidiary roads feeding into major highways, and sometimes short cuts that can bypass rate-limiting intersections.

The understanding of how signal transduction networks function during development is still in its infancy, but a picture is beginning to emerge of how connectivity is built up in some types of cells. In this chapter we discuss the approaches that can be used to study signalling pathways, the signalling 'currencies' that are used by cells, the types of molecules and complexes that exchange these

A

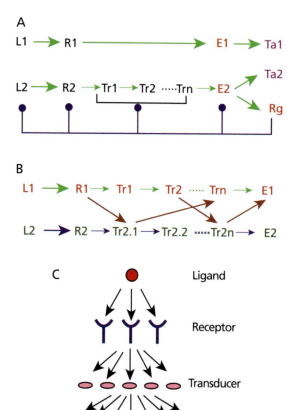

Fig. 5.1. Structure and properties of signalling pathways. (A) General structure of signal transduction pathways. Ligands (L1 and L2) interact with specific receptors (R1 and R2). In some cases (*top*) this interaction is directly linked to the effectors (E1) which act on the targets (Ta1) whereas in others (*bottom*) there is a variable number of intermediaries or transducers (Tr1–Trn) leading to an effector (E2) and targets (Ta2). Pathways are directional: the information flows from the receptor to the effector. However, effectors often modulate the activity of regulators (Rg) which can act on the pathway to modulate the flow of information at any level. The bullets indicate that the effects of the regulators can be positive or negative. **(B)** Pathways are sometimes functionally connected in such a way that certain points act as nodes or connectivity points which integrate information from different pathways and thus create networks. **(C)** An important property of signal transducers and relays is that they allow the amplification of signals. A single ligand molecule can interact with more than one receptor molecule, but more often a single receptor leads to the activation of a number of transducer molecules which thus amplify the initial input.

fined processes and the characterization of the wild-type function of the genes identified by these mutations have been an essential ingredient in this analysis.

When developmental events are analysed genetically it is often observed that mutations in a collection of different genes, or loss of activity of a particular set of proteins, generate the same or related phenotypes, suggesting that the functions of the genes or proteins are related (Fig. 5.2). This is usually taken as a hint that the products of the genes are involved in a common process or event and the task is to sort out the functional relationships among the gene products. Sometimes sequence information can provide clues. For example, if the sequences of gene X and gene Y suggest that X is a transcription factor and Y a cell-surface receptor, and if targets of gene X are not expressed in mutants for gene Y even though gene X itself is expressed, one could conclude, with reasonable confidence, that X functions downstream from Y in a signalling pathway. However, things are not always so clear-cut, for example, in the case of a set of enzymes, all of which yield the same phenotype when their genes are mutated. At first sight this observation does not indicate whether the individual elements form part of a linear pathway or function in parallel pathways.

The analysis of epistasis is the most common approach to clarifying such situations. At its simplest level, epistasis is a genetic technique that allows the inference of functional relationships between gene products from the analysis of double, or sometimes multiple, combinations of mutations. One mutation, and by extension a gene or its

currencies and convert them into specific signals, and the pathways along which signalling information flows. Many of the basic signalling pathways were first elucidated not from the analysis of developmental systems but in studies on the responses of cells in culture to a variety of stimuli such as growth factors. The realization that these pathways were central to developmental systems, and their study in these systems, often by genetic analysis, revealed complex interconnections and a multitude of ways in which their activity can be regulated. In the last part of the chapter we shall explore some aspects of the regulation and integration of signalling pathways.

Epistasis and the analysis of signalling pathways

The genetic analysis of development has provided important insights into the molecules that mediate these processes and the molecular interactions they are involved in. In particular, the isolation of mutations affecting de-

A Metabolic pathways

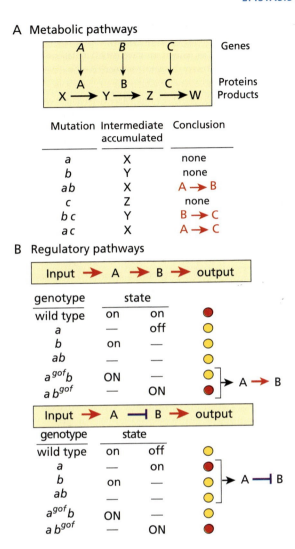

Fig. 5.2. Analysis of epistasis to establish regulatory relationships between gene products. (A) In a metabolic pathway, genes *A*, *B*, and *C* encode enzymes A, B, and C which catalyse the progressive transformation of X into Y into Z into W. Every enzyme catalyses one step and produces one intermediate. Loss of function of A (*a* mutant) will lead to the accumulation of X, whereas loss of function of B (*b* mutant) will lead to the accumulation of Y. However, a double mutant *ab* will look the same as the single mutant *a*, because it will lead to the accumulation of X. This observation indicates that gene *A* is epistatic over *B* and therefore that its product, A, operates upstream of B. The same analysis can be applied to B and C, allowing the ordering of A, B, and C in a linear chain. **(B)** The situation is different in a regulatory pathway, in which an input produces an output through the regulation of the activity of two proteins A and B, encoded by genes *A* and *B*. The interactions between A and B can be of two kinds: positive, that is, A activates B (*top*) which means that the input produces the output (red bullet); or negative, that is, A supresses B (*bottom*) and thereby suppresses the output (indicated with a yellow bullet). Epistasis can be used to unravel the functional order of A and B in these situations. In the case where A activates B, mutations in *A*, *B*, or both *A* and *B*, produce the same output so it is not possible to establish a functional relationship or order of action between A and B. One way to bypass this problem is by a combination of gain of function (gof) mutations in *A* or *B* with loss of function mutations. This shows that A needs B but B does not need A and that therefore A is upstream and promotes the activity of B. In the case where A suppresses the activity of B, the phenotypes of single and double mutants are different and therefore significant. Thus, a mutation in *A* allows B to operate and produce the outcome, whereas mutations in *B* do not alter the outcome; the double mutant *ab* has a similar phenotype to the *b* mutant. Together these three observations suggest that B is negatively regulated by A. This is confirmed by the use of gof mutations, as shown.

product, is said to be epistatic over another if its effects are dominant in combination with that mutation. Thus, for mutations *a* and *b* in genes *A* and *B*, if the double mutant *ab* has the 'a' phenotype, *A* is said to be epistatic over *B*.

Figure 5.2 illustrates how the analysis of epistasis can be used to work out the functional relationships among components of both biochemical (e.g. metabolic) and regulatory pathways. Note that the logic of the analysis (or the interpretation of the results) is different in these two types of situations. In a metabolic pathway, if a mutation in gene *A* is epistatic over a mutation in gene *B*, it means that the product of the step carried out by A is needed for the step carried out by B, and therefore that A precedes, or is upstream of, B in the pathway. The interpretation is aided by the fact that genetics and biochemistry work hand in hand

in the elucidation of metabolic pathways, in which a chain of enzymes operates to generate a final product. For example, mutants for individual genes may accumulate identifiable biochemical intermediates, so tying specific genes to specific steps in the pathway.

In a regulatory pathway, such as a potential signalling pathway, if *A* is epistatic over *B*, it means that A is necessary for the function of B, and therefore is downstream of B. This is the reverse of the situation in a metabolic pathway. It is a very common situation in the analysis of developmental systems but its interpretation is not straightforward. There are two reasons for this. First, the available data are genetic rather than biochemical; that is, usually there are no molecular intermediaries to be assayed and mutant phenotypes are all that is available. The mutant phenotype reflects the outcome of the last regulatory

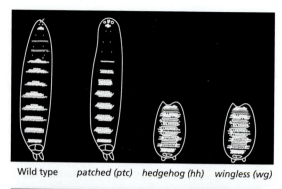

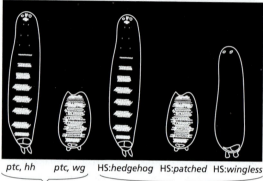

Fig. 5.3. Analysis of epistasis: A practical case. The top panel shows cartoons of ventral cuticles of first instar *Drosophila* larvae that are wild type or mutant for three genes, mutations in which alter the pattern of every segment. Mutations in *hh* and *wg* produce similar phenotypes, whereas mutations in *ptc* have a very different appearance. Analysis of double mutant combinations reveals possible functional interactions between these genes. Thus, the double mutant *ptc,hh*, rather than being intermediate between the single mutant phenotypes or entirely different from both, resembles one of them, in this case *ptc*, indicating (see Fig. 5.2) that *ptc* is downstream of *hh*. In the case of the double mutant *wg,ptc* the phenotype resembles that of *wg*, suggesting that *wg* is downstream of *ptc*. Functional relationships among the three genes can be inferred from these results, as indicated. Use of gain of function mutations provides further information. Gain of function is achieved by placing these genes under the control of the heat shock promoter (HS), which directs high and indiscriminate levels of gene expression when the temperature is raised. As shown, HS:*hh* produces a phenotype similar to that of *ptc* mutants. Since *hh* is upstream of *ptc*, the observation that overexpression of *hh* generates a loss of function of *ptc* suggests that a function of *hh* is to negatively regulate *ptc*. On the other hand, HS:*ptc* produces a phenotype that looks like *wg* or *hh*. Because *wg* is downstream of *ptc*, this suggests that *ptc* regulates *wg* negatively. Finally, HS:*wg* produces a novel phenotype, indicating that it acts on other genetic elements that are not, in principle, related to *hh* and *ptc*.

interaction, which will always be dominant over others that precede it (Fig. 5.2). Second, there are two possibilities (positive and negative) for any interaction between two elements; this will also affect the outcome of any experiment (Fig. 5.2).

Sometimes the situation is clear-cut. For example, mutations in the genes *patched* (*ptc*) and *hedgehog* (*hh*) affect the patterns of the cuticle in *Drosophila*, and the single mutants are distinguishable from each other and from wild type (Fig. 5.3). The double mutant *ptc,hh* is always like *ptc*, and so *ptc* is epistatic over, and therefore functionally downstream of, *hh* (Fig. 5.3). In other cases, the single and double mutant phenotypes look the same or very similar (e.g. for *wingless* or *hedgehog*) and further information is needed to order the genes in a regulatory hierarchy. The use of gain-of-function mutations, whether naturally occurring or engineered, becomes a useful tool for this task. To order two genes, one can use a gain of function mutation in one gene combined with a loss of function mutation in the other (Fig. 5.2 and 5.3). If a loss of function mutation suppresses a gain of function mutation, the gene

corresponding to the loss of function mutation is said to be epistatic over the other, and therefore 'downstream' from it in a regulatory sense.

In summary, analysis of epistasis is a powerful tool to find out the genetic relationships between genes, but it should not be overinterpreted in molecular terms. Epistasis can be used to build up a set of genetic relationships, but this does not necessarily mean that they also reflect functional or molecular relationships. Too often, a series of cloned genes with certain properties inferred from sequence data is arranged in a functional order on the basis of epistatic relationships and said to form a signal transduction pathway. However, without a proper molecular analysis of the relationships this conclusion is not warranted. All the genetic analysis tells us is that a phenotype is upstream or downstream of another, without saying much about the biochemical relationship between the products of the corresponding genes. A gene involved in the early specification of the top half of the body plan is clearly upstream from those involved in forelimb development—but this does not mean that it regulates them directly. The same applies to the analysis of phenotypes at a much finer level of detail. Branching pathways, parallel pathways, and regulatory loops in which two elements will alternate being upstream or downstream of each other, can also confound epistatic analysis, which can only be translated to molecular pathways if the pathways are linear.

Signalling currencies

The transactions involved in signalling interactions within cells are carried out using a fairly limited repertoire of molecular 'currencies'. These currencies fall into three main groups (Fig. 5.4)

The first type of currency (though not strictly a molecular one) is potential differences between cellular compartments, which are generated by movements of ions across membranes. These ion flows are regulated as a result of ligand–receptor interactions or the propagation of chemical signals. Ion flows occur in most cell types and have been shown to play a specific functional role in the propagation of action potentials during neural activity. So far, their role in other processes remains to be explored.

Intracellular second messengers constitute a second type of currency. These are small molecules whose function is to relay signals within the cell. Second messengers include the divalent cation Ca^{2+}, cyclic AMP, cyclic GMP, inositol phospholipids, diacylglycerol, and certain free radicals. In signal transduction pathways that involve second messengers, the binding of ligand to receptor at the cell surface typically activates a biochemical pathway that stimulates production of the second messenger, which then transmits the signal by binding to a specific protein target, thereby inducing structural changes that alter its activity.

Molecular substituents, such as phosphate groups that modify proteins covalently, are a third type of signalling currency. For example, as described in Chapter 4, many cell-surface receptors have kinase or phosphatase activity, adding or removing phosphate groups from tyrosine, serine or threonine residues of their protein targets. The phosphorylation or dephosphorylation of the receptor's target molecule, and often also of the receptor itself, may trigger a conformational change that enables binding of the next molecular component in the signal transduction pathway. As well as cell-surface proteins, a vast number of intracellular signalling proteins also have kinase or phosphatase activity: it has been estimated that in human cells there may be as many as 4000 different enzymes of this type.

Adaptor motifs

The interactions between signalling elements inside the cell are not left to depend on random associations between the components of the different signalling systems, but are highly regulated through specific molecular interfaces on these components. As we have seen in previous chapters, there are some molecular motifs that appear over and over again in proteins with specific functions, such as DNA-binding transcription factors, ligands, receptors, and extracellular proteins. Characteristic motifs are also found in the intracellular molecules involved in the transduction and networking of signals. Certain structural domains in these proteins act as 'adaptor motifs' that mediate defined protein–protein connections and thus create specific functional interactions (Fig. 5.5).

Protein–protein interaction domains are found on many proteins in all cellular compartments and are not restricted to proteins involved in signalling. For example, a motif known as cdc10/ankryin repeats appears in proteins with many different functions including, as the name implies, the ankyrin protein (a cytoskeletal protein of the erythrocyte membrane), the cytoplasmic domain of the receptor Notch, and the NFκB transcription factor. Another motif, the WD40 motif, is found in proteins involved in processes as diverse as the activation of programmed cell death, protein degradation, transcriptional regulation, and mitotic spindle function. Thus, rather than being associated with a particular function, these motifs provide interfaces that will allow only restricted interactions, that is, motif A will interact with motif B, but not with C, D or E.

The number of motifs, though finite, is large and on the increase as genome sequences are completed and functional interactions between proteins are established. However, from the available information, it is already clear that certain adaptor motifs do reappear often in proteins that function in signalling systems, where their role appears to be to create networks of signalling components. As we have seen, phosphate groups on tyrosine, serine, and threonine residues are essential currencies in the exchange of molecular information. Phosphorylation of these groups in the context of a particular amino acid sequence creates interfaces that are recognized by specific structural motifs. Some of these motifs recognize phosphotyrosine and some recognize phosphoserine and phosphothreonine. In other situations, certain amino acid sequences in signalling pathway proteins, for example stretches of proline, are used as recognition interfaces for adaptor motifs. Other motifs

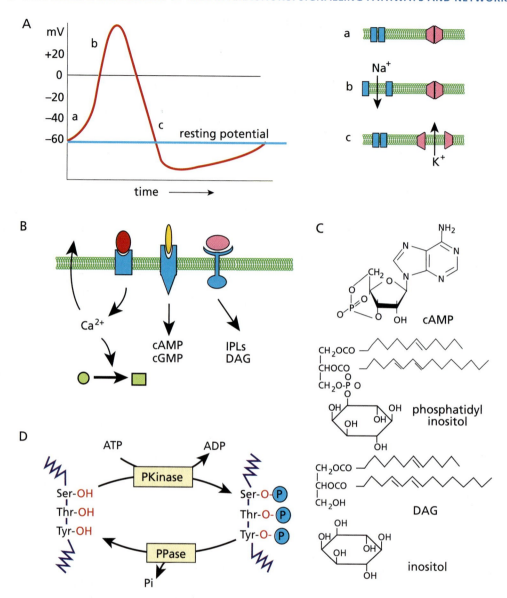

Fig. 5.4. Signalling currencies. (A) Potential differences between the outside and the inside of the cell can be used to generate information, as in the conduction of neural impulses. Changes in the potential are indicated on the left and the molecular events associated with these changes are indicated on the right. A neuron has a typical resting potential (a) which results from the balance of the fluxes of ions across the membrane; this potential is usually negative, approximately −60 mV. An electrical input opens Na^+ channels (b) and triggers an influx of Na^+ ions into the cell inducing a very rapid change in the membrane potential: the action potential. At a certain value, this change in potential induces the closing of the Na^+ channels and the opening of K^+ channels (c), allowing K^+ ions to rush out of the cell, restoring the resting potential. The propagation of the action potential along an axon results in cellular communication.

(B) Second messengers are released or generated after the interaction between specific types of signals and receptors. Some examples of second messengers are shown. IPLs, inositol phospholipids; DAG, diacylglycerol. In some cases the second messenger effects a functional change in cytoplasmic proteins; the figure shows Ca^{2+} having such an effect. **(C)** Structures of some second messenger molecules. **(D)** Phosphorylation and dephosphorylation of specific proteins results in conformational changes that modulate their activity. For this reason, the phosphorylation balance of a cell may determine many of the variables of signalling pathways. The three amino acids most commonly targeted by phosphorylation reactions are serine, threonine, and tyrosine. These reactions are mediated by specific proteins that we have seen in Chapter 4: protein kinases (PKinase) and protein phosphatases (PPase).

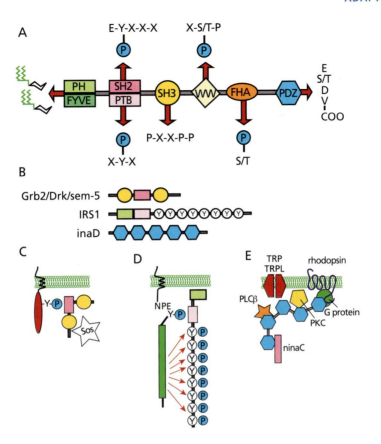

Fig. 5.5. Signal transducing adaptor motifs. (A) Some of the main adaptor motifs in intracellular signalling proteins (coloured blocks) and their substrates (indicated as amino acid motifs, using the one-letter code for representing amino acids). The phospho-inositide substrates for PH and FYVE domains are indicated by cartoons of their molecular structures. PH (pleckstrin homology domain); FYVE (Fab1p/YOTB/Vac1p/EEA1 domain); SH2 (Src homology type 2 domain); PTB (phosphotyrosine-binding domain); SH3 (Src homology type 3 domain); WW (tryptophan–tryptophan domain); FHA (forkhead-associated domain); PDZ (PSD-95/SAP90/DLG/ZO-1). **(B)** Examples of adaptor proteins. Grb2 is a protein made up of two SH3 domains (yellow circles) and one SH2 domain (pink box). The domain structure and function of the protein are conserved in *Caenorhabditis elegans* (sem-5) and *Drosophila* (Drk). The insulin receptor substrate 1 (IRS1) protein contains a PH domain at its amino terminus (green box), a PTB domain (pale pink box) through which it interacts with the insulin receptor, and a domain containing multiple tyrosine residues which are potential substrates

for phosphorylation. InaD is a *Drosophila* protein that contains five PDZ domains (blue polygons). **(C–E)** Functional interactions of the Grb2, IRS1 and inaD proteins. **(C)** The role of the Grb2 proteins is to recruit cytoplasmic proteins such as the guanine nucleotide exhange factor Sos (also see Fig. 5.8 and 5.15) to some RTKs, establishing a scaffold for signal transduction. **(D)** The IRS1 protein is associated with the membrane through its PH domain, and with one of the subunits of the insulin receptor (IR) through a PTB domain which recognizes a specific phosphotyrosine residue in the IR. Activation of the IR activates its tyrosine kinase domain (vertical green box) which phosphorylates tyrosine residues in the IRS1. There are up to 18 potential tyrosine substrates in IRS1 and each one becomes the docking site for a specific SH2-containing protein (not shown). **(E)** InaD is involved in signal transduction in the *Drosophila* eye. Each of the five PDZ domains interacts with a specific protein, as indicated; in this way, inaD assembles a specific signalling complex at the cell surface. In the absence of inaD most of these components are not properly localized within the cell.

mediate more specific interactions between proteins containing characteristic amino acid sequences, while yet others mediate interactions between proteins and the membrane. A selection of the most important of these motifs and how they are used *in vivo* is summarized in Fig. 5.5.

The Src homology type 2 (SH2) domains highlight many of the features of adaptor motifs. SH2 domains are domains of about 100 amino acids that are involved in the recognition and binding of phosphotyrosine (pY) residues. Each SH2 domain has a conserved pocket that recognizes pY and a more variable pocket that binds to a site 3–6 residues carboxy-terminal to the pY site. The nature of the amino acid residue at this site determines which type of SH2 domain will recognize and bind to the phosphorylated protein and so confers specificity on the interaction. Activated RTKs, for example, have several different phosphotyrosine residues in their intracellular domain, each with a different constellation of surrounding amino acids, which create a battery of docking sites for different SH2 domain-containing proteins. The specific set of SH2 proteins recruited to an activated receptor determines the nature of the downstream signalling events.

The amino acid context of the pY residue is also important for the binding specificity of a second type of pY-binding domain, the phosphotyrosine binding (PTB) domain. Like SH2 domains, the function of PTB domains appears to be to recruit other signalling proteins to an activated receptor, but the amino acid context of the pY residues they recognize differs from that recognized by the SH2 domains.

Serine and threonine residues are phosphorylated by some protein kinases, and there is a group of proteins that recognize these modifications in specific amino acid contexts (Fig. 5.5). Tryptophan–tryptophan (WW) domain-containing proteins have been shown to target phosphoserine residues within a specific amino acid context containing proline. Forkhead-associated domain (FHA) proteins can bind phosphothreonine and phosphoserine residues directly in much the same way as SH2 domains bind phosphotyrosine. FHA domains were first found associated with the forkhead transcription factors but then were also discovered in other proteins that are not transcription factors. Other phosphoserine/threonine binding motifs include some WD40 motifs. In some cases whole proteins, rather than just specific motifs, function as adaptor domains. The 14-3-3 proteins are examples of this type. Whereas phosphotyrosine binding proteins tend to be associated with linking molecules to the cell surface,

those that bind phosphoserine/threonine are found associated with a wider variety of functions including recognition of proteins targeted for ubiquitin-dependent degradation, and regulation of protein entry into the nucleus.

Src homology type 3 (SH3) domains are present in many proteins that participate in signalling pathways involving tyrosine phosphorylation, and also in some cytoskeletal proteins. These domains bind to proline-rich peptides (Fig. 5.5). Once again, the context of the these residues determines the specificity of the interaction and therefore the selectivity of the signalling output. Proteins can have several SH3 domains, enabling them to cluster other proteins. For example, the *Caenorhabditis elegans* sem-5 protein and its mammalian and *Drosophila* homologues Grb2 and Drk are made up entirely of SH2 and SH3 domains and therefore act as true adaptor/linkers of different active modules. The function of these proteins appears to be to couple activated RTKs to their downstream targets, as we shall discuss in more detail later in this chapter.

PDZ (acronym derived from the founding members of the family: PSD-95/SAP90/DLG/ZO-1) domains are structurally similar to PTB domains but rather than binding phosphotyrosine they recognize short peptides with the specific carboxy-terminal sequence E(S/T)DV, a motif that is characteristic of several transmembrane ion channel and receptor proteins. Many PDZ proteins have multiple PDZ domains, through which they can cluster interacting proteins (Fig. 5.5). A striking example, the *Drosophila* inaD protein, contains five linked PDZ domains. By virtue of the different binding specificity of these domains, inaD acts as an assembly platform, or scaffold, for a multicomponent signalling complex that regulates light-activated signalling events in the *Drosophila* eye.

Some proteins containing adaptor motifs become associated directly with the membrane. This association is mediated by motifs like the pleckstrin homology (PH) domain or the FYVE (for the 'founding members' of the motif: Fab1p, YOTB, Vac1p, and EEA1) domain, which appear to localize proteins to the membrane by binding to phosphoinositides (Fig. 5.5). This allows the translocated proteins to interact more efficiently with other effectors that are located in the membrane. Proteins with these domains also contain other motifs, such as SH2 domains, through which they can recruit additional proteins and thereby couple signalling complexes and activities to specific sites on the membrane.

Adaptor motifs serve several purposes in the assembly of signalling networks. They act as connectors between different proteins, they help localize signalling proteins to specific regions of the cell where they can interact with activators and targets, and they increase the efficiency of signalling by creating molecular microenvironments with the appropriate concentrations of elements. Proteins that contain adaptor motifs usually contain more than one type of motif, and more than one copy of each motif, each with its own binding specificity. This creates a broad repertoire of combinatorial possibilities from which specific networks can be created.

A tool for detecting molecular interactions

One way of identifying proteins that interact with the motifs shown in Fig. 5.5 is to use a method known as the 'two hybrid screen'. This method makes use of the modular nature of transcription factors, that is, the fact that their DNA binding and transactivation domains are separable and that their interactions depend on a critical concentration of the binding domain and the transactivation domain. The two hybrid screen is used more generally as a way of identifying and characterizing protein–protein interactions (Fig. 5.6).

The original basis for the two hybrid screen was the Gal4 transcriptional activator from *Saccharomyces cerevisiae*. If the DNA binding and transactivation domains of this protein are not linked in some way, it will not promote transcription of genes under the control of the UAS to which the Gal4 protein binds. However, if the link between the two domains is restored using a molecular 'bridge' created by two interacting proteins A and B, transcription of Gal4 targets will ensue. This is achieved by creating fusion proteins between the Gal4 DNA-binding domain and protein A, and the Gal4 transactivation domain and protein B—hence the name 'two hybrid'. The effectiveness of the transcriptional activity will depend on the strength of the interaction between A and B. This is measured by incorporating a reporter, such as the gene for β-galactosidase or another enzyme, downstream of the UAS that binds Gal4 (Fig. 5.6).

The principle of the two hybrid method can be used in screens to search for interacting proteins. If one wants to find proteins that interact with protein X, for example, one can fuse X to either the DNA-binding domain or the transactivation domain of Gal4, and then use a library of proteins fused to the other domain to search for partners of X that will restore the activity of Gal4. A reporter determines the strength and nature of the interaction (Fig. 5.6).

Modular protein kinases and phosphatases

In the previous chapter we described some receptors that do not have intrinsic enzymatic activity but acquire it by associating, after activation, with other proteins such as cytoplasmic kinases and phosphatases. The enzymatic domains of these cytoplasmic proteins are associated with adaptor motifs that determine the specificity of their association.

Cytoplasmic tyrosine kinases, which include the Src and Abl families of proteins or the phosphatase Shp (Fig. 5.7), mediate signalling by a wide range of different membrane-bound receptors and participate in the regulation of cellular events ranging from cell adhesion and migration to cell cycle progression, programmed cell death and cell differentiation. Some of these kinases (e.g. those of the Abl family), shuttle between the nucleus and the cell surface. The Src kinases have been particularly well studied and provide a good model for understanding the mechanism of activation of cytoplasmic kinases (Fig. 5.7). Src proteins contain both an SH2 and an SH3 adaptor motif, as well as a catalytic kinase domain, an amino-terminal domain that is a target for lipid modification to anchor the protein to cellular membranes, and a carboxy-terminal domain involved in regulation of the protein's catalytic activity. In the same way that receptor-linked kinases only become activated upon ligand binding to the receptor, cytoplasmic kinases also remain inactive until the appropriate stimulus reaches them. In the absence of a stimulus, Src proteins are autoinhibited by remaining in a structurally tight configuration that sequesters the catalytic domain. They have multiple activation mechanisms, including dephosphorylation of a regulatory tyrosine residue in the carboxy-terminal domain; ligand-binding to the SH2 or SH3 domains, which disrupts inhibitory interactions between the SH and kinase domains; and phosphorylation of a specific tyrosine residue in the catalytic domain. Different combinations of these events can be triggered by interaction with different activated receptors, endowing the Src proteins with their considerable flexibility.

Many phosphatases, such as the Shp phosphatase, have the same modular nature (Fig. 5.7), enabling them to interact transiently with their targets and thus mediate the dynamic regulatory interactions that are essential for signalling during development.

G proteins and intracellular switches

The flow of information through a signal transduction pathway can be controlled at steps that function as gates or switches which can be turned on or off. There are several kinds of molecules that can fulfil this criterion, for example, allosteric enzymes whose functional state, associated with structural changes, is regulated through phosphorylation. A particularly important group of proteins that work as molecular switches are the G proteins, all of which have intrinsic GTPase activity. These proteins are activated by the binding of GTP and deactivated by hydrolysis of bound GTP to GDP (Fig. 5.8). Transitions between these states

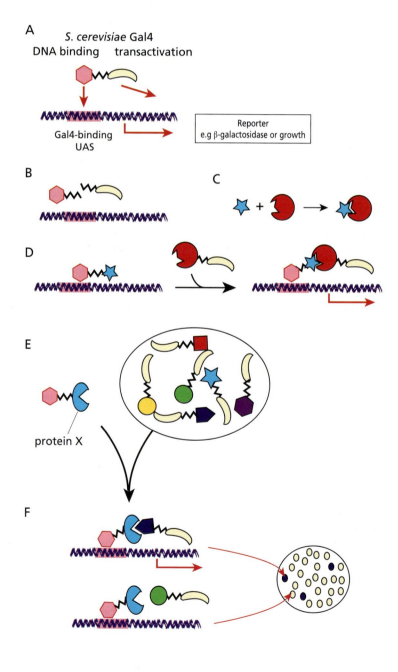

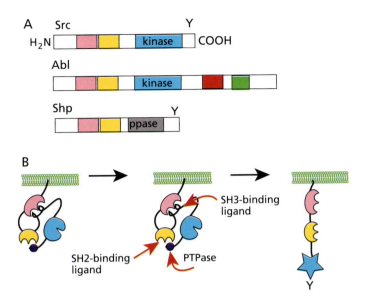

Fig. 5.7. Cytoplasmic protein kinases and phosphatases. (A) Domain structure of non-receptor cytoplasmic protein tyrosine kinases and phosphatases. The structure of the Src tyrosine kinase provides a framework for others. At the amino terminus there is a sequence to which a lipid substituent is added, attaching the protein to the cell membrane. An SH2 (pink) and an SH3 (yellow) domain allow the kinase domain (blue) to associate with and act on its targets. At the carboxy terminus there is a tyrosine residue whose phosphorylation state determines the overall configuration of the protein. While some members of this family of proteins are associated with cell-surface proteins others, like Abl, can be either nuclear or cytoplasmic and contain additional domains such as DNA-binding (red) or actin-binding (green) domains. Cytoplasmic phosphatases, such as Shp, have a similar basic structure but contain a phosphatase domain in place of the kinase domain. **(B)** Proposed mechanism of activation of Src. *Left*: Binding of the SH2 domain to the pY at the carboxy terminus (black circle) and of the SH3 domain to some internal proline residues keeps the kinase in an inactive configuration. *Centre*: Activation occurs by a variety of reactions: displacement of the intramolecular association of the SH2 and SH3 domains by the binding of high-affinity ligands, the phosphorylation of a Y residue in the catalytic domain (not shown) and the removal of the carboxy-terminal tyrosine residue by a phosphatase. *Right*: The activated kinase domain (star) can now bind to and phosphorylate substrates. (After Thomas, S. M. and Brugge, J. S. (1997) Cellular functions regulated by Src family kinases. *Annu. Rev. Cell Dev. Biol.* **13**, 513–609.)

Fig. 5.6. The yeast two hybrid screen for identifying interacting proteins. (A) The DNA-binding domain of the Gal4 transcription factor recognizes a specific sequence (Gal4-binding UAS) and the transactivation domain contacts the transcriptional machinery. Gal4 will drive the expression of any sequence that is downstream of the Gal4-binding UAS; this can be, for example, a reporter, or a gene that stimulates growth. **(B)** Separation of the two domains abolishes the activity of the protein. **(C)** Two proteins, which need not necessarily be associated with transcription (e.g. they could be two molecules involved in signal transduction, or two transcription factors that interact at an enhanceosome) can interact physically. **(D)** The interaction between these molecules can be used to reconnect the DNA binding and transactivation domains of Gal4. When the two hybrid proteins are inserted into yeast, the interaction reconstitutes the activity of Gal4, which now promotes transcription. **(E)** The type of experiment outlined in D can be used to test interactions between two specific proteins, to delineate the interaction interfaces of two proteins, or, as shown here, to screen for proteins that interact with a protein of interest (X, blue). X (which by itself does not promote transcriptional activation) is linked to the DNA-binding domain of Gal4. A library of potentially interacting proteins is prepared as fusion proteins with the Gal4 transactivation domain. Some members of this library will interact with X, though most will not. A reporter gene is inserted downstream of the Gal4-binding UAS. **(F)** The library of fusion proteins is transferred into a yeast strain that contains the fusion between X and the Gal4 DNA binding domain as well as the reporter under the control of the Gal4-binding UAS. This is done in such a way that, on average, every yeast cell will have one molecule from the library and one of X fused to the Gal4 DNA-binding domain. The resulting mixture is plated on to growth medium and colonies in which Gal4 activity is restored will be detected by the expression of the reporter. These colonies will have, fused to the transactivation domain of Gal4, a protein (or protein fragment) that interacts with X.

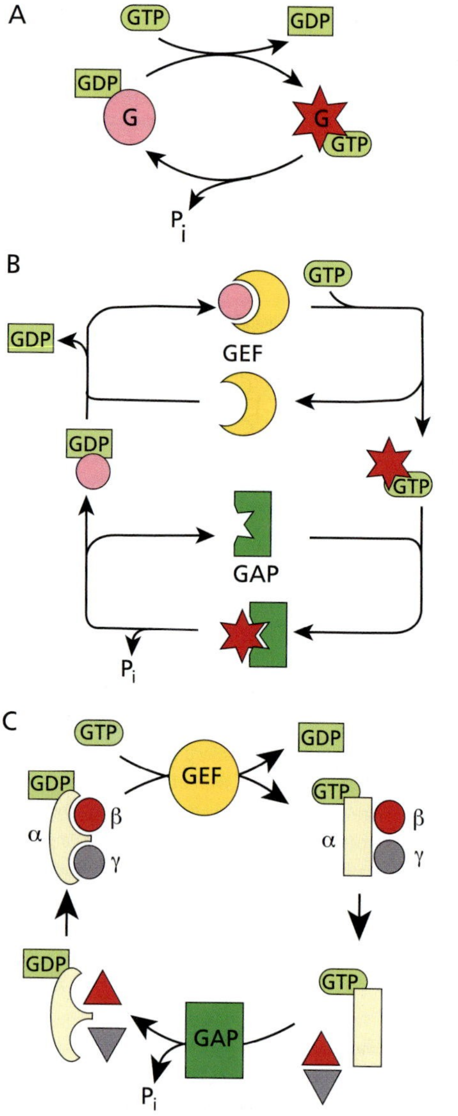

Fig. 5.8. G proteins as intracellular switches. (A) Members of the family of GTP binding proteins (G proteins) cycle between two states: an activated GTP-bound state (star) and an inactive GDP-bound form (circle). The transition from the active to the inactive state is brought about by the hydrolysis of GTP to GDP, and from the inactive to the active state by the exchange of GDP for GTP. It is the conformational changes in the G protein associated with these transitions that modulate its ability to interact with other proteins and modify their activity. **(B)** Monomeric G proteins, as exemplified by Ras, cycle between active and inactive states under the influence of G protein exchange factors (GEF) and GTPase-activating proteins (GAP). GEFs and GAPs catalyse transitions between GTP- and GDP-bound states in a G-protein-specific manner, that is, each G protein has its own GAP and its own GEF. **(C)** Heterotrimeric G proteins are composed of three subunits, α, β, and γ, but only the α subunit binds GTP. As in the case of the monomeric G proteins, the GTP-bound form is active and the GDP-bound form is inactive. When bound to GTP, the α form separates from the $\beta\gamma$ dimer and becomes active. The GDP-bound form of α has affinity for $\beta\gamma$, enabling the trimeric form to be reconstituted. Specific GAPs and GEFs catalyse the reactions as in the case of the monomeric G proteins.

cellular receptor to be amplified into a massive intracellular response.

There are two classes of G proteins: the heterotrimeric G proteins and the small (monomeric) GTP-binding proteins. The monomeric G proteins (Fig. 5.8B), exemplified by the protooncogene Ras, encompass an expanding group of proteins involved in linking extracellular signals to a variety of cellular responses including cell movement (Rac, Rho, and Cdc42), nuclear transport (Ran), and trafficking of intracellular vesicles (Rab). Most monomeric G proteins are anchored to a membrane via a covalently attached prenyl moiety. In the case of Ras, the protein is attached to the cytoplasmic surface of the plasma membrane, where it is involved in relaying signals from activated RTKs.

Heterotrimeric G proteins (Fig. 5.8C) are involved in relaying signals from the large seven transmembrane domain group of cell-surface receptors. These GTP-binding proteins contain three subunits: α, β, and γ. The trimeric form of the protein is inactive and contains GDP bound to the α subunit. Upon activation, the α subunit exchanges GDP for GTP and dissociates from the complex to exert its signalling function. Signalling is switched off by hydrolysis of GTP to GDP and re-formation of the trimeric complex with the β and γ subunits.

The on/off cycling of G proteins is regulated by a balance between negative and positive regulators known as GAPs (GTPase-activating proteins) and GEFs [G protein exchange factors, also called guanine nucleotide exchange

turn signal flow through a pathway on and off, allowing rapid and sensitive response to extracellular signals. In the 'on' state, these proteins catalyse protein–protein interactions that allow information to flow through a pathway. In some instances, part of the information is provided by the rate of cycling between the on and the off state.

A wide variety of cell-surface receptors are coupled to intracellular effectors via G protein switches of this type. A single activated receptor can often activate many G protein molecules, and these in turn can activate many target molecules before switching themselves off by GTP hydrolysis, enabling a small change in the concentration of an extra-

factors (GNEFs) or guanine nucleotide releasing-proteins (GNRPs)] respectively, which in turn respond to the activation state of the receptor. The integrated activities of the G proteins and their regulators provide signalling gateways to all the internal compartments of the cell.

Signalling pathways

The assembly and linking of a specific collection of receptors, adaptors, transducers, and effectors generates a signalling pathway (Fig. 5.1) which allows the transfer of information between different cellular compartments. In this way, extracellular signals can be transformed into specific inputs into the identity and behaviour of a cell that will contribute to specific patterns during development. Although future research may reveal novel pathways that we have no hint of as yet, it appears so far that cells, either in culture or within an organism, actually use quite a limited number of different types of pathways. In the remainder of this chapter we will outline some of the major types of pathways that have been elucidated so far.

These pathways can be grouped, somewhat artificially, into four different classes that we shall discuss in turn. In the first class, a ligand–receptor interaction regulates the direct transit of a transcription factor to the nucleus. The second class of pathways also regulates the activity of transcription factors, but does so through a relay containing several steps and culminating with the entry of a transducer, rather than the effector, into the nucleus. The third class involves the generation of second messengers which can act on a variety of processes. Finally, Wnt and hedgehog signalling, first uncovered in studies in developing embryos, might participate in networks involved in the coordinated regulation of various cellular processes.

Routes triggering nuclear translocation of the effector

The activity of the nucleus is a key variable in the generation of cell diversity and so it is not surprising that many signalling events result in changes in the transcriptional activity of the cell, with dramatic consequences during development. A signal can reach the nucleus in many ways. The most direct way is for the signal itself to trigger changes in transcriptional activity: as outlined in Chapter 4 (Fig. 4.25), steroids can diffuse through cytoplasmic and sometimes also nuclear membranes to reach their receptors, which act as transcription factors that directly regulate the activity of target genes.

There are other signalling systems in which very few intermediates separate the signal and its effector in the nucleus. Some examples will be discussed in this section.

Delta–Notch signalling

In the case of the Notch receptor (Chapter 4, Fig. 4.22; Fig. 5.9) it is the activated receptor/sensor itself, rather than the

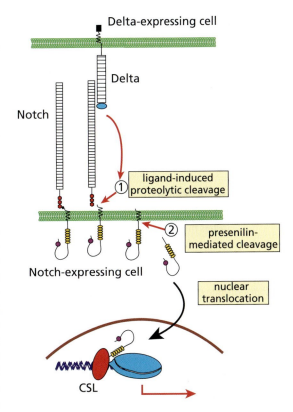

Fig. 5.9. Notch signalling. The binding of Delta to specific EGF-like repeats in the extracellular domain of Notch promotes two cleavage events. The first (1) occurs in the extracellular domain near the transmembrane domain and is mediated by extracellular proteases of the TACE family. This event is completely dependent on the binding of Delta to Notch. The second (2), takes place close to the membrane but in the intracellular domain and releases a fragment (Nintra) containing the intracellular domain. This event is dependent on the previous cleavage and is mediated by the transmembrane presenilin proteins. Once released, the intracellular domain of Notch is able to enter the nucleus and interact with the transcription factor CSL [CSL is named from the initial letters of the corresponding proteins in vertebrates (CBF), *Drosophila* (suppressor of Hairless) and *Caenorhabditis elegans* (lag-1)], resulting in the transcriptional activation of target genes. The cdc10/ankyrin repeat domain of Notch/Nintra provides a transcriptional activation function as well as an interface for interactions with CSL.

ligand, that makes the journey to the nucleus. Upon binding its ligand Delta, the Notch receptor undergoes two cleavage events, one extracellular and the other intracellular. These events generate a soluble intracellular fragment of the receptor, often referred to as Nintra (or NICD for Notch intracellular domain), that can translocate to the nucleus. Nintra cannot bind DNA by itself but in combination with the nuclear protein CSL it acts as a transcription factor targeted to specific promoters (Fig. 5.9). The first cleavage at the cell surface is mediated by proteases of the TACE family and is completely ligand-dependent. The second proteolytic cleavage depends on the first and requires some intracellular membrane proteins called presenilins. In most cases, the amounts of Nintra required for signalling are very small—in many systems only just within the present limits of detection (Fig. 5.10)—indicating that this signalling system has a very low threshold for activity.

NFκB/dorsal signalling

In other cases, the receptor itself does not travel to the nucleus, but transfers the signal to an effector molecule that is located in the cytoplasm but is mobilized to the nucleus upon signalling. A classical example of this signalling mode is provided by the NFκB transcription factor, which mediates immune and inflammatory responses in mammalian cells and has also been shown to be involved in certain aspects of pattern formation during embryogenesis (Fig. 5.11). NFκB is a dimer made up of two subunits, known as p50 and p65, which are both members of the Rel family of transcription factors. In the absence of ligand, NFκB is complexed with an inhibitor protein, IκB, which keeps it in the cytoplasm. Activation of the TNFα or IL1 receptor recruits to the activated receptor a large multiprotein complex that includes a cytoplasmic protein kinase, NIK, and an I-κB kinase. This association activates NIK, which in turn phosphorylates IκB kinase. Phosphorylation of IκB targets it for ubiquitin-dependent degradation. As a result, NFκB enters the nucleus where it modulates transcription of target genes (Fig. 5.11).

In *Drosophila*, the *dorsal* gene encodes an NFκB homologue that associates with an IκB-like protein encoded by the *cactus* gene (Fig. 5.11). In this case, the receptor that is activated is a protein, encoded by the *Toll* gene, that has an intracellular domain similar to that of the IL1 receptor. This signalling system participates in the definition of the dorsoventral axis in the embryo (see Fig. 4.26) and—in an interesting parallel to mammals—a related pathway functions in the insect's immune response. Despite the similar-

ities between the receptor, major transducer, and effectors, the intermediates in the pathway appear to be different. In *Drosophila*, the connection between Toll and cactus is mediated by two proteins, tube and a kinase, pelle, which have no clear homologues in the IL1/NFκB signal transduction pathway (Fig. 5.11).

JAK–STAT signalling

Two further examples of effectors that are directly activated by interaction with an activated receptor are provided by the STAT (Fig. 5.12) and Smad (Fig. 5.13) families of proteins. The STATs (signal transducers and activators of transcription) are SH2-domain-containing transcription factors that normally reside in the cytoplasm or at the inner surface of the plasma membrane. Interaction between a receptor and ligand (usually a cytokine) leads to phosphorylation of the STATs which then form heterodimers with other STATs and are translocated to the nucleus. Nuclear entry is mediated not by the removal of an inhibitor, as in the case of NFκB, but by a conformational change. Once in the nucleus, STATs bind to DNA and activate transcription, often acting synergistically with other transcription factors that are themselves regulated by signalling pathways (Fig. 5.12).

STATs were first identified in the interferon signalling pathway but have subsequently been found to be involved in responses to more than 35 different polypeptide ligands and to play a role in a wide variety of cellular responses and developmental processes. Receptors such as those for EGF and PDGF, which have tyrosine kinase activity, can when activated recruit STATs to the membrane and phosphorylate them. In other cases, including the interferon pathway, the receptor that activates STATs has no intrinsic kinase activity, and STAT phosphorylation is accomplished by a member of the Janus kinase (JAK) family, which is noncovalently attached to the receptor and becomes activated upon ligand binding. The specificity of the signalling response is determined by the interaction of a specific heterodimeric pair of STAT molecules with a specific receptor: the specificity can be artificially altered by changing either the STAT or the phosphotyrosine site in the receptor. Figure 5.12 illustrates further details of STAT activation in interferon signalling.

Signalling by cytokines is usually transient, and several mechanisms have been found that act to switch off STAT-mediated signalling. These include inhibition of JAK activity by a group of negative regulators called suppressors of cytokine signalling (SOCS) proteins, which downregulate the receptors, dephosphorylation of activated

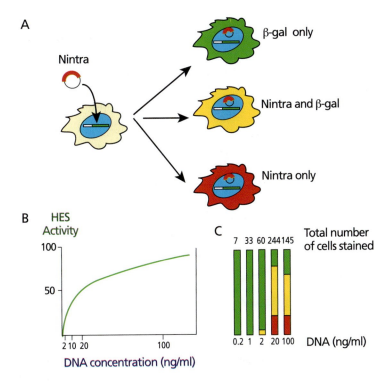

Fig. 5.10. Thresholds of protein detection and the activity of Notch. Nintra can effect transcriptional activation at very low ('catalytic') concentrations that are dfficult to detect by existing methods. This experiment illustrates this point. **(A)** A piece of DNA encoding a soluble form of Nintra that can be detected with specific antibodies is transfected into cells that have a reporter gene (β-galactosidase) linked to a Notch-responsive element (HES). After transfection, antibodies are used to visualize Nintra and the activity of the reporter. There are three possible outcomes of the transfection: cells in which the reporter is activated but Nintra cannot be detected (green); cells in which both the reporter and Nintra are detected (yellow), and cells in which Nintra but not the reporter can be detected (red). Because the reporter is absolutely dependent on Nintra, the green cells indicate that Nintra can work at very low con- centrations that cannot be detected. The red cells indicate that Nintra has not been able to activate the reporter. All three possible outcomes are observed. **(B)** Nintra is transfected in the form of DNA, so the DNA concentration can be used as a measure of Nintra concentration in an experiment to titrate the Nintra requirement. The graph shows the activity of the reporter in response to different concentrations of Nintra-encoding DNA. As shown, the response of the reporter is very sharp and reaches near saturation at very low concentrations of input DNA. **(C)** Analysis of the numbers of cells with reporter activity and/or detectable Nintra reveals, when compared with B, that Nintra will activate the reporter at a concentration below its detection level. (After Schroeter, E. H. et al. (1998) Notch-1 signalling requires ligand-induced proteolytic release of intracellular domain. *Nature* **393**, 382–6.)

JAKs and receptors by the Shp phosphatases, and negative regulation of STAT DNA binding by the protein inhibitors of activated STATs (PIAS) group of proteins.

TGFβ–Smad signalling

Another family of proteins that exert their function by translocating to the nucleus are the Smad proteins (the name comes from a combination of the *sma* genes in *C. elegans* and the *Drosophila Mad* gene). These nuclear proteins mediate the effects of the TGFβ/BMP family of signalling molecules (Fig. 5.13). Smad proteins are asso- ciated with the membrane through their interaction with the FYVE domain-containing proteins known as SARA (Smad anchor for receptor activation). Members of the activating class of Smads, the R-Smads, interact directly with the activated ligand–receptor complex and are phosphorylated at carboxy-terminal serine residues by the Type I receptor (see Chapter 4, Fig. 4.17). Phosphorylation is thought to counteract intramolecular binding between the carboxy- and amino tails of the protein, opening the structure and allowing the phosphorylated protein to form a heteromeric complex with a cytoplasmic Smad protein of

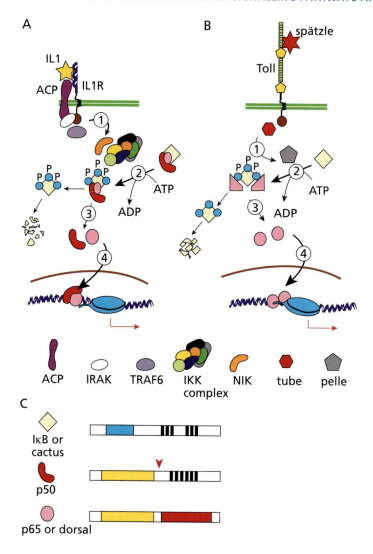

Fig. 5.11. Regulated nuclear translocation of NFκB and dorsal. (A) Upon binding its ligand IL1, the IL1 receptor associates with the transmembrane accessory protein (ACP) and this complex recruits and activates (1) two kinase proteins, TRAF6 and IRAK (IL1R activated kinase). Together, these two proteins activate a very large protein kinase complex (IKK complex; approximately 500–900 kDa) that contains the two subunits of the IκB kinase and is associated with the protein NIK (NFκB inducing kinase). Upon activation, the IκB kinase phosphorylates IκB which is complexed with the two subunits of the NFκB transcription factor (p50 and p65) (2). In the inactive state, IκB retains the two subunits of the NFκB transcription factor, p50 and p65, in the cytoplasm. The phosphorylation targets IκB for degradation, releasing NFκB (3) and allowing it to enter the nucleus (4) and activate transcription. **(B)** Activation of the Toll receptor in *Drosophila* results in the nuclear entry of dorsal (an NFκB family member) and the activation of transcriptional targets. Upon binding the ligand spätzle, Toll is activated. Toll activation leads, through the cell membrane associated protein tube, to the activation of the pelle serine/threonine kinase (1), which phosphorylates the IκB-like protein cactus (2). The effect of this phosphorylation is to release dorsal from its association with cactus (3), allowing dorsal to enter the nucleus and activate transcription as a dimer (4). **(C)** Structure of IκB and cactus, the two components of NFκB (p50 and p65) and dorsal. Cactus and IκB are characterized by an acidic region that is the substrate for phosphorylation (blue) and cdc10/ankyrin repeats (black bars) which mediate protein–protein interactions. p50, p65, and dorsal are members of the Rel family of transcription factors characterized by the 'Rel domain' (yellow), which contains DNA binding and dimerization domains. This domain also contains the interfaces for interaction with IκB. p65 and dorsal contain transactivation domains at the carboxy terminus (red), whereas p50 contains some cdc10/ankyrin repeats (black bars) which inhibit its activity and are removed by proteolytic cleavage at the position indicated (arrow head).

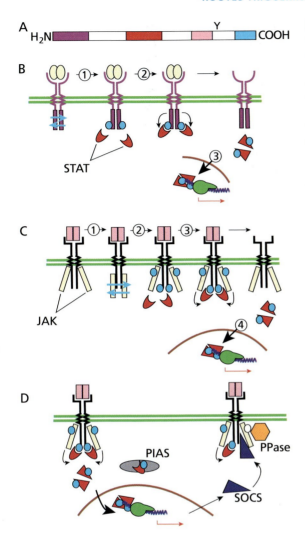

Fig. 5.12. Structure and function of STATs. (A) Domain structure of members of the STAT family of transcription factors. These proteins are composed of several domains: a dimerization domain (purple) at the amino terminus and a centrally located DNA-binding domain (red); on the carboxy-terminal side of this there is an SH2 domain (pink) which allows the protein to recognize the pY site of the receptor, and close to the SH2 domain a tyrosine (Y) residue whose obligatory phosphorylation results in dimerization and nuclear translocation. At the carboxy terminus there is a transcriptional activation domain (blue). The STATS recognize a palindrome with the sequence: TTCNNNGAA. **(B)** When STAT activation is mediated by receptor tyrosine kinases (RTKs), ligand binding induces dimerization of the receptor and transphosphorylation of its subunits (1, blue arrows), resulting in the creation of pY docking sites (blue spots) for SH2 domains. In specific sequence contexts, these sites are recognized by the SH2 domains (red) of members of the STAT family, which are thereby recruited to the receptor (2) and phosphorylated by the RTK. As a consequence, STATs change conformation, dimerize, and enter the nucleus where they modulate transcription (3). **(C)** Sometimes (e.g. in the case of the cytokine and interferon receptors), the receptor that relays the signal to the STAT has no enzymatic activity of its own, but is associated with a cytoplasmic kinase of the Janus kinase (JAK) family (yellow box). Upon ligand binding, a conformational change activates the JAK (1), which phosphorylates certain Y residues in the receptor and on the JAK itself. Those in the receptor act as docking sites (blue spots) for the STAT (2). When the STAT is recruited to the receptor (3), the JAK catalyses a second phosphorylation on the STAT protein which results in its dimerization and translocation to the nucleus. **(D)** There are several ways in which STAT signalling can be down-regulated, most of which have been identified by analysing cytokine signalling. The JAK or the receptors may be dephosphorylated by specific phosphatases (Shp1 and Shp2, orange hexagon) or the STATs may be dephosphorylated by specific tyrosine phosphatases (not shown). In addition, there are SH2-containing proteins called suppressors of cytokine signalling (SOCS), whose expression is induced by STATs. SOCS compete with the STATs for binding to the receptors. Protein inhibitors of activated STATs (PIASs) are proteins that can bind and inactivate STATs and therefore down-regulate their transcriptional activity.

the Co-Smad class. Co-Smads, such as Smad4, are not phosphorylated by the activated receptor but are obligatory components of all Smad heterodimers. The heterodimeric complex migrates to the nucleus where it binds DNA. Different members of the activating R-Smads are associated with different receptors. For example, BMP receptors effect signalling through Smad1, whereas TGFβ and activin do so through Smads 2 and 3. In addition to this activating class of Smads, there is also a class of inhibitory Smads (I-Smads) that can interact with an activated receptor and compete with R-Smads for binding. However, the I-Smads lack the transcriptional activation domain and, in a manner reminiscent of the 'mock ligands' or 'mock receptors' that negatively regulate some ligand–receptor interactions at the cell surface (see Chapter 4, Fig. 4.28),

titrate the signalling ability of the activating Smads. Some I-Smads are themselves transcriptionally induced by TGFβ family signalling, acting as feedback inhibitors of the signalling pathway.

Although they can bind DNA, Smad proteins are not very effective in generating transcriptional activity on their own. Instead, they interact with other transcription factors, creating a context-dependent complex that activates transcription (see below). At different promoters, different Smads interact with different transcription

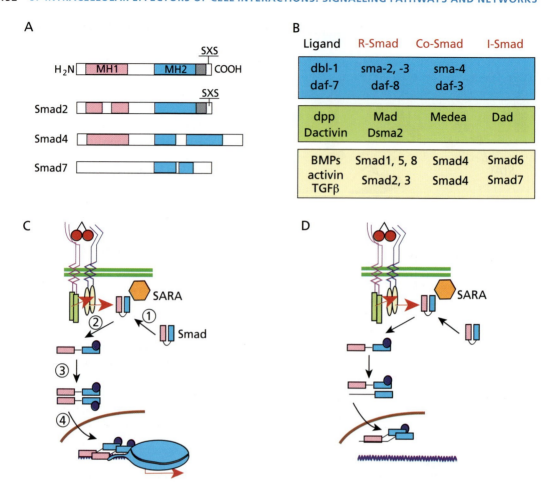

Fig. 5.13. Structure and function of Smads. (A) The Smad family of transcriptional regulators act as heterodimers. They are characterized by two MAD-homology (MH) domains: an amino-terminal domain, MH1 (pink), which is associated with DNA binding, and a domain, MH2 (blue), that is closer to the carboxy terminus. The MH2 domain allows Smads to interact with each other or with other transcription factors. In some Smads, this domain also provides transactivation functions. There are three kinds of Smads. The R-Smads are exemplified by Smad2 [R stands for 'restricted' since each member of this class is associated with a particular ligand/receptor complex (see B)]. In these Smads, the MH1 domain is split into two subdomains. Near their carboxy terminus R-Smads contain a series of serine residues (grey box) which are the substrate for binding to the receptor, and on the carboxy-terminal side of this region there is a serine/threonine phosphorylation site (shown as SXS). The second class of Smads, the Co-Smads, exemplified by Smad4, are obligatory partners for the pathway-restricted Smads (i.e. they are required for the activity of the R-Smads with which they form heterodimers). They contain a bipartite MH2 domain and a single MH1 domain, but lack the receptor-binding carboxy-terminal serine-rich domain. Members of the third Smad class, the I-Smads, exemplified by Smad7, have an MH2 domain but lack the MH1 domain and the car-

boxy-terminal phosphorylation site; these Smads are also pathway-restricted. Because they lack the MH1 domain, these Smads cannot bind DNA, and have an inhibitory function. **(B)** Examples of the relationships between different Smads and particular TFFβ/BMP family members in different organisms, to highlight the specificity of the R-Smads and I-Smads. Blue, *Caenorhabditis elegans*; green, *Drosophila melanogaster*; yellow, vertebrates. **(C)** Signalling through Smads. Before receptor stimulation, R-Smads are found either in the cytoplasm or associated with the cell surface through interactions with SARA (Smad anchor for receptor activation) proteins. The two MH domains are tightly bound to each other (1). Activation of the Type I receptor (see Chapter 4, Fig. 4.17) leads to phosphorylation (arrows) of residues at the carboxy terminus of R-Smads (2). This induces a conformational change which dissociates the R-Smad from SARA and allows it to interact with a Co-Smad (3). The heterodimer enters the nucleus where it contributes to transcriptional activation of target genes (4). **(D)** I-Smads, such as Smad6 and 7, can counteract productive Smad signalling in two different ways: by blocking the productive interaction of R-Smads with the receptor (not shown), or by forming sterile heterodimers with R-Smads or Co-Smads.

factors, contributing to the specificity of the transcriptional response.

In all of these examples, the nuclear translocation step itself is an important point of control, and cells have evolved mechanisms to regulate this event. NFκB may be an exception in having an intrinsic ability to enter the nucleus. Most other proteins that follow this route are probably enabled to do so by forming complexes with other proteins that have an escorting function.

Membrane–nucleus relays: Mitogen-activated protein kinase (MAPK) signalling cascades

In some pathways, the outcome of the signalling event at the cell membrane does not act directly on the effector but triggers a sequence of molecular interactions that lead to the activation of a key element of the cascade that in turn acts on the effectors to modulate their activity. An important aspect of such relay-based systems is that the more elements they have the more points of interaction and regulation they provide. Mitogen-activated protein kinase (MAPK) pathways function in this manner in all eukaryotic cells and are one of the best characterized intracellular signalling relays (Fig. 5.14). MAP kinases were first identified as a set of proteins that are phosphorylated in response to certain mitogens or stress signals. They are serine/threonine kinases that can activate effectors in either the nucleus or the cytoplasm. The kinases are themselves activated upon phosphorylation of specific serine, tyrosine, and threonine residues by MAPK kinases (MAPKK), which in turn become activated by serine phosphorylation mediated by kinases dubbed MAPKK kinase, or MAPKKK. Each of these phosphorylation events activates the kinase by inducing a conformational change, and the chain of mutually dependent protein kinases forms a functional cassette that has been conserved across a wide range of phyla (Fig. 5.14).

MAPK signalling pathways are activated by many different ligand–receptor systems, including RTKs and heterotrimeric G-protein-coupled receptors. There are many different MAPK proteins (Fig. 5.14); for example in mammalian cells six MAPKs, seven MAPKKs and seven MAPKKKs have been identified so far but this number will undoubtedly increase as the genome is analysed. These proteins can be grouped into three classes—ERK, JNK, and p38—each corresponding to a pathway that responds to a particular stimulus and has a defined set of effectors (Fig.

5.14). Within a given group the MAPKs show some sequence homology, as do the MAPKKs. However, the defining feature of a particular group is a combination of two elements: the stimulus that activates the pathway, and the site of action of the MAPKK on the MAPK (Fig. 5.14). The second of these features is particularly important because it defines a MAPKK–MAPK pair as a specific functional module. The MAPKKKs can be more promiscuous (Fig. 5.14).

The pathway that activates the ERK (extracellular signal-responsive kinase) MAPKs in vertebrates is one of the best characterized so far and reveals a functional network containing many of the signalling system components that we have described in this chapter (Fig. 5.1. and 5.15). In addition to an RTK, such as the EGF receptor, the pathway involves the small GTPase protein Ras, whose activation/inactivation cycle represents an important control point. For this reason, this pathway is often referred to as 'the Ras signalling pathway'.

Binding of ligand (often a growth factor in the case of Ras) by the RTK leads, after dimerization, to transphosphorylation of its intracellular domain. This event generates phosphotyrosine residues, some of which act as docking sites for a complex of the SH2 domain-containing protein Grb2 and its associated guanine nucleotide exchange factor Sos (Fig. 5.15). Recent genetic evidence suggests that there may also be several other regulatory factors involved in this step of the pathway. The activity of Sos switches Ras, at the inner surface of the membrane, from its inactive (GDP-bound) to its active (GTP-bound) state and this results in the recruitment to the membrane of the serine/threonine kinase Raf, which acts as a MAPKKK, phosphorylating and activating the dual-specificity kinase MEK (a MAPKK). MEK in turn phosphorylates and activates the MAPK ERK, which can then enter the nucleus where it phosphorylates various transcription factors, modulating their activity (Fig. 5.15). Alternatively, activated ERK can interact with cytoplasmic targets, one of which, the serine/threonine kinase Rsk, can itself enter the nucleus and phosphorylate transcription factors.

Numerous experiments show that the outcome of Ras signalling is different in different cells even though the relay system is the same. For example, studies of cell fate determination during development in *Drosophila* and *C. elegans* provide numerous examples in which different cell fates are specified through the activity of the same MAPK pathway. In *Drosophila*, for example, Ras signalling is involved in specifying the termini of the embryo early in development, cell fates in the peripheral and central nervous system, the epidermis and the mesoderm, and the

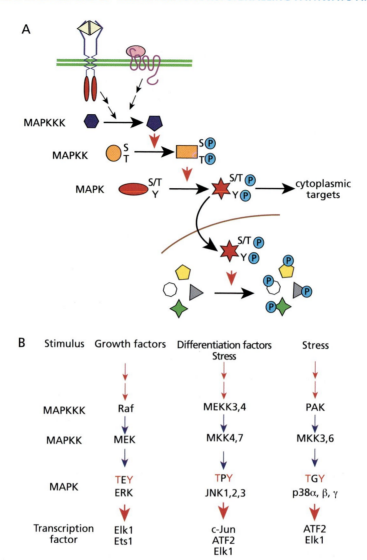

Fig. 5.14. Mitogen-activated protein kinases. (A) A MAPK cascade results from the functional assembly of a series of protein kinases that act sequentially upon each other until the last link of the chain phosphorylates a number of effector substrates. The relay is triggered by the activation at the cell surface of either receptor tyrosine kinases or seven transmembrane domain receptors. A series of steps, often mediated by G proteins, activates MAPKKK proteins. Active MAPKKK phosphorylates target serine and threonine resides in a MAPKK and activates its kinase activity. Active MAPKK can then in turn phosphorylate serine, threonine and tyrosine residues in a MAPK. The activated MAPK enters the nucleus where it phosphorylates effectors such as transcription factors. Alternatively, it may act on cytoplasmic targets. **(B)** Sequence of kinase activity in the three known types of MAPK module. In some modules, more than one kinase can function at each step. Different MAPK proteins fall into three classes depending on the structure of their activation site: TEY for MAPK, TPY for JNK, and TGY for p38. Some targets of each pathway are shown; some of these targets, such as the transcription factor Elk1, can be substrates for more than one pathway.

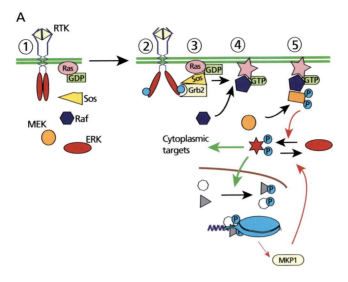

B	S. cerevisiae	C. elegans	D. melanogaster	Vertebrates
		sem-5	Drk	Grb2
		let-60/Ras	D-Ras	Ras
GEF		CRas-GEF	Sos	mSos
MAPKKK	Ste11	lin-45	D-Raf	Raf
MAPKK	Ste7	mek-2	D-Sor	MEK
MAPK	Fus3	mpk-1	rolled	ERK

expression of many genes whose products are involved in the growth and patterning of appendages such as the wings and the legs. The reason for this functional diversity of outcomes is that the constellation of targets of MAPK is different in different cells and this, rather than the signalling event itself, turns out to be the important element in defining the specificity of the response.

MAPK signalling pathways also function in responses to stimuli other than growth factors, such as hyperosmolarity, mechanical stress, and UV irradiation. These responses, as well as responses to some specific growth factor families, are mediated by a group of MAPKs that includes the p38 and Jun N-terminal kinase (JNK) proteins, sometimes known as stress activated protein kinases, or SAPKs (Fig. 5.16). These pathways operate as a highly interconnected network. Their targets include the transcription factors c-Jun, Elk1, and ATF2, and their upstream components include individual MAPKK, each specific for a member of either the p38 or the JNK subfamily of MAPK proteins. Further upstream, there are Raf-like MPKKK. Experiments in tissue culture suggest that the small GTPase proteins Rho, Rac, and Cdc42 may function in

Fig. 5.15. The Ras/ERK (MAPK) pathway. (A) The components of the Ras/MAPK pathway (1). Upon activation of an RTK (2), tyrosine phosphorylation of the intracellular domain recruits the Grb2 adaptor protein to the receptor and this in turn recruits to the membrane the GEF protein Sos (yellow) through its SH3 domain, and Ras (3). As a result, Sos catalyses the conversion of Ras (pink) from an inactive GDP bound form to an active GTP bound form (star). Ras then triggers the MAPK cascade by recruiting the MAPKKK Raf (dark blue) to the membrane and activating it (4). Raf then phosphorylates and activates MAPKK (MEK) (orange), which in turn activates MAPK (red). (5). MAPK then interacts with its targets either in the cytoplasm or after translocation to the nucleus. One of the nuclear targets activates the expression of a cytoplasmic tyrosine phosphatase, MKP1, that dephosphorylates and inactivates MAPK and thus provides a feedback control on the intensity of signalling. **(B)** Elements of the canonical Ras signalling pathway in *Saccharomyces cerevisiae, C. elegans, Drosophila,* and vertebrates.

these pathways in an analogous way to Ras in the ERK pathway. However, in contrast to the Ras pathway, there is at present no clear mechanistic link between these small GTPases and the downstream MAPKs (Fig. 5.16).

Genetic studies in *Drosophila* and *C. elegans* have

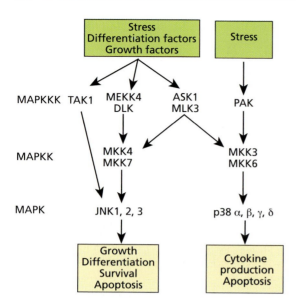

Fig. 5.16. Signalling by stress activated protein kinases. (A) Physical stress and some cytokines and growth factors, usually involved in programs of cellular differentiation, elicit the activity of a family of MAPK proteins called JNK (c-Jun N-terminal kinase) and p38 families. These kinases are sometimes referred to as SAPK (stress activated protein kinases). The initial steps in the activation of these pathways are not well understood at the moment but they are known to be triggered by a number of receptors which include members of the seven transmembrane domain family. Several different MAPKKKs have been identified in these pathways, and small G proteins, such as Rac and Cdc42, may also be involved. In ways that are also not yet clear, signalling through these intermediates leads to the activation of one of several possible MAPK cascades. There are many members of each class of kinases in these cascades, indicating that these proteins operate as a network.

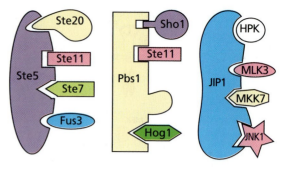

Fig. 5.17. MAPK modules and scaffolding. Examples of proteins that act as scaffolds for the assembly of MAPK signalling complexes. *Left*: yeast Ste5 is a scaffold protein that assembles four components of a MAPK pathway (the Ste11 kinase Ste20, the MAPKKK Ste11, the MAPKK Ste7, and the MAPK Fus3) and facilitates their functional interactions during the response to mating pheromone. *Centre*: Pbs1 is a MAPKK from yeast which also acts as a scaffold for other members of a MAPK cascade during responses to osmotic shock; in this response, Sho1 is an osmotic sensor, and Hog1 a MAPK of the JNK subclass. The MAPKKK Ste11 is required for signal transduction both during the pheromone response and in response to osmotic shock. Its efficient use in one or the other process is facilitated by its association with the appropriate scaffolding protein. *Right*: in vertebrates, the JNK interacting protein 1 (JIP1) organizes the haematopoietic progenitor kinase 1 (HPK1) together with the MAPKKKK mixed lineage kinase 3 (MLK3), and JNK activator MKK7 to activate JNK1 during the stress response.

demonstrated that the JNK signalling pathway takes part in a variety of morphogenetic processes and the embryonic lethality of mutations in some JNK pathway components in the mouse suggests that this may also turn out to be true for other animals. For example, in *Drosophila*, Rac/Cdc42 mediated activation of JNK signalling has been implicated in regulating epithelial changes in the embryo that bring about a morphogenetic process known as dorsal closure, by regulating the spatial expression of the TGFβ family member dpp.

Individual cells of eukaryotes typically contain the components of several different MAPK pathways. How then does the cell ensure that a MAPK relay specific for the required response is assembled? Very little is known about how this is achieved. In some cases the specific interactions

between each pair of members of a particular cascade (MAPKKK and MAPKK or more often MAPKK and MAPK) ensure a specific response. In others, specific 'scaffold' proteins assemble a specific MAPK module. The prototypes for these scaffolding proteins are the *S. cerevisiae* Ste5 and Pbs1 proteins, which assemble specific elements of MAPK pathways to mediate distinct responses (Fig. 5.17). The MAPKKK Ste11 can function not only in the pheromone response pathway that mediates mating between α and **a** cells, but also in MAPK pathways that respond to hyperosmolarity or starvation. The protein Ste5 acts as a scaffold for assembly of the MAPK cassette that functions in response to mating pheromone ensuring that, during this process, Ste11 associates with the appropriate MAPKK and MAPK partners and that the signal is routed to the correct transcription factor targets. The Pbs1 protein performs a similar function in the hyperosmolarity response and associates Ste11 with other MAPKs. Evidence has been obtained that there are mammalian proteins that function in a similar way to Ste5 and Pbs1 (Fig. 5.17). Some, such as JIP1, which is involved in the JNK pathway, seem to cluster multiple proteins. These associations are likely to play an active role in the routing of specific proteins to specific responses and it will be important to understand their regulation.

Second-messenger signalling pathways

Many cell-surface receptors, including the huge family of seven transmembrane domain receptors, and several RTKs, do not access their target through a direct route but employ intracellular 'second messenger' molecules, such as cyclic AMP, calcium, diacylglycerol, and inositol phospholipids, to transduce the signal within the cell (Fig. 5.4). These second messengers lack the specificity of the kinases,

phosphatases, and adaptors that we have discussed earlier in this chapter and often interact with various targets. The pattern of activity is then determined by other variables that set or pre-set the activity of those targets. Two very common second messengers are cyclic AMP and phospholipids.

Cyclic AMP and protein kinase A

The intracellular second messenger cyclic AMP is used in all animal cells, participating for example in hormone responses in many different tissues. Receptor activation is coupled to activation of the enzyme adenylyl cyclase by a trimeric G protein. Cyclic AMP is produced by adenylyl cyclase and activates the enzyme cyclic AMP-dependent protein kinase (PKA, also known as protein kinase A or A-kinase). PKA, a tetramer composed of two regulatory subunits and two catalytic subunits, phosphorylates a wide variety of intracellular proteins, from other kinases and receptors to transcription factors that act as effectors. There are several different regulatory subunits which associate with a smaller repertoire of catalytic subunits. Thus, the cyclic AMP response in a given cell is determined by its complement both of target proteins and of regulatory subunits of PKA.

In some cells, production of cyclic AMP leads to the transcription of particular sets of genes. These genes are characterized by regulatory sequences known as cyclic AMP response elements (CREs), which bind the cyclic AMP response element binding protein (CREB). The signal transduction pathway leading to activation of gene transcription by cyclic AMP is shown in Fig. 5.18: cyclic

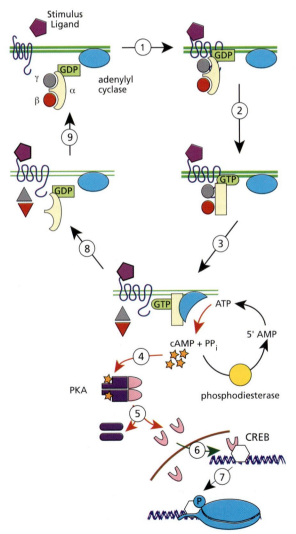

Fig. 5.18. Cyclic AMP and protein kinase A. cAMP is produced as a response to a variety of stimuli which activate the inner membrane protein adenylyl cyclase (blue). In the example shown the activation is mediated by a seven transmembrane domain receptor through a heterotrimeric G protein. In response to ligand binding (1) the GDP-bound form of the G protein is converted to the GTP-bound form (2). The GTP bound α subunit activates adenylyl cyclase (3), which produces cAMP from ATP. cAMP diffuses through the cytoplasm and binds the regulatory subunits of the tetrameric protein kinase A (PKA) (purple/pink) (4) inducing a conformational change which releases the catalytic kinase subunits (pink) (5). These enter the nucleus (6) and phosphorylate DNA-bound CREB, triggering a conformational change that allows the DNA-bound CREB (white hexagon) to activate transcription (7). The activity of CREB is very sensitive to the levels of cAMP and therefore lowering the cAMP concentration, either by activation of the phosphodiesterase or by changes in the activity of the heterotrimeric G protein, will alter the activity of the pathway. The cycle of GTP/GDP exchange (8,9) allows the signalling system to return to the initial state.

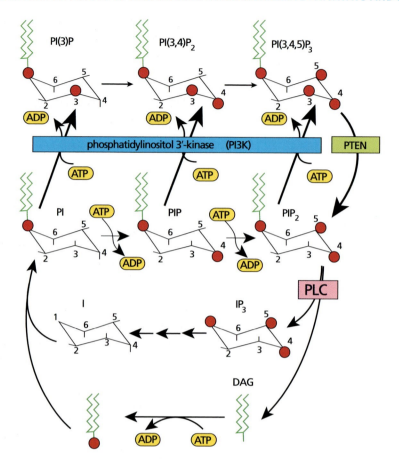

Fig. 5.19. The phospholipid cycle and second messengers. Structures and biochemical relationships between different phospholipids, some of which act as second messengers in signalling reactions. Phosphorylated residues are indicated in red; diacylglycerol is indicated by double green zig-zag lines. The best starting point for following the phospholipid cycle is the phospholipid phosphatidylinositol (PI), which together with two phosphoderivatives, phosphatidylinositol 4-phosphate (PIP) and phosphatidylinositol 4,5-bisphosphate (PIP$_2$), is found in the inner leaflet of the plasma membrane. PI, PIP, and PIP$_2$ can be converted to 3'-phosphoderivatives by the phosphatidylinositol 3'-kinase (PI3K). The phosphatase PTEN converts PI(3,4,5)P$_3$ into PIP$_2$. Members of the phospholipase C (PLC) family can cleave PIP$_2$ into the soluble inositol-1,4,5-trisphosphate (IP$_3$) and a lipophilic molecule diacylglycerol (DAG). These molecules are precursors for PI.

AMP binds to the regulatory subunits of PKA and leads to the dissociation of the catalytic subunits which can enter the nucleus and phosphorylate DNA-bound CREB, enabling it to activate transcription.

Lipid second-messenger pathways

Even more widely used than cyclic AMP are the lipid second messengers: inositol phospholipids and diacylglycerol. These molecules are derived from phospholipids in the inner leaflet of the plasma membrane, by phosphorylation and dephosphorylation reactions (Fig. 5.19). Their structural variability provides the basis for a wide repertoire of interactions with other molecules. As we have seen earlier in this chapter, there are adaptor motifs that mediate the interactions of proteins with these membrane lipid derivatives. There are two main pathways involved in the processing of phospholipids: one is mediated by the phospholipase C enzyme and the other by the phosphatidylinositol 3'-kinase, PI3 kinase (PI3K).

Activation of some G-protein-coupled receptors and of some RTKs induces the activity of the enzyme phospholipase C (PLC) which hydrolyses phosphatidylinositol 4,5-bisphosphate (PIP$_2$) to produce soluble inositol 1,4,5-trisphosphate (IP$_3$) and the membrane lipid diacylglycerol (DAG) (Fig. 5.20). There are different forms of PLC and each is linked to a different type of receptor; for example, RTKs use PLCγ while G-protein-coupled receptors use PLCβ. IP$_3$ is soluble and can diffuse through the cytoplasm. Its receptor, a multichannel protein, lies within the

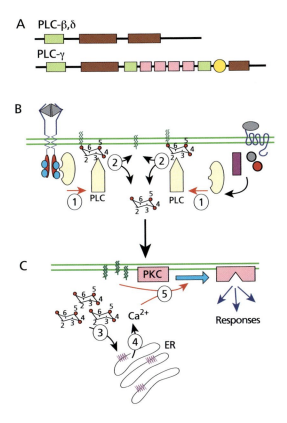

Fig. 5.20. IP₃-dependent activation of PKC. (A) Phospholipase C (PLC) is a key enzyme in the generation of IP_3. There are different PLCs, all of which share a bipartite catalytic domain (brown) and one or more PH domains (green) which associate the protein to the membrane. In addition, some members of the family contain SH2 (pink) and SH3 (yellow) adaptor domains, which allow interaction with various other proteins. **(B)** Activation of PLCs (1), for example, by seven transmembrane domain receptors (*right*) or, in the case of PLCγ, by RTKs (left), leads to the hydrolysis of PIP_2 to generate IP_3 and DAG (2). DAG remains associated with the membrane while IP_3 diffuses through the cytoplasm in search of its receptor, which is usually located in the ER membrane. **(C)** Binding of IP_3 to the ER membrane (3) activates receptor-linked channels that release Ca^{2+} from stores within the ER (4). Ca^{2+} has several effects on the physiology of the cell and gene expression. One of its actions is to cooperate with DAG, which has remained near the membrane, to activate the membrane-associated enzyme protein kinase C (PKC) which also remains associated with the inner side of the cell surface (5). PKC activation is linked to a wide variety of cellular responses.

intracellular membrane network, in vesicles which accumulate Ca^{2+}. The receptor is coupled to a channel and, upon IP_3 binding, releases Ca^{2+} into the cytoplasm. Ca^{2+} release has a variety of effects. One is, in cooperation with DAG, to activate protein kinase C, a Ca^{2+}-dependent protein kinase that has a large number of physiological substrates. There are at least eight different isoforms of protein kinase C, linked to a plethora of different signalling pathways. These pathways participate in eliciting the waves of Ca^{2+} that accompany fertilization and other effects during early embryogenesis (see Fig. 4.9), a process that appears to be elicited in response to activation of some key elements of the pathways by nitrosylation by NO. The release of Ca^{2+} also affects the activity of certain transcription factors.

In a second pathway—more recently discovered—the lipid second messengers are membrane-bound inositol phospholipids produced by the action of the PI3K enzymes. There are several kinds of PI3Ks, which are generally composed of two subunits (Fig. 5.21); in the best characterized PI3K, the two subunits, p85 and p110, form an active heterodimer at the membrane (Fig. 5.21). The products of PI3Ks (Fig. 5.19) are not substrates for PLCs, and are instead involved in the activation of distinct signalling pathways (Fig. 5.21). Within the last few years, PI3K signalling pathways have been implicated in a variety of growth factor responses, as well as metabolic changes, cell survival, cytoskeletal rearrangements, and cell adhesion (see Chapters 6 and 7).

Many of the products of PI3Ks recruit target proteins to the membrane by binding adaptor motifs such as PH and FYVE domains. These target proteins are activated either directly by binding the phosphatidylinositol phosphate or by interaction with other membrane-associated proteins. In other cases they act by recruiting the exchange factors (GEFs) of some monomeric G proteins such as those of the Rac protein.

These signalling systems also contain negative regulators in the form of phosphatases that participate in the metabolism of phospholipids. One enzyme that has been well characterized is the phosphatase PTEN, which catalyses the transformation of phophatidylinositol 3,4,5-trisphosphate ($PI(3,4,5)P_3$) into $PI(4,5)P_2$ and thus down-regulates the signalling mediated by $PI(3,4,5)P_3$. In animals mutant for PTEN, some PI3K-dependent processes are deregulated.

Hedgehog and Wnt signalling

Many of the pathways that we have outlined above were initially delineated from experiments on cultured cells, exploring the responses of cells to various pathological and physiological conditions. Later their elements were found to play a role in developmental processes. Other pathways, in particular those involving the hedgehog and Wnt families of signalling molecules introduced in Chapter 4, have

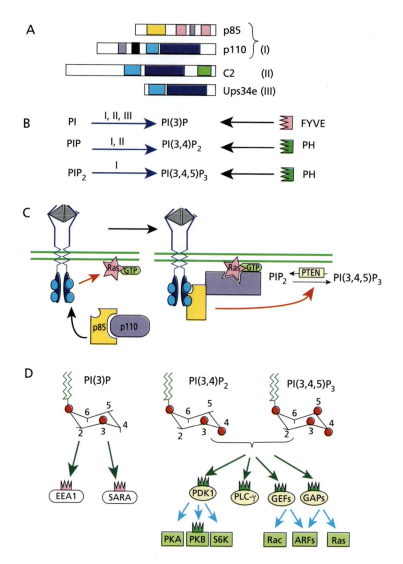

Fig. 5.21. PI3 kinases. (A) There are three classes of PI3 kinases (PI3Ks) all of which share a kinase domain (dark blue) and a PI3K-specific conserved domain (light blue) of unknown function. The first class (I) comprises the p110 protein, which acts as a heterodimer with p85, a protein that has no enzymatic activity but which associates physically with p110 through a specific domain (purple in both proteins) and regulates its activity. p85 contains SH2 (pink) and SH3 (yellow) domains that allow the complex to interact with a variety of proteins in a regulated manner. In addition to the kinase and p85 interaction domains, p110 contains a region (black) that allows it to interact with Ras. The second class of PI3K (II) contains a C2 domain (green) at the carboxy terminus which allows it to associate with the membrane. It is not clear whether these proteins act through adaptors, or whether their membrane association allows them to interact directly with their substrates. The third class of PI3Ks (III), exemplified by Ups34e, are very short proteins with specialized functions. **(B)** Each class of PI3Ks generates specific 3'-phosphorylated phosphatidylinositides as shown. In turn, each of these phospholipids can interact with specific adaptor motifs: PI(3)P interacts with the FYVE motif, while PI(3,4)P_2 and PI(3,4,5)P_3 interact with specific PH domains. **(C)** Example of the mode of action of a class I PI3K during activation of an RTK. Activation of the receptor creates docking sites for the SH2 domains of the p85 subunit which recruits the complex to the membrane, where the p110 subunit can interact with Ras. This leads to the activation of the kinase, which phosphorylates the phosphatidylinositide and creates the effector phospholipid, PI(3,4,5)P_3 in this case. A phosphatase, PTEN, antagonizes the effects of PI3K by transforming PI(3,4,5)P_3 into PI(4,5)P_2, thus regulating the activity of this phospholipid. **(D)** Examples of targets of various 3'-phosphoinositides. Through this wide variety of targets, the activation of PI3K can be used to modulate a number of signalling pathways. EEA1, early endosomal antigen 1; SARA, Smad anchor for receptor activation; PDK1, phosphatidylinositide dependent kinase 1; PLCγ, phospholipase Cγ; GEF, guanine nucleotide exchange factor; GAP, GTPase activating protein; PKA, protein kinase A; PKB, protein kinase B; S6K, S6 kinase; ARF, small GTPase involved in vesicular trafficking.

emerged from the genetic analysis of development and have subsequently been shown to function in basic cell biological processes. These pathways play important roles in the assignment and spatial organization of cell fates and although the details of their organization are still being elucidated, they have already revealed some characteristics that mark them out as different from the sorts of pathways we have described so far. Although the elements of these pathways can be arranged in linear arrays through genetic analysis, this linearity breaks down at the molecular level. Instead, these pathways appear to act as networks that link and coordinate apparently unrelated molecules with a variety of cellular activities.

The Wnt and hedgehog pathways were first uncovered through genetic analysis of development in *Drosophila* and much, but not all, of the understanding of their function has come from these studies. However, the strong evolutionary conservation of all the elements of the pathways suggests that the lessons learned from *Drosophila* are generally applicable. The experimental evidence that is available so far supports this contention.

Hedgehog signalling

The hedgehog (hh) proteins are members of a large family of signalling proteins (see Chapter 4, Fig. 4.14) found both in vertebrates and invertebrates (although, interestingly, not so far in *C. elegans*). The key element of hh signalling is the modulation of the activity of members of the ci/Gli family of zinc-finger-containing transcription factors (Fig. 5.22).

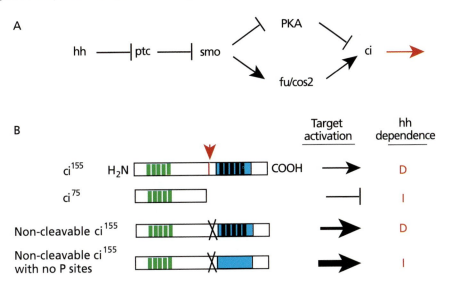

Fig. 5.22. Elements of hedgehog signalling. (A) The elements of the hedgehog (hh) signalling pathway are the receptors patched (ptc) and smoothened (smo), the proteins costal2 (cos2) and Suppressor of fused (Su(fu)), the kinase fused (fu), protein kinase A (PKA), and the C_2H_2 zinc-finger transcription factor ci/Gli. Genetic analysis of gain- and loss-of-function mutations in genes involved in hh signalling has enabled the establishment of a number of functional relationships which can be assembled into a pathway as shown. The target of these interactions is the activity of ci, which acts as the effector of hh signalling. **(B)** In *Drosophila*, ci can be found in two different forms: a 155 kDa form (ci^{155}) and a shorter 75 kDa form (ci^{75}) derived from the 155 kDa form by proteolysis at a precise site (red arrow). The zinc fingers (green lines) are located in the amino-terminal half of the protein and therefore are present in both forms. The carboxy-terminal half, which is absent from the 75 kDa form, contains a transactivation domain (blue), a site of interaction with the transcriptional coactivator protein CBP (see Chapter 3) and five putative sites for phosphorylation by PKA (dark bars). ci^{155} can mediate transcriptional activation, whereas ci^{75} mediates transcriptional repression constitutively, that is, in a way that is independent (I) of hh. The effect of hh signalling on the activity of ci has been elucidated by ectopically expressing various forms of ci as shown here. The thickness of the arrows indicates the strength of transcriptional activation. Overexpression of ci^{155} can activate various targets but in a way that is still dependent (D) on the presence of hh. The same is true of a form of ci^{155} that cannot be cleaved (cleavage site destroyed). This protein, however, has a higher activity than that the wild type. If the PKA phosphorylation sites are removed, the activity of ci^{155} increases (not shown) and if the cleavage site is mutated as well, the activity of this protein also becomes independent of hh. These experiments suggest that hh regulates the cleavage and the phosphorylation of ci, and that these modifications play a role in its transcriptional activation function.

The hh protein is secreted. After activation by proteolysis, its amino-terminal moiety is modified by cholesterol. Although tightly associated with the cell surface, the protein is nevertheless capable of diffusing through the extracellular space where it binds to the cell surface via the receptor patched (ptc). In the absence of hh, patched destabilizes the cell-surface association of the seven transmembrane domain protein smoothened (smo). Binding of hh to ptc stabilizes the association of smo with the cell surface, resulting in a signal. At present it is not clear whether there is also a ligand for smo, but it is clear that the release of smo from ptc regulation leads to the activation of members of the ci/Gli family, which are the effectors of the hh signal (Fig. 5.22 and 5.23).

A great deal of the work on the activity of ci and its regulation has been done in *Drosophila* and the conclusions derived from these experiments have been shown to hold for the vertebrate homologues too. There are two different forms of ci: a 155 kDa form (ci^{155}) that is a transcriptional activator and a 75 kDa form, (ci^{75}) derived from ci^{155} by proteolysis, that is transcriptional repressor. The current view of hh signalling is that in the absence of hh, regulated proteolysis of ci^{155} leads to the generation of the ci^{75} form, which represses some hh target genes. This process is associated with the phosphorylation of ci^{155} in its carboxy-terminal domain by PKA, an event that appears to have two effects: to promote the cleavage of ci^{155} into ci^{75}, and to inactivate ci^{155}. Hedgehog signalling inhibits the proteolysis and phosphorylation of ci^{155} and so promotes ci^{155} activity (Fig. 5.22). Experiments in *Drosophila* indicate that hh signalling can lead to an increase in the expression of target genes without changing the detectable concentration of ci^{155}, which can be observed mostly in the cytoplasm. This is a situation similar to that of the Nintra fragment during Delta–Notch signalling (see Fig. 5.10); it suggests that ci is active at very low concentrations.

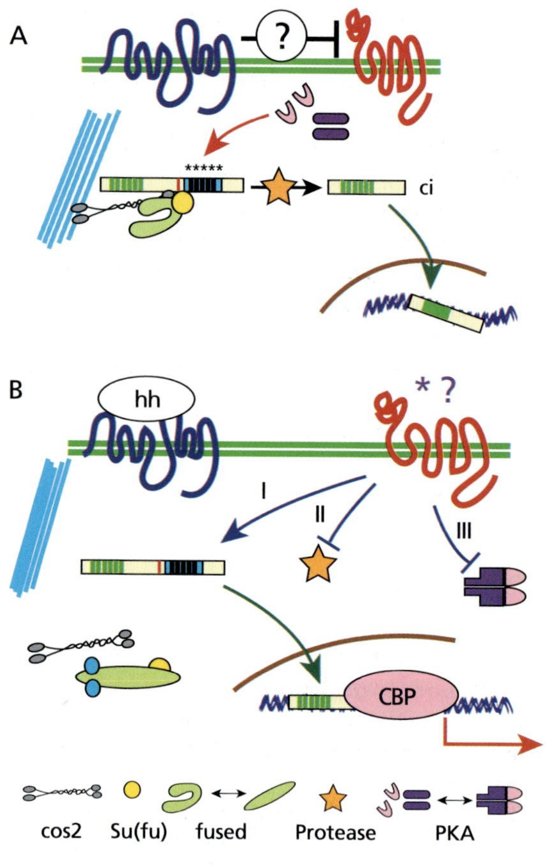

Fig. 5.23. **Hedgehog signalling.** Symbols for elements of the cascade are shown in the key in part B. **(A)** In the absence of hedgehog signalling, patched (dark blue) prevents the accumulation of smoothened (red) at the cell surface by an, as yet, unknown mechanism. In the cytoplasm, ci^{155} is bound by the cos2/fused/Su(fu) complex which is associated with microtubules (blue lines). Within this complex, ci^{155} is phosphorylated in its carboxy-terminal domain (asterisks) by PKA and cleaved into the ci^{75} form which enters the nucleus and represses the expression of some targets. The nature of the protease that catalyses this cleavage remains unknown. **(B)** Binding of hedgehog to patched blocks its effects on smoothened and allows smoothened to accumulate at the cell surface where it becomes active (asterisk), apparently without a ligand. The activation of smoothened leads to three outcomes: (I) the phosphorylation and release of the cos2/fused/Su(fu) complex from the microtubules, (II) the inactivation of the ci protease, thus decreasing the amount of ci^{75}, and (III) inactivation of PKA. As a result of these activities ci^{155} enters the nucleus and interacts with CBP to promote the transcription of hh targets. **(C)** Requirements of events I–III for the different elements of the pathway. Hedgehog is absolutely necessary for the events described in B but some events show differential dependence on the fu kinase and on PKA, suggesting that hh signalling activates a network of events rather than a linear pathway.

The balance between ci[155] and ci[75] is controlled in a complex way by an interplay between hh, PKA and a complex of proteins that includes a kinase encoded by the *fused* (*fu*) gene, a microtubule-associated protein encoded by the *costal 2* (*cos2*) gene, and a small cytoplasmic protein encoded by *Suppressor of fused* (*Su(fu)*) (Fig. 5.23).

The current model of hh signalling can account for many of the results of genetic experiments investigating interactions between the different genes, but still leaves detailed questions about mechanisms and the role of individual components unanswered. In particular, the relationship between patched and smoothened, the role and substrates of PKA, and the precise activity of the fused kinase remain to be elucidated.

Wnt signalling

Signalling by Wnt proteins also involves modulation of the activity of specific transcription factor effectors through the modulation of multiprotein complexes (Fig. 5.24). Also like hedgehog, Wnt proteins acting at the cell surface can trigger a variety of different responses. However, in contrast to hedgehog which appears to trigger these effects through its interaction with one receptor, patched, and through a chain in which most of the components are devoted to hh signalling, the elements that mediate Wnt signalling are connected with various proteins and processes that diversify the response to particular Wnt proteins (Fig. 5.24).

Wnt triggers different responses through interactions with different kinds of receptors (Fig. 5.24 and 5.25). Interactions with complexes between members of the frizzled family of seven transmembrane domain cell surface receptors and the LDL receptor-related protein (LRP) arrow have been shown to mediate many of the signalling activities of Wnt proteins. In *Drosophila*, the Wnt family member wingless has been shown to bind to the Notch receptor and modulate its activity. Many organisms contain several Wnt proteins and frizzled receptors. For example, *Drosophila* has seven Wnt proteins and at least four members of the frizzled family, whereas vertebrates have over ten members of each kind. It is therefore likely that functional specificity arises from the matching of particular Wnt proteins to particular frizzled receptors. Interactions between Wnt proteins and other cell surface receptors probably contribute to the specificity of these interactions.

Genetic analysis in *Drosophila* identified a cytoplasmic pool of armadillo, a *Drosophila* homologue of the cytoskeleton-associated protein β-catenin, as the major effector of Wnt signalling. This relationship was established largely through experiments with loss- and gain-of-function mutations. As in the case of hedgehog, the generality of this model has been shown by experiments in vertebrates, which also point to β-catenin as the crucial element in Wnt signalling. Work in vertebrates, however, has identified a number of Wnts which can signal through a β-catenin-independent pathway that modulates Ca²⁺-dependent intracellular signalling pathways and activates PKC (Fig. 5.25). This pathway also requires the frizzled receptors but perhaps not arrow.

There are two pools of β-catenin: a stable membrane-associated pool involved in cell adhesion through its interactions with specialized cell-surface proteins called cadherins (see Chapter 6, Fig. 6.16) and an unstable, cytoplasmic pool. In the absence of Wnt signalling the cytoplasmic pool of β-catenin is efficiently degraded by the activity of a large complex assembled on a scaffolding protein (axin) in which a key element is the serine/threonine kinase glycogen synthase kinase 3 (GSK3). GSK3 phosphorylates specific residues in the amino terminus of β-catenin that promote its degradation by the proteasome, and thus antagonizes Wnt signalling. Upon binding of Wnt to cell-surface receptors (Fig. 5.25), a cytoplasmic adaptor protein, dishevelled, inactivates the axin-based complex and leads to the accumulation of β-catenin in the cytoplasm. As a result of this event, β-catenin enters the nucleus where it promotes transcription by interacting with members of the Tcf family of HMG box containing proteins.

In the absence of Wnt signalling, Tcfs repress transcription by associating with corepressors such as the groucho protein. The activation of β-catenin disrupts this association and turns Tcf into a transcriptional activator. It is possible that the surge of β-catenin in the nucleus competes with groucho for binding to Tcf, but it is also possible that Wnt signalling destabilizes the groucho–Tcf interaction in a way that is independent of the interaction between Tcf and β-catenin. The net effect of either process would be the formation of the complex between Tcf and β-catenin that promotes transcription.

β-catenin also interacts physically and functionally with other nuclear proteins in addition to Tcf. Some of these (e.g. members of the Sox family of HMG box containing proteins), are related to Tcf while others, such the zinc-finger protein teashirt, or members of the retinoic acid receptor family, are involved in apparently unrelated processes (Fig. 5.24 and 5.25).

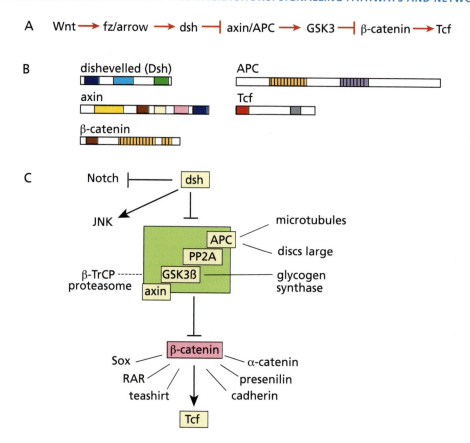

Fig. 5.24. Elements of Wnt signalling. (A) The elements of the Wnt signalling pathway are members of the frizzled (fz) family of seven transmembrane domain receptors, the arrow/LRP receptors, the dishevelled (dsh) protein, axin, the adenomatous polyposis coli (APC) protein, glycogen synthase kinase 3 (GSK3), β-catenin (armadillo in *Drosophila*) and the HMG box-containing T-cell-specific factor (Tcf). Genetic analysis of gain- and loss-of-function mutations in genes involved in hh signalling has enabled the establishment of a number of functional relationships which can be assembled into a pathway as shown. Wnt signalling modulates the interaction between β-catenin and Tcf, promoting transcription of target genes. **(B)** Structures of some of the elements involved in Wnt signalling. These proteins all contain motifs that mediate their interactions with other proteins and cellular activities. Dishevelled contains an amino-terminal domain (DIX) (dark blue) shared with axin, a central DHR domain (light blue) and a carboxy-terminal domain involved in the activation of JNK (DEP domain, green). Axin is a scaffolding protein which interacts with and holds together many of the proteins involved in the transduction of Wnt signalling: within the amino terminus there is a domain involved in the interaction with APC (yellow); two adjacent domains are involved in interactions with GSK3 (brown) and β-catenin (pale yellow), then there is a domain which mediates an interaction with the phosphatase PP2A and dishevelled (pink) and a DIX domain (dark blue). β-catenin contains an amino-terminal domain (brown) through which it interacts with α-catenin, and an intermediate region with several repeats known as armadillo repeats (light brown) which mediate different protein–protein interactions. APC contains several interactive domains, only two of which are highlighted here: armadillo repeats (light brown) and a tract of characteristic repeats which bind β-catenin (purple). Tcf contains an amino-terminal domain (red) involved in its interaction with β-catenin and a HMG box (grey) at the carboxy terminus. **(C)** Proteins involved in Wnt signalling are associated with a variety of different functions. Arrows indicate activating interactions whereas blunt-ended lines indicate repressive ones. Connecting lines indicate physical interactions. The green box represents the β-TrCP proteasome (see Fig. 5.25). It is not clear if the multiple interactions of some of the proteins (e.g. APC or β-catenin) reflect interactions of pools of these molecules that are in equilibrium inside the cell, or interactions of different pools that are compartmentalized.

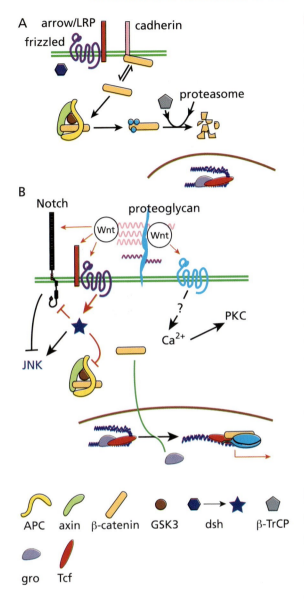

Fig. 5.25. Wnt signalling. Genetic analysis suggests that a central element of Wnt signalling is the modulation of the state of β-catenin and its associations with other molecules. The elements of the pathway are indicated in the key at the foot of the Figure. **(A)** In the cell there are two pools of β-catenin that are in equilibrium: a cytoplasmic pool and a pool that is associated with the membrane through binding to the intracellular domain of cadherins. In the absence of Wnt, cytoplasmic β-catenin associates with a complex that includes axin, APC, and GSK3. Within this complex GSK3 phosphorylates specific residues at the amino terminus of β-catenin, promoting its association with β-TrCP, which targets it for degradation. In the nucleus, complexes containing Tcf and the groucho (gro) co-repressor repress transcription of Wnt target genes. **(B)** Wnt proteins interact at the cell surface with a number of receptors (members of the frizzled family, LRP/arrow, Notch) and cell-surface proteoglycans. The interaction of some Wnts with a complex of frizzled and arrow leads, through the activation of dishevelled, to the inactivation of GSK3 in the axin/APC complex and the stabilization and nuclear localization of β-catenin, which interacts with Tcf and promotes transcriptional activation of specific targets. It is likely that when it enters the nucleus, β-catenin displaces Tcf from its association with the co-repressor groucho. In the nucleus β-catenin also interacts with other proteins (see Fig. 5.24). Some Wnts interact with some frizzled receptors to trigger a β-catenin-independent signalling event that leads to a rise in intracellular Ca^{2+} and activation of PKC. Dishevelled, in addition to destabilizing the axin/APC complex, can also activate JNK and this interaction might be significant for Wnt signalling. Both the Wnt family member wingless and dishevelled have been shown to bind the Notch receptor and modulate a CSL-independent activity of this receptor which down-regulates JNK activity.

Dynamics and regulation of signal transduction during cell interactions

Continuous signalling does not provide information. For a signal to have meaning it must be turned on from the off state. It follows that the interpretation of a signal during cell interactions must rely on the ability of a cell to read transitions between the on and the off states. This in turn suggests that there must be mechanisms that down-regulate and modulate the intensity, frequency, and duration of signalling once it has been initiated.

Indeed, in some cases, different intensities of signalling can be interpreted differently by the cell. For example, stimulation of PC12 cells with EGF causes transient stimulation of the Ras pathway, and cellular proliferation results. On the other hand, stimulation of the same cells with NGF results in prolonged stimulation of MAPK signalling, leading to neuronal differentiation without proliferation

The interactions of Wnt signalling with other pathways are not restricted to the multiple interactions of β-catenin in the cytoplasm and the nucleus. Wnt and dishevelled have been shown to regulate the activity of JNK in all organisms, and this raises the possibility that in addition to the β-catenin-dependent and independent pathways, there might be a third mode of Wnt signalling from the cell surface (Fig. 5.25). Whether all these pathways respond to a single signalling event or, if not, how the cell manages to decide which pathway to deploy, remains to be clarified.

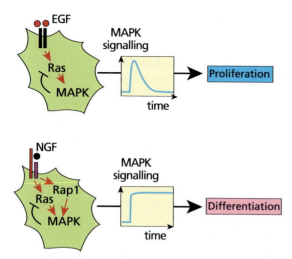

Fig. 5.26. Different intensities of signalling can lead to different cellular responses. Treatment of PC12 cells with either epidermal growth factor (EGF) or nerve growth factor (NGF) results in activation of the Ras pathway, but with very different kinetics and physiological outcomes. EGF causes the cells to proliferate and this is associated with a transient activation of MAPK (indicated digrammatically in the graph). In contrast, treatment with NGF induces the cells to differentiate into neurons, an effect that is associated with a persistent activation of MAPK. The difference between the two situations appears to be that NGF, but not EGF, can activate the Rap1 G protein, bypassing some of the feedback regulation of Ras and resulting in a sustained activation of MAPK.

(Fig. 5.26). In another example, different frequencies of 'global' calcium waves, set up by the coordination of local calcium signalling events at the plasma or endoplasmic reticulum membranes, have different effects. Waves of calcium signalling across the fertilized egg, oscillating with a period of 1–35 minutes, activate the developmental program of the zygote, whereas much slower calcium oscillations, with a period of 10–20 hours, trigger cell division.

Unfortunately, very little is known about the real dynamics of signalling *in vivo*, the exception being signalling by ions—initially Na^+ and K^+ in the conduction of nerve impulses and, more recently, the many physiological functions of Ca^{2+}. However, it is likely that with the development of new tools to follow signalling pathways in real time our knowledge of these important parameters will increase rapidly. In the meantime, more static analysis of signalling pathways has revealed a number of ways in which the flow of information can be regulated. Three of these in particular stand out as likely to be operative in all signalling systems. They involve regulation of the activity,

concentration, and localization of key components of signalling pathways.

Changes in activity of signalling pathway components

A sensitive way of regulating the flow of information through a pathway is by regulated changes in the activity of its components. This can be achieved by post-translational modification or by interaction with regulatory proteins. We have already seen some of the most important post-translational modifications of signalling proteins, in particular phosphorylation and dephosphorylation. The importance of these changes in the regulation of signalling can be seen, for example, in the control of the activity of small GTPases and its regulation by exchange factors (see Fig. 5.6). The alternation between GTP- and GDP-bound states is a source of information. In the case of Rac, Cdc42, and Rho, which form a parallel functional array, it is not just the state of the factors, but the frequency of the alternation between states, that is important.

Phosphorylation and dephosphorylation are obviously important regulatory steps in the many signalling pathways involving kinases and phosphatases. Often, one of the outcomes of a signalling event involving a kinase is the recruitment, synthesis or activation of a phosphatase that targets and down-regulates a key component of the signalling pathway and thus provides an important feedback control (Fig. 5.27). This is very common in pathways triggered by the activity of RTK and Ras, where the receptor itself may recruit a phosphatase that regulates the rate of phosphorylation of specific residues. In the MAPK pathway, there are MAPK phosphatases that are activated or synthesized in response to the activity of the pathway and provide a finely tuned control of the signalling event (Fig. 5.27).

This type of inbuilt negative feedback also operates at the level of the cell surface. In many cases, the signalling event triggers the expression of a molecule that dampens signalling at the source, either by quenching the signal or by competing with it for binding to the receptor. An example of quenching can be found in hedgehog signalling, where the signalling event results in the expression of the non-signalling receptor element patched, which will bind the ligand and so reduce the extent of signalling (Fig. 5.28).

Changes in the concentration or stability of signalling pathway components

Many of the individual reactions in a signalling chain are equilibrium reactions and therefore they can be pro-

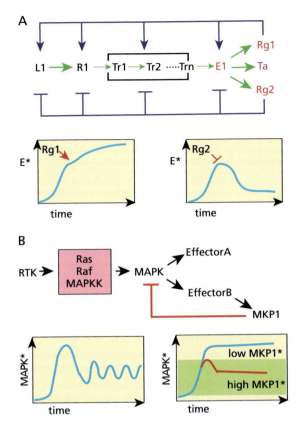

Fig. 5.27. Changes in the activity of the components of signalling pathways. (A) The effector (E) of a signal transduction pathway can activate target molecules (Ta) that will implement the function induced by the signalling molecule (L), and regulators (Rg) that can act, positively or negatively, at various points in the transduction chain. The graphs show the effects of a positive (Rg1) or a negative (Rg2) regulator on the activity of an effector (E*) over time. **(B)** Negative regulation of the activity of a MAPK protein by a MAPK-specific protein phosphatase (MKP1) induced by one of the effectors of the pathway. Signalling through MAPK cascades leads to the expression of proteins, such as MKP1, that provide a feedback control on the signalling. The graphs at the bottom illustrate the effects that the activity of MKP1 will theoretically have on the activity of MAPK. *Bottom left panel*: Under normal conditions, the activity of the phosphatase will dampen the activity of MAPK (MAPK*) and create fluctuations that maintain it within a narrow band of activity. *Bottom right panel*: Small deviations in these fluctuations might have consequences for the activity of MAPK in different cells: two effectors could be activated at different thresholds of MAPK activity and therefore the activity of the kinase will determine which target or targets are expressed. If MKP1 is very effective, only targets within the green band will be activated; if the activity of MKP1 is low, targets which require high MAPK activity will respond.

foundly affected by the concentrations of the molecules involved. We have already seen that many effectors (e.g. Nintra, ci, and β-catenin) can be present at such tiny concentrations that they are difficult to detect in their active location. When crucial effectors are present in the active state at limiting concentrations, the system can respond very sensitively, in a sigmoidal or switch-like fashion, to small changes in either their activity or their concentration.

This exquisite sensitivity means that there is a possibility that activation could occur spontaneously, and so controls are needed to take account of this. For example, the mechanism of RTK activation—ligand-induced dimerization and autophosphorylation—depends crucially on the concentration of receptor. Increasing the receptor concentration above a certain threshold could trigger dimerization and signalling in the absence of ligand. For this reason, the concentration of the receptor must be integrated as part of the signalling event, so that it is regulated simultaneously with that of its ligands. This is why a common response to a signalling event is often down-regulation of the receptor

itself—not just of the number of molecules at the cell surface by ligand-induced internalization—but also down-regulation of the transcription of the gene encoding the receptor.

In other cases, signalling can require the assembly of a complex with a certain stoichiometry: too much ligand can have a dominant negative effect on signalling by titrating away one of the components. For example, in the Notch signalling system the ligands can pass from an activation mode to a dominant negative mode over a phsyiological range of concentrations and this is used, *in vivo*, to modulate the signalling event. In a similar vein, the stoichiometry between NFκB and IκB is essential for the output of signalling through the IL1 or TNFα receptors (Fig. 5.11).

A very common way to regulate the concentration of a cytoplasmic or membrane-associated signalling protein is to target it for degradation by a complex of proteases, the proteasome (Fig. 5.29). This process is regulated and it is becoming clear that it plays an important role in development. Proteins are targeted for regulated degradation by the addition of a series of ubiquitin residues through a tightly regulated process. First ubiquitin is prepared for addition to the target protein by a ubiquitin-activating enzyme E1, and this is followed by the addition of

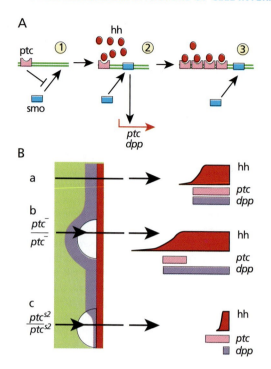

Fig. 5.28. **Receptor concentration as a way of regulating signalling intensity.** In some cases, signalling results in an increase, at the cell surface, in the concentration of a receptor that can reduce the concentration of effective ligand. When the signalling event involves a spatially fixed source of signal, this can have the effect of reducing the overall spatial range of the signalling event. **(A)** Binding of hedgehog (hh) to its receptor patched (ptc): (1) allows accumulation of the smoothened (smo) receptor at the cell surface where it signals (2) and promotes the expression of hh target genes, such as *dpp* and *patched* itself. This results in an increased concentration of patched at the cell surface (3) which will bind hedgehog and reduce its effective concentration. **(B)** When hh acts from a fixed source over a wide range, an increased concentration of patched at the cell surface can affect the range of hh signalling. *Left*: The Figure shows, diagrammatically, sections through a region of the wing primordium of *Drosophila*, with a source of hh (red) and the response to it indicated as *dpp* expression (purple). *Right*: The corresponding expression profiles of the hh protein, and the *ptc* and *dpp* genes. (a) In a wild-type primordium, hh is produced from a fixed source and diffuses anteriorly into cells that do not express hh, creating a short-range concentration gradient that induces the expression of *dpp* and *ptc* over a certain spatial domain (see Chapter 9, Figs. 9.23, 9.24 for details). (b) By use of a genetic mosaic (see details in Chapter 4, Fig. 4.6), the expression of *ptc* can be eliminated from a group of cells (white area) just anterior to those that express hh. In this situation, the expression of *dpp* extends further than in the wild type, to a point just anterior to and around the group of cells that cease expressing the ptc protein. Expression of *ptc* can also be observed in cells that have the *ptc* gene and are anterior to those that have lost *ptc*. This extension can be explained by the absence of patched at the surface of the cells adjacent to the source of hh, which can now reach further because it is not trapped by patched (i.e. the absence of ptc extends the range of diffusion of hh). **(C)** A mutation in the *ptc* gene, *ptc*[s2], produces a patched protein that has an increased ability to interact with hh and is therefore more effective in removing hh. This reduces the range of action of hh. This can be shown by generating a group of cells that express the ptc[s2] protein (white), resulting in a reduction of the spatial extent of the response to hh as measured by the expression of *dpp*. (Adapted from Chen, Y. and Struhl G. (1996) Dual roles for patched in sequestering and transducing Hedgehog. *Cell* **87**, 553–63.)

ubiquitin to the protein through the combined activity of a ubiquitin-conjugating enzyme (E2) and a ubiquitin ligase (E3), which create a covalent bond between ubiquitin and a chosen lysine residue in the target protein. Addition of other ubiquitin monomers to form a polyubiquitin tail marks out the protein for degradation.

The E2/E3 ubiquitin ligase complexes are multiprotein complexes that are present in low abundance and that provide the rate-limiting step in the ubiquitination process (Fig. 5.29). A key component of these complexes is a protein that has a 'ring finger' motif. This motif was for some time thought to be a transcription factor motif but has recently been shown to be characteristic of E3 ubiquitin ligases, mediating the interaction between the ligase and the E2 conjugating enzyme. Another important element in E2/E3 ubiquitin ligase complexes are F-box-containing proteins, which help to present the proteins selected for degradation to the E2 enzymes, and which function in a phosphorylation-dependent manner (Fig. 5.29). The founding members of the F-box family were identified in *C. elegans* and yeast, where mutations in the genes encoding these genes were found to have profound developmental effects.

Large numbers of F-box-containing proteins are now known (e.g. more than 60 in *C. elegans*); in different members of the family, the F-box is associated with specific mo-

tifs that allow specific interactions with different target proteins. As a result of mutations in F-box proteins, specific target proteins are not degraded and so their activity increases. For example, mutations in the F-box protein Slimb/β-TrCP result in an increase in the active form of β-catenin and increased Wnt signalling. Similarly, in *C. elegans* mutations in the sel-10 protein, a homologue of yeast Cdc4, increase the efficiency of the Notch protein lin-12, by increasing its concentration in the cell.

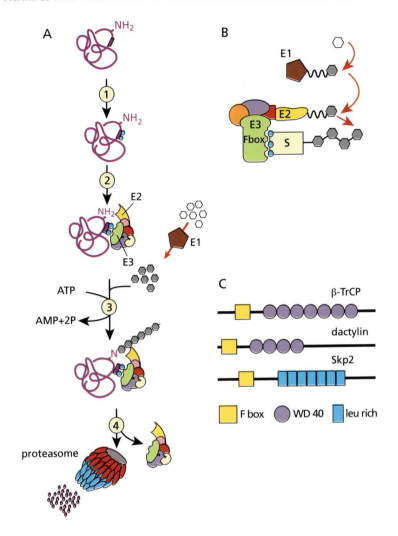

Fig. 5.29. Regulation by changes in the concentration of components: ubiquitin-dependent protein degradation. An efficient way of regulating the activity of a signalling protein is to regulate its degradation. The ubiquitin-dependent proteasome pathway provides a mechanism to do this. **(A)** The protein that is to be degraded is targeted first by phosphorylation (blue spots) of particular residues (1), triggering its recognition by a component of a large protein complex with two catalytic elements: a ubiquitin-conjugating enzyme (E2), and a ubiquitin ligase (E3) (the E2/E3 ubiquitin ligase complex) (2). The assembled complex can then catalyse the ATP-dependent addition of a ubiquitin tail (grey hexagons) to the amino groups of specific lysine residues (3). Ubiquitin has previously been activated by the activity of the ubiquitin-activating enzyme

(E1). Marked by a tail of ubiquitin, the protein is recognized by the proteasome and degraded (4). **(B)** The E3 moiety of the ubiquitination complex is made up of three subunits which ensure that the ubiquitin tail is added to the substrate(s). An important component is the F box protein component, which links the substrate to the complex. **(C)** There are many different F box proteins. All of them share a conserved motif of about 45 amino acids near their amino terminus, linked to a variety of motifs that often include WD40 motifs and leucine-rich repeats, through which they interact with and thereby recruit the substrates to be degraded. Some examples of F-box-containing proteins are β-TrCP, which is involved in the degradation of β-catenin, dactylin, and Skp2, which is involved in the degradation of the cell-cycle regulator p27 (see Chapter 7).

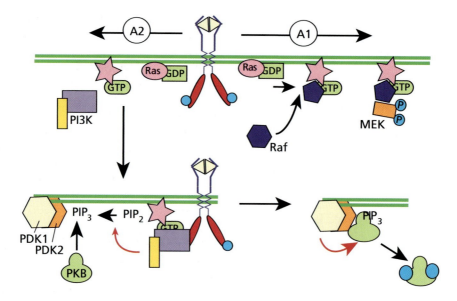

Compartmentalization

A third general way to regulate signalling is to link the activity of specific pathway components to their localization to specific compartments or structures within the cell. For example, as described earlier in this chapter, translocation to the membrane is an essential element of Ras/Raf signalling and for many of the proteins that have PH domains. The recruitment allows the assembly of catalytic complexes in which individual elements are activated and so become competent to act on their substrates (Fig. 5.30).

The nucleus is a central target of many signalling pathways, and the transport of transducers, effectors, and regulators into and out of the nucleus is potentially an important regulatory step (Fig. 5.31). Some cytoplasmic effectors are held in specific compartments where they receive the input that activates them and allows their entry into the nucleus. In other instances, as in the case of PKA-regulated genes, a pathway component, in this case a pool of PKA itself, is positioned near the nucleus to facilitate its entry. This is achieved through its association with a protein, A kinase anchoring protein (AKAP), which binds to the regulatory subunit and tethers the complex to the nuclear membrane. Signalling then allows rapid access of the catalytic subunit to its target, CREB.

AKAP is a member of a family of proteins that interact with protein kinases and target them to different cellular compartments (Fig. 5.32). This type of compartmentaliza-

Fig. 5.30. Summary of compartmentalization. There are three intracellular locations that are important for signalling: the membrane, the nucleus and the cytoplasm. Recruitment of signalling components to the membrane is a common part of many signalling mechanisms. The recruitment is usually nucleated by one or more components that are associated with the inner plasma membrane and that become activated upon signalling. In the case of receptor tyrosine kinases, the receptors are already at the membrane and their activation can recruit other proteins to the membrane in a variety of ways. For example (A1), Ras (and related proteins) are associated with the membrane by a chain of myristic acid at their amino terminus. Activation of an RTK leads to the activation of Ras (pink star) by recruitment to the membrane of a G protein exchange factor (not shown but see Fig 5.15 for details of this step). The activation of Ras results in the membrane localization of Raf (blue), which then becomes activated by interactions with Ras and triggers the MAPK cascade. If the Raf protein is engineered so as to make it associate spontaneously with the membrane, the requirement for Ras is bypassed. In other cases (A2), an intracellular protein is attracted to the membrane by a specific anchor (e.g. a phospholipid). In the example shown, Ras acts in concert with specific docking sites in some RTKs to recruit the p110/p85 PI3K to the membrane where it generates PI(3,4,5)P$_3$ (PIP$_3$). PIP$_3$ attracts proteins, such as protein kinase B (PKB), that contain PH or FYVE domains, enabling them to be activated or regulated. In the case of PKB (green), interaction with PIP$_3$ and translocation to the membrane allows interaction with the protein kinases PDK1 (yellow) and PDK2 (orange), both constitutively active kinases that reside in the membrane. These interactions activate PKB, which then moves to the cytoplasm to find its substrates.

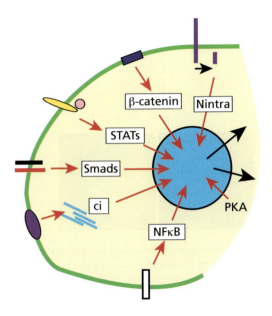

Fig. 5.31. Nuclear translocation of effectors. Nuclear translocation of effectors is an important step in many signalling pathways. This cartoon summarizes some of the nuclear translocation events associated with various signalling pathways that are shown in more detail in other figures. Transport into (red arrows) and out of (black arrows) the nucleus can be regulated in a variety of ways and contributes to the regulation of signalling pathways.

tion can create different pools of the same molecule and so allow it to be involved in different processes. Within the cell the different pools of PKA are likely to be in equilibrium and so the concentration of PKA may be a factor in its ability to signal differentially in different processes. We have seen another example of this in the two pools of β-catenin: one associated with the cell-surface cadherin proteins and involved in cell adhesion, and the other involved in Wnt signalling (Fig. 5.25). Experimentally increasing the concentration of cadherins reduces Wnt signalling (see Chapter 6, Fig. 6.20), indicating that these pools are indeed in equilibrium, and that differential compartmentalization can be used as a way of connecting different cellular functions through a common intermediate.

Integration of signalling pathways

Signalling pathways, such as those described in this chapter, do not operate as separate, parallel processors but are elements of complex networks that process information from the external environment and coordinate internal changes in the state of the cell and its nuclear activity. The constellation and concentrations of signal transduction pathway proteins that a cell contains by virtue of its developmental history define both the range of signals to which it responds, and its repertoire of responses. *C. elegans* provides a particularly striking example: the *C. elegans* EGF receptor homologue, encoded by the *let-23* gene, regulates cell fates in many cell types, but whereas in most tissues its activities are implemented by the Ras/MAPK pathway, in the germ line they are mediated by the IP_3 receptor which, together with a kinase for IP_3, are specifically expressed in the germ line (Fig. 5.33).

In general, however, a signal triggers a complex array of responses. Not only can activation of a single receptor mobilize a number of different pathways (Fig. 5.34), but a single signal may also have the effect of generating a number of secondary signals, which will increase the complexity of the response over a very short period of time. An experiment using DNA microarrays (introduced in Fig. 2.15) to probe the expression of a large number of genes illustrates many of the features of signalling networks operating in a specific cell type (Fig. 5.34). In this experiment, an array of cDNAs corresponding to about 8600 different human genes was used to follow the temporal expression pattern of these genes in a culture of human fibroblasts stimulated with serum. One striking observation was that an important part of the cells' response was the expression of signalling molecules and other factors involved in communication with other cells: each cell mounts a response not as an isolated unit, but as part of a coordinated group. A second observation was that many of the genes that were expressed are known to be involved in a specific process—wound healing—indicating that the cells' response was not a generalized one of growth and division but a specific physiological activity determined and delimited by the cell type.

The complexity of the response even in this 'simple' situation suggests that unravelling the dynamics of signalling and signalling responses during developmental processes will be a daunting task. It is important to remember that a signalling pathway is not a linear series of non-equilibrium, irreversible chemical reactions, as diagrams tend to imply. Rather, signalling pathways are non-linear (because of their high degree of interconnectivity), they are in dynamic equilibrium (which means that at certain nodal points small changes in the concentration of active components can have profound effects on the flow of

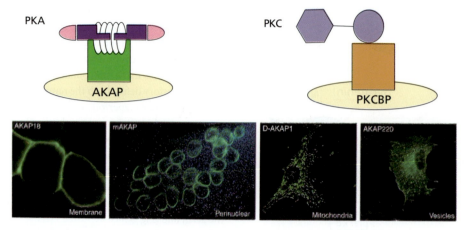

Fig. 5.32. Targeting of intracellular effectors of signalling pathways. Some intracellular mediators of cell signalling are involved in multiple biochemical interactions. For example, GSK3 is involved in glycogen metabolism and Wnt signalling, and protein kinases, such as PKA or PKC, have a very wide range of activities and interactions. One way to ensure that specific proteins are involved in specific activities in some cell types is to locate them in the intracellular compartment where they are needed, in the neighbourhood of their activators and effectors—in other words, to create local pools of these proteins. This can be achieved through the mediation of in-teracting proteins that provide intracellular 'addresses'. Examples are shown here for PKA and PKC. PKA can interact with a variety of A-kinase anchoring proteins (AKAP), which contain a helix through which they interact with the PKA regulatory subunit, and a targeting region, which locates the complex to particular cellular compartments. PKC-binding proteins (PKCBP) perform a similar role. The photographs show immunofluorescence images of PKA targeted to different compartments through specific AKAPs. (Adapted from Colledge, M. and Scott, J. (1999) AKAPs: from structure to function. *Trends Cell Biol.* **9**, 216–21.)

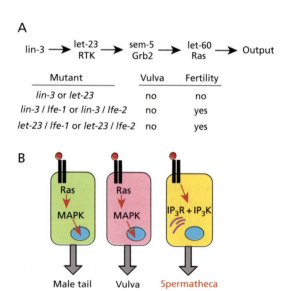

Fig. 5.33. Tissue-specific signalling through the EGF receptor in *Caenorhabditis elegans*. (A) In *C. elegans* the Ras/MAPK cascade, activated by the ligand lin-3 acting on the RTK receptor let-23, plays an important role in cell fate decisions in many tissues, including the development of the vulva (see Chapter 11). The scheme shows the *C. elegans* proteins corresponding to the components of the canonical Ras/MAPK signalling pathway. Both lin-3 and let-23, but not Ras, are also required for proper functioning of the spermatheca (the organ in the *C. elegans* hermaphrodite that contains the sperm) during egg laying: animals mutant for either lin-3 or let-23 are both infertile and vulvaless. The spermatheca is a germ-line tissue. A genetic screen was designed to identify the downstream components of lin-3–let-23 signalling required for egg laying. The screen relied on the ability to screen simultaneously for mutants in vulval development and mutants in egg laying (measured as fertility). Mutations were sought that could rescue egg-laying defects generated by loss of the let-23 RTK function, but not vulval development defects. This screen revealed two genes, *Ife-1* and *Ife-2*, mutations in which could restore fertility but not vulval development in either lin-3 or let-23 mutants. One of these genes, *Ife-1*, encodes an IP$_3$ receptor, and the other, *Ife-2*, an IP$_3$ kinase. **(B)** The cartoons represent the lin-3–let-23 signalling pathways operating in different developmental or physiological processes in *C. elegans*: male tail development, vulval development, and egg laying (spermatheca). The *Ife-1* and *Ife-2* genes are expressed exclusively in the germ line, explaining their ability to mediate RTK signalling in the spermatheca. There also appears to be a germ-line-specific SH2 adaptor that links the activated let-23 receptor to the IP$_3$ signalling pathway. The outcome of activation of the IP$_3$ receptor is the release of intracellular Ca^{2+} which can control a variety of physiological processes including the contractions of the spermatheca during ovulation.

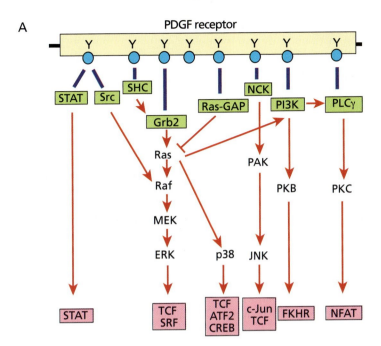

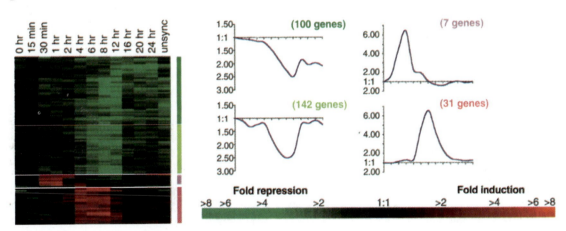

Fig. 5.34. Connectivity and diversification in the response to single signalling events. (A) The intracellular domain of the RTK receptor for PDGF contains several tyrosine (Y) residues that become phosphorylated upon ligand binding and activation of the kinase. Each of these pY residues acts as a docking site (circles) for a cognate SH2-containing protein, linking the PDGF receptor to any of several signalling pathways. The response to the activation of the receptor will depend on the set of elements available in a given cell, as well as the state of the cell and the thresholds for activation of the targets of each signalling pathway. Proteins boxed in green are associated with the membrane. Proteins boxed in pink are in the nucleus (Adapted from Pawson, T. and Saxton, T. (1999) Signalling networks—do all roads lead to the same genes? *Cell* **97**, 675–8.) **(B)** The pattern of genes expressed in human fibroblasts after stimulation with serum. Cells were made quiescent by the removal of serum for 48 hours. Serum was then added and RNA was extracted from the cells at different times after serum addition. cDNA was prepared from the RNA

sentative of 8613 genes (see Chapter 2, Fig. 2.15), and probed with the cDNA from each time point. Most of the genes showed changes in expression profile during the first 24 hours. The changing expression pattern of some of the genes is shown in the panel on the left. Each gene is represented by a horizontal strip, and the colour of the signal at each time point represents the fold repression (green) or induction (red) of the gene compared with its expression level at time zero. (The 'unsynch' column represents the expression levels in exponentially growing cells.) The genes were grouped on the basis of similar expression profiles; the groups are indicated by the narrow vertical coloured bars immediately to the right of the panel. The graphs on the right show the average time courses of levels of induction or repression for each of the four groups shown in the left panel. The graphs are colour-coded to correspond with the coloured bars representing the groups. Note that both early and delayed expression responses are seen. (After Iyer, V.R. *et al.* (1999) The transcriptional program in the response of human fibroblasts to serum. *Science* **283**,

information) and any individual step is in principle reversible (a feature that is fundamental to regulation and homeostasis). Different pathways are connected by internal 'wiring', that is, regulatory molecules which by acting on elements of more than one pathway provide links between them; this means that activating one receptor, say an RTK, will not simply set in motion one discrete pathway but will have an effect on other linked pathways.

Although analysis of the regulation and integration of signalling networks is still in its infancy, some principles are beginning to emerge that are crucial for the efficient operation of signalling systems in development. There are two basic ways in which signalling pathways can be regulated: at the level of the flow at specific steps in a pathway, and in the connectivity between different pathways operating in the same cell. There is a further tier of control in the integration of the incoming information by a given type of cell to generate a cell-type-specific response. In the remainder of this chapter we will highlight some examples of how these types of control might be achieved.

Connectivity and integration of signalling networks

The enormous repertoire of cell fates and behaviours generated during development contrasts with the relatively small stock of proteins responsible. Combinatorial control, as discussed in Chapter 3 in the context of transcription factors, is the key to understanding how both diversity and specificity are generated. Just as in transcriptional regulation, combinatorial control in signalling is a function of the cell's developmental history: its complement and concentrations of signalling components and regulatory molecules. Of course, the control of signalling networks and gene regulation are inextricably linked: a cell's responses to signalling input and transduction are determined by its specific complement of transcription factors and regulators, and the state of its chromatin. Even signalling responses that bypass the nucleus are still determined ultimately by the cell's developmental program, for example, signals that cause changes in cell motility via effects on the cytoskeleton (see Chapter 6) will only have this effect in cells whose developmental history has programmed them to produce the necessary intracellular apparatus for movement.

It is by no means a simple task to analyse and understand the connectivity of signalling pathways and to predict how variations in this connectivity will affect the response. Many signalling pathways are known to be connected because they share components. For example, signalling through both EGF and FGF receptors deploys Ras as a downstream effector. Cells that express both types of receptor presumably have ways of ensuring that they can distinguish whether the signal has originated with FGF or EGF, and of responding appropriately. In this case it is known that there are components that will stimulate Ras activity in an FGF-dependent, EGF-independent manner. Another way in which a pathway may be 'insulated' from other, potentially interconnecting pathways, is by the assembly of scaffolded complexes such as those we have described earlier in this chapter (Fig. 5.17). Such complexes can effectively 'route' signalling from a given activated receptor through a specific pathway. The clearest examples come from yeast, where different scaffolding proteins assemble shared MAPK pathway components to form specific complexes that can differentially route signalling by mating pheromones, hyperosmolarity, and nutrient starvation.

In contrast, in other circumstances activation of one signalling pathway may not be an 'insulated' event but may lead to the activation of parallel pathways. For example, Ras can activate some of the JNK pathways in addition to the ERK MAP kinase pathway. This 'connectibility', which is probably cell-type-specific, is very important in determining the outcome of signalling.

Recently, some attempts have been made to model mathematically the behaviour of biological signalling networks, using experimentally determined values for the parameters of biological signalling reactions (Fig. 5.35). Two, three or four pathways were linked in the modelling process by common second messengers, such as diacylglycerol, or by shared intermediates, such as activated kinases. Perhaps the most interesting outcome of these studies was the finding that signalling networks are very much more than the sum of their parts: they exhibit 'emergent' properties that are not found in the individual pathways. For example, mathematical modelling predicts how interactions between pathways can be necessary to activate regulatory feedback loops, and how feedback controls in a network can lead to the establishment of a 'bi-stable' system which produces an output that is not continuously variable but switches between distinct steady-states at different signalling thresholds. Moreover, the output can persist for some time even after the initial signal has stopped, a phenomenon that has been observed in some biological responses to signalling. Although these mathematical models necessarily simplify the system, for example they do not take into account the spatial localization or compartmentalization of signalling network components,

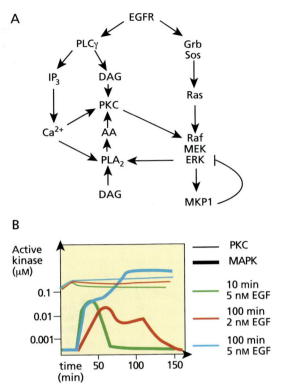

Fig. 5.35. Modelling of signal transduction integration in the response to EGF. (A) Functional interactions between signal transduction components in pathways activated by ligand-bound EGFR. The different elements of the pathways have been introduced in other figures. The biochemical parameters of these interactions (K_m of each reaction and input/output relationships) were measured in studies *in vitro* and used to test the properties of the system in various simulations. In one study, conditions were sought that would lead to MAPK (ERK) remaining stably activated after removal of the input, EGF. Note that a feedback loop that would achieve this could be established in two ways: through the activation of Raf by PKC which would sustain activation of the pathway; and by the activa-tion of phospholipase A_2 (PLA_2) by ERK itself, since the arachidonic acid (AA) produced will activate PKC, leading, again, to the activation of Raf. **(B)** Result of simulations of the response of the system shown in A. The concentration of active PKC (thin lines) or MAPK (thicker lines) is plotted as a function of time after the onset of stimulation with EGF. Three different conditions of stimulation were tested (represented by different colours): only stimulation for 100 minutes with 5 nM EGF can activate the feedback loop and achieve sustained activation of MAPK. (After Bhalla, U. and Iyengar, R. (1999) Emergent properties of networks of biological signalling pathways. *Science* **283**, 381–7.)

they represent a promising beginning in the attempt to understand how biological signalling networks behave (Fig. 5.35).

Signal integration in the nucleus

How do signalling pathways impinge on the activities of the transcriptional regulators that determine the changes in gene expression that occur in response to signalling? Many examples are beginning to emerge of how combinatorial interactions determine the cell-type-specific response to signalling input.

At the simplest level, signal-directed changes in the nature of a single transcription factor can determine a transcriptional response. As an example, the 75 kDa form of the ci protein represses transcription of target genes, while the 155 kDa form produced in response to hh signalling activates these genes (Fig. 5.22). At a slightly higher level of complexity, in Wnt-responsive cells in *Drosophila* the transcription factor Tcf forms a transcriptional repressor complex with the groucho protein; in response to Wnt signalling, β-catenin/armadillo forms an activating complex with Tcf (Fig. 5.25).

An example with further layers of complexity is provided by the variety of transcriptional responses to signalling through the TGFβ/BMP pathways (Fig. 5.36).

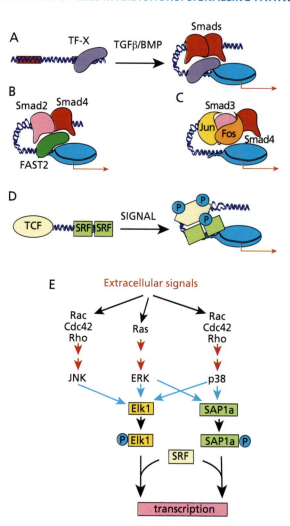

Fig. 5.36. Integration of signalling pathways at the level of transcription factors. Examples of integration of signalling pathways at the level of DNA. **(A)** TGFβ/BMP signalling results in the translocation of Smads to the nucleus where they interact both with DNA and with other transcription factors (TF-X). These interactions provide a means of integrating different signals and modulating transcription, and are essential for Smad-mediated signalling. **(B)** At the promoter of the *goosecoid* gene of *Xenopus*, the Smad2/4 heterodimer interacts with FAST2 to promote transcription. **(C)** At some TGFβ-responsive promoters that contain AP1-binding sites, it is the interaction of Smad3/4 with Jun and Fos, through different Smad domains, that contributes to complex formation and transcriptional activation. **(D)** Serum response elements (SREs) are the recognition sites for a ternary complex consisting of a serum response factor (SRF) dimer, and a ternary complex factor (TCF), which is usually a member of the Ets family of transcription factors. In the example shown, the TCF can be either Elk1 or SAP1a. In the absence of an external stimulus, the ternary complex is assembled at the enhancer but does not promote transcription. Signalling through one or more of the different MAPK pathways results in the phosphorylation of the TCF and the activation of transcription. **(E)** Different MAPK have preferences for different transcription factors and so the interplay between the nuclear and cytoplasmic signalling repertoires of the cell determines its response.

We have already described how combinatorial mechanisms involving heterodimers of the Smad signalling proteins distinguish between signalling by activated TGFβ/activin receptors and BMP receptors (Fig. 5.13). Further combinatorial interactions in the nucleus, involving Smads and other transcriptional regulators, ensure a specific transcriptional response in a given cell type. For some time, it was a puzzle that, although Smads can bind DNA, analysis of DNA sites required for ligand-activated transcription showed no readily apparent common features. For example, binding sites for the 'winged helix' transcription factor FAST1 were required for the definition of the mesoderm by activins in *Xenopus* (see Chapter 9) whereas other TGFβ responses required, for example, AP1-binding sites (which bind c-Jun/c-Fos) or cyclic AMP response element (CRE) sites. And yet Smads are necessary for ligand-activated transcription from these promoters. The explanation is that, at target promoters, specific Smads interact both with Smad-specific binding sites and with other DNA-bound transcription factors to form gene-specific transcriptional-regulatory complexes (Fig. 5.36). Thus in *Xenopus*, Smad2/4 interacts with FAST1 at the *Mix2* promoter and with FAST2 at the *goosecoid* promoter. At promoters with AP1-binding sites that respond to TGFβ signalling, Smad3/4 interacts via different domains with both c-Jun and c-Fos, forming a tripartite complex. The interactions of individual Smads with different transcription factors are mediated by distinct sequences on the Smads. At a given gene, the unique combinatorial assembly of Smads and other transcription factors determines the response.

In a different twist on the combinatorial theme, there are other promoters at which the cell-type-specific assembly of a transcription factor complex at a target promoter is modulated by the signal transduction machinery via a post-translational modification. In MAPK signalling pathways, members of the Ets family of transcription factors are the key link between the signalling pathway and transcriptional regulation (Fig. 5.36). Ets proteins are phosphorylated at one or more sites by activated MAPKs. At different target promoters, Ets proteins are associated with specific combinations of other transcription factors, such as c-Jun/c-Fos, serum response factor (SRF) or the tissue-specific factor Pit1, and the complex is rendered active or inactive by phosphorylation or dephosphorylation of the Ets component. Different Ets proteins provide examples both of broad and narrow specificity in their interactions with MAPK partners: some are specific for a single MAPK whereas others, such as Elk1, can interact with several different MAPKs and thus can potentially integrate MAPK-mediated responses to different stimuli (Fig. 5.36).

Detailed studies of the functions of serum response elements (SREs)—sequences present in the promoters of many genes whose expression is rapidly induced in mammalian cell cultures stimulated with serum (the so-called 'immediate-early' genes, such as c-*fos*)—give an idea of the potential complexity and versatility of the interactions that determine a cell's response to external stimuli (Fig. 5.36). SREs respond to signalling mainly via Ras pathways, and are bound by regulatory complexes containing members of the Ets family. As an example, the SRE of the c-*fos* gene constitutively binds a complex consisting of a dimer of the SRF transcription factor and a member of the TCF subfamily of Ets proteins. (Here, TCF stands for ternary complex factor—not to be confused with the Tcf proteins that function in the Wnt signalling pathway.) Which TCF proteins can bind is determined by the precise sequences and spacing of both the SRF- and Ets-binding sites in the promoter. In the case of the c-*fos* SRE, the relevant factors include Netb, Elk1, and SAP1a. The transcriptional output depends on the relative amounts of the TCF proteins in the cell (a function of its history) and what signalling pathways(s) have been activated (determined both by the cell type and its environment). As mentioned previously, Elk1 is a substrate for all the classes of MAPKs; in contrast, SAP1a is preferentially activated by ERK or p38. The Netb protein, which contains an inhibitory domain that represses transcription, and lacks a MAPK activation domain, may function as a constitutive repressor that switches transcription off after signalling. In summary, the SRE acts as a sensitive and sophisticated integrator that determines the transcriptional output from the gene in response to the prevailing signalling and cellular conditions.

The expression of the *even skipped* gene in certain muscle precursors of the *Drosophila* embryo (Fig. 5.37) provides an example of how different signalling pathways and their effectors converge on a promoter to direct spatially and temporally specific expression in a developing embryo. The assembly of the different transcription factors on the DNA must occur incrementally over time and is likely to be similar to the assembly of enhanceosomes that we have seen in Chapter 3. The gene is transcribed when all of the components assemble in the complex. Each component is essential: if any one is missing, the gene is not transcribed.

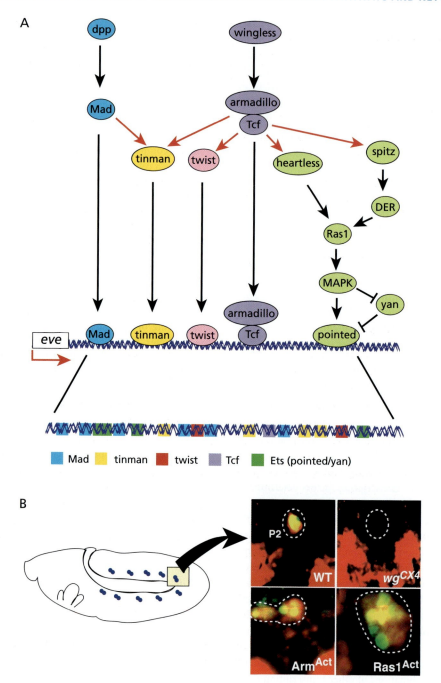

Fig. 5.37. **Signal integration during pattern formation in the *Drosophila* embryo. (A)** The expression of the *even skipped* (*eve*) gene in muscle founder precursors in the mesoderm of the *Drosophila* embryo (see also Chapter 10) requires the assembly, at a specific enhancer (on the 3′ side of the gene), of the effectors for several signalling pathways. These effectors accumulate as a result of the convergence and coordination of the activities of the signalling pathways. In some cases, a signalling pathway not only affects the expression of *eve* through the effector of its own pathway, but also by inducing the expression of another transcription factor or of a ligand for another pathway. The enhancer is active only when all the components are assembled. **(B)** Example of the integration of signalling pathways in the expression of the *eve* gene of *Drosophila*. *Left*: Diagram of an embryo showing expression of eve in some mesodermal cells (for further details of this see Chapter 10). *Right*: Visualization of the expression of eve (green) and of the activity of

MAPK (through an antibody specific for the phosphorylated form of MAPK, red), in the mesoderm of a *Drosophila* embryo. *Top left* panel: in the wild type, the yellow colour indicates that there is an overlap between the expression of *eve* and the activity of MAPK. *Top right* panel: in a *wingless* mutant (wg^{cx4}), *eve* is not expressed and there is no activation of MAPK. *Bottom left*: Expression of an activated form of armadillo (the *Drosophila* homologue of β-catenin) throughout the mesoderm leads to constitutive wingless signalling, and eve expression and MAPK activity are seen in more cells than in the wild type. *Bottom right*: in an embryo in which an activated form of Ras is expressed in the mesoderm, eve expression and MAPK activity are turned on in a broad domain. (From Halfon, M. S. *et al.* (2000) Ras pathway specificity is determined by the integration of multiple signal-activated and tissue-restricted transcription factors *Cell* **103**, 63–74.)

SUMMARY

1. Signalling molecules, receptors, and transcription factors are key players in the information-processing that underlies the unfolding of the developmental program, and in the coordination of the program in different cells.

2. These elements function as components of signal transduction pathways that are linked and integrated to form information-processing networks.

3. The components of signal transduction networks use molecular currencies (such as second messengers and phosphorylation) to exchange information, and their activities are linked through conformational changes and interactions involving specific molecular interfaces. Some of these molecular interfaces are used to group sets of signal transduction elements into functional complexes, or to localize them to specific cellular compartments.

4. Some signalling routes from the cell surface to the nucleus employ only a few signal transducers, while others involve multiple intermediaries or use second messengers, offering many opportunities for interactions with other pathways.

5. Signal transduction networks are subject to complex regulatory mechanisms that can be effected, for example, by changes in the concentration, activity or stability of signalling components, or in their compartmentalization within the cell. Feedback control systems are particularly important.

6. Integration of signalling pathways can occur in the nucleus but also at the level of the signal transduction networks and even at the cell surface.

COMPLEMENTARY READING

The remarks made at the end of Chapter 4 also apply to this chapter. Our understanding of signalling pathways and neworks, their regulation and their role in development, will undoubtedly see important advances in the next few years, and it would not be helpful to single out references on particular topics. The journals listed at the end of Chapter 4 are useful sources of up-to-date reviews on individual signalling pathways, their components and interactions. The following references are to reviews that deal with more general aspects of signal transduction and its regulation, or to resources that catalogue the growing number of proteins and other molecules that participate in signalling networks.

Bader, G. D., Donaldson, I., Wolting, C., Ouellette, B. F., Pawson, T., and Hogue, C. W. (2000) BIND (BIND; http://binddb. org): The biomolecular interaction network database. *Nucleic Acids Res.* **29**, 242–5.

Bray, D. (1995) Protein molecules as computational elements in living cells. *Nature* **376**, 307–12.

Ciechanover, A., Orian, A., and Schwartz, A. L. (2000) Ubiquitin-mediated proteolysis: biological regulation via destruction. *Bioessays* **22**, 442–51.

Freeman, M. (2000) Feedback control of intercellular signalling in development. *Nature* **408**, 313–19.

Halfon, M. S. *et al.* (2000) Ras pathway specificity is determined by the integration of multiple signal-activated and tissue-restricted transcription factors. *Cell* **103**, 63–74.

Hill, C. and Treisman, R. (1999) Growth factors and gene expression: fresh insights from arrays (www.stke.org/cgi/content/full/OC_sigtrans;1999/3/pe1).

Pawson, T. and Nash, P. (2000) Protein–protein interactions define specificity in signal transduction. *Genes Dev.* **14**, 1027–47.

Weng, G., Bhalla, U., and Iyengar, R. (1999) Complexity in biological systems. *Science* **284**, 92–6.

Chapter 6

Cells and their interactions

The protein networks we have described in the preceding chapters operate within and across molecularly confined domains: cells. Cells, created by and from proteins, are the units of development. It is their proliferation, diversification, specialization, and spatial organization that form all the tissues and body structures which make up the organism.

All cells are variations on a basic structural theme: a package of highly structured cytoplasm contained within a plasma membrane whose fundamental component is a lipid bilayer. The cytoplasm is subdivided by internal membrane systems into several compartments specialized for particular functions (Fig. 6.1): nucleus for transcription and replication; endoplasmic reticulum for translation; Golgi apparatus for protein sorting and allocation, and organelles, such as mitochondria, peroxisomes, and lysosomes, for specialized biochemical functions. The insulating properties of membranes enable the biochemical activities of these compartments to be segregated. For example, enclosure of the highly acidic contents of the lysosome within the lysosomal membrane protects the rest of the cell from their effects on protein degradation, and sequestration of the powerful second messenger Ca^{2+} in specialized stores allows its regulated release in response to specific stimuli (see Chapter 5, Fig. 5.20). Membranes can also function as platforms for the organization of protein complexes, such as ribosomes and the signal transduction complexes that assemble at the inner face of the plasma membrane (see Chapter 5). There may be further substructures within individual cellular compartments, for example, recent evidence suggests that the nucleus has a dynamic structure of microdomains in which different functions, such as transcription and splicing, may be localized. A similar theme is emerging at the plasma membrane where there appear to be microdomains with a specialized composition of lipids and proteins tuned to particular functions. The mechanical strength and capacity to change the shape of a cell are largely provided by an intracellular network of specialized protein polymers, the cytoskeleton, which also forms an internal solid phase for many of the biochemical interactions that take place inside the cell.

The different compartments of the cell do not exist in isolation but are in constant communication, with a highly regulated traffic of molecules passing between them. We are beginning to get glimpses of this dynamic organization which, as we have seen in Chapter 5, makes important contributions to the regulation of cell signalling.

However, cells rarely exist on their own. As they become functionally specialized, they elaborate a wide array of adhesion and junctional complexes that enable them to interact in a dynamic fashion with other cells and with their environment—in the case of tissues, these interactions must also be coordinated with those of their neighbours to create a harmonious product. The outside of a cell contains a microenvironment that modulates the interactions of cells with each other. As we have mentioned in Chapter 4 (Figs. 4.30, 4.31), the interactions of cells with their environment are mediated by a complex external meshwork of proteins called the extracellular matrix (ECM), which also serves as a milieu for the modulation of cell interactions.

The genetic analysis of development has revealed that many proteins and processes involved in basic cellular functions, from cell adhesion to vesicular transport or the cell cycle, contribute to the generation of multicellular patterns. Therefore, in order to understand development we need in turn to understand these processes and how they are modified to regulate the assignment and spatial arrangements of cell fates. In particular, we have to understand the activities of the proteins within cellular contexts, that is, how they operate in individual cells or, more important during development, groups of cells.

For many years it has only been possible to study the behaviour of cells and of proteins within cells in fixed samples or through experiments *in vitro*. However, the advent of techniques that can be used to visualize cell-biological processes *in vivo* has revolutionized the study of cell

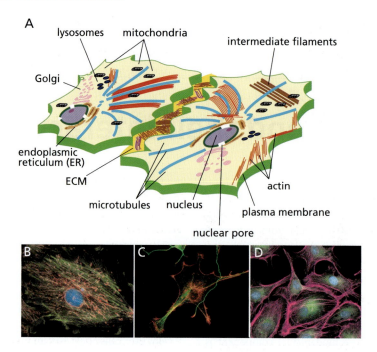

Fig. 6.1. Basic cell structure and some variations. (A) A typical animal cell is bounded by a plasma membrane. Its cytoplasm contains a number of functionally distinct structures that in some cases are also structurally independent. A prominent structure within the cell is the nucleus, the repository of the chromosomes which contain the DNA; within the nucleus, various cytochemical stains reveal a dense structure, called the nucleolus, which is made of rDNA. The nucleus is surrounded by a double membrane, pierced by pores that allow regulated transport of various molecules. The interior of a cell is traversed by a complex network of membranes which includes the endoplasmic reticulum (ER) and the Golgi apparatus. The ER, topologically continuous with the outer nuclear membrane, is rich in ribosomes and is the place where the RNAs for membrane and secreted proteins are translated. The Golgi is a stack of flattened membrane vesicles where many proteins made in the ER are sorted, processed, and transported to their destinations. The intracellular membrane network also contains vesicles which serve as specialized stores for different kinds of molecules, from small ions, such as Ca^{2+}, to large signalling peptides (see Chapter 4). Lysosomes contain specialized enzymes whose function is the controlled digestion of proteins and other macromolecules. In addition, the interior of a cell is criss-crossed by a dynamic meshwork of polymerized proteins (mostly actin and tubulin): the cytoskeleton. The various structures that make up the cytoskeleton, including intermediate filaments and microtubules, regulate the shape of the cell and provide an internal scaffold for protein–protein interactions. The cytoplasm also contains mitochondria, which have their own DNA and act as the powerhouses for the cell. Cells are seldom in isolation and the space between them contains a highly specialized proteinaceous structure, the extracellular matrix, through which they interact. **(B)** Actin cytoskeleton (green) and mitochondria (red) in an endothelial cell revealed by specific probes conjugated to fluorescent dyes. The nucleus is stained with DAPI (blue). **(C)** Confocal image of the cytoskeleton of a population of neurons and glial cells. The meshwork of actin is revealed by Texas red conjugated phalloidin (red) and the microtubules with an antitubulin antibody (green). **(D)** Actin cytoskeleton and membranous structures of an endothelial cell, revealed with specific probes: fluorescently labelled phalloidin for actin (magenta), and a green-fluorescent DiOC6(3) for intracellular membranes (green). The nuclei were revealed with a DAPI stain. (Images B–D, courtesy of Molecular Probes Inc.)

architecture and its dynamics. In particular the jellyfish green fluorescent protein (GFP) (Fig. 6.2) has proven useful and versatile in this respect: to mention just one of many applications, the *GFP* gene can be fused to a gene of interest—a cytoskeleton component, for example—and so used as a reporter to follow its temporal and spatial pattern of expression in a living wild-type or mutant embryo. More than one non-invasive fluorescent tag can be used, enabling several proteins to be followed simultaneously.

In this and the next chapter, we discuss the basic cellular operations that contribute in a general way to developmental events. Here, we first describe the properties and functions of the extracellular matrix and the cytoskeleton and then discuss how these functions relate to basic cellular processes such as cell adhesion, movement, and polar-

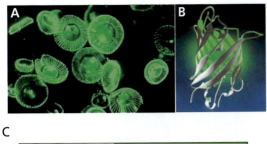

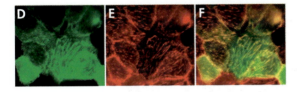

Fig. 6.2. Green fluorescent protein (GFP). (A) The jellyfish *Aequoria victoria* contains a protein which emits green fluorescence when stimulated by ultraviolet light. This protein is known as green fluorescent protein (GFP). **(B)** Model of the three-dimensional structure of GFP. **(C)** GFP can be fused to a protein of interest (X). If the fusion protein is introduced into cells or organisms, the location of the original protein can be followed *in vivo* through the fluorescence emitted by the GFP. **(D–E)** Fibroblasts transfected with GFP fused to the cytoskeletal component actin. **(D)** Fluorescence emitted by the actin : GFP protein. **(E)** Endogenous actin revealed by staining the same cells with phalloidin coupled to a fluorescent dye. **(F)** Merging of the images from D and E. The differences between D and E are explained by the fact that whereas phalloidin only detects actin that has polymerized into filaments, the actin : GFP protein reveals actin monomers as well (see Fig. 6.5). (Images D–E, courtesy of J. Sutherland.)

ity. In the next chapter we discuss cell proliferation and growth, as well as cell death. Altogether these functions can be seen as a set of basic 'routines' that affect and regulate many of the elements and molecular networks we have discussed in previous chapters.

The extracellular matrix

Whether in an embryo or an adult organism, cells are not alone, but are organized into groups with precise and reproducible spatial arrangements. The space between cells is filled with a variety of molecules that are secreted by the cells and create a narrow microenvironment consisting of two layers. One of these layers is very close to the cells and is made up of molecules directly associated with the cell surface. The other, which is broader but more variable in thickness, is made up of molecules that are indirectly associated with the cell surface and act as a physical buffer between cells, as well as a substrate to which to adhere and over which to move. This layer is called the extracellular matrix (ECM) and in some cases its composition provides information about other cells in the group and their environment.

Most of the component elements of the ECM are very large proteins whose properties are determined by their structure and a variety of post-translational modifications, chiefly glycosylations and sulphations. These components include collagens, elastins, proteoglycans, fibronectins, and laminin (Fig. 6.3).

Proteoglycans are highly hydrated proteins responsible for the gel-like character of tissues such as connective tissue. They surround cells and protect the tissue against pressure damage, a function that is as important for differentiated cells as it is during development; in some cases, they contribute to the overall shape of a particular cell assembly. Also, as we have seen in Chapter 4, proteoglycans are not just structural components of the intercellular space but can also modulate some signalling events. They do this by associating with other proteins that can affect the activity of ligands and receptors, or by binding specific growth factors and creating local niches where they become very concentrated and can be used effectively (Chapter 4, Fig. 4.30).

Fibronectins (Fig. 6.3) are large glycoproteins with a variety of structural motifs that allow them to interact specifically with other components of the ECM and, most importantly, with the specialized cell adhesion proteins, integrins, on the cell surface. The regulation of these interactions in space and time provides a way of modulating the exchange of information between cells. As we have seen in Chapter 2 (Fig. 2.11), the fibronectin gene is alternatively spliced in a cell- and tissue-specific manner to produce subtly different variants with different functional properties tailored to particular cells. In the chick, for example, an exon that is present in fibronectin in mesenchymal cells is specifically spliced out of the fibronectin mRNA in developing cartilage cells. Because fibronectin can interact both with cells (through the integrins) and with other proteins of the ECM, it can play a role in mediating the attachments of cells to other surfaces. Migrations of individual cells and other global cellular movements, such as displacements of tissues over or under one another, are often associated with the presence of fibronectin in the ECM. An important element in the interactions between fibronectin and cells is

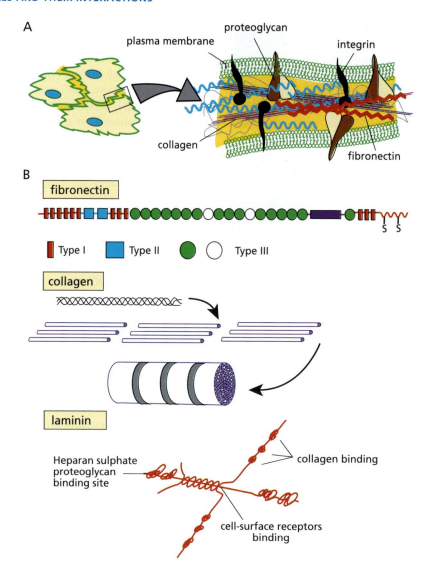

Fig. 6.3. Major components of the extracellular matrix. (A)
The space between cells is filled with a dense array of proteins
known as the extracellular matrix (ECM). The proteins of the ECM
are large and heavily post-translationally modified, largely by different degrees of glycosylation. Glycosaminoglycan chains of proteoglycans provide an extreme example of this modification (see
Chapter 4, Fig. 4.30). In addition, ECM components tend to form
aggregates that are very important for generating elastic properties,
as in the attachment of muscles to bone, or the epidermis. Cells interact with the components of the ECM through specialized cell-surface receptors, such as integrins, which recognize specific proteins
of the ECM. **(B)** Structures of some of the main components of the
ECM. **Fibronectin** is a modular protein, of about 2440 amino acids,
made up of three types of structural modules: fibronectin Type I,
Type II, and Type III repeats. There are two kinds of Type III repeats indicated in white and green; those in white are alternatively spliced
in different cell types, contributing to differences in the properties of

the protein (see Chapter 2, Fig. 2.11). These modules create a series
of binding sites for cell-surface receptors (integrins) or other molecules of the ECM such as collagens and heparin proteoglycans. In
the ECM, fibronectin is composed of two identical chains, joined by
disulphide bonding near their carboxy termini. **Collagens** are fibrous proteins that are very common in the ECM and can constitute
up to 25% of the total protein mass of a vertebrate. The basic unit
of the large collagen fibre is a triple α helix made up of interwound
collagen chains called α chains, each 1050 amino acids long. The
triple helical structures are organized into higher-order bundles.
There are several kinds of α chains each tailored for a specific tissue
need. **Laminin** is a trimeric molecule made up of three subunits
each composed of about 1500 amino acids. It has a central trihelical
region and three globular domains through which it can interact
with different proteins in the ECM and the cell surface to form complex linked networks.

the tripeptide arg–gly–asp (RGD), which is present not only in fibronectin but also in other proteins involved in cell–matrix attachment. The RGD peptide provides the anchorage point of these ECM proteins to their receptors. An isolated RGD tripeptide added to cells in culture can displace fibronectin from the cell surface.

An example of the importance of fibronectin for the organization of certain cell types is provided by its effects on the movement of neural crest cells, a multipotent migratory cell population that in vertebrate embryos gives rise to a number of cell types including much of the peripheral nervous system, epidermal pigment cells and facial cartilage and bone. In culture, neural crest cells show a clear affinity to fibronectin and will move along tracks of this protein. *In vivo*, however, this property appears to be more permissive than essential, since blocking the interactions between fibronectin and neural crest cells in the embryo impairs the movement of the cells in the head region but not in the trunk.

Collagen is a large fibrous and insoluble protein that makes up the bulk of the ECM in tissues such as cartilage, bone, and skin. It is highly glycosylated and organizes itself into large cable-like bundles that criss-cross the ECM. There are many kinds of collagens whose structure and composition are characteristic of different tissues (Fig. 6.3) but that are all characterized by their ability to withstand stretching and other tensions. There are specialized collagens for different adult tissues and even for some embryonic ones. Elastins are also highly hydrophobic molecules which form cross-linked meshes that impart elasticity to skin, blood vessels, and other tissues but, in contrast to the collagens, they are not glycosylated.

Laminin is one of the first ECM components produced during embryonic development. It is a major constituent of the basal lamina, a thin carpet of matrix located at the base of many epithelia that can perform a variety of roles such as separating different tissues and providing a substrate over which cells can move (Fig. 6.3). Laminin can interact with many other extracellular proteins and may act as a link between these proteins, forming networks that contribute to determining the properties of the extracellular space.

The ECM is neither a static nor a passive structure. It is a product of cells that changes as they change in response to their programs of gene expression. As development proceeds, the ECM that is deposited increases in amount and complexity and undergoes constant tissue-specific remodelling. Many of these changes are achieved directly by changes in composition but they can also be brought about

by the activity of specific proteases, matrix metalloproteases, that cleave ECM components, changing the properties of pre-existing molecules and generating substrates for novel interactions between ECM proteins and between the cells and the ECM. Because the ECM harbours many signalling molecules, these modulations of the activity of the ECM provide an alternative means of regulating cell signalling (see Chapter 4). Altogether, these processes contribute to building up the dynamic landscape of cell interactions and movements that characterizes the construction of tissues and organs.

The cytoskeleton

The maintenance and dynamic regulation of cell shape, and the organization of biologically efficient compartments within the cytoplasm, are the functions of a proteinaceous intracellular framework known as the cytoskeleton. In addition, the cytoskeleton provides a framework for many molecular processes, including intracellular transport and regulated responses to mechanical stress (Fig. 6.4).

The cytoskeleton is made up of three classes of protein filaments with different physical properties and functions: intermediate filaments (about 10–12 nm in diameter), microtubules (about 25 nm in diameter), and actin filaments (about 7–9 nm in diameter) (Fig. 6.5). Whereas intermediate filaments are for the most part static and concerned with providing support against mechanical stress, microtubules and actin filaments are dynamic structures that adapt to new environments and are responsible for the motile and elastic properties of cells. All three types of filament are polymers of protein units. In each case, other proteins known as accessory proteins are responsible for cross-linking cytoskeletal filaments and attaching them to other cellular structures, in particular the plasma membrane, and thus controlling filament assembly.

Intermediate filaments

These form a structural meshwork most prominent in cells that undergo mechanical stress, such as epithelial and muscle cells. The protein components of intermediate filaments are fibrous proteins that assemble to form strong, rope-like bundles which are very stable and provide mechanical support for the structure of the cell (Fig. 6.5). There are several kinds of intermediate filaments, each with a different protein composition, that are specialized for different roles in particular cell types. For example, the felt-like nuclear lamina that maintains the structure of the

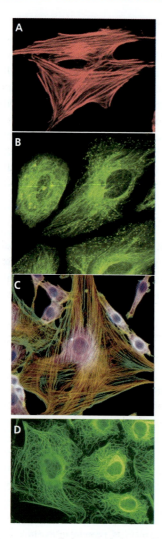

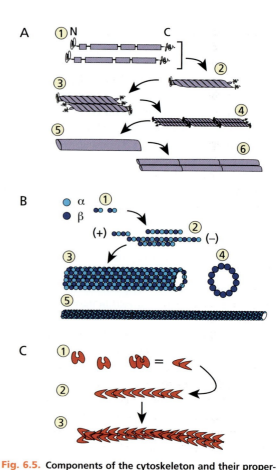

Fig. 6.4. Components of the cytoskeleton. (A) Filamentous actin in a 3T3 cell revealed with fluoresceinated phalloidin (red). **(B)** Microtubules of a HeLa cell in interphase revealed with Oregon green 488 paclitaxel. **(C)** Mouse fibroblast stained for actin (green) and tubulin (red). Nuclei are revealed by DAPI. **(D)** Fibroblasts stained with an anti keratin-8 antibody to reveal the meshwork of intermediate filaments. (Images A–C, courtesy of Molecular Probes Inc. Image D courtesy of D. D. Owens.)

depicting it with 'pointed' (minus) and 'barbed' (plus) ends—a shape that is observed when the protein is complexed with myosin in actinomyosin filaments. The actin monomer is assembled into thin fibres of variable length and stability which provide a flexible scaffold for many of the activities of the cell (2). Actin fibres are sometimes woven (3) into different architectures which can create supramolecular structures within the cell (see below) through interactions with cross-linking proteins.

Fig. 6.5. Components of the cytoskeleton and their properties. (A) Intermediate filaments are polymers of fibrous proteins which assemble into higher-order rope-like fibres that are very strong. The basic unit of the fibre is a dimer of two subunits which have a central α-helical region that is used to form the dimer, and globular amino- and carboxy-termini (1 and 2). The monomer subunits have molecular weights between 50 and 100 kDa. While the central domain is conserved between different monomers, the termini are not and it is these that give different intermediate filaments their distinctive properties. Heterodimers assemble into long rods through interactions between the globular ends (3 and 4). These rods then form higher-order fibres (5, 6). It is the stacking of these arrays that creates filaments and confers their strength and stability. **(B)** Microtubules are polymers of a dimer of two subunits of tubulin: α and β (1) (50–60 kDa). This heterodimer polymerizes into longitudinal structures (2) which through lateral interactions form the hollow structure that is the microtubule (3). The perimeter of a microtubule is made up of 13 longitudinal polymers (4). The structural differences between the α and β subunits provide a polarity to the dimer and to the tubule itself, which has a plus end and a minus end (5). Microtubules are dynamic structures that play important roles in mitosis, transport, and scaffolding within the cell. **(C)** Actin is a polymer of a single polypeptide, G actin (approximately 40 kDa) (1). The polarity of the actin molecule is often represented by

See Bottom Left

nucleus (and perhaps also creates microenvironments for specialized nuclear activities) is composed of intermediate filaments; epithelia contain keratin filaments, the major component of feathers, hair, and scales; muscle, cells from the haematopoietic system and connective tissue contain vimentin, and linked neurofilaments provide tensile strength to nerve cell axons. The shapes and spatial arrangement of many of these cells depend on the spatial organization of the different kinds of intermediate filaments. The basis of the filaments is, in most cases, a heterodimer of two subunits with different composition tailored to the function of the filament (Fig. 6.5).

Microtubules

As their name implies, microtubules are tubular structures whose walls are composed of parallel filaments of polymerized tubulin (Fig. 6.5). The basic tubulin subunit protein is a dimer of α and β subunits which polymerize into long fibres. The fibres interact laterally, creating sheets that are converted by the structure of the dimer into the walls of the tubule. The orderly stacking of these units during polymerization means that the microtubule also has a global polarity, with the α-tubulin moiety of each tubulin oriented towards one end (− end), and the β-tubulin moiety oriented towards the other (+ end). There are several different forms both of α- and β-tubulin monomers, allowing the cell to form microtubules with subtly different properties adapted to the function of different cells. In many vertebrates, for example, developing neurons often contain a neural-specific tubulin that has different properties from the tubulin types in other cells and in *Drosophila*, there is a sperm-specific tubulin.

Microtubules, though structurally strong, are dynamic structures in a constant state of flux. They expand and contract in length, and sometimes depolymerize completely. The rate of polymerization at one end is rapid (the + end), while polymerization at the other end is slow (− end). GTP plays a vital role in this 'dynamic instability': β-tubulin carries a bound molecule of GTP, which is hydrolysed to GDP when a new molecule of tubulin binds to the end of a microtubule. The incorporation of the GDP-bound molecule destabilizes the tubule structure and can lead to its disassembly. However, tubulin binding and GTP hydrolysis are not simultaneous: in a rapidly growing microtubule, there may be a build-up of GTP-bound tubulin at the growing end which tends to stabilize the microtubule. When

growth is slower, this GTP 'cap' disappears or does not form, and the plus end is destabilized, making it more likely to depolymerize (Fig. 6.6).

In the cell, microtubules generally grow out in random directions from nucleating structures called microtubule organizing centres. The most important of these centres is the centrosome (Fig. 6.6). Nucleation of microtubules at the centrosome requires the participation of several centrosome-associated proteins including a special form of tubulin, γ-tubulin, which acts as an anchor for the polymerization of the tubules. The minus ends of the microtubules are anchored in the centrosome and the plus ends may reach as far as the plasma membrane. This polarization is used to direct many cellular activities. For example, during cell division microtubule arrays nucleated by the centrosomes organize to form the mitotic spindle. After their separation at the end of metaphase, the two new sets of chromosomes travel along these arrays (in a plus-to-minus direction) to opposite poles of the cell. Controlling the orientation of the mitotic spindle through the orientation of the microtubules from the centrosome provides a way of orienting cell cleavages. This can enable the differential allocation of factors that influence the assignment of cell fates (see Chapters 8 and 10). Another example in which the polarity of microtubules is used in the generation of pattern is provided by the axons of many neurons, where the oriented organization of the microtubules in tracks from the cell body to the growth cone provides routes for the movement of many proteins that contribute to the laying down of neural pathways.

Microtubule stability is also influenced by specific proteins known as microtubule-associated proteins, or MAPs (not to be confused with the MAP kinase signalling proteins discussed in Chapter 5). MAPs are ATP-binding proteins that stabilize microtubules and link them to other elements of the cytoskeleton or, in some cases, to membrane-associated proteins. An important class of MAPs are the microtubule motor proteins: kinesins and cytoplasmic dyneins (Fig. 6.7). These proteins contribute to the stability of the microtubules but more importantly use the energy of ATP hydrolysis to translocate along the microtubule, carrying with them a bound 'cargo' that may be, for example, an organelle, a protein complex or an RNA molecule. Dyneins transport their cargo towards the minus ends of microtubules, while most kinesins move towards the plus end. By this mechanism the microtubules act as tracks for the energy-dependent, directional transport of components within the cell—a process

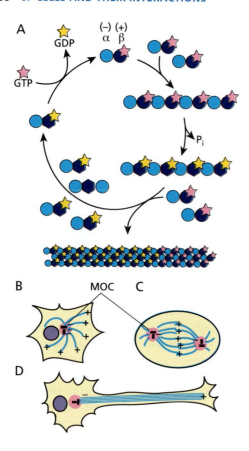

Fig. 6.6. Microtubule dynamics. (A) The β subunit of the α, β dimer of tubulin binds GTP. Under conditions in which a microtubule has nucleated, it grows by addition of GTP-bound dimers to the plus end of the tubule. However, after binding to the polymer, GTP can be hydrolysed, inducing a conformational change in the dimer; this bends and destabilizes the polymer, which will break unless new GTP-bound dimers are added. The balance between GTP- and GDP-bound monomers at the end of the tubule determines the stability of the structure. GTP-bound subunits favour linear configurations. In stabilized tubules, the GDP-bound subunits are kept in a linear configuration by lateral interactions with other subunits. Because of the effect of the GTP-bound tubulin on the stability of the filament, its concentration in the medium is an important parameter. **(B–D)** Examples of arrays of microtubules in different kinds of cells. **(B)** In an interphase cell, microtubules radiate randomly from the microtubule organizing centre (MOC) which focuses the minus end of each microtubule. **(C)** In a dividing cell, microtubules make an important contribution to the formation of the mitotic spindle and provide a platform for the movement of the chromosomes to opposite poles of the cell. Special MOCs called centrosomes are associated with the minus ends of the microtubules. **(D)** In a neuron, arrays of microtubules run along the length of the axon, contributing to its polarity and providing tracks for the movement of proteins from the cell body to the growth cone at the tip of the axon.

that is vital during the organization and activity of the mitotic spindle and a number of developmental events involved in setting up and maintaining asymmetry or polarity (Fig. 6.7).

In many organisms, microtubules are used early in embryogenesis during the establishment of the main body axes. In *Drosophila* the anteroposterior axis of the embryo is derived from the polar organization of specific mRNAs at two opposite poles of the oocyte during oogenesis: the mRNA for the transcription factor bicoid at the anterior and the mRNA for the posterior determining protein oskar at the posterior end. In the oocyte, microtubules are organized with their minus ends at the anterior end, and their plus ends at the posterior. The mRNAs for oskar are associated with kinesins and require kinesin for their posterior localization (Fig. 6.7) (see also Chapter 3 Figs. 3.22 and 3.24 for the effects of localized mRNAs). Also during this period, the oocyte nucleus undergoes stereotyped move-

ments along microtubules which are essential for the definition of both the anteroposterior and the dorsoventral axes (Fig. 6.7).

In vertebrate embryos, microtubules also play an important role in the establishment of the main body axes. For example, during the early stages of the development of fishes and frogs, an array of microtubules organizes parallel to the cortex of the embryo and provides tracks for the movement of protein complexes, particularly components of the Wnt signalling pathway (see Chapter 5), that after translocation define the future dorsal side of the embryo (Fig. 6.8).

In addition to providing a system of tracks for the movement of molecules that have the potential to alter cell fates very early in development, there is evidence that microtubules play a role in some signal transduction pathways that are used at later developmental stages, though the reason for their participation in these processes remains unclear. As mentioned in Chapter 5 (Fig. 5.23), microtubules are implicated in the hedgehog signalling pathway via the function of the microtubule-associated protein cos2. The sequence of cos2 suggests that it is related to kinesins, but it does not appear to have any motor or translo-

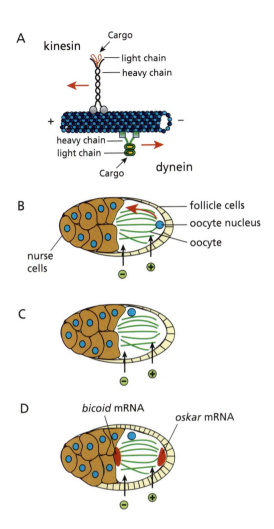

Fig. 6.7. Microtubule-associated proteins. (A) The two main motor proteins associated with microtubules are kinesins and dyneins. Kinesins are dimers of two heavy chains (about 130 kDa) plus two light chains (64 kDa). Each heavy chain consists of a large globular domain that binds to the microtubule, and a long stalk that connects the head to a small globular tail. The tail segments are associated with the light chains, which bind the cargo (usually a vesicle). The kinesin moves over the microtubule in the direction of the plus ends, by means of conformational changes in the head that are mediated by ATP hydrolysis. Dyneins are large proteins (>1000 kDa) composed of two or three chains. Like the kinesins, they have a head segment that binds to microtubules and hydrolyses ATP, and a tail that binds the cargo. The monomers are associated with a variable number of light chains that might be involved in the binding to the cargo. Dyneins move along microtubules in the minus direction. **(B–D)** Function of microtubules (green lines) in the polarization of the oocyte in *Drosophila*. During oogenesis, an array of microtubules defines a polarity in the oocyte with the plus end at the prospective posterior and the minus end at the prospective anterior end. **(B)** As the oocyte grows, its nucleus becomes located at the plus end of the microtubules, where it plays a role in the definition of the posterior end of the oocyte through a signalling event that involves the oocyte and the follicle cells. **(C)** The oocyte nucleus, using the microtubule tracks, moves to one side; the position at which the oocyte nucleus comes to rest defines the dorsal region of the embryo. **(D)** The anteroposterior axis of the oocyte is set up by the localization of specific mRNA and protein determinants: at the anterior end, the RNA for bicoid and, at the posterior end a complex of the protein staufen and the mRNA for oskar (Fig. 3.22). The localization of the complex of staufen and *oskar* mRNA is mediated by a kinesin that uses the microtubular tracks. These asymmetries will be transformed into different fates after fertilization.

catory function. Rather its role seems to be to bind the ci^{155}-containing protein complex to microtubules in the absence of hh signalling, thereby flagging it for proteolysis, perhaps by a microtubule-associated protease. Components of the JNK signalling pathway have also been found to be associated with microtubules, which might have a scaffolding function somewhat like that of the Ste5 and JIP proteins (see Chapter 5, Fig. 5.17). In most cases, however, the reasons for this association remain to be explored.

Actin filaments

The third major component of the cytoskeleton, actin filaments, are helical polymers of a monomer, the globular G actin, which associate in bundles or networks to form strong but (in contrast to microtubules) flexible structures (Figs 6.5 and 6.9). Actin is a very abundant protein that in some cell types can account for up to 10% of the total protein. An actin meshwork below the plasma membrane—the cortex—is involved in regulating cell shape and movement in all cells (discussed further later in this chapter), while in other cells actin associates with specialized proteins to perform particular functions. In muscles, for example, it associates with myosin to form the contractile myofibrils.

Like microtubules, actin filaments are polar, with a plus and a minus end defined by the basic structure of the monomer and the shape of the polymer, as well as by the relative rates of growth of the two ends: the minus ('pointed') end is slow-growing, whereas the plus ('barbed') end grows very rapidly. In contrast to microtubules, however, actin filaments are very unstable and display a much more dynamic behaviour.

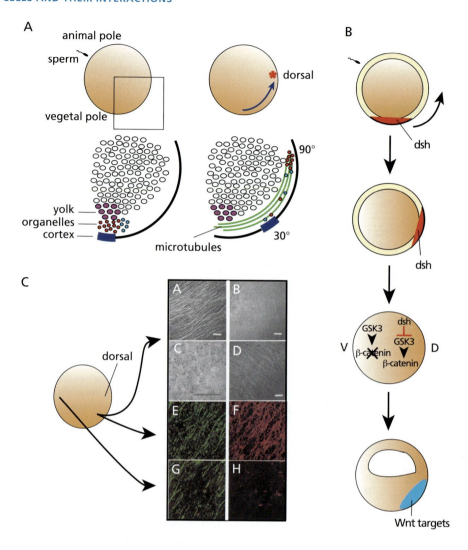

Fig. 6.8. Microtubules and the establishment of the dorsoventral axis in *Xenopus*. (A) The *Xenopus* oocyte contains well-defined animal and vegetal regions. The vegetal region is characterized by large amounts of yolk and the presence of some organelles and large protein particles. Fertilization induces a 30° rotation of the cortex with respect to the cytoplasm and, at the same time, the formation of arrays of microtubules parallel to the cortex (green lines). These tracks serve to move a number of organelles and protein complexes, but not yolk, about 90° towards the animal pole, leading to the definition of the dorsal pole of the embryo at a point opposite to that of sperm entry. **(B)** The establishment of the dorsal pole of the embryo correlates with the translocation, along the microtubular tracks, of a pool of dishevelled (dsh) protein located at the vegetal pole, and the activation of Wnt signalling over a region opposite the sperm entry point (as represented by the activation of Wnt targets). This is achieved by blocking the GSK3 dependent degradation of β-catenin (see Chapter 5, Fig. 5.25). **(C)** Photomicrographs of the formation of the microtubular arrays and the differential activation of Wnt signalling in amphibian embryos. (A) Parallel arrays of microtubules induced by fertilization in an egg of *Rana pipiens*, (B) As A, but after treatment with colchicine which eliminates the microtubules. (C) Same region of an egg treated with ultraviolet (UV) irradiation before the formation of the microtubular arrays. This treatment inhibits the formation of the arrays. (D) In this case, the egg was treated with UV after the formation of the arrays. Note that they have been disrupted (compare with A). (E) Microtubular arrays on the future dorsal side of a *Xenopus* embryo. (F) Same region as in E stained for β-catenin; note its accumulation over the microtubules. (G) Microtubules over the future ventral side of the embryo; they are not arranged in clear tracks like those in F. (H) β-catenin in the same region as G. Note that in this region there are low levels of cytoplasmic β-catenin. (Images A–D from Rowning, B. *et al.* (1997) Microtubule-mediated transport of organelles and localization of β–catenin to the future dorsal side of *Xenopus* eggs. *Proc. Natl Acad. Sci.* **94**, 1224–9. Images E–H from Elinson, R. and Rowning, B. (1988) A transient array of parallel microtubules in frog eggs: potential tracks for a cytoplasmic rotation that specifies the dorsoventral axis. *Dev. Biol.* **128**, 185–97.)

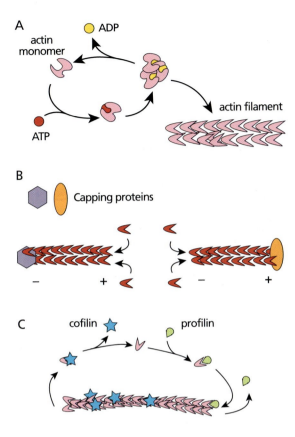

Fig. 6.9. Actin. (A) The actin monomer has a pocket, created by two almost symmetrical domains each composed of two lobes, which can bind ATP. Hydrolysis of actin-bound ATP promotes the assembly of actin into helical filaments and release of ADP promotes dissociation of monomers from the filament. The initial stages in the assembly of the filament require a nucleation event mediated by specialized proteins. The asymmetry of the monomer, together with the fact that all monomers bind with the same orientation in the polymer, create a polarity in the filament with a minus or 'pointed' end, which grows slowly, and a plus or 'barbed' end, which grows rapidly. The names are derived from observations of actinomyosin filaments under the electron microscope. **(B)** *In vivo*, actin filaments are in a state of dynamic instability largely because the ends of a filament are not very stable and are prone to break down. Stability is achieved because the rate of ATP binding and hydrolysis, as well as the ability of monomers to assemble into the filament, are regulated by a number of actin-binding proteins. Actin-capping proteins, such as profilin, bind specifically to the plus or minus end and prevent incorporation of new monomers into the filament, thus promoting stability of that end and directional growth. **(C)** Actin-severing proteins, such as cofilin, destabilize filaments. The combined activities of capping (green) and severing (blue) proteins result in the dynamic instability of actin filaments, allowing them to grow or shrink depending on the concentration of actin monomers. This dynamic process is known as treadmilling. (Adapted from Borisy, G. and Svitkina, T. (2000) Actin machinery: pushing the envelope. *Curr. Opin. Cell Biol.* **12**, 104–12.)

The balance between the polymerization and depolymerization of actin is dependent on nucleotide triphosphate hydrolysis—this time of ATP rather than GTP—and the activities of a variety of actin-binding proteins that regulate filament stability by determining whether they grow or break down (Fig. 6.9). These proteins include profilin, which binds to and helps to regulate the concentration of free monomeric actin and its addition to the growing polymers, and capping protein, which binds to the plus ends of actin filaments and prevents the addition of further

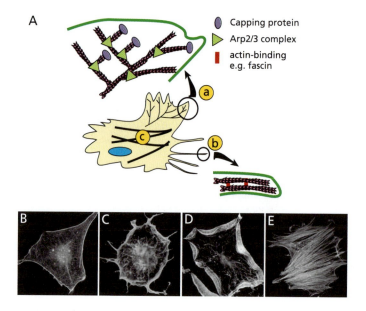

monomers. Severing proteins, including gelsolin and cofilin, can bend and break actin filaments. Other actin-binding proteins, such as fimbrin, α-actinin, and filamin, are responsible for cross-linking actin filaments into different functional assemblies. The balance of actin assembly and disassembly mediated by these factors is highly dynamic, allowing the cell to change its actin architecture rapidly (Fig. 6.9).

Actin-based assemblies form the structural basis for intracellular stress fibres, and underlie the characteristic shape and behaviour of cell protrusions known as lamellipodia and filopodia. These assemblies, which can also be elicited in tissue culture, play important roles in morphogenesis *in vivo*, determining the shape, mobility and flexibility of cells (Fig. 6.10). Small monomeric G proteins regulate the formation of these assemblies, making their behaviour rapidly responsive to extracellular stimuli. Activation of, or inteference with, different proteins of this family reveals a strikingly specific link between particular molecular switches and specific types of actin assemblies: Rho induces stress fibres, Cdc42 filopodia, and Rac lamellipodia. The interplay between the activities of these proteins is almost certainly essential for coordinating many of the cell interactions that mediate the emergence of shape and cell motility during development (Fig. 6.11).

Actin polymerization can be triggered locally by a number of stimuli including activation of RTK or seven-transmembrane-domain receptors, or phosphatidylinositide-dependent activation of Rho/Rac/

Fig. 6.10. Actin networks. (A) Diagram of a cell showing different types of actin networks: (a) the branched networks underlying lamellipodia; (b) the cross-linked parallel networks underlying filopodia; and (c) stress fibres. Each of these architectures depends on the cross-linking of actin filaments by specific proteins. Filopodia contain parallel actin filaments with the minus end at the tip, cross-linked by actin-binding proteins such as fascin. Lamellipodia contain branched actin filaments, with plus ends at the tips of the branches and with both ends capped. Branching is promoted by the activity of a complex of actin-related proteins (Arp2/3) at the branch points. Within both filopodia and lamellipodia, the actin filaments have to be anchored to the membrane: just once in filopodia but several times in lamellipodia to provide the sheet-like appearance characteristic of these structures. Little is known about the structural basis for these links. Stress fibres are bundled filaments which provide strength and rigidity. **(B)** Actin cytoskeleton in a Swiss 3T3 cell. **(C–E)** Actin cytoskeleton in Swiss 3T3 cells transfected with activated small GTPases. **(C)** Actin cytoskeleton organizing filopodial extensions in a cell that has been transfected with a dominantly active Cdc42 GTPase. **(D)** Actin cytoskeleton organizing large lamellipodial extensions in a cell that has been transfected with a dominantly active Rac GTPase. **(E)** Actin cytoskeleton organized into compact stress fibres in a cell that has been transfected with a dominantly active Rho GTPase. (Images B–E courtesy of K. Nobes.)

Cdc42 (Fig. 6.12 and see Chapter 5). The link between these events and the polymerization of actin is mediated by adaptor proteins: WASP (Wiskott-Aldrich syndrome protein) for RTK and phosphatidylinositide-dependent Cdc42, and SCAR (suppressor of cAMP

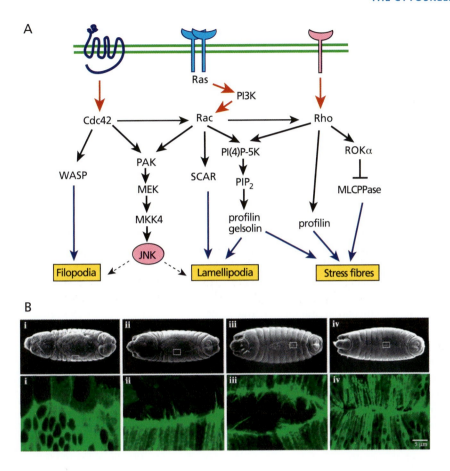

Fig. 6.11. Regulation of the activity of the cytoskeleton by cell-surface receptors. (A) Interactions of cell-surface receptors and signalling pathways in the regulation of the actin cytoskeleton. PI3K (PI3 kinase); WASP (Wiskott–Aldrich syndrome protein); PAK (p21-activated protein kinase); MEK (MAPKK kinase); MKK4 (MAPK kinase 4); JNK (c-Jun kinase); ROKα (Rho-associated kinase α); MLCPPase (myosin light-chain phosphatase) SCAR (suppressor of cAMP receptor). (After Schmidt, A. and Hall, A. (1998) *Annu. Rev. Cell Dev. Biol.* Signalling to the actin cytoskeleton. **14**, 305–38.) The dashed arrows indicate regulatory interactions which might be indirect. **(B)** Formation of filopodia and lamellipodia during the morphogenetic process of dorsal closure in *Drosophila*. The upper row shows a dorsal view of a *Drosophila* embryo during the final stages of embryogenesis when a gap in the dorsal region of the embryo is closed by epidermal sheets. This process does not involve cell division and is mediated by changes in cell shape and cytoskeletal activity of the epidermal cells. The lower panels show details of the leading edge of the epidermal sheet from the regions indicated in the top panels. The cells express actin : GFP and this allows the activity of the cytoskeleton to be visualized. During the closing of the gap, the epidermal cells stretch and the dorsal-most cells send out filopodial and lamellipodial extensions which contribute to the closure. (From Jacinto, A., Wood, W., Turmaine, M., Rosenblat, J., Balayo, T., Martinez Arias, A., and Martin, P. (2000) Dynamic actin-based epithelial adhesion and cell matching during *Drosophila* dorsal closure. *Curr Biol.* **10**, 1420–6.)

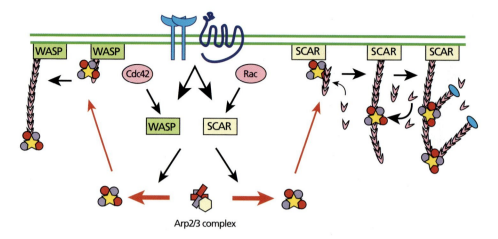

receptor) for the seven-transmembrane-domain receptors and Rac. Recruitment of WASP and SCAR to defined membrane sites nucleates actin polymerization by activating a complex of seven proteins, known as the Arp2/3 (for actin related protein 2/3) complex, which functions as an organizing factor for the assembly of actin filament networks at the edge of cells (Fig. 6.12). In this way, information from extracellular signals leads to spatially controlled assembly of the machinery for cell shape change and movement, which we discuss in more detail later in this chapter. As the filament grows, the Arp2/3 complex remains associated with the pointed end of the polymer; in the case of lamellipodia, it is present at the branching sites.

Cell adhesion and cell junctions

Within an organism, cells associate with other cells to form tissues and organs, often mixing with like cells and avoiding integrating with different kinds of cells. Blood cells, cartilage, and bone are examples of differentiated tissues in which these properties are obvious, but other cell populations, such as the undifferentiated cells of the embryonic ectoderm, mesoderm, and endoderm, are also able both to form coherent groups and to interact within and across these groups. In order to achieve this organization, cells must have mechanisms that favour contacts with some cells, while repelling contact with others. This is evident from cell-sorting experiments in which mixtures of cells from different origins spontaneously sort themselves out, like with like. This was first demonstrated by the experiments of Johannes Holtfreter and his colleagues on cells from early amphibian embryos (Fig. 6.13). In these experiments, cells from different origins were mixed and

Fig. 6.12. Regulation of the assembly of actin networks. (A) The nucleation of actin polymerization is facilitated by a complex of seven proteins, the Arp2/3 complex, which is differentially activated by Cdc42 and Rac to organize filopodia or lamellipodia. The activation is mediated by different adaptor proteins: WASP for Cdc42 and SCAR for Rac. Activation of the monomeric GTPases by specific receptors leads to the recruitment of cell-type-specific WASP and SCAR proteins to the membrane and to the localization and activation of the Arp2/3 complex which can then begin to nucleate the formation of actin filaments. In this way, influences that regulate the activity of the monomeric GTPases regulate the activity of actin, and thereby the shape of the cells. Arp2/3 acts in a different way in filopodia and lamellipodia. In filopodia (*left*), Arp2/3 binds to the minus end and protects it from rapid degradation. In lamellipodia (*right*), Arp2/3 binds to growing filaments and promotes branching by nucleating branched filaments on existing ones. In this case the plus end is capped with capping proteins and so is very stable. In the branched architectures, the stability of the interactions between Arp2/3 and the minus end determines the length of the branch and this, in collaboration with treadmilling at this end, produces the broad membrane flaunts that are lamellipodia. (After Borisy, G. and Svitkina, T., (2000) *Curr. Opin. Cell Biol.* **12**, 104–12; Svitkina, T. and Borisy, G. (1999) *Trends Biochem. Sci.* **24**, 432–6.)

allowed to settle. After a while, cells sort out according to their origin. For example, if presumptive epidermal cells are mixed with mesoderm, the mesodermal cells sort out from the epidermal cells. Furthermore, the sorting out is not random and the epidermal cells surround the masses of mesodermal cells as they would in the organism (Fig. 6.13). This sorting must depend on the recognition of cell-surface components that are different in different cell types.

Differential affinities are also evident during develop-

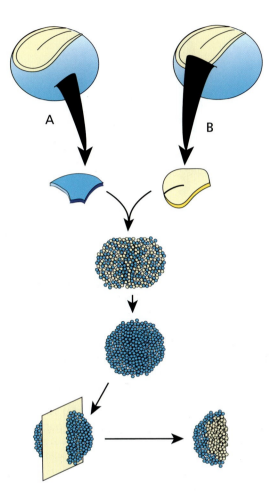

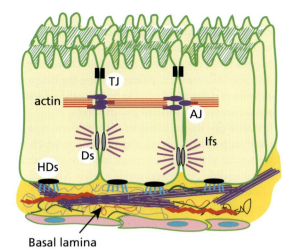

Basal lamina

Fig. 6.14. Cell–cell and cell–matrix junctions. In an epithelium, cells adhere to each other and to the ECM through specialized protein complexes. Cells are linked to other cells through specialized complexes which fall into three classes: tight junctions (TJ), adherens junctions (AJ), and desmososmes (Ds). Adherens junctions are linked to the actin cytoskeleton, and desmosomes to intermediate filaments (Ifs). These associations provide rigidity and strength to each cell and, in the case of the adherens junctions, a mechanical link between their cytoskeletons. The interactions between cells and the ECM are mediated either through specialized structures, hemidesmosomes (HDs), or through direct interaction between specialized cell-surface proteins and components of the ECM. In an epithelium such as the one shown here, the ECM over the basal side of the cell forms a layer called the 'basal lamina'. The basal lamina is rich in collagen fibres and proteoglycans, which are good substrates for cell–ECM interactions. Non-polarized cells also show adhesive interactions that are often mediated by the some types of proteins but they lack the striking, stereotyped architecture of epithelial cells.

Fig. 6.13. Differential cell adhesion in embryos. This experiment tests the ability of cells to recognize other cells of the same type or from the same region of an embryo. Pieces from defined regions of a *Xenopus* embryo, neural plate (yellow) and presumptive epidermis (blue) are dissociated into individual cells which then are mixed together. After a while, the cells reorganize themselves and sort out in such a way that cells from the neural plate and presumptive epidermis form separate groups with one wrapped around the other. Thus, to see the neural plate derived cells it is necessary to cut through the ball of tissue. This experiment suggests that there are molecules that enable cells to recognize other cells of the same class or kind.

ment: as cells diversify and change their properties, they alter their associations and form groups that behave as interactive units. These changes are highly regulated, accompanying, and in some cases mediating, the transformations in shape and form that characterize embryogenesis. In many situations, cells of one kind must remain grouped together in the course of large-scale movements, such

as those involved in gastrulation, and avoid interactions with surrounding cells. When cells undergo these processes, they do so in groups that are connected through specialized cell–cell attachments that are only established between cells from the same kind, to keep them as a coherent unit. There are two non-exclusive mechanisms by which cells connect with each other and make coherent groups: indirectly by interaction with the ECM, and by direct cell–cell contact or adhesion (Fig. 6.14). For these attachments to be effective in governing the properties of cells, they must be linked to the cell interior, where the information-processing machinery lies. As we shall see later in the chapter, this is usually dependent on a relay of the adhesive properties of cells to the cytoskeleton.

Cell–cell and cell–matrix contacts are characterized by

the development of molecular structures, or junctions, that promote intimate association between adjacent cells or between cell and substrate (Fig. 6.14). Because they are connected to the cytoskeleton, cell–cell junctions, in particular, provide a means of interconnecting the structural framework of different cells and so facilitate the coordination of behaviour in associated groups of cells.

Epithelial cells display examples of the different types of cell–cell junctions (Fig. 6.14). Tight junctions, which are characteristic of these cells, form an impermeable barrier that enables different surfaces within a cell to be functionally isolated from one another. These junctions contribute to the apico-basal polarity of these cells as discussed later in this chapter. Adherens junctions, which can mediate both cell–cell and cell-matrix interactions, contribute to the ability of cells to withstand mechanical stress. Adherens junctions are associated with the actin cytoskeleton and can provide a way of coordinating the activity of the cytoskeleton through several cells. In most epithelial cells, the adherens junctions, linked by actin fibres, form a tight belt around the cell which is referred to as the 'zonula adherens'. Desmosomes, which are cell–cell junctions, and hemidesmosomes, which connect cells with the ECM, are both connected to intermediate filaments. Other types of junctions include the nearly ubiquitous gap junction, which is formed by channel-forming proteins and allows small molecules, such as ions and water, to pass between cells.

Molecular mediators of cell adhesion

Several families of cell-surface molecules are involved in both cell–cell and cell–matrix adhesion. One of the most important and versatile of these is the integrin family (Fig. 6.15), whose members participate mainly in interactions between cells and the ECM. These transmembrane proteins are heterodimers of two noncovalently associated glycoprotein subunits, α and β, both of which are essential for divalent-cation-dependent binding of ligands such as laminin, fibronectin, vitronectin, and collagen. There are different kinds of α and β subunits which can associate with each other in a combinatorial manner to generate a large number of integrin subfamilies specialized for binding different combinations of ECM components, and so providing a versatile selection of specificities for cell–matrix association (Fig. 6.15). Ligand-binding specificity may also be modulated by association of specific integrins with other transmembrane proteins such as members of the immunoglobulin superfamily.

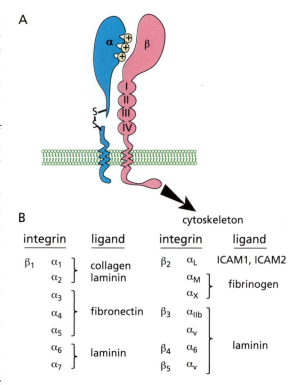

Fig. 6.15. Integrins and their interactions. (A) Basic structure of an integrin dimer at the cell surface as inferred from structural and biochemical data. Integrins are dimers of an α and a β subunit. The α subunit is processed to yield two elements joined by a disulphide bond. The globular domain, which binds cations, is involved in the ligand interactions which mediate attachment to the matrix. The β subunit also contains an amino-terminal globular domain, which cooperates with the α subunit globular domain in ligand binding, and a set of four cysteine-rich repeat domains (I–IV). Each of the subunits has a molecular weight over 100 000 kDa. The intracellular domain of the β subunit can establish contacts with talin and α-actinin, and links the integrin molecule to the cytoskeleton. Formation of the heterodimer is essential for the expression of integrins at the cell surface; in cells of animals deficient for one of the subunits, the other subunit does not appear at the cell surface either. **(B)** Different cells and tissues have different complements of integrin subunits and can therefore assemble combinatorial arrangements with different properties and different affinities for ligands in the ECM.

The intracellular domain of the integrin dimer interacts, either directly or indirectly, with a variety of different cytoskeletal proteins as well as with components of intracellular signalling pathways. Like the interactions of integrin cell-surface domains, many intracellular interactions of integrins are specific to different types of integrin subunits. For example, in adhesion plaques—adherens junctions

A

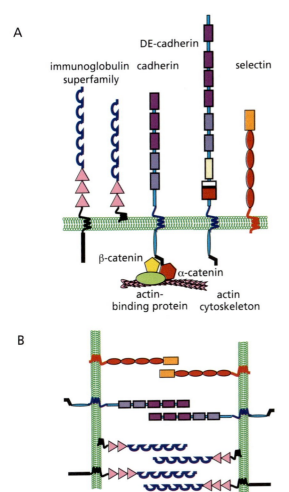

immunoglobulin cadherin
superfamily

DE-cadherin

selectin

β-catenin

α-catenin

actin-
binding protein

actin
cytoskeleton

B

Fig. 6.16. Molecules involved in the generation of cell–cell attachments: selectins, IgG superfamily and cadherins. (A) Representative members of the different families of proteins involved in cell–cell adhesion. Members of the immunoglobulin superfamily (those participating in cell adhesion are also known as IgCAMs) contain a number of IgG domains at the amino terminus (dark blue) that are usually combined with fibronectin Type III repeats in the membrane-proximal region (pink triangles). Their attachment to the cell surface can vary and sometimes the same gene will produce a form with a transmembrane domain and another that is anchored to the membrane through a lipid moiety. Cadherins are members of a large family of proteins characterized by globular extracellular domains, some of which bind calcium and mediate homophilic protein–protein interactions. Vertebrate cadherins have five of these domains, three of which (deep purple) can bind calcium but the number varies in invertebrates. The intracellular domain is linked to the actin cytoskeleton through a complex that involves β- and α-catenin. Calcium binding is absolutely essential for the homophilic interactions between cadherin molecules. Invertebrate cadherins have, in addition to the classical cadherin domains, other structural motifs in the extracellular domain. The example shown is from *Drosophila* E-cadherin which has an atypical cadherin domain, an EGF-like repeat (yellow) and a structural motif that is present in laminin A (the red box). Selectins contain an amino-terminal lectin domain (orange) that allows them to bind carbohydrates on the cell surface and a number of repeated motifs (red). **(B)** Interactions between the different cell-surface molecules involved in cell–cell adhesion. In most cases they are homophilic.

through which cells may attach to the extracellular matrix—binding of specific integrins to the actin associated proteins α-actinin and talin mediates connection of integrin to the actin cytoskeleton. On the other hand, at hemidesmosome junctions, which attach the basal surface of epithelial cells to the underlying basal lamina, different types of integrin are connected, via attachment 'plaque' proteins at the cytoplasmic surface of the membrane, to the intermediate filament network.

Direct cell–cell interactions are mediated by three families of proteins: members of the immunoglobulin (Ig) superfamily, cadherins and selectins, all of which are transmembrane proteins with the ability to interact, directly or indirectly, with the cytoskeleton (Fig. 6.16). These proteins can also participate in cell–matrix interactions, but are more often involved in interactions between cells. Other types of proteins also play a role in specific situations, such as the tight junctions at the apical ends

of epithelial cells; little is currently known about these proteins.

IgCAMs are proteins of the Ig superfamily, so called because the extracellular part of the proteins contains one or more domains similar to a set of characteristic, repeated domains first described in immunoglobulin molecules (see also Chapter 4). These proteins take part in calcium-independent interactions with other proteins of the same family. Such interactions are termed homophilic, while interactions between different types of molecules are called heterophilic. The best known of the Ig superfamily of cell adhesion molecules is neural cell adhesion molecule (NCAM), which exists in a variety of forms including membrane-bound, extracellular, and membrane-anchored forms. NCAM is important during many developmental processes, in particular, during the development of the nervous system, where it promotes cell interactions that hold cells together at specific developmental stages such as the development of the neural tube. Other members of this family are involved in mediating short-range attractive cues that participate in guiding axon growth during the laying down of the pathways of the nervous system. In the chick, for example, axons that are destined to

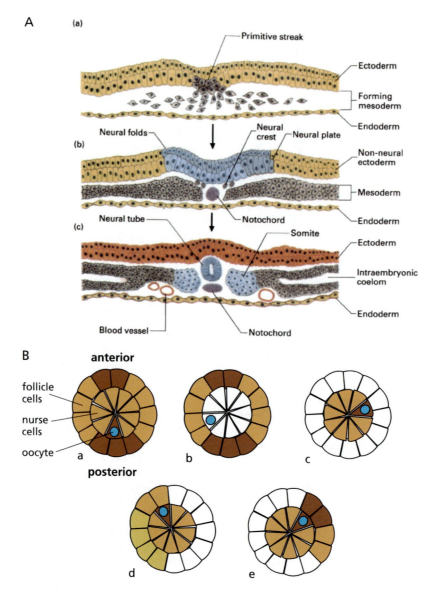

Fig. 6.17. Developmental dynamics of cadherin expression.
(A) Changes in the cadherin composition in cells during the early stages of chick development. (a) Early in development, cells in the embryo have an epithelial appearance and express E-cadherin (yellow); mesodermal cells (grey) become mesenchymal, stop expressing E-cadherin and migrate towards the interior of the embryo. (b) Gastrulation generates a three-layered embryo with ectoderm, mesoderm, and endoderm. Within the ectoderm, a group of cells (blue) loses expression of E-cadherin, begins to express N-cadherin and folds to form the neural tube. (c) Later, different groups of cells express combinations of cadherins. For example, the epidermis, which remains epithelial, expresses E-cadherin and P-cadherin (red) and a mesodermally derived structure, the notochord, expresses N- and P-cadherin (purple). Thus, tissues, and presumably their properties, are labelled by different kinds of cadherins. (After Takeichi, M. (1987) Trends Genet. (1987) **3**, 213, adapted from Lodish, H. et al. (1995) Molecular cell biology. (3rd edn). Scientific American Books.)

(B) Differential concentrations of Drosophila E-cadherin (DE-cadherin) control the positioning of the oocyte in Drosophila. The Figure shows diagrammatic cross-sections through an early developing Drosophila egg follicle. A wild-type follicle contains 16 germline cells (of which eight are visible in the cross-section), surrounded by a follicular epithelium (see also Fig. 6.7). The different concentrations of cadherin are indicated by the intensity of the brown colour. (a) DE-cadherin is expressed at high levels at the anterior and posterior poles of the epithelium (dark brown) and the oocyte always positions itself at the posterior. Mosaic analysis (see Chapter 4, Fig. 4.6) can be used to test if there is a correlation between the levels of DE-cadherin and the positioning of the oocyte. If all of the germ-line cells (b) or all of the follicle cells (c) lack expression of cadherin, the oocyte positions itself at random. If some of the epithelial cells lack DE-cadherin and others have varying levels (d, e), the oocyte positions itself next to the cells with higher levels of DE-cadherin. (After Tepass, U. (1999) Curr. Opin. Cell Biol. **11**, 540–8.)

cross the midline (commissural axons) are characterized by expression of an IgCAM called axonin-1. Attractive interactions between axonin-1 and another IgCAM family member called NrCAM, which is expressed by midline neurons, are involved in mediating correct pathfinding. The membrane-bound forms of IgCAMs may participate, via their intracellular domains, in signal transduction pathways and/or interactions with the cytoskeleton.

Selectins are carbohydrate-binding proteins that mediate transient calcium-dependent interactions between different types of blood cells during such processes as inflammatory responses. Carbohydrate binding is the function of an extracellular domain that recognizes glycoproteins exposed on the surface of cells such as platelets, endothelial cells, and leucocytes (Fig. 6.16).

Cadherins mediate homophilic cell–cell adhesion, but this time in a strictly calcium-dependent fashion, and are linked both to the cytoskeleton and the ECM (Fig. 6.16). A prototypic cadherin has an extracellular domain composed of five motifs that bind Ca^{2+} and which are absolutely restricted to members of the cadherin family. In invertebrates, there are multiple repeats of these motifs and the molecules also contain other motifs characteristic of extracellular proteins, such as EGF-like repeats or domains also present in laminins (Fig. 6.16). The cytoplasmic domain of cadherins is linked to the actin cytoskeleton via the proteins α- and β-catenin and plakoglobin (also known as γ-catenin), which together form a flexible anchor between the cadherin and the cytoskeleton (Fig. 6.16).

Cadherins are essential for many of the interactions between cells in the early embryo and there are both cell-type and stage-specific forms of these proteins. Changes in the set of cadherins that are expressed often correlate with major changes in cell associations that occur during development, for example, during gastrulation and in neural crest formation and migration (Fig. 6.17). At a very early stage of development, E-cadherin is required for the compaction of cells that occurs at the eight-cell stage of the mouse embryo and triggers the segregation of embryonic and extraembryonic tissues. Embryos that are mutant for E-cadherin cannot form a polarized external epithelium and die before implantation.

There is some evidence that the quantitative level of cell adhesion mediated by cadherin can be an important developmental parameter and that differential cell adhesion can play a role in the assignment of cell fates. An experiment illustrates this suggestion. As we have seen in Chapter 4 (Fig. 4.1), during oogenesis in *Drosophila* the oocyte is selected from a set of 16 sister germ-line cells on the basis of its number of contacts with other cells of the group. Then it is positioned within a population of surrounding somatic cells, the follicle cells, that will provide important cues for the organization of the oocyte later in development. The oocyte is always located at the posterior end of this group of cells (see Fig. 6.7). The oocyte always expresses higher levels of cadherin than the surrounding cells and always positions itself next to the follicle cells that have high levels of DE-cadherin. Mosaic experiments indicate that there is a causal correlation between the levels of DE-cadherin and the positioning of the oocyte (Fig. 6.17). Removal of DE-cadherin from the germ line, the surrounding cells or groups of cells in either population shows that the oocyte will always position itself adjacent to cells with higher DE-cadherin expression and therefore more cell adhesion (Fig. 6.17).

Cell adhesion and signalling

Some cell-surface adhesion molecules such as the integrins and cadherins do not act merely as an inert 'glue' to stick cells together or to a substrate, but as components of complex signalling systems that sense the cell's environment and use this information to regulate its patterns of gene expression, cell shape, and cell movement. This has been shown by experiments in which overexpression of mutant integrin or cadherin molecules that lack an intracellular domain results in dramatic alterations of both the adhesive and the morphogenetic properties of the cells. This is a surprising result if the only function of these proteins is to promote adhesion between cells, as the intracellular domain would be expected to be dispensable for this function. In these experiments, the mutant molecules appear to act as dominant negative receptors (see Chapter 4, Fig. 4.16G) and suggest that the process of adhesion is linked to an exchange of information between the environment and the cell that requires the intracellular domain of these adhesion molecules.

When cells adhere to a substrate (a dish when they are in culture or the extracellular matrix of an embryo) they develop foci of strong adhesion known as 'focal adhesion plaques'. At these sites, local cytoskeletal rearrangements lead to the formation of intracellular actin-based stress fibres linked to a transmembrane complex of proteins that in turn is strongly attached to the ECM. As we shall see later in this chapter, the dynamics of formation of focal adhesions is a central element in the process of cell movement.

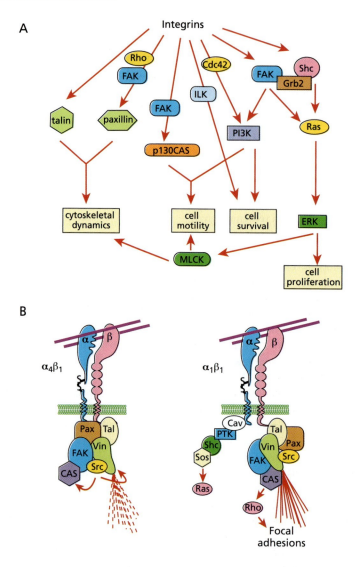

Fig. 6.18. Integrin signalling. (A) Summary diagram of signalling pathways accessed by activated integrins, and their functional consequences. Integrin activation has direct effects on the cytoskeleton and, less directly, on cell survival and proliferation. FAK, focal adhesion kinase; ILK, integrin linked kinase; MLCK, myosin light chain kinase; CAS, Crk associated substrate. **(B)** Signalling through different integrin complexes can have different intracellular effects. When combined with the α_4 subunit (*left*), the β_1 subunit nucleates the formation of a complex of proteins in which focal adhesion kinase (FAK) is activated and combines with Src or other Src

family protein kinases to phosphorylate some of the components of the complex. The substrates recruited by the complex lead to the destruction of actin filaments and the local breakdown of the actin cytoskeleton. In contrast, when β_1 combines with the α_1 subunit (*right*), the complex assembled by β_1 is not phosphorylated by FAK and acts, in a way that includes participation of the small GTPase Rho, to promote the formation of actin filaments. The α_1 subunit also contributes to the specificity and kinetics of the activity of the complex through a connection to the Ras pathway via an interaction with the protein caveolin.

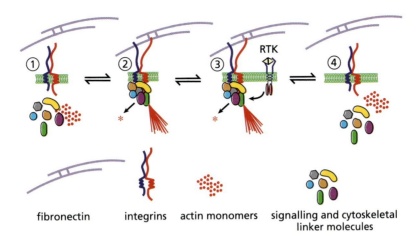

fibronectin   integrins   actin monomers   signalling and cytoskeletal linker molecules

Activation of a kinase, focal adhesion kinase (FAK) by integrin is an important requirement for the establishment of adhesion plaques. The steps that link activation of integrin to the activation of FAK are not yet well defined but must involve an interaction of the cytoplasmic tail of integrins with adaptor proteins which in turn can access other elements of signal transduction pathways. Small G proteins, such as Rho, Rac, and Cdc42, which are involved in the modulation of the actin cytoskeleton, are likely targets for the integrin-activated signalling pathways. Intermediates such as these are also involved in nuclear signalling pathways, so that integrin 'signalling' could also contribute to the transcriptional activation of genes (Fig. 6.18). This type of networking enables information about the cell's environment, transmitted by integrins, to be input into, for example, the cell cycle control system (see Chapter 7). By signalling to the interior of the cell that appropriate extracellular contacts are in place, specific integrins also appear to be involved in preventing cell death; when these signals are absent, indicating that the cell is in the 'wrong' place, the programmed cell death pathway (discussed in more detail in Chapter 7) is activated.

A wide-ranging role in mediating the functions of integrins has been suggested for another intracellular kinase, known as integrin-linked kinase (ILK) because it has been shown to interact directly with the β subunit of integrin. ILK has also been implicated as an effector in growth factor signalling (Fig. 6.18). Roles for ILK have been suggested in the regulation of processes as diverse as anchorage-dependent cell growth and cell-cycle progression, and expression and assembly of ECM components such as E-cadherin and fibronectin, as well as Wnt signalling.

In addition to acting as sensors of the extracellular envi-

Fig. 6.19. Integrin signalling. Interactions between integrins and their ligands in the ECM (e.g. fibronectin, in purple) triggers a conformational change which results in the assembly of a complex of proteins on the intracellular domain of the integrin, accompanied by intracellular signalling events and modifications of the cytoskeleton (1 and 2). The complex of proteins results in signalling (*) and in the regulation of cytoskeletal activity, indicated here by the formation of actin filaments (red lines). These complexes are typical of focal adhesion plaques. In turn, intracellular signalling events, like those triggered by the activity of RTK (3) can alter the conformation of the integrins and affect their interactions with the ligands (3 and 4). This can result in the dissolution (but sometimes also in the strengthening) of the complexes at the focal adhesion.

ronment, integrins additionally provide a link from the inside to the outside of the cell because their binding activities in the ECM can be modulated by intracellular signalling events: local signalling pathways that do not involve the nucleus influence integrin activation and thus the cell's adhesive properties. For example, signalling pathways such as those triggered by activation of RTKs have been shown to regulate the ligand-binding affinity of integrins (Fig. 6.19), while the cytoskeleton itself can influence integrin's functional interactions by regulating its diffusion or clustering in the plane of the membrane. So far, unfortunately, little is known about the mechanics of these events although they are clearly important in the processes leading to cell shape change and movement.

Cadherins are also linked to a variety of signalling pathways that play a role in the regulation of cell adhesion and cell shape during early embryonic development. The intracellular domain of cadherins binds β-catenin and plakoglobin, which, through an association with α-catenin, are linked to the actin cytoskeleton. Both β-catenin and plako-

A

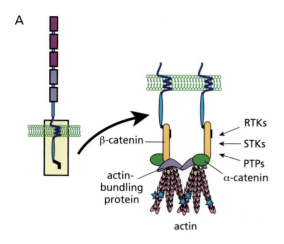

B

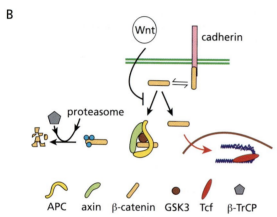

Fig. 6.20. Cadherins and signalling. (A) The intracellular domain of cadherins is linked to the cytoskeleton through an ordered series of protein–protein interactions: the cytoplasmic tail interacts with either β-catenin or its relation plakoglobin, which bind α-catenin. α-Catenin links the cadherin/β-catenin complex to the cytoskeleton through interactions with actin bundling proteins (e.g. α-actinin or ZO-1). β-Catenin is a substrate of various RTKs, ser/thr kinases (STKs), and protein phosphatases (PTPs), which can affect the extracellular interaction by modulating the conformation of β-catenin and its interactions with the cadherin molecule. Often, more than one cadherin molecule is used to create strong adhesion. **(B)** Cadherin-associated β-catenin is in dynamic equilibrium with a cytosolic pool of β-catenin that plays a central role in Wnt signalling (see Chapter 5, Fig. 5.25). Cytoplasmic cadherin is degraded but Wnt signalling inhibits this degradation so that the concentration of β-catenin rises and it enters the nucleus where it acts as an effector of transcription with Tcf.

globin are substrates for a variety of signalling pathways, enabling cadherin activity to influence signalling events and vice versa. Although many of these interactions have been observed in tissue culture and their significance *in vivo* remains to be proven, it seems likely that they are used during development. Activation of RTKs has been shown to result in the phosphorylation of β-catenin and plakoglobin and the modulation of cadherin-mediated adhesion. The observation that some protein tyrosine phosphatases target these proteins—in other words that reversal of the phosphorylation is a regulated event—indicates that the phosphorylation is likely to be physiologically significant (Fig. 6.20).

Like the integrins, cadherins also interact with members of the Rho/Rac/Cdc42 family of GTPases. For example, experiments in tissue culture show that an effector of these proteins, the GTPase-activating protein IqGAP1, interacts with β-catenin at the adherens junctions and disrupts its assembly with α-catenin and cadherin. Activation of Rac or Cdc42 has been shown to stabilize cadherin-based cell

adhesion *in vitro*. This is likely to be achieved in part by recruitment and inactivation of IqGAP1, thus preventing its interaction with β-catenin at the cytoplasmic tail of cadherin. It seems likely that such mechanisms are also employed *in vivo* in the dynamic control of cell–cell associations during development.

The relationship between β-catenin and Wnt signalling (Chapter 5, Fig. 5.25 and Fig. 6.20 and 6.21) provides a clear example of a link between adhesion and signalling that is likely to be of importance in many developmental contexts. There are three cellular pools of β-catenin: one associated with the plasma membrane, one in the cytosol, and a third, in the nucleus, that is dependent on Wnt signalling. In the absence of Wnt signalling there is an equilibrium between the membrane/cadherin-associated pool and the pool in the cytoplasm, where β-catenin is efficiently degraded (Fig. 6.20). As we have seen in Chapter 5, Wnt signalling prevents this degradation and leads to translocation of β-catenin to the nucleus. Interestingly, an increase in the concentration of cadherin at the cell surface antagonizes Wnt signalling, presumably by depleting the cytoplasmic pool of β-catenin. Similarly, expressing β-catenin molecules that are membrane bound increases Wnt signalling by releasing β-catenin from adherens junctions (Fig. 6.20 and 6.21). Although these are experimental situations, they raise the possibility that similar relationships operate *in vivo*, so that changes at the cell surface in response to cell interactions could influence Wnt signalling. As β-catenin is also able to migrate into the nucleus during signal transduction, this type of regulation may be a way of coordinating the activities of three major cell compartments: cell membrane, cytoskeleton, and nucleus.

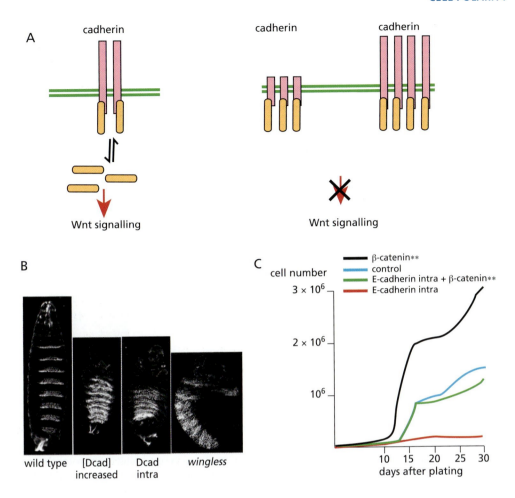

Fig. 6.21. Cadherin-mediated adhesion and Wnt signalling.
(A) The existence of a dynamic equilibrium between the cytosolic and the cadherin-associated pools of β-catenin is demonstrated by the observation that overexpressing the intracellular domain of cadherin, or increasing the overall concentration of cadherins at the cell surface in cultured cells or embryos, results in a lowering of Wnt signalling by titrating cytoplasmic β-catenin. **(B)** If, during *Drosophila* embryogenesis, the concentration of *Drosophila* cadherin ([Dcad]) is increased, or the intracellular domain (Dcad intra) is artificially expressed, the phenotype that is generated looks like the phenotype of loss-of-function mutants for *wingless*, a member of the *Wnt* gene family. (Modified from Sanson, B. *et al.* (1996) Uncoupling cadherin-based adhesion from wingless signalling in *Drosophila*. *Nature* **383**, 627–30.) **(C)** The rate of proliferation of keratinocytes in culture is increased when Wnt signalling is activated by expressing a stable form of β-catenin (β-catenin**). Expression of the intracellular domain of E-cadherin (E-cadherin intra), on the other hand, reduces the proliferative ability of these cells. The effects of the intracellular domain of E-cadherin are neutralized by coexpression with β-catenin**. This indiates that the intracellular domain of E-cadherin affects the proliferation of the keratinocytes through its ability to in-

teract with β-catenin. (Modified from fig. 6. in Zhu, A.J. and Watt, F. (1999) β-catenin signalling modulates proliferative potential of human epidermal keratinocytes independently of intercellular adhesion. *Development* **126**, 2285–98.)

These examples begin to show how key proteins such as the integrins and β-catenin allow intricate interactions between signalling networks involved in apparently disparate functions, thereby enabling the cell to modulate its behaviour in response to its neighbours and its environment.

Cell polarity

Cells have obvious asymmetries in structure, shape, orientation, internal composition, and surface properties that are of fundamental importance for the generation of patterns during development. For example, in many early embryos such asymmetries are a prerequisite for the differential cell divisions that initiate the allocation of different cell fates and ultimately lead to the generation of cell diversity. In the case of insect eggs, these asymmetries, which are laid down during oogenesis, serve to define the basic anteroposterior and dorsoventral axes of the embryo and thus have a direct impact on the establishment of the body plan (see Fig. 6.7). In other embryos, such as those from amphibia or fish, they serve to localize proteins that will trigger specific developmental events when mobilized to specific locations following zygotic stimuli (see Fig. 6.8).

These asymmetries sometimes result in a stable polarization of the cell and can be an essential element of the differentiated function of some cell types. For example, the polarity of neurons is clearly visible in their structure, with one side of the cell body bearing a long projection, the axon, and the other a set of much shorter extensions, the dendrites. This structural polarization is essential for the function of the cell in receiving, conducting and transmitting nerve impulses which travel from the cell body down the axon. In epithelial cells, such as those in the intestine or the skin, polarization of the cell is essential for its function as a mediator between the 'outside' and the inside of the tissue or the body.

Epithelial cells are a very common cell type which makes up the sheets and tubes of cells that line external or internal body surfaces. An epithelial cell is an extreme form of polarized cell that illustrates two kinds of polarity: apico-basal and planar (Fig. 6.22). Apico-basal polarity is polarity in a plane perpendicular to the epithelium: the apical surface of the cell is exposed to the exterior (or the lumen of organs such as the intestine), and the basal or inner surface is exposed to other types of cells within the animal. Planar polarity—polarity within the plane of the epithelium—allows cells to orient themselves within the epithelium and to read and react to positional signals from sources within the plane (Fig. 6.22). The polarity of epithelial cells is reflected in the asymmetric locations of many different proteins and subcellular structures. For example, proteins functioning in information exchange with neighbouring cells in the epithelium are often localized to the lateral membranes, whereas the signalling molecules found on the apical and basal faces of the cell participate in communication with the outside environment and with different cell types, respectively. Also, differential positioning of signalling ligands and receptors at specific locations within the epithelial plane allows cells to orient themselves within the field and defines the direction of many processes linked to cell signalling (Fig. 6.22).

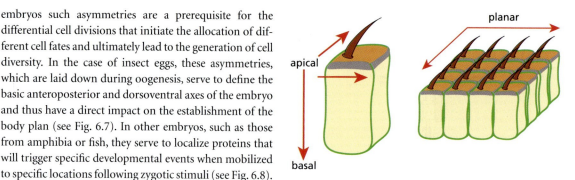

Fig. 6.22. Cell polarity. An epithelium illustrates the two planes in which cells can be polarized. The apico-basal plane, perpendicular to the main plane of the epithelium, demarcates functional domains within the cell. An obvious manifestation of apico-basal polarity in this diagram is the formation of a hair on the apical face of the cell. A plane parallel to the plane of the epithelium defines planar polarity, which plays a role in the development of pattern across a field of cells (as shown here by the orientation of the hairs).

Apico-basal polarity

Apico-basal polarity is an exclusive characteristic of epithelial cells, which are often also referred to as polarized cells (Fig. 6.22). In most epithelia, the apical surface of the cells is essentially sealed off from the basolateral surface by two structures: in vertebrates these are the zonula adherens (an actin belt linked by adherens junctions) and, on the apical side of the zonula adherens, tight junctions (Fig. 6.14 and 6.23). These specialized structures, in particular the zonula adherens, subdivide the cell surface into two domains, an apical domain and a basolateral domain, and prevent the free movement of molecules across the epithelium through the intercellular space. The apical and basolateral domains are each characterized by a specialized protein composition tailored to the function of the cell type of a particular epithelium. For example, the apical membrane of the cells of the vertebrate intestine contains protein complexes that transport glucose across the membrane from the intestinal lumen. The basal domains of many epithelia are characterized by hemidesmosomes, the junctions through which integrin-containing complexes connect the epithelium to an underlying mat of ECM, the basal lamina.

This regional specialization has to be set up and maintained by the sorting and trafficking apparatus of the cell, which must direct subcellular components, particularly vesicles loaded with proteins, to the right places. Even in

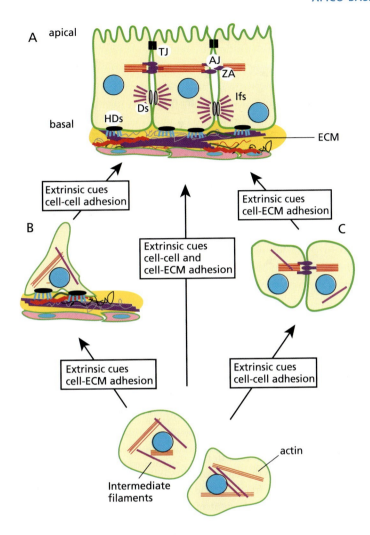

Fig. 6.23. **Mechanisms for the establishment of apico-basal polarity. (A)** A typical epithelium is characterized by the organization of the plasma membranes into domains in a plane (the apico-basal plane) perpendicular to the epithelium. The apical–basolateral boundary is demarcated by two macromolecular structures: the tight junction (TJ) and the adherens junction (AJ). Tight and adherens junctions often form belts around the cell that are visible in the electron microscope as defined structures known as the zonula adherens (ZA) in the case of AJ and zonula occludens in the case of the TJs (not shown). A number of protein complexes in this region of the cell are necessary for establishing and maintaining the apical and basal domains. The most basal surface is often characterized by specialized junctions between the cell and the ECM secreted by the underlying cells. **(B, C)** Apico-basal polarity may be induced by extrinsic cues acting on non-polarized cells and defining points of cell–ECM **(B)** or cell–cell **(C)** adhesion which nucleate and organize the epithelium. (Adapted from Yeaman, C., Grindstaff, K., and Nelson, J. (1999) New perspectives on mechanisms involved in generating epithelial cell polarity. *Physiol. Rev.* **79**, 73–98.)

non-polarized cells the secretory apparatus sorts proteins to particular subcellular domains; the polarization of epithelial cells imposes an extra set of spatial constraints on these sorting processes. How the asymmetries of polarized cells are established is not known but it is possible to envisage two types of 'nucleating events' (Fig. 6.23): adhesion between two non-polar cells, creating an interface for the stabilization of cell–cell adherens junctions or cell–matrix hemidesmosomes, or a response to some local polar signalling event that begins to localize the maintenance components to particular domains within the cell. These processes must occur in a regulated manner during

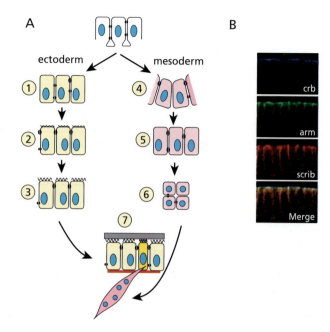

development when they contribute to the organization of tissues and organisms.

Although the establishment of the apical and basolateral domains in epithelial cells is still largely a mystery, we are beginning to learn how differences between these domains are maintained in some cell types. Invertebrates cell lack the tight junctions typical of vertebrates but next to the zonula adherens there is a domain known as the sub-apical region that performs essentially the same function. The sub-apical domain is characterized by specific protein complexes which appear to act as organizers for the polarity of the cell. In *Drosophila* and *Caenorhabditis elegans* these complexes include a number of adaptor proteins that are involved in holding together large protein complexes. Examples are the PDZ proteins bazooka, scribble, and discs large (*Drosophila*), and par-3 (*C. elegans*). These complexes also appear to include enzymes, such as the atypical protein kinase C, aPKC. Genetic analysis has revealed that these proteins are required for maintaining the apico-basal polarity of many epithelial cells: in mutants for any of these proteins, cells lose their polarity and the other proteins lose their localization (Fig. 6.24). There are homologues for some of these proteins in vertebrates. However, it is too early to say if the observations from *Drosophila* can be extrapolated to other organisms. Nor is it clear whether there is one set of proteins that generates and maintains apico-basal polarity in most epithelia, or different complexes specific to different epithelia.

Fig. 6.24. Assembly of junctions during *Drosophila* development. (A) Representation of cell interactions between epidermal cells and mesodermally derived muscle cells during *Drosophila* embryogenesis. Apical is towards the top. At the blastoderm stage (*top*) all cells are epithelial and display a number of immature adherens junctions (small buttons). In the ectoderm, cells develop adherens junctions as their epithelial characteristics become more pronounced (1–3). During the late stages, the apical side secretes cuticle and the basal side attaches to the basal membrane (7). The process of gastrulation gives rise to the mesodermal cells (4) which lose their epithelial and adhesive characteristics (5, 6). Mesodermal cells never form strong adherens junctions. Late in development they fuse to form multinucleate syncytia which develop into muscle fibres. These attach to the epidermis at specialized cells in a process that involves the function of integrins. (After Tepass, U. and Hartenstein, V. (1994) The development of cellular junctions in the *Drosophila* embryo. *Dev. Biol.* **161**, 563–96.) **(B)** Distribution of proteins involved in the apico-basal organization of epithelia in blastoderm cells from *Drosophila*. Apical is to the top and basal at the bottom. Crumbs (crb) is localized apically with a sharp boundary of expression in the subapical region. Armadillo/β-catenin (arm) is localized in the region of the adherens junction, just below the subapical domain. Scribble (scrib) is an adaptor protein, essential for the apico-basal organization of the cell, which is localized to the basolateral region and is particularly concentrated, with armadillo, at the adherens junction. In the panel at the bottom the images are merged, showing the composite distribution of the different proteins. (From Wodarz, A. (2000) Tumor suppressors: linking cell polarity and growth control. *Curr. Biol.* **10**, R624–R626.)

In some situations during normal development, epithelial cells lose their apico-basal polarity and such transitions can also be associated with changes in cell fates. For example, after the first three divisions of the mouse embryo, the cells (blastomeres) become tightly associated and display a clear apico-basal polarity, with an outer/apical surface covered with microvilli. From this mass, divisions in the external region of the embryo generate an internal non-polarized cell mass from which the embryo will develop, and an outer, highly polarized epithelium that will give rise to the extraembryonic tissue. In this case, polarization of the cells is likely to be associated with their fate.

In all organisms, gastrulation provides many examples of a more dramatic loss of polarity that has developmental consequences. During this process, groups of cells that up to this point have displayed a clear apico-basal polarity, lose this polarity synchronously as they move inside the embryo. The term 'mesenchyme' is often used to refer to such a mass of non-polarized cells and transitions of this type, called epithelial–mesenchymal transitions, figure in many large-scale topological transformations of developing embryos. They involve extensive remodelling of the cytoskeleton, as we shall discuss later in this chapter. Epithelial–mesenchymal transitions rely on a cell being able to sense its position with respect to other cells, and on a complex interplay between the machinery that mediates the cell's interaction with its neighbours, and the allocation of cell fate. These transitions are not irreversible and cells that have become mesenchymal can still in some circumstances revert to an epithelial character.

Planar polarity

The plane of the epithelium defines the second axis of polarity for a cell: the planar or tissue polarity. All cells are asymmetric along this plane to some degree, in terms of the distribution of proteins and structures within them. This polarity can sometimes be discerned from the orientation of surface structures elaborated by the apical membrane of the epithelial cells, such as the sensory hairs and bristles on the body and limbs of an adult *Drosophila* (Fig. 6.25), the scales and feathers in fish or birds or the microvilli of the 'hair' cells of the vertebrate inner ear. However, planar polarity is not restricted to specialized situations; rather, it is probably a general property of all groups of cells, reflecting the way the cells interact with their neighbours as the field of cells grows and is patterned.

In *Drosophila*, mutations in the seven-transmembrane-domain receptor frizzled and in some of the Rho GTPases have effects on planar polarity. In mutants for genes encoding these proteins, cell-surface structures such as hairs or bristles that normally reflect the planar polarity of an epithelium lose the correct orientation (Fig. 6.25). Although the polarity of an individual cell is an autonomous property that depends on the ability of a cell to orient itself within a field, it requires an input from the nearest neighbours: that is, once a cell is polarized within the plane of the epithelium it contributes to the polarity of its neighbours. This is demonstrated by analysis of the effects of mutations in *frizzled* on the hairs and bristles of the fly (Fig. 6.25). Genetic mosaic experiments in which clones of *frizzled* mutant cells are located within a wild-type epithelium have shown that the absence of frizzled affects the polarity of wild-type cells around the clone (Fig. 6.25). In contrast, clones mutant for genes downstream in the signalling pathway behave autonomously, consistent with the idea that the effects of these genes are restricted to intracellular processes. It is not yet clear, however, whether the signal carried by the ligand for frizzled is serially propagated from cell to cell in a wave-like fashion, or diffuses from a source to create a gradient across the tissue that is interpreted in a concentration-dependent fashion. All these activities must, eventually, be reflected in the organization of the cytoskeleton. An issue that remains to be resolved is whether the effects are mediated through the nucleus or directly on the cytoskeleton.

Members of the frizzled family are receptors for Wnt proteins and are coupled in an unknown manner to the adaptor protein dishevelled. Mutations in *dishevelled*, a component of the Wnt signalling pathway, also affect planar polarity, but they do so in a strictly cell-autonomous manner (Fig. 6.25) indicating that dishevelled is a component of the mechanism that relays the signal to the cell. It appears that this function of dishevelled is not mediated by β-catenin, the usual transducer of Wnt signalling.

Both apico-basal and planar polarity play major roles in pattern formation during development. It is still early days in the analysis of these processes, but it is clear that many of the answers will come from identifying the cellular machinery that provokes particular responses and then observing how the major signalling pathways impinge on these effectors in particular situations.

As mentioned above, gastrulation in most embryos provides an example in which a concerted epithelial–mesenchymal transition of a large mass of cells contributes to a morphogenetic event: the placement of the mesodermal

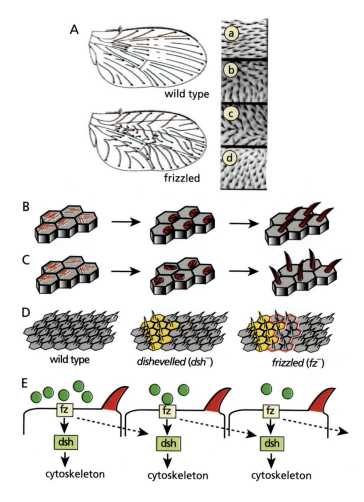

Fig. 6.25. **Planar polarity.** **(A)** *Left*: Every cell of the *Drosophila* wing secretes a hair. The arrows over the surface of the wing indicate the overall polarity of these hairs. In the wild type (top) the polarity is uniform and orientated towards the edge of the wing. In flies mutant for the gene encoding the frizzled receptor (bottom) the overall polarity is very different, much less uniform and with region-specific patterns. (From Gubb, D. and Garcia Bellido, A. (1982) A genetic analysis of the determination of cuticular polarity during development in *Drosophila*. *J. Embryol. Exp. Morphol.* **68**, 37–57.) *Right*: Details of the hair pattern on wings from wild type (a) or mutants (b–d) affecting the polarity of the hairs: *frizzled* (b), *dishevelled* (c), and *prickle* (d). In the wild type the hairs have a uniform polarity consistent with the diagram of the wild-type wing on the left. In the mutants the pattern is altered. A particular mutation alters the orientation in a reproducible way. (From fig. 4 in Wong, L. and Adler, P. (1993) Tissue polarity genes of *Drosophila* regulate the subcellular location for prehair initiation in pupal wings. *J. Cell Biol.* **123**, 209–21.) **(B, C)** Cartoons representing the activity of the actin cytoskeleton during hair formation in cells from the *Drosophila* wing. In the wild type **(B)** actin polymerizes in a defined position

within the plane of the epithelium in all cells, and hairs appear with a uniform orientation. In mutants in which the polarity is affected **(C)**, the polymerization of actin that gives rise to the hair occurs at random within the plane of the epithelium. **(D)** Fields of wing epidermal cells with hairs. The properties of the mechanisms that establish planar polarity can be assessed by generating small clones of cells mutant for genes that affect the process, in otherwise wild-type backgrounds. The clones of mutant cells are marked (yellow in the cartoon) in a way that is independent of the mutation and that allows their identification (see Chapter 4, Fig. 4.6). In these experiments, the effects of mutations in *dishevelled* are completely cell-autonomous, but in *frizzled* mutants, mutant cells can alter the polarity of wild-type cells (outlined in red) that are adjacent and in contact with them. **(E)** Experiments with genetic mosaics like those shown in D suggest that the mechanisms that regulate planar polarity have two components: a cell autonomous one that regulates the activity of the cytoskeleton, and a second component, possibly requiring the activity of the frizzled receptors, which coordinates the activities of neighbouring cells.

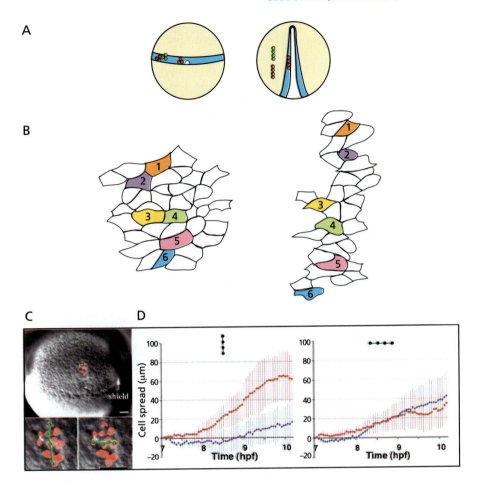

Fig. 6.26. Effects of dishevelled on cell movements during gastrulation in the zebrafish. (A) Movement of cells from the marginal zone (presumptive mesoderm) of a zebrafish embryo during gastrulation. Cells that are in groups before the invagination become organized in rows. This process is called convergent extension. **(B)** During the process of convergent extension, cell shape changes and movements of cells relative to one another lead to cell intercalation and to a reorganization of the cells along the anteroposterior axis of the embryo. The cartoon shows cells at differest stages of gastrulation, labelled by number and colour. **(C)** Lateral view of a zebrafish embryo in which a group of cells labelled with rhodamine as a lineage tracer (red; see Chapter 8, Fig. 8.2) have been transplanted to the mesodermal region. As indicated in the inset, cells could be counted and measured across two perpendicular axes (green lines). **(D)** Cells as in C were followed over time to determine the degree of spread along the anteroposterior (*left*) and mediolateral (*right*) axes. The distance between pairs of cells is referred to as 'cell spread'. The cells tracked in this experiment were wild-type (blue) or mutant for the *silberblick* (*slb*) gene which encodes a member of the dishevelled family of signalling proteins (red). Wild-type cells spread widely along the anteroposterior axis, whereas *slb* mutant cells do not. (Heisenberg, C.P. *et al.* (2000) Silberblick/Wnt11 mediates convergent extension movements during zebrafish gastrulation. *Nature* **405**, 76–81.)

group of cells on the inner side of the embryo. This implies that the transition is associated with a signal of some sort that instructs the cells to move in a particular direction. This signal might be planar, as suggested by the observation that changes in the activity of zebrafish dishevelled disrupt the pattern of gastrulation (Fig. 6.26).

Cell shape, movement, and migration

During development there are many situations in which individual cells or groups of cells undergo changes of shape

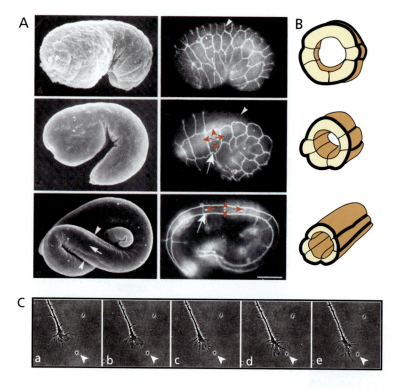

that have a specific spatial orientation and which therefore contribute to the generation of the spatial asymmetries that form the basis of pattern formation. Some obvious examples include the concerted movements that drive gastrulation. In some cases, these shape changes follow a characteristic sequence that results in a long-range movement of cells. Broadly speaking, three kinds of cell movements can be distinguished in developing embryos. In the first, groups of cells change shape in a coordinated manner and drive large-scale changes in the shape and orientation of tissues (Fig. 6.27). Examples of this kind can be found in the folding of many structures, such as the folding of the neural plate to form the neural tube (see Chapter 10), or the global activities of many epithelia, for example during the differentiation of the epidermal cells of *C. elegans* (Fig. 6.27). In a second type of movement, the cell as a whole does not relocate but rather sends out growth processes which change the overall location of the cell. The most familiar example of this type of movement is the growth of neurons, which project their axons over vast distances in cellular space (Fig. 6.27). A different kind of movement results when a cell or group of cells moves to a distant location. The clearest example of this kind is provided by migratory cells, which include the germ cells of many or-

Fig. 6.27. Cell shape changes and cell movement. (A) Cell shape changes in epidermal cells of *Caenorhabditis elegans* during morphogenesis promote an increase in body size. *Left column*: Scanning electron micrographs of *C. elegans* embryos during the late stages of embryogenesis. *Right column*: Epidermal cell outlines from embryos at stages comparable to those shown on the left, revealed with an antibody against a cell surface antigen. The shape changes in a single hypodermal cell are indicated by the frame of small red arrows. Note the dramatic elongation that, when repeated over a number of hypodermal cells, drives the extensive elongation of the embryo. **(B)** Schematic representation of the elongation process, as seen in an ideal transverse section. Much of the driving force is generated by contraction of the apical surface of the cells (thick line). (A and B from Priess, J. and Hirsh, D. (1986) *Caenorhabditis elegans* morphogenesis: the role of the cytoskeleton in elongation of the embryo. *Dev. Biol.* **117**, 156–73.) **(C)** Promotion of axonal extensions by the activity of a growth cone followed over time. Note the extension towards the point indicated by the arrow and the associated emission of filopodia and lamellipodia from the growth cone. (Image courtesy of D. Bray.)

ganisms, the cells of some early vertebrate embryos, and the neural crest cells (Fig. 6.28).

Cell shape changes reflect regulated alterations in the cytoskeleton and in some instances this is driven transcrip-

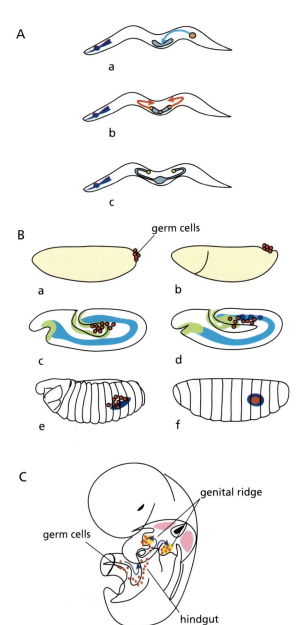

Fig. 6.28. Migrations in development. (A) Examples of cell migrations during the development of the vulva of *C. elegans*. (a) A specific muscle precursor cell, the sex myoblast (orange), is born in the posterior region of the animal during the first larval stage and migrates anteriorly towards the vulva (pale blue) where it generates uterine and vulval musculature. (b) As the vulva develops, the gonadal tip cells undergo a complex migration (red arrows) which generates the U shape of the gonad (c). **(B)** In *Drosophila*, the germ cells are determined as a subset of the posterior blastoderm cells (red in (a) and (b)). As development proceeds, the embryo undergoes cell rearrangements (c and d) and the mesoderm (blue) and the endoderm (green) are defined. The germ cells migrate through the posterior endoderm (c) and come to lie against the posterior mesoderm (d). After proliferation, the embryo undergoes the final stages of differentiation (e and f) and the germ cells become incorporated into a subset of mesodermal cells, the gonadal mesoderm (dark blue), in which they will form the gonad. **(C)** Like *Drosophila* germ cells, those of vertebrates such as the mouse are born in the posterior part of the embryo and undertake a migration (blue arrows) to the site of gonad formation (the genital ridge).

assemblies, and the spatial control of these activities determines many of the properties of the shape change. The generation of force by actin assemblies involves their interaction with myosins. Myosins use energy generated by ATP hydrolysis to induce conformational changes that translocate the molecule directionally along an actin filament. The myosins, of which there are at least ten different types, all share the same basic structure (Fig. 6.30). A globular head region at the amino terminus binds actin and has ATPase activity. The neck region mediates interactions with regulatory subunits usually known as the light chains. The tail region, which is of variable length, mediates many of the interactions with other proteins. Myosin II, which is responsible for much of the contractile activity of muscles, has a long tail that allows it to form multimers and fit into the muscle fibres like long rods. In contrast, the monomeric myosin I, common in non-muscle cells, has a very short tail through which it interacts with lipids or, in some cases, actin. The orchestration of the interactions between actin, myosin and actin binding proteins can be used to generate both intracellular movement and movement of the whole cell.

A sequence of spatially organized cell shape changes mediates cell movement (Fig. 6.31). As it starts to move, a cell defines a leading edge by emitting protrusions that appear as thin spikes (filopodia) or broad expansions of the membrane (lamellipodia). These dynamic structures are pushed out and supported by oriented bundles of actin

tionally. For example, the transcription factors slug and snail down-regulate the expression of cadherins during embryogenesis, in areas in which cells move (Fig. 6.29). Loss of cadherin results in a loss of adhesion, allowing the cells to move.

Another important target of the signals and activities that promote cell movement is the assembly and disassembly of actin filaments. A large number of actin-binding proteins and members of the family of small monomeric GTPases are involved in the dynamic regulation of actin

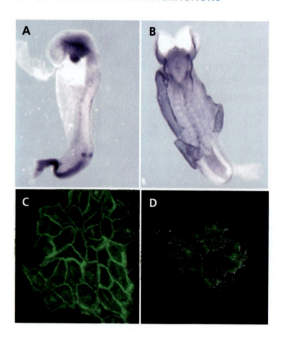

Fig. 6.29. The activity of the transcription factor snail inhibits the expression of E-cadherin and promotes epithelial–mesenchymal transformations. (A, B) The patterns of expression of mRNAs for E-cadherin (blue in A) and for the zinc-finger-containing transcription factor snail (blue in B) in 8.5-day-old mouse embryos. In regions where snail is expressed, no E-cadherin expression is seen. **(C)** E-cadherin expression in MCA3D cells. **(D)** MCA3D cells expressing the transcription factor snail. The expression of E-cadherin is strongly reduced and, as a consequence, cells lose their epithelial character (compare with c). (From Cano, A. *et al.* (2000) The transcription factor Snail controls epithelial-mesenchymal transitions by repressing E-cadherin expression. *Nature Cell Biol.* **2**, 76–83.)

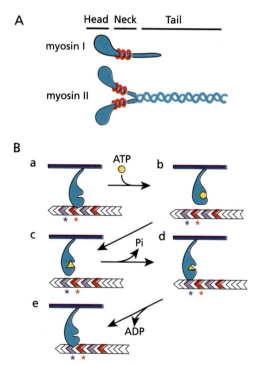

Fig. 6.30. Myosin and its dynamic interactions with actin. (A) The two best-studied myosin molecules are myosin I and myosin II. Both have a modular structure, with a globular head containing an ATP-binding site and an actin-binding site, a neck which binds light chains (red) that regulate the activity of the head, and a tail, whose length differs in the two proteins and which is responsible for interactions with other molecules. In myosin II, the long tail enables the protein to interact with itself and form multimers, which are important for its function during muscle contraction. Myosin I has a shorter tail through which it can interact with lipids in the membrane or other elements of the cytoskeleton including actin itself. **(B)** The head of myosin can bind actin to form a contractile machine. The binding to actin is very specific. The tail can bind to membranes to provide an anchor for the activity of the head. The movement of myosin over actin is powered by cycles of conformational changes as myosin binds ATP, hydrolyses it and releases ADP and Pi. In this figure, different actin monomers are indicated with different colours and asterisks to provide landmarks for the relative movement of the myosin molecule. Binding of ATP to a myosin molecule weakens its interaction with actin and activates its ATPase activity (a, b). ATP hydrolysis causes a conformational change in the myosin head which places it in a different position over the actin filament (c). In this configuration myosin binds actin as it releases Pi (d) and ADP (e) and undergoes a second conformational change that returns it to its initial configuration. Because the myosin molecule is bound to the actin filament, this conformational change exerts a force known as the 'power stroke' which displaces the myosin relative to the actin, creating movement.

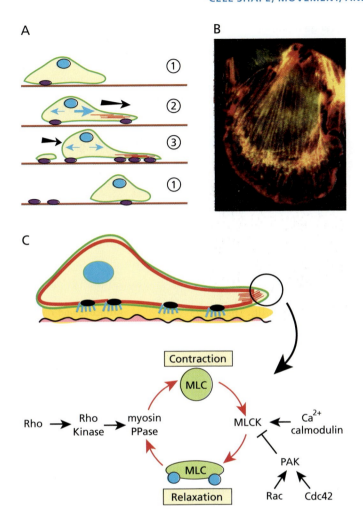

Fig. 6.31. Cell movement. (A) The movement of a cell can be described by a sequence of three stages. First (1, 2), filopodia and lamellipodia protrude at the leading edge, sample the environment and extend the realm of the cell in a polarized manner. These extensions are based on rapid rearrangements of the actin architecture mediated by actin-binding proteins and their regulators (see also Fig. 6.10, 6.11). Second (2, 3), many of these protrusions are stabilized through the formation of focal adhesion plaques between the cell and the substrate, at which integrins provide stability for the interactions between the ECM and the cytoskeleton. This adhesive state is consolidated by the formation of actin-based stress fibres across the cell which will provide the tension (dark blue arrows) needed for the net movement in the direction of the protrusions (black arrow). The final phase (3, 1) is the retraction of the 'feet' at the rear of the cell. This is sometimes accompanied by the loss of some of the material from the adhesion plaques. **(B)** Myosin (green) and actin (red) in a motile cell. Where both proteins coincide they appear yellow. In this cell, actin-based structures can be seen near the leading edge (bottom of photograph), where they play a role in the protrusion of the membrane. Immediately behind the leading edge, thick cables of actin and myosin generate the force for movement. (Image courtesy of K. Cramer and T. Mitchison.) **(C)** Regulation of myosin activity during cell movement. There are several classes of myosin and one of them myosin I, accumulates near the leading edge of many motile cells, where it can contribute to movement in two ways: by creating a force-generating network with actin and by carrying a cargo of new membrane to be added to the extending edge of the cell, moving over the actin filaments. Myosin light chains (MLC) regulate the activity of the myosin molecule (see Fig. 6.30) and their activity responds to myosin light chain kinase (MLCK) and phosphatase (MLCPPase) which, in turn, are regulated by external stimuli and various signalling pathways. The diagram indicates the effects of some of these regulatory inputs on the activity of MLC. (Adapted from Horwitz, A. and Parsons, J. (1999) Cell migration. Moving on. *Science* **286**, 1102–3.)

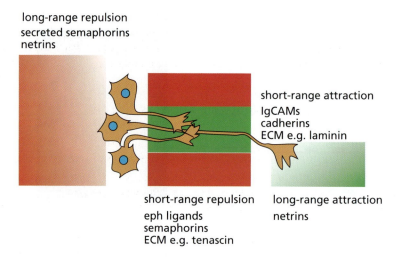

long-range repulsion
secreted semaphorins
netrins

short-range attraction
IgCAMs
cadherins
ECM e.g. laminin

short-range repulsion
eph ligands
semaphorins
ECM e.g. tenascin

long-range attraction
netrins

filaments in the cell cortex (see also Fig. 6.10). The Arp2/3 complex, which we have introduced earlier in this chapter (Fig. 6.12), nucleates the directional growth of the filaments, and the architecture of the assembly in a particular motile cell is a response to signalling information transmitted via specific GTPase and adaptor proteins. Thus, signalling by Cdc42 recruits the WASP adaptor protein and leads via the nucleating activity of Arp2/3 to the formation of parallel actin filaments that grow towards and push out the cell membrane, forming filopodia. Lamellipodia form when Arp2/3 responds to signalling via Rac and the SCAR adaptor. In this case, Arp2/3 nucleates the formation of dense arrays of diagonal actin branches from the sides of existing filaments. As these grow towards the membrane they push it out along a broad front (Fig. 6.10 and 6.12). Growth of the filaments and the localization of the growing end near the membrane involve the participation of many different actin-binding proteins (Fig. 6.31).

The protrusions lead to the formation of focal complexes that attach the leading edge of the cell to the substrate, and of actin stress fibres anchored by one end at the focal contact and by the other end either at another focal contact or in the intermediate filament network. As we have mentioned earlier, integrins and their regulation play an important role in this step, while stress fibre formation involves the regulatory activity of the small GTPase Rho (Fig. 6.10). As the cell moves, the focal contacts are strengthened and act as traction points for contractile forces generated by myosin, enabling the cell body to move forward. At the lagging edge of the cell, the focal adhesions weaken and detach. Some of the protein components of

Fig. 6.32. Attractive and repulsive cues that guide growth cones. During the development of the nervous system, axonal projections from neurons sample the environment and navigate towards their targets to create precise patterns of connectivity. Analysis of this process in vertebrates and invertebrates has revealed a number of molecular cues, common to both systems, that are present on the surface of the cells surrounding the navigating axons. There are both repulsive (red) and attractive (green) cues and they can act over a short or a long range. Short-range cues are usually cell-surface molecules, some of them involved in cell–cell adhesion; long-range cues are secreted molecules. Examples are indicated in the figure. In the absence of these cues the axons are unable to follow their pathways and an abnormal nervous system results. (Adapted from Tessier-Lavigne, M. and Goodman, C. (1996) The molecular biology of axon guidance. *Science* **274**, 1123–33.)

these complexes may be recycled through the cell for use in newly forming adhesion complexes at the leading edge, and some may be left behind, forming 'plaques' on the surface over which the cell is migrating (Fig. 6.31).

The type of cell movement in which the cell stays stationary but emits projections that grow in defined directions is a variation on the same theme and also appears to involve regulation by members of the family of small G proteins. The projection path of an axon, for example (Fig. 6.27), is the result of actin polymerization within a structure called the growth cone in response to both attractive and repulsive local cues that it encounters as growth proceeds.

Cues for cell movement

Genetic analysis has complemented cell biology approaches in confirming the involvement of actin, myosin, and the Rho/Rac/Cdc42 family of small GTPases in the fundamental mechanics of cell shape change and movement. In addition, it has also yielded some insight, at the molecular level, into how movement is initiated and spatially directed in response to environmental cues. FGF signalling, for example, has been shown to provide an orienting influence for several types of motile cells in *C. elegans*, *Drosophila*, and vertebrates (see, for example, the discussion of *Drosophila* tracheal development and vertebrate lung development in Chapter 12). Perhaps surprisingly, screens for genes involved in the movement of cells have often implicated transcription factors as mediators of the movement, for example, the Ets transcription factor pointed during the development of the tracheae (the breathing apparatus) of the fruit fly, and some GATA factors in vertebrates. The identity of other genes implicated by these screens has revealed the importance of remodelling the ECM as the cells move; for example, the *gon-1* gene of *C. elegans*, which encodes a secreted metalloprotease, is essential for the migration of the leader cells of the nematode gonad. Not surprisingly, the results of similar screens in *Drosophila* suggest that lipid metabolism is also likely to play a prominent role in these processes. The genes *wunen* (*wun*), which encodes a protein related to lipid phosphatase 2A, and *columbus* (*clb*), which encodes an enzyme (HMGCoA reductase) that in mammals is involved in the biosynthesis of cholesterol, have both emerged from such screens. In *wun* mutants, germ cells migrate into regions of the embryo that are normally forbidden to them, whereas in *clb* mutants they do not follow the normal migration path through the midgut.

One of the situations in which cues are most important is directing the movement of the growth cones of axons, which ultimately generate neural networks. Genetic and biochemical analysis has played an important role in the deciphering of these cues and has yielded a large number of molecules that exert either attractive or repulsive influences on the growth cone. The include a variety of signalling molecules of the ephrin, semaphorin, NCAM, and netrin families (see Chapter 4) whose receptors are linked to the activity of the small GTPases (Fig. 6.32).

SUMMARY

1. Cells are the units of development. Surrounded by a lipid membrane, they contain several different types of organelles and are traversed by complex internal membrane systems and a proteinaceous structual framework called the cytoskeleton.

2. Cells secrete a microenvironment that surrounds them and provides a platform for their interactions. This microenvironment contains two layers: one, very close to the cell surface, is made up of proteins tightly associated with the membrane of the cell and provides a link between the outside and the inside of the cell. The second, the extracellular matrix (ECM), is broader, not linked directly to the cell surface and creates a medium for the regulation of cell interactions, whether adhesive or signalling.

3. The composition, internal architecture, and organization of cells change during development. These changes and their regulation constitute a set of basic developmental 'routines'.

4. This microenvironment is in a dynamic state as cells change their fates and composition during development.

5. Intermediate filaments, actin and tubulin are protein polymers that form the core of the cytoskeleton, an intracellular protein network that buttresses the structure of the cell and mediates changes in its shape and behaviour.

6. Cells associate with each other to create ensembles that form the basis for their organization into tissues and organs. These associations rely on specialized cell-surface molecules that mediate direct interactions between cells (cadherins and immunoglobulin-like molecules) or between cells and the ECM (integrins).

7. The activity of the cytoskeleton is physically coupled to the activity of receptors that mediate cell interactions and this allows the shape of the cell to be regulated in response to changing environment and signalling events.

8. Individual cells have an asymmetric organization that imposes spatial constraints on cell interactions. That is, physical associations or the exchange of information with other cells are limited to certain interfaces. The propagation of these asymmetries is fundamental to the patterning of fields of cells.

9. Epithelial cells are a common type of cell, which by virtue of their structural polarization, contribute to multiple morphogenetic processes. These cells are polarized both in a plane perpendicular to the epithelium (apico-basal polarity) and in the plane of the epithelium (planar polarity). Mesenchymal cells do not display this overt polarization. Epithelial–mesenchymal transitions play an important role in the spatial reorganization of cells during development.

10. External signalling events regulate the activities of the cytoskeleton and cell–cell interactions, determining patterns of cellular associations and movement that contribute to the generation of form in multicellular organisms.

COMPLEMENTARY READING

The understanding of cell biology, and of the contribution that regulated changes in cell shape, polarity, adhesion and movement make to developmental processes, is growing rapidly. Once again, the best sources of up-to-date information are review journals, in particular *Trends in Cell Biology* and *Current Opinion in Cell Biology*, as well as reviews published in primary journals such as *Nature* and *Molecular and Cell Biology*. The following books cover, in greater detail, some of the basic cell biology topics discussed in this chapter.

Alberts, B. *et al.* (1994) *The molecular biology of the cell* (3rd edition) Chapters 16 and 19. Garland.

Bray, D. (2000) *Cell movements: From molecules to motility.* Garland.

Edelman, G. (1993) *Topobiology. An introduction to molecular embryology.* Basic Books.

Lodish, H. *et al.* (2000) *Molecular cell biology* (4th edition), Chapters 18, 19 and 20. Freeman.

Chapter 7

Basic cellular routines: Division, differentiation, and death

The generation of a large number of cells is an essential factor in the development and patterning of multicellular organisms. It is achieved through many rounds of cell division which are often associated with the assignment of different programs of gene expression to the emerging cell populations. The spatial and temporal control of cell division are important elements in the generation of shape and form. However, as the number of cells increases, the overall size of the animal also increases, indicating that the size of the cell is regulated in concert with its division.

The death of a cell can be thought of as the opposite of its division and it is interesting that the development of an organism relies on a balance between these two processes. Cell death is known to be common in embryos, and experiments during the last decade or so have shown that this is not a random phenomenon, but a regulated form of cell differentiation that is essential for normal development. Death appears to be a 'default' choice for cells in many types of multicellular organisms and both survival and division are actively triggered by signals from the cell's environment.

In this chapter, we introduce the components of the molecular machinery that drives cell proliferation and cell death and we discuss what is known about the control of this machinery and its relationship to development and pattern formation. We look at these processes as routines that emerge from molecular interactions in a cellular context; when these routines come under coordinated control through cellular interactions, they play a fundamental role in the shaping of cellular ensembles.

The cell cycle

During development, some animals have to produce as many as 10^{15} cells from the single-celled zygote. As shown by the cloning experiments discussed in Chapter 2 (Fig. 2.2), most cells do not lose information during this process, so every division has to be preceded by the faithful replication of DNA, and the division process itself must involve the precise partitioning of the two copies of the genome between the daughter cells. In general, cells do not get smaller as proliferation proceeds so, in addition, each division is usually followed by a period of replenishment of cellular constituents.

From the point of view of the genetic material, the behaviour of the chromosomes illustrates the cycle of events that underlies each cell division (Fig. 7.1): every replicated chromosome has two chromatids, which segregate to each of the daughter cells during every somatic division or mitosis (M). After mitosis, each of the chromatids is replicated during a period of DNA synthesis (S), thereby creating the templates for the next round of division. During development there is a cyclical alternation between the M and S phases. In some situations, such as the early fast cleavages of some embryos, there is no interval between M and S. However, as development proceeds the rhythm of cell division slows down and gaps (G) are interspersed—in a cell- and stage-specific manner—between M and S. The period between M and S, called G1, is usually associated with the replenishment of cell contents and growth in size. The period between S and M is called G2. Together with S, G1 and G2 are known as the interphase.

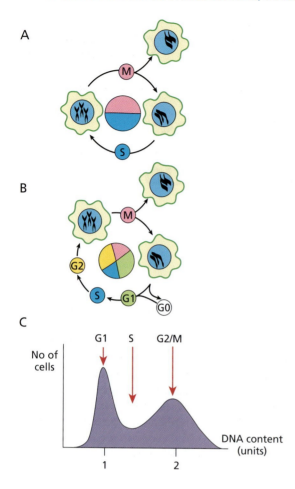

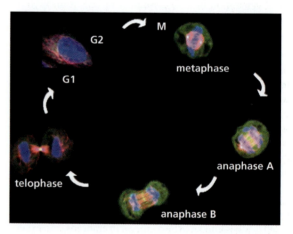

Fig. 7.1. **The cell cycle. (A)** The simplest cell cycle, common in the early stages of many embryos, alternates cell division or mitosis (M) with the replication of the genetic material (S phase). **(B)** The most common variation of the alternation of M and S phases is the introduction of two gaps or G phases, one before the S phase (G1) and another before mitosis (G2). Much of what a cell does during development depends on the length and regulation of these gaps. During G1 cells have the option of arresting the cycle in a state known as G0, from which they can re-enter the cycle under appropriate conditions. Chromosomes are indicated in the nuclei before (single chromatid) and after (double chromatid) replication. **(C)** The cell-cycle phase of individual cells in a growing population can be assessed by staining the DNA with a fluorescent dye and passing the cells through a fluorescence-activated cell sorter to determine the DNA content of each cell in the population. Plotting DNA content against the number of cells gives a bimodal distribution with two peaks, indicating that cells in G2 (before replication) have twice as much DNA as cells in G1 (before replication).

Fig. 7.2. **Cellular dynamics during the cell cycle.** Micrographs of HeLa cells at different stages of the cell cycle. Microtubules are highlighted by a red fluorescent antibody while the DNA in the nucleus is stained with DAPI (blue). The green reveals the presence of an enzyme (polo kinase) that is present during mitosis. (Image courtesy of D. Glover.)

The cell-cycle phase of cells in a population can be determined by measuring the amount of DNA of individual cells (Fig. 7.1). The reiterated succession of these four phases generates the cellular routine known as the cell cycle (Fig. 7.1 and 7.2).

The overall timing of the cell cycle and the relative lengths of its four phases vary enormously in different cell types and at different stages in development. In a complete cycle, M and S tend to be the shortest phases and cells spend most of their time in the period of interphase, where they carry out most of their activities. Within this period, the gap phases may lengthen, shorten, or be skipped altogether depending on the physiological or developmental circumstances. For example, in some embryos the rapid cycles during the early cleavages last on the order of 15 to 30 minutes; this requires very short interphases which are skipped altogether in many embryos. At the other end of the spectrum, mature eukaryotic cells have average cycles of 10 to 20 hours with extended gap phases. Some cells may arrest their cell cycle, either temporarily or permanently. For example, starved yeast cells or mammalian cells deprived of growth factors arrest during G1 at a point sometimes called Start (in yeast) or the 'restriction point' (in animal cells). Non-dividing cells may entering a quiescent state known as G0: adding back mitogenic factors can stimulate the cell to re-enter G1 and progress through the

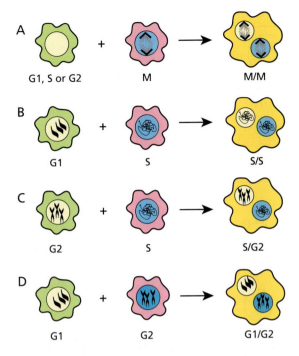

Fig. 7.3. Cell fusion experiments reveal the existence of factors promoting different activities at different phases of the cell cycle. (A) Cells in G1, S or G2 when fused with cells in M are induced to undergo chromosome condensation and other events associated with mitosis, such as spindle formation and nuclear breakdown. This shows that mitosis is dominant to other states of the cell cycle and that mitotic cells contain factors that mediate this dominance. **(B)** Fusing cells in G1 with cells in S induces premature replication of the G1 nuclei. This suggests that replication is dominant over G1, and that nuclei in G1 are competent to replicate their DNA. **(C)** Fusing cells in G2 with cells in S does not alter the activity of either nucleus: cells in S do not speed up their replication, and cells in G2 do not re-replicate their DNA. **(D)** Fusing cells in G1 with cells in G2 does not alter the activity of either of these nuclei. (After Rao, P. and Johnston, R. (1970) Mammalian cell fusion studies on the regulation of DNA synthesis and mitosis. *Nature* **225**, 159–64; Johnson, R. and Rao, P. (1970) Mammalian cell fusion: induction of premature chromosome condensation in interphase nuclei. *Nature* **226**, 717–22.)

cell cycle. Cells in G0 slow down their metabolism, and often use this state as an opportunity to undergo growth that is not linked to division.

There is, of course, an important exception to the cell division process we have described so far: the specialized cell division, known as meiosis, that takes place during the generation of germ cells. During meiosis, the diploid genome is divided into its two haploid constituents, en-suring that fertilization maintains rather than duplicates the amount of genetic material. This is achieved by a sequence of two specialized divisions that occur without an intervening round of DNA replication: during the first, a non-conventional mitosis, the number of chromosomes is reduced by half, and in the second, the haploid chromatids are apportioned to each of the daughter cells. Although meiosis is undoubtedly an important event in the development of an organism, we shall concentrate in the rest of this chapter on the molecular engine that drives the mitotic cell divisions that are responsible for cell proliferation.

Uncovering the cell-cycle regulators

The variety of cell cycles that can be observed in nature suggests that there must be regulators that control the lengths of the cell-cycle phases and the transitions between them. Studies on heterokaryons (see Chapter 3, Fig. 3.1) provided some of the first experimental evidence for such regulatory factors and showed that the presence and/or activities of these factors fluctuate during the cycle (Fig. 7.3). Fusing cells in either S, G1 or G2 with cells in M led the interphase cells to condense their chromosomes and enter M, indicating that cells in M phase have components that can drive cells in other stages of the cell cycle into mitosis. In other experiments (Fig. 7.3), cells in either G1 or G2 were fused with cells in S. In the first case (G1 + S), the G1 nuclei immediately began to replicate their DNA. In the second (G2 + S), the nuclei from the cells in S continued to replicate but the nuclei in G2 did not enter mitosis until the nuclei in S had completed replication. These experiments indicate not only that there are regulatory activities that can drive cells into M or S phase (so-called M- and S-phase-promoting factors), but also suggest that there are controls to ensure that cells cannot enter the next phase of the cycle unless all the events of the previous phase have been properly completed. They also suggest that there is a mechanism to ensure that the DNA, once replicated, cannot be replicated again until mitosis has occurred. This is why cells in S can drive G1 cells but not G2 into replication.

The first advance in uncovering the molecular nature of cell-cycle regulators came from experiments in which *Xenopus* oocytes were induced into meiosis—or 'maturation'—by application of the hormone progesterone (Fig. 7.4). It was observed that once maturation of the oocyte to

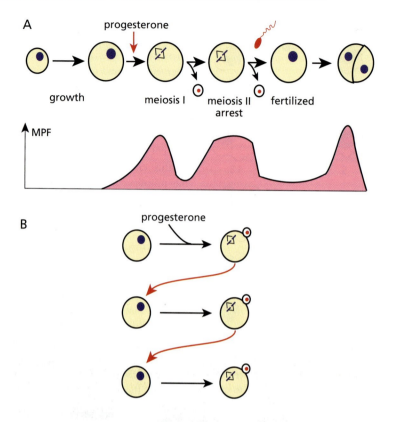

an unfertilized egg had been triggered, its cytoplasm could be used to trigger the maturation of other oocytes, in the absence of hormone. These experiments complemented those performed in heterokaryons and suggested the existence of a cytoplasmic factor which regulated the entry of cells into division; this factor was called MPF (initially for maturation promoting factor but, helpfully, also standing for mitosis-, meiosis- or M-phase promoting factor). MPF activity is not restricted to the meiotic process since after fertilization, the embryo was shown to contain MPF activity that rose and fell with the rhythm of the cell cycle (Fig. 7.4). In these experiments, the activity of MPF did not require DNA synthesis, since it could be obtained from enucleated eggs or eggs in which DNA synthesis had been blocked. The activity was also found in other cell types and other organisms.

Experiments that were performed first with clam eggs showed that the oscillation in MPF activity coincided with the appearance and disappearance of a protein that was dubbed cyclin because of its periodic pattern of expression (Fig. 7.5). Eventually, purification of MPF showed that it is composed of two subunits: a catalytic kinase subunit, which is present throughout the cell cycle, and an oscillat-

Fig. 7.4. MPF and its fluctuations during the cell cycle. (A) After being born, *Xenopus* oocytes replicate their DNA, arrest in G2 and grow to a large size (up to 1 mm in diameter). A fully grown oocyte can be induced by the hormone progesterone to undergo the first meiotic division and to initiate the second, where it is arrested at metaphase until fertilization. The entry of the sperm triggers the completion of the second meiosis and initiates the early embryonic cell cycles. **(B)** Oocytes that have been treated with progesterone contain a cytoplasmic factor that can be retrieved and transplanted into other oocytes that are arrested in G2. These oocytes are induced to enter meiosis without progesterone by the transplanted factor. The inducing activity is called 'maturation promoting factor' (MPF) because it induces the maturation of an oocyte into a mature egg. Measurement of the concentration of this activity during the maturation of the oocyte (see MPF in A) shows that it is found only in cells that are in mitosis.

ing regulatory cyclin subunit that determines the activity of the kinase. The kinase component is usually referred to as cyclin-dependent kinase, or Cdk, to reflect the absolute dependence of its activity on its association with cyclin.

Related experiments revealed the existence of S-phase promoting factors (SPF), which also proved to be composed of a cyclin and a Cdk. In agreement with the results

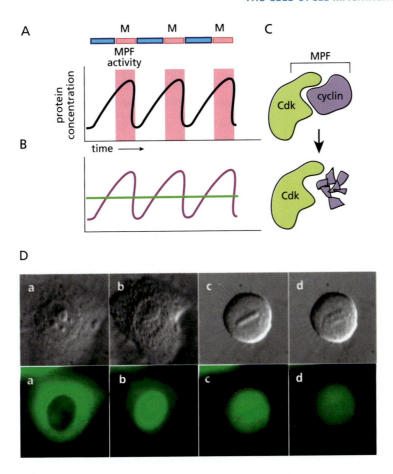

Fig. 7.5. Composition of MPF. (A) Measurements of protein profiles during the development of many embryos show that there are some proteins whose concentration cycles over time. The peaks of concentration of these proteins are associated with the mitotic phases of the cell cycles (M) and coincide with the activity of MPF (pink boxes). These proteins are called cyclins. **(B)** Biochemical analysis shows that MPF is composed of two subunits, a cyclin and a protein kinase (cyclin-dependent protein kinase, or Cdk) whose activity is strictly dependent on its association with the cyclin. In contrast to the cyclic variations in the concentration of cyclins during the cell cycle (purple line), the concentration of Cdk is constant (green line). **(C)** During mitosis the cyclin is destroyed, coinciding with the sharp decline in A, and this inactivates the associated Cdk. **(D)** Changes in the intracellular localization of cyclin B1 during the cell cycle in a HeLa cell. The images are from a live video of cells expressing a cyclin : GFP fusion protein. The upper row shows Nomarski images and the bottom row shows fluorescent images of the same cell. At interphase (a), cyclin is cytoplasmic and as mitosis begins (b), it enters the nucleus where it remains until, during the transition from metaphase (c) to anaphase (d), it is degraded. (Photographs courtesy of J. Pines.)

from heterokaryons made from cells in different phases of the cell cycle (Fig. 7.3), whereas MPF could induce chromosome condensation and mitosis in cells at any stage of their cell cycle, SPF induced cells in G1, but not G2, to replicate their DNA.

The cell-cycle machinery at work

In parallel with biochemical approaches using eggs from frogs and marine invertebrates, genetic experiments in both budding and fission yeast identified a number of genes whose mutant phenotypes suggested that their products play roles in cell-cycle regulation. In the course of these experiments it became apparent that many of these genes and their products are highly conserved, not just structurally, but also functionally, in all eukaryotes. This was most dramatically demonstrated by the ability of certain human genes to complement cell-cycle defects in yeast mutants, an experiment that suggested it might be possible

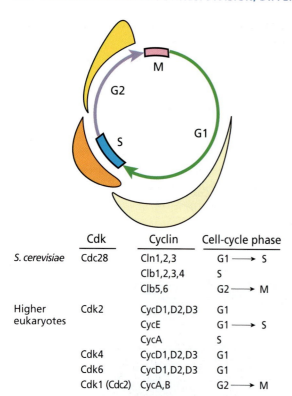

	Cdk	Cyclin	Cell-cycle phase
S. cerevisiae	Cdc28	Cln1,2,3	G1 ⟶ S
		Clb1,2,3,4	S
		Clb5,6	G2 ⟶ M
Higher eukaryotes	Cdk2	CycD1,D2,D3	G1
		CycE	G1 ⟶ S
		CycA	S
	Cdk4	CycD1,D2,D3	G1
	Cdk6	CycD1,D2,D3	G1
	Cdk1 (Cdc2)	CycA,B	G2 ⟶ M

Fig. 7.6. Overview of cell-cycle regulation. (A) Progress through the different stages of the cell cycle is determined by the presence and activity of phase-specific cyclin–Cdk complexes. The concentration and intracellular location of the cyclin subunit is a key regulatory factor of these complexes. The cartoon illustrates an 'average' cell cycle, with the fluctuations in the activity of specific cyclin–Cdk complexes associated with different phases illustrated by the thickness of their profiles. S, synthesis; M, mitosis. Note that at the end of each phase the activity of a particular cyclin–Cdk complex disappears as its cyclin component is degraded, and a new one appears. Each cyclin–Cdk complex acts on a set of downstream effectors which mediate the biochemical and cell biological functions associated with that stage: DNA synthesis and replication in S, or the mechanics of cell division during M. **(B)** Table indicating a representative sample of the different cyclins and Cdks in yeast and higher eukaryotes, as well as the cell-cycle phase regulated by particular cyclin–Cdk complexes.

Fig. 7.7. Mechanics of the cell cycle. (A) Progression through the cell cycle is promoted by the activity-coupled assembly and disassembly of cyclin–Cdk complexes. The activity of these complexes is regulated by combinations of factors that act on the components of the complex, causing their synthesis and destruction, in the case of cyclins, or post-translational modification in the case of Cdks. In addition, the activity of the complex is regulated by association with specific proteins, in particular Cdk inhibitors (Cki) whose activity is under tight regulation. **(B)** The events of the cell cycle, generalized from the results of research on yeast and higher eukaryotes. The cycle can be described from any point; here, we begin in G2. The activity of MPF, consisting of the cyclin-dependent kinase Cdk1/Cdc2 (pink) and the mitotic cyclin A or B (blue), is determined both by the synthesis and degradation of its cyclin component, and by the phosphorylation state of the Cdk at position 15 (Y15) and position 161 (T161). As the cyclin B concentration rises in G2, the Cdc2/Cdk1–cyclin B complex forms, and is a substrate for the Wee1 kinase, which phosphorylates Y15. The Y15-phosphorylated complex is in turn a substrate for Cdc2/Cdk1-activating kinase (AK), which phosphorylates T161 and renders the Cdc2/Cdk1–cyclin B complex ready for activation. This is mediated by the activity of the Cdc25 phosphatase, which specifically dephosphorylates Y15. The active complex (phosphorylated on T161 and dephosphorylated on Y15) has properties that favour maintenance of its own activity: it promotes the activity of Cdc25 phosphatase and inhibits Wee1 kinase; as a result, the amount of active complex rises through G2. The complex acts upon several proteins that carry out the mechanics of mitosis (spindle assembly, chromosome condensation, and nuclear envelope breakdown). Above a certain threshold, however, the complex begins to promote its own deactivation and the cyclin is proteolytically degraded. A specific phosphatase (PPase) dephosphorylates Cdc2/Cdk1, leaving it inactive until the cyclin B concentration rises again in G2. After mitosis and cell division are completed, cells enter G1 where, depending on the molecular environmental conditions of the cell, they have an opportunity to leave the cell cycle and arrest in a state known as G0. Progression through G1 and the transition between G1 and S are regulated in a similar way to the regulation of the G2 to M transition, by the activity of complexes between Cdks and cyclins (D or E in this case), but with the additional involvement of Ckis whose proteolytic inactivation is essential for the activity of the cyclin–Cdk complexes. Important targets of the cyclin-Cdk complexes involved in the G1 to S transition include genes involved in DNA replication. During S phase, there is also a rise and decline in the activity of specific cyclin–Cdk complexes (not shown).

to identify cell-cycle regulators by complementation experiments across species.

The functional conservation of the key components of the cell cycle (Fig. 7.6) enables us to outline a universal cell cycle whose basic features, described in more detail in Fig. 7.7, are common to all eukaryotic cells. The overall picture is one in which regulatory complexes specific to successive phases of the cell cycle are sequentially activated and deactivated. The basic units of these complexes, phase-specific cyclin–Cdks, are responsible both for initiating events specific to a particular phase of the cycle (for example, activation of replication enzymes in S, and assembly of the

spindle apparatus in M), and for triggering the regulatory events that control the transition to the next phase of the cycle.

The specificity of cyclin–Cdks for different phases of the cell cycle is conferred by their cyclin subunit: there are mitotic cylins, S phase cyclins, and G1 cyclins (Fig. 7.6). Some Cdks can associate with more than one cyclin. For example, in yeast, the Cdk encoded by the *CDC28* gene in *Saccharomyces cerevisiae* and *cdc2* in *Schizosaccharomyces pombe* associates with mitotic cyclins to promote entry into M phase, with S phase cyclins to trigger DNA replication, and with G1 cyclins to drive the cell through G1 and

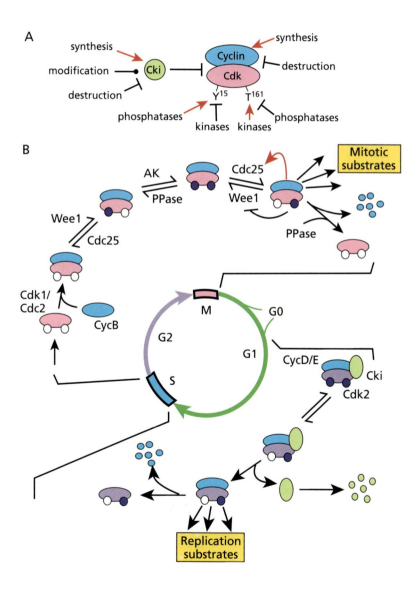

control the transition from G1 to S. Higher eukaryotes, however, have multiple Cdks, and the cell cycle phases and transitions between them are associated with specific sets of cyclin–Cdk complexes.

The activity of the cyclin–Cdks is controlled by multiple mechanisms (Fig. 7.7), in particular the regulated synthesis and destruction of their cyclin subunits. Because the activity of Cdks is absolutely dependent on cyclins, this control mechanism is important for ensuring that the cell cycle is directional: the destruction of a key regulator of a particular phase will both bring that phase to an end and allow the events that have been set in motion by its activity to take over and initiate the next phase. Cyclin–Cdk complexes are also regulated by other mechanisms, including post-translational phosphorylation/dephosphorylation of

the kinase subunit, and the activity of a set of inhibitory proteins called cyclin-dependent kinase inhibitors (Ckis) (Fig. 7.7).

Much of the detailed work on cell-cycle regulation has been done using yeast and there is no doubt that in multicellular organisms there are both more proteins that participate in cell-cycle reulation and more regulatory pathways that control the transitions between the different phases of the cycle. However, the basic principles are common to all eukaryotic organisms. Within this overall framework, the set of substrates for cyclin–Cdk activity—that is the effectors for each phase—are likely to be specific for each organism and probably even for different cells within an organism.

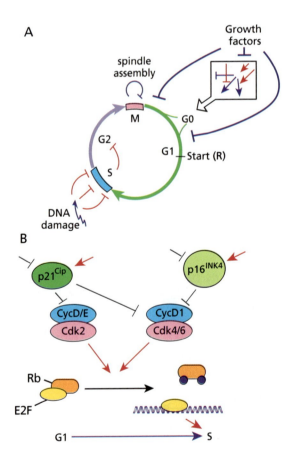

Fig. 7.8. Checkpoints. (A) Checkpoints are regulatory pathways that operate at several steps in the cell cycle to ensure the fidelity of the process. During G1, a checkpoint control ensures that cells will only proceed to replicate their DNA if they are ready for mitosis. The elements of this checkpoint integrate a number of growth and developmental signals and act at the Start or restriction point (R). Beyond this point the only quality controls cells will be subjected to concern the integrity of their DNA and mitotic machinery and thus only cells that can and should divide will proceed beyond it. During S phase another checkpoint ensures that only cells with undamaged DNA replicate their chromosomes. S phase, in turn, has a regulatory effect on G2, which ensures that only cells that have replicated their DNA correctly proceed onwards to the G2–M transition. During mitosis, a further checkpoint at the point of spindle assembly ensures that mitosis only proceeds when the mitotic apparatus is properly assembled. (Adapted from Elledge, S. (1996) Cell-cycle checkpoints: preventing an identity crisis. *Science* **274**, 1664–72.) **(B)** An important target of the cyclin–Cdk complexes regulating G1 and the transition from G1 to S (cyclin D- and E-containing complexes in higher eukaryotes) is the transcription factor E2F, which regulates the expression of genes involved in replication. The activity of E2F is inhibited by its association with the retinoblastoma (Rb) protein. The cyclin–Cdk complexes phosphorylate Rb, inducing its dissociation from E2F and thus allowing the transcription of S-specific genes. The activity of the cyclin–Cdk complexes is in turn regulated by two types of Ckis: the Cip/Kip (p21) class, which regulate the activity of both types of cyclin–Cdk complexes, and the Ink4 (p16) class, which regulate only the cyclin D complexes. The activity of these Ckis is subject to controls from the general signalling machinery of the cell as well as by transcription factors, thereby integrating developmental and metabolic signals with progression from G1 to S phase.

Cell-cycle checkpoints: Integration and surveillance

A complex set of surveillance mechanisms ensures not only the internal coordination of the cell cycle, but also that the cell-cycle transitions are in tune both with the biochemical state of the cell and with the signals it is receiving from its environment (Fig. 7.8). The stages of the cell cycle at which there is a quality control of the physiology of the cell and/or its DNA, are called 'checkpoints' and ensure that the cell cycle will not proceed unless the right conditions for progress are met. For example, blocking DNA replication leads to arrest of the cell cycle before entry into mitosis, while DNA damage leads to arrest in the G1, S, and G2 phase, and induction of the DNA repair apparatus. In both cases it is clear that it would not be to the advantage of the cell to proceed, since it would propagate an incomplete or damaged set of genes that would lead to abnormal cell activities in the descendant cells.

An important checkpoint from the developmental point of view is at the transition between G1 and S. As we have seen, this is a point of no return in the cycle which, when crossed, commits a cell to a full cycle (barring major mistakes in its DNA). This checkpoint is not only important as a sensor for cellular damage, but also as a regulatory point that enables the cell cycle to be coordinated with the cell's developmental program. Many cell types, for example, exit the cell cycle when they begin to differentiate, and do not divide again. There are several parameters that can feed into control at this point, such as the size of the cell, its metabolic state, and the presence of transcription factors mediating differentiation. Together, these parameters determine whether a cell is ready to divide or whether its developmental program requires one further division.

Checkpoints controls are implemented when 'sensing proteins' lead to the activation of signal transduction pathways whose ultimate effectors or targets are the key regulators of the cell-cycle oscillator and the mitotic apparatus itself, in particular the cyclin–Cdk complexes (Fig. 7.8). Checkpoint signalling pathways regulate cyclin–Cdks, both positively and negatively, by a variety of mechanisms including phosphorylation and dephosphorylation by specific kinase and phosphatase enzymes, and the action of Ckis. The progression from G1 to S in flies and mammals, for example, is dependent on a number of Ckis (members of the Cip/Kip family, Fig. 7.8) which regulate the activity of cyclinD– and cyclinE–Cdk complexes. Because the activity of these Ckis is under the control of growth factors and cell interactions, this checkpoint provides a way of linking progress through the cell cycle to the processes that determine cell diversity during development, since many of these growth factors are also involved in the regulation of transcription and thereby the fate of the cells. Division is as much part of a fate as the acquisition of cell-surface, cytoskeletal, or signalling properties.

The cell cycle in early development

The modular nature of the cell cycle and the mechanisms for its regulation at intrinsic control points enable the cycle to be adapted to the needs of different cells during development. For example, the lengths of the different phases can be adjusted by linking the activity of the cyclin–Cdk regulators to the cell signalling network that mediates cell interactions, thus ensuring that a cell will divide only if its fate requires it to.

The early stages in the development of many animals provide several examples of the regulative flexibility of the cell cycle (Fig. 7.9). In animals that develop from large eggs rich in maternal components (such as *Drosophila* and *Xenopus*), the initial cell divisions, which essentially partition the zygote into a set of smaller cells, are rapid, lack gap (G) phases, and are not constrained by the need for cell growth between divisions. Probably because of the fast pace of division, there is no transcription during this period. Eventually, one or more maternally supplied cell-cycle components become limiting, arresting or slowing the cell cycle until zygotic transcription replenishes the factor and a slower rhythm of cell division takes over. The fact that this point (sometimes known as the midblastula transition or MBT) is reached after a defined number of divisions characteristic of each species suggests that it is regulated by a mechanism capable of sensing the number of times that cells have divided.

The ratio of nuclei to cytoplasm is one parameter that could be measured by the cells during the early cleavages (Fig. 7.9). Some experiments support this possibility (Fig. 7.10). For example, it is possible to obtain haploid embryos of *Drosophila* and observe the cell cycles of their nuclei early in development. These embryos have a lower nuclear : cytoplasmic ratio than wild-type diploids and their MBT is delayed compared to wild type. Conversely, polyploid embryos, or embryos in which the DNA content has been increased artificially by injection, undergo MBT earlier than

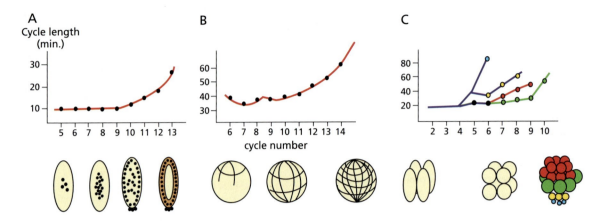

Fig. 7.9. Strategies in cell-cycle regulation during early development. Variations in the rates of cleavage (i.e. cell-cycle length) in early *Drosophila* **(A)**, *Xenopus* **(B)**, and sea urchin **(C)** embryos. The lower panels show, diagrammatically, the patterns of the early divisions in each species. (Note that in *Drosophila* the early nuclear divisions generate a syncytium.) The mid-blastula transition (MBT) signals the point where the rate of the cell cycle slows down and G phases are introduced. The MBT might not be a universal inflection point, but rather a species-specific process of change in the rhythm of cleavage that is often accompanied by the onset of zygotic transcription. In sea urchin embryos (C), in which early cleavages generate tiers of cells that are very different in size, the dynamics of division are different in each tier, with bigger cells dividing more times than smaller ones. In the figure, the colours of the different tiers correspond to the colours of the lines on the graph. The correlation between size and number of divisions suggests that the ratio of nucleus to cytoplasm is an important determinant of this number: a low ratio leads to more divisions. (Adapted from Yasuda, G. and Schubiger, G. (1992) Temporal regulation in the early embryo: is MBT too good to be true? *Trends Genet.* **8**, 124–7.)

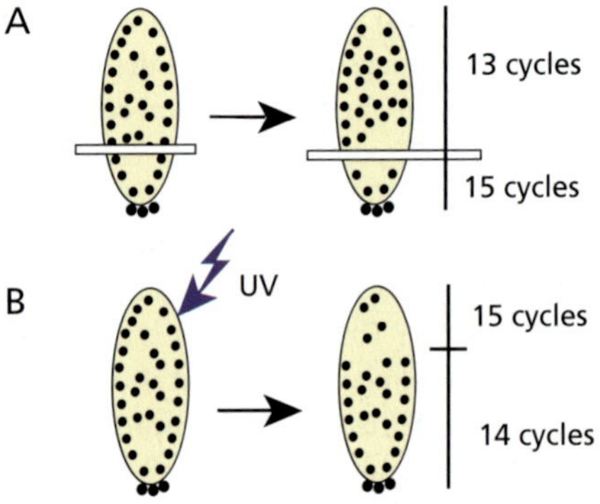

Fig. 7.10. The nucleocytoplasmic ratio and the dynamics of cell division. The relationship between the nucleocytoplasmic ratio and the dynamics of cell division can be tested experimentally. In *Drosophila*, where the blastoderm is a syncytium, the number of nuclei in a given cytoplasmic region can be changed by ligating **(A)** or locally irradiating **(B)** the embryos. These procedures generate local domains with high or low nucleocytoplasmic ratios, respectively. Nuclei in the blastoderm normally divide 14 times. A low ratio of nuclei to cytoplasm increases the number of divisions that those nuclei undergo, while a high ratio reduces this number.

wild type. In all cases, the resumption of the cell cycle after the MBT results in a slower cell cycle; in some cases, G phases appear.

Changes in the cell cycle during development have been studied in detail in *Drosophila*. These studies illustrate both the range of cell-cycle modes and tempos that cells can employ, and the variety of regulatory controls (Fig. 7.11). In the fly embryo, 13 rapid synchronous nuclear divisions, which lack G1 (and, for the most part G2) phases and are not accompanied by cytokinesis, create the syncytial blastoderm. During the first 13 cycles, progressive depletion of a maternal factor, which recent work suggests may be a factor required for DNA replication, leads to a gradual lengthening of interphase and, by a so-far unknown mechanism, to an associated activation of zygotic transcription at the MBT. These new zygotic gene products promote the degradation of the maternally

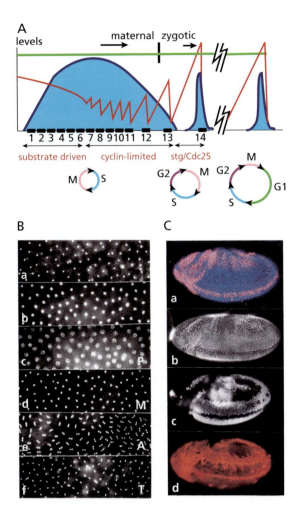

Fig. 7.11. **Regulation of the cell cycle during *Drosophila* development. (A)** Fluctuations in the concentration of cyclin B (red line), string/Cdc25 (blue), and Cdc2 (green line) during early *Drosophila* embryogenesis. The horizontal axis indicates the lengths of the 14 successive cell cycles. The first 13 cycles rely on maternally deposited stores of proteins and RNA that are progressively depleted, resulting in a continuous alternation of M and S phases. The early cycles are very fast and only when cyclin becomes limiting do they begin to slow down (cycles 11, 12, and 13). Cycle 14 and those that follow are driven by activation of the cyclinB–Cdc2 complex by dephosphorylation mediated by zygotically transcribed string/Cdc25 phosphatase; this regulatory input introduces the G2 phase which defines the timing of string/Cdc25 activity. Later in embryogenesis the G1 phase is introduced. **(B)** Photomicrographs showing nuclei undergoing synchronous mitoses in the syncytial blastoderm of *Drosophila*; P, prophase; M, metaphase; A, anaphase; T, telophase. **(C)** The cell cycle during the first postblastoderm mitosis in *Drosophila* stage 9 embryos, showing that cells in different regions of the embryo enter this mitosis asynchronously. (a) Pattern of expression of the RNA for string/Cdc25 in groups of cells that will undergo mitosis. (b) Pattern of mitosis revealed with an antibody against tubulin. The pattern differs from that shown in a. (c) Expression of cyclin B in a pattern that is very similar to that of the cells that are undergoing mitosis in (b). (d) Pattern of DNA replication (S phase) revealed by incorporation of BrdU. (Images A and B adapted from Edgar, B. *et al.* (1994) Distinct molecular mechanisms regulate cell cycle timing at successive stages of *Drosophila* embryogenesis. *Genes Dev.* **8**, 440–52; Image C from Foe, V., Odell, G., and Edgar, B. (1993) Mitosis and morphogenesis in the *Drosophila* embryo: point and counterpoint. In *The development of* Drosophila melanogaster. (M. Bate and A. Martinez-Arias (eds). Cold Spring Harbor Laboratory Press.)

laid-down product of the *string* gene, which encodes the *Drosophila* homologue of the yeast Cdc25 phosphatase that activates mitotic cyclin–Cdks. As a result, the embryonic cells arrest in G2 of the 14th division cycle, until regulated zygotic production of string/Cdc25 introduces control of the G2–M transition. The *string* gene is under transcriptional control and, at the 14th cycle, displays a complex pattern of expression that reflects the complexity of its transcriptional control elements. As a consequence, the spatial and temporal distribution of cells entering the 14th mitosis is also very complex, and appears to bear no simple relation either to the visible embryology of the system or to the patterns of expression of many genes (Fig. 7.11).

For the remainder of the embryonic divisions of *Drosophila*, there is still no G1 phase in the cycle. The divisions are regulated by the transcription of *string*, at the G2–M transition. In the last division that cells undergo during embryogenesis, inactivation of cyclin E–Cdk2 complex by a transiently expressed Cki encoded by the *dacapo* gene arrests the cells in G1 (Fig. 7.11). The regulation of *dacapo* must therefore define both the end of the embryonic divisions and the G1 arrest, and highlights the intricate relationships that must exist between the regulators of the cell cycle, the transcriptional program of the cell, and the global developmental program that is determined by cell interactions.

How cells 'know' when to introduce G1 is not an easy question to answer. Forced expression of string during this period will not introduce another division, probably because a stringent G1 checkpoint control has come into operation.

Mammalian embryos do not undergo the dramatic cleavage cycles that are characteristic of insects or amphibia, and from the outset have long cell cycles with well-established G1 phases. The existence of a G1 phase provides a control point for cell division, at the Start or restriction point. Crossing this point will commit a cell to a full cycle, so before doing so it must ensure that it is ready to undertake division; at the very least it needs to have doubled its mass and be capable of reading mitogenic signals. The first of these conditions ensures that the cell does not lose any of its components as a consequence of cell division (except, as discussed in Chapters 6 and 8, some specific components that are asymmetrically distributed in order to establish polarity in the embryo). The second condition ensures that it can progress through the cycle without stalling.

Coordinating cell division with cell growth

An increase in total cell mass is an essential component of embryonic development because it provides the substrate for the generation of cell diversity. In general, cells grow as they proliferate, but there are situations, such as the early cleavages of many embryos, in which proliferation occurs without growth. This can happen because the rapid oscillations between DNA synthesis and mitosis run on maternal stores of protein and RNA. Once these stores are empty, the cells have to rely on regulatory events, usually accompanied by the introduction of a G1 phase (see above), to monitor their state. Interestingly, this relationship between growth and proliferation does not work both ways: for example, many cells, such as neurons, continue to grow after they have withdrawn from the cell cycle and begun to differentiate.

The relationship between cell division and growth during proliferation must involve regulatory links between the protein biosynthetic machinery and the cell cycle (Fig. 7.12). Once again, the G1–S transition is a key regulatory point, since during G1 the cell has the option of entering into the G0 phase where it can arrest the cycle, integrate other inputs and grow. In proliferating cells, experimentally increasing the concentration of G1 cyclins can bring forward the timing of cell division. This suggests that the concentration of G1 cyclins could be used as a readout for the performance of the protein biosynthetic machinery

and thus provide a link between growth and the cell cycle: the more biosynthetic activity, the more cyclin and the more efficient the transition. Some observations support this view. One such observation (which also has the merit of suggesting a mechanism) is that the mRNA for the yeast G1 cyclin Cln3 is unstable and not very efficiently translated. Under these conditions, the more ribosomes the higher the chance that it will be translated. Therefore, variations in the number of ribosomes as a result of cell growth can affect the concentration of Cln3 and thus the timing of S phase (Fig. 7.12). In a similar manner, the concentration of cyclin D in vertebrates is increased dramatically by overexpression of the translation initiation factor eIF4E, whose activity is stimulated by growth factors via MAP kinase and PI3-kinase signalling pathways (see below). Mutants in eIF4E have an altered body and cell size, as well as altered rates of protein synthesis. In both yeast and animals, proteolysis of the cyclins at the end of G1 imposes another layer of control that will affect the sensitivity of the system to fluctuations in general metabolism.

Not surprisingly, coordination between growth and proliferation also employs the regulatory activities of Cdk inhibitors and their links with the activitiy of growth factors (Fig. 7.8). Transgenic mice lacking the Cki known as p27/Kip1 have larger than normal organs. A similar phenotype is produced by mutations in another Cki, p18/Ink. Interestingly, mice mutant for both genes have organs that are bigger than those of either single mutant, but in all cases the organs do eventually stop growing (Fig. 7.13). The partial redundancy revealed by these and many other genetic experiments indicates that the control system that coordinates cell size and cell division has a number of back-up safety mechanisms.

Experiments with *Caenorhabditis elegans* and *Drosophila* also underscore the role of Ckis in developmental coordination of growth with cell-cycle progression. For example, in *C. elegans* expression of the protein cki-1 (a member of the Cip/Kip family of Ckis) has been shown to coincide with G1 arrest in differentiating cells, while ectopic expression of the protein causes extra cell divisions. Cell-cycle exit (G1 → G0) occurs during dauer larva development under starvation conditions, a process that has been shown to require both *cki-1* and a G1 phase Cki regulator belonging to the cullin family.

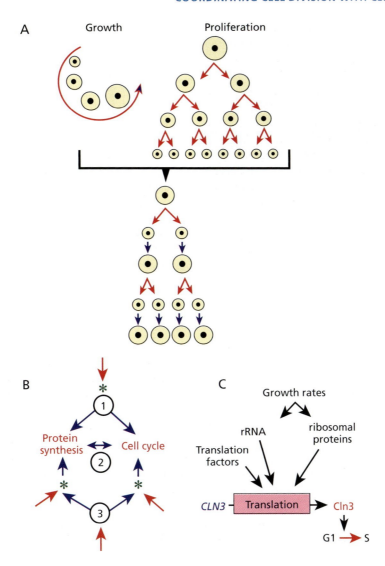

Fig. 7.12. Coupling between the cell cycle and cell growth.
(A) Cell growth can occur without proliferation. However, proliferation without growth leads to ever-smaller cells, suggesting that in proliferating cells there needs to be some coupling between cell division and growth. **(B)** There are several possible ways to couple growth (represented in the figure by protein synthesis) and the cell cycle. For example, there might be common regulators for both processes (1), or there might be regulatory interconnections between them (2). Alternatively, there might be a common regulator that affects specific regulators for both processes (3). Regulators are indicated by green asterisks and red arrows indicate influences on the activity of the regulatory molecules. **(C)** One strategy for linking growth to proliferation is exemplified by the translational regulation of the expression of the G1 cyclin Cln3 in yeast. The *CLN3* mRNA requires a high number of ribosomes to be translated efficiently and therefore imposes a growth-dependent constraint on cell division (see text for details).

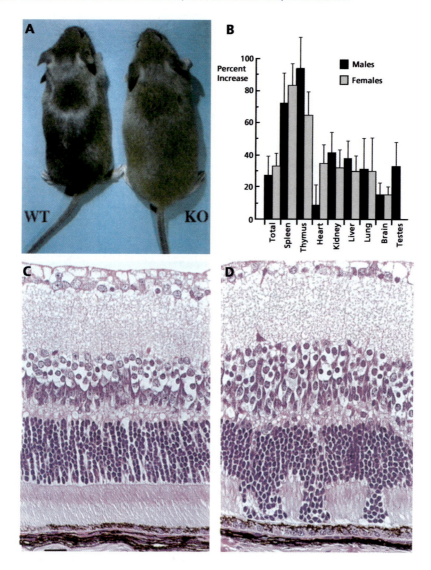

Fig. 7.13. Mutations in the Cki p27^Kip1 increase body and organ size. (A) A mouse that lacks the Cki p27^Kip1 (KO) is much bigger than a wild-type litter-mate (WT). **(B)** Plot showing the percentage increase in the weight of various organs from *p27* mutant mice relative to wild type. **(C, D)** Histological section through the retina from a wild-type **(C)** and a *p27* mutant **(D)** mouse showing the layers of the different cell types. The mutant mice show an increased cell number and cell size relative to the wild type. (Images A and B from Fero, M. *et al.* (1996) A syndrome of multiorgan hyperplasia with features of gigantism, tumorigenesis, and female sterility in p27^Kip1-deficient mice. *Cell* **85**, 733–44; Images C and D from Nakayama, K. *et al.* (1996) Mice lacking p27^Kip1 display increased body size, multiple organ hyperplasia, retinal dysplasia and pituitary tumors. *Cell* **85**, 707–20.)

Extracellular regulation of the link between growth and proliferation

There is ample evidence that cell growth and proliferation are also subject to extracellular controls. Systemic regulators play a central role in the determination of overall shape and size and ensure coordination in different parts of the animal. Direct evidence in support of this idea comes from simple experiments in butterflies, in which removal of the hind wing primordia at the larval stages results in adults with abnormally large forewings and forelegs, which develop by way of compensation. This suggests that during growth, cells compete for limiting amounts of growth and survival factors and that removal of one developing organ increases the availability of these factors to the cells of other remaining organs and tissues.

The behaviour of *Minute (M)* mutants in *Drosophila* illustrates this phenomenon. *Minute* mutations affect a variety of genes associated with the basic growth machinery (e.g. ribosomal proteins and translation initiation factors), slowing cell growth in a dosage-sensitive manner. Mosaics of mutant and wild type cells in imaginal discs (see Fig. 7.14 for an explanation of this approach) reveal that *M/M* cells cannot survive when proliferating together with wild-type cells, and that *M/+* cells grow more slowly than *+/+* cells (Fig. 7.15 and 7.16). One explanation for this observation is that cells are in constant competition for exogenous factors that promote growth. Any impairment in their ability to respond effectively to these factors will prevent them from dividing rapidly enough.

At least in some situations, the factors that cells compete for might be locally restricted. In experiments on the developing *Drosophila* wing, clones of *+/+* cells can be generated in a *M/+* environment. The *+/+* cells grow faster and in general out-compete the *M/+* clones, but surprisingly they respect certain boundaries that do not correspond to any visible anatomical structure (Fig. 7.15). The units revealed by these experiments are called compartments and the boundaries between them compartment boundaries. In the wing, a prominent boundary of this type subdivides the wing disc into an anterior and a posterior compartment (Fig. 7.15). The existence of apparent compartments, defined by these boundaries, suggests that the factors triggering cell growth and division may be spatially localized. One way to think about these compartments is that they are like organs which will reach a certain size

independently of the size and number of the cells that make them up. This analogy would suggest that some of the factors that cells compete for in order to divide and grow might be different in different organs and tissues.

Many of the systemic factors regulating cell growth and/or proliferation are likely to be general ones that signal through well-known pathways. During the development of the wing of *Drosophila*, dpp (a member of the BMP growth factor family) is required for growth: in the absence of this signal, cells do not grow and fail to produce clones (Fig. 7.17). Insulin and insulin-like growth factors (IGFs) also play a role, while the MAP kinase signalling pathway that is activated by signalling through RTKs has been shown to stimulate both cell growth and cell division (Fig. 7.18 and 7.19). MAP kinase acts through several different routes, some of which are illustrated in Fig. 7.18: it directly stimulates synthesis of nucleotides required for DNA and RNA synthesis, it enhances translation by stimulating the activation of translation initiation factor eIF4E, it stimulates cell proliferation by increasing transcription of genes encoding cyclin D and the transcriptional regulator Myc, and it may also aid transcription in general by facilitating chromatin remodelling.

PI3K has also been implicated in promoting cell growth. Upon activation, potentially through a variety of stimuli but particularly by IGF molecules, PI3K stimulates translation in two ways (Fig. 7.18): by modulating the activity of S6 kinase, which phosphorylates ribosomal protein S6, and by contributing to the activation of eIF4E. The role of the PI3K pathway in the overall control of cell size has been shown by genetic approaches in *Drosophila*, where mutants in any component of the pathway display phenotypes of small organ and cell size and concomitant altered growth rates, but no changes in the total number of cells.

The cell cycle and cell differentiation

Differentiation is the process whereby a cell acquires its final phenotype—in other words, it is the final expression of cell fate. The differentiated state is characterized by a specific pattern of gene expression that produces a characteristic set of proteins, and is often associated with visible differences in cell architecture and behaviour. This pattern of gene expression is set up as a result of the cell's history (or lineage, see Chapter 8) as well as by the signals it receives from other cells.

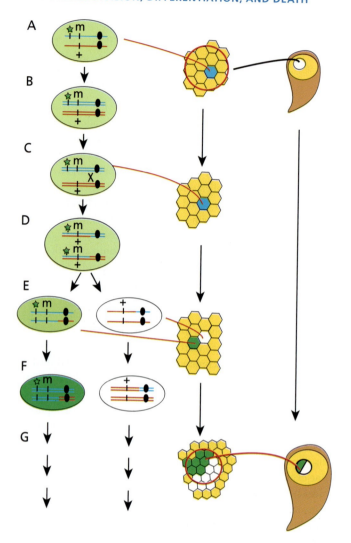

Fig. 7.14. Mitotic recombination and clonal analysis in *Drosophila*. Clones of cells mutant for specific genes (m) can be generated within a group of wild-type cells by inducing mitotic recombination in animals that are heterozygous for m (i.e. m/+). In heterozygous cells, mitotic recombination generates two types of daughter cells after division: homozygous for the mutation (m/m) and wild type (+/+). The m/m cells can be easily identified if they contain a phenotypic marker (star) that is genetically linked to m and that does not interfere with the development of the cells. The marker can be a pigment that will colour the cell. The marker is, like the mutation, recessive and therefore will only be displayed in homozygous cells. Mitotic recombination enables the effects of mutations—most importantly those that might be lethal at the level of a whole organism—to be studied at the cellular level. The diagram illustrates details of the technique in an imaginal wing disc of *Drosophila* (represented diagrammatically on the right). In the imaginal disc, the cells that will give rise to the wing undergo several rounds of division. The middle column highlights a patch of cells from the region that will give rise to the wing and the column on the left describes the events that accompany and follow mitotic recombination in the nucleus of a cell. In an imaginal disc that is heterozygous for m and for the phenotypic marker (*), each chromosome contains two sister chromatids after DNA replication **(A, B)**. Mitotic recombination leads to an exchange of chromatids before mitosis **(C, D)**. This occurs at random in some cells of the imaginal disc, blue in this case. As a consequence of this event, segregation of sister chromatids of the newly hybrid chromosomes generates two cells with different genotypes, which are also different from those of the surrounding cells **(E)**: one cell is homozygous for m (m/m) and the other for the wild-type allele of m (+/+). Because the mutation m is linked to a phenotypic marker that labels mutant cells, these can now be readily identified (green). Sometimes the phenotypic marker is sensitive to dosage, making it possible to distinguish +/+, */+, and */* cells. **(F, G)** Proliferation of these cells generates clones of mutant and wild-type cells that can be identified by virtue of the phenotypic marker. Once these have been identified, it will be possible to investigate the effects of the mutation m in those cells.

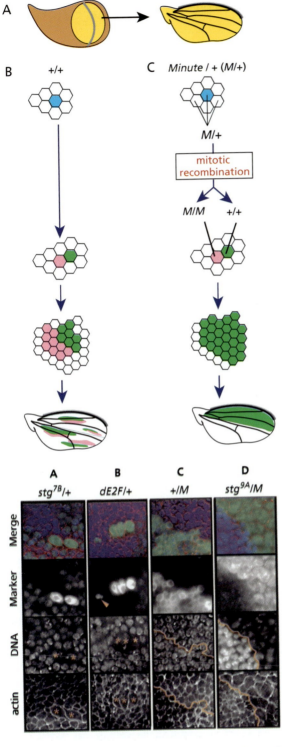

Fig. 7.15. Growth of *Minute* mutant cells in the wing of *Drosophila*. (A) The *Drosophila* wing is derived from a defined region of the wing disc (yellow). As the wing develops, the epithelium of the disc folds and extends outwards (out of the plane of the page) so that the dorsal and ventral surfaces of the wing become flattened and closely apposed. The wing margin is derived from the boundary (grey) between the dorsal and ventral halves of the disc. **(B)** Mitotic recombination clones of wild-type cells can be induced within the developing wing and recognized through a phenotypic marker, as shown in Fig. 7.14. These clones generate compact and continuous patches of cells within the wing that, if generated at the same time, are about the same size. Clones induced early reach a larger final size than those generated late. **(C)** If mitotic recombination is induced in flies that are heterozygous for *Minute* mutations (*M*) which influence the rate of growth of the cells in a dose-sensitive manner, three types of cells will coexist in the developing wing. The majority of the cells will be heterozygous for *M* (*M/+*), some will be wild type (*+/+*) (green), and their sisters will be homozygous for *M* (*M/M*) (pink). Allowed to proliferate, the *+/+* cells generate very large clones, probably because they grow much faster than the heterozygous *M/+* clones that surround them. In contrast, the *M/M* cells do not produce any observable progeny, probably because they grow and divide too slowly and are out-competed by the *M/+* and *+/+* cells. In a background of *M/+* cells, the *+/+* cells colonize very large areas of the wing in the adult fly (as shown) but surprisingly, stop proliferating when they occupy a defined territory that covers about half of the wing in the anterior or the posterior region. The territories revealed in these experiments are called compartments and the wing is said to be divided into an anterior and a posterior compartment.

Fig. 7.16. Clonal analysis of mutations affecting cell growth and proliferation in *Drosophila*. Effects of mutations that affect cell growth and the cell cycle in clones of cells induced by mitotic recombination (see Fig. 7.14). The genotypes on the top line indicate the heterozygous backgrounds in which the clones are induced.

First row: Merged images of cell outlines revealed by actin staining and a marker for the mutant cells. The black and white panels in rows 2–4 correspond to these images. *Second row*: The mutant cells are labelled with a phenotypic marker. The labelled cells correspond to the green cells in the first row. *Third row*: DNA is labelled with Hoechst dye. *Fourth row*: Cell outlines are revealed by staining for actin. **(A)** Loss of *string* (encoding the *Drosophila* homologue of Cdc25) leads to an arrest in cell division. Mutant cells are bigger than the surrounding cells. Note that the *+/+* cells that result from the recombination event (identifiable because they have lost the green marker) have divided a large number of times and generated a clone of cells that lies adjacent to *string* mutant cells (which are indicated by asterisks in the third and fourth rows). **(B)** Cells mutant for the gene encoding the *Drosophila* homologue of E2F also show a mitotic arrest and, asociated with this, a larger size with respect to surrounding cells. **(C)** Mosaic of *Minute* cells showing that despite their differential growth properties, *M/+* cells (green in first row) and *+/+* cells are the same size. The boundary between the clones is traced by a red line in the third and fourth rows. *M/M* cells are not seen in this picture because they die. **(D)** *String* mutant cells in a *stg/M* background. The mutant cells (green in first row) are large. (From Neufeld, T. *et al.* (1998) Coordination of growth and cell division in the *Drosophila* wing. *Cell* **93**, 1183–93.)

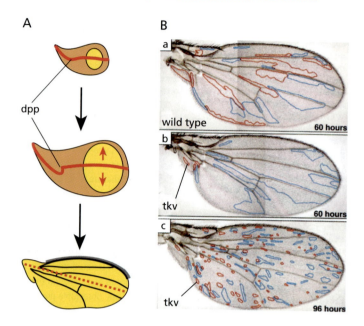

Fig. 7.17. Dpp promotes proliferation in the *Drosophila* wing. (A) A gene encoding a BMP family member, *dpp*, is expressed in a narrow stripe of cells (red line) across the developing wing disc of *Drosophila*. The expression of *dpp* lies just anterior to the antero-posterior compartment boundary (Fig. 7.15) and allows diffusion of dpp into each compartment. The diagram of the differentiated wing indicates the location of the AP compartment boundary (dashed line) which roughly coincides with the expression domain of *dpp*. **(B)** Effects of dpp signalling on cell proliferation. (a) Mitotic recombination was induced in a wild-type wing disc at about 60 hours of development and clones of cells were identified through suitable markers. The picture shows clones from such an experiment: adjacent blue and red outlines highlight clones from sister cells. (b) Mitotic recombination clones were induced at 60 hours of development in a fly heterozygous for a mutation in the dpp receptor thick vein (*tkv/+*) (see Chapter 4, Fig. 4.17). This generates (+/+) cells (blue outline) and (*tkv/tkv*) cells (red outline). Note that clones of *tkv* mutant cells are not visible in the wing and that the +/+ clones are large. This is likely to be because in the absence of the dpp signal, cells do not grow well and, as in the case of the *Minute* mutants, are out-competed by surrounding cells. (c) Mitotic recombination induced in *tkv/+* flies later than 60 hours (96 hours here) produces smaller clones and in this case it is possible to see clones of *tkv* mutant cells (red outlines). These are smaller than the wild-type clones, highlighting their problems in proliferation. In this case, the mutant clones are not out-competed by the wild-type ones. (Modified from Edgar, B. and Lehner, C. (1996) Developmental controls of cell-cycle regulators: a fly perspective. *Science* **274**, 1646–52.)

Although it is often assumed that proliferation and differentiation are mutually exclusive states of a cell, some cells, such as those in the intestinal epithelium, continue dividing after becoming highly specialized, passing on their phenotype (and by implication their pattern of gene activity) faithfully to their daughter cells. However, in other cell types, for example, nerve and muscle cells, differentiation is accompanied by irreversible withdrawal from the cell cycle, and the fully differentiated cell does not divide. In this situation, many of the aspects of cell-cycle regulation that we have highlighted earlier in this chapter come into play.

Studies on the differentiation of mammalian muscle cells in culture (Fig. 7.20) have shed some light on how differentiation and cell-cycle arrest can be coordinated. Muscle precursor cells, known as myoblasts, are already committed to differentiate into myocytes (mononucleated muscle cells that subsequently fuse to form myotubes), but will carry on proliferating in culture without differentiating as long as growth factors are supplied in the medium. Withdrawal of growth factors leads to induction of myogenic-specific proteins, such as the bHLH transcription factors MyoD, Myf5, and myogenin and to the subsequent expression of the MADS box transcription factor MEF2, followed by irreversible cell-cycle arrest and then phenotypic differentiation (production of contractile proteins and myotube formation) (see Chapter 10 for details of the role of these proteins in myogenesis). Measurements

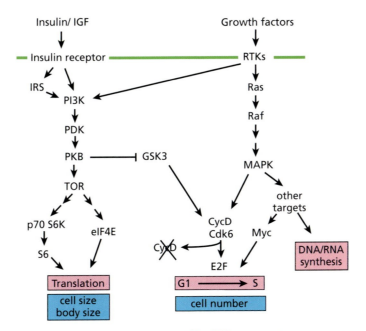

of the levels of key regulators of cell-cycle progression during muscle differentiation in tissue culture indicate that this process correlates with up-regulation of the Cki proteins p21 and p27. Activation of p21 has been shown to arrest the cell cycle at the G1–S transition by blocking the inhibitory effect of cyclin–Cdk complexes on Rb (see Fig. 7.8B) which, in turn, inhibits the activity of E2F transcription factors. Rb is also required, in an indirect way that is not completely understood, for the activity of MEF2, an

Fig. 7.18. Extracellular control of growth and proliferation. Signalling pathways regulated from cell-surface receptors converge coordinately on cell growth and proliferation. The rate of protein synthesis and the associated cell size are regulated by the activity of the S6 ribosomal protein and the initiation factor eIF4E through a phosphorylation cascade that includes: IGFs, insulin-like growth factors; IRS, insulin receptor substrate; PI3K, PI3 kinase; PKB, protein kinase B; TOR, target of rafampycin. The activation of RTKs links the regulation of the cell cycle with the processes of cell growth, as shown.

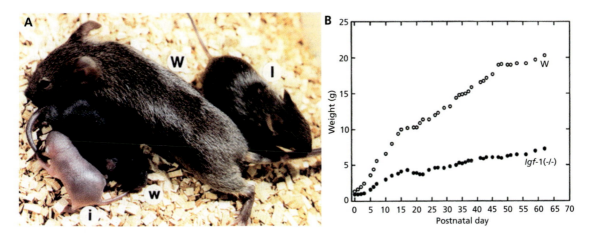

Fig. 7.19. Insulin and the control of growth. (A) Pairs of 6-day-old and 2-month-old litter mates that are wild type (w and W) or mutant (i and I) for the *Igf1* gene. Note the smaller size of the mutants. **(B)** Postnatal growth curves of two litter-mates that are wild type (W) or mutant (*Igf1*(–/–)) for the *Igf1* gene. (From Baker, J. *et al.* (1993) Role of insulin-like growth factors in embryonic and postnatal growth. *Cell* **75**, 73–82.)

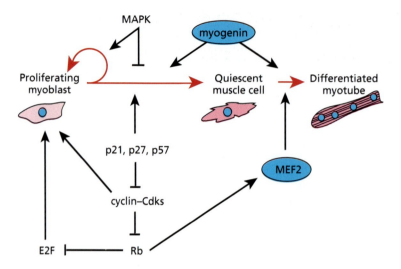

Fig. 7.20. Differentiation and the cell cycle. Diagram of the events that regulate the transition between proliferation and differentiation during muscle development in a mouse embryo. The proliferation of the myoblasts is promoted by the activity of E2F and by cyclin–Cdk complexes acting on the cell cycle and inhibiting the activity of Rb. Withdrawal of growth factors from the medium (or, *in vivo,* complex interactions between signalling molecules), elevate the levels of Cki proteins like p21, p27, and p57, which inactivate the cyclin–Cdk complexes that mediate the G1 to S transition. This allows Rb to act, inhibiting E2F and assisting differentiation factors such as MEF2. All these events are coordinated with the induction and activity of differentiation specific transcription factors. (Based on Walsh, K. and Perlman, H. (1997) Cell cycle exit upon myogenic differentiation. *Curr. Opin. Genet. Dev.* **7**, 597–602; Naya, F. and Olson, E. (1999) MEF2: a transcriptional target for signalling pathways controlling skeletal muscle growth and differentiation. *Curr. Opin. Genet. Dev.* **11**, 683–8; Zhu, L. and Skoultchi, A. (2001) Coordinating cell proliferation and differentiation. *Curr. Opin. Genet. Dev.* **10**, 91–7.)

essential element of the process of differentiation of myocytes into myotubes (Fig. 7.20). Thus, the effect of Cki proteins, coordinated with the activity of muscle-differentiation-specific transcription factors, imposes a cell-cycle arrest in G1 and promotes the expression of differentiation proteins. This is confirmed by the observation that, although mice singly mutant for p21 or p27 develop normally, myogenesis is impaired in p21; p27 double mutant mice.

The general features of the muscle example can probably be extrapolated to other systems: during differentiation, induction of the expression of cell-type-specific transcription factors is linked to the transcriptional and post-transcriptional activation of key regulators which halt the cell cycle at G1.

Cell death: The ultimate 'differentiation'

During the last twenty years it has become clear that 'programmed cell death' (PCD) is an integral part of development and homeostasis in all multicellular animals. It appears that during development, more cells are produced than those that contribute to the final organism. Death of excess cells is a vital process during the growth and shaping of organs and tissues such as the brain and the digits, while the proper functioning of the mature mammalian immune system depends on the controlled destruction of many immune-cell types. One way to think about this is that, from the point of view of the embryo, it is better to have a mechanism that produces more cells than are necessary and then eliminate the excess than to produce too few cells and thus run the risk of making an incomplete organism. Also, as we have already mentioned, cells have intrinsic surveillance mechanisms to detect intracellular damage, particularly to DNA. If the damage cannot be repaired, the cell essentially commits suicide, thus preventing the proliferation of cells with defective genomes. Given the enormous numbers of cells involved in the construction of many organisms and the error rate of many of the enzymes involved in DNA replication, errors do occur and cells do die.

Both developmentally programmed cell death and cell

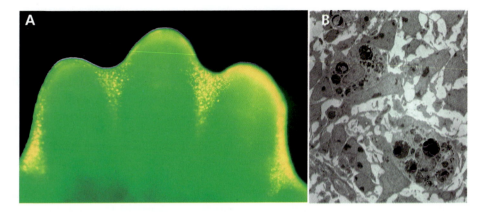

Fig. 7.21. **Cell death by apoptosis. (A)** Foot of a mouse embryo stained with acridine orange to reveal apoptotic cells (yellow staining) in the epithelium between the digits. **(B)** Electron micrograph of two macrophages, from the apoptoic interdigital region of A, that have engulfed apoptotic cells and appear as electron-dense objects (black). (Images courtesy of P. Martin and W. Wood.)

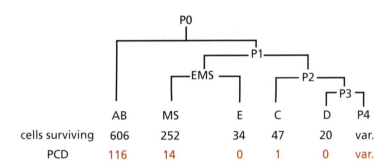

	AB	MS	E	C	D	P4
cells surviving	606	252	34	47	20	var.
PCD	116	14	0	1	0	var.

suicide under stress conditions involve activation of a specific cell death pathway that initiates a process called apoptosis: the cells shrink and are rapidly engulfed by scavenging cells, such as macrophages, without leaking their contents (Fig. 7.21). In contrast, cells killed by many disease processes die by necrosis, involving massive degradation of cell membranes followed by leaking of cell contents and often leading to an inflammatory response.

Programmed cell death can in some ways be thought of as a form of differentiation, in which activation of the differentiation pathway leads to the acquisition of a specific cell phenotype: in this case apoptotic death. As in other forms of differentiation, both signals from other cells and cell-intrinsic factors can be involved in initiating the pathway. Unlike 'true' differentiation, however, it appears that the protein machinery required for expression of the apoptotic phenotype is constitutively present within all cells, and all that is needed is its activation. Thus cells are per-

Fig. 7.22. **Programmed cell death in *Caenorhabditis elegans*.** Outline representation of the total number of cells produced in the major lineages during the development of the *C. elegans* hermaphrodite, and of the ones that die: a total of 131 cells undergo programmed cell death (PCD), distributed as shown between the different lineages.

manently poised between life and death, ready to commit suicide in the appropriate circumstances.

Genetic studies on *C. elegans* first began to uncover the molecular basis of programmed cell death. Normal development of the worm involves the programmed death of a precise set of 131 cells out of a total of about 1000, but it was possible to find mutants (named *ced*, for cell death abnormal) in which these cell deaths failed to occur, leading to the development of adult animals containing extra cells (Fig. 7.22). Cloning and characterization of the genes that were mutated in these animals revealed that the biochemical basis of programmed cell death was a regulated cascade of proteolytic events that acted upon specific intracellular targets to produce the tightly controlled phenotype of apoptosis.

Surprisingly, the *C. elegans* cell death genes were found to be highly conserved between the worm and vertebrates, including humans; for example, a human homologue (*Bcl2*) of the *C. elegans ced-9* gene can functionally substitute for *ced-9* in the worm. Despite strong conservation of the basic molecular mechanism of cell death, however, there are some differences between its role in animals such as *C. elegans* that have a precise number of cells, and in those such as vertebrates where total numbers of cells are not regulated so exactly. Inactivation of the cell death pathway is almost always lethal in animals with indeterminate cell numbers, whereas cell death mutants in *C. elegans* are generally viable. The reason for this difference is not entirely clear. In animals like *C. elegans*, total numbers of cell divisions are fairly tightly controlled, with cell death fulfilling a tidying up function to eliminate the 131 extras. Worms that are defective in cell death show tissue hyperplasia, but not to a lethal extent, perhaps because the presence of the extra cells does not interfere too much with the lineage-driven interactions that characterize the development of this animal (see Chapter 8). On the other hand, in animals with indeterminate cell numbers, such as vertebrates, cell division on its own appears to be less tightly constrained and may continue to a lethal degree unless counteracted by the active killing of extraneous cells.

The cell death machinery

In *C. elegans* four proteins, encoded by the *egl-1, ced-3, ced-4*, and *ced-9* genes, form the core machinery of cell death. Genetic experiments, mainly based on epistasis (see Chapter 5, Figs. 5.2 and 5.3), have suggested an order for the functions of these proteins in the cell death pathway, placing *egl-1* 'upstream' of *ced-9*, which is in turn upstream of *ced-3* and *ced-4* (Fig. 7.23). These experiments also suggest that *ced-9* protects cells from apoptosis by negatively regulating *ced-3* and *ced-4*, while *egl-1* acts as a negative regulator of *ced-9*. Biochemical characterization revealed that ced-3 is a protease of the caspase family, so-called because these enzymes contain a cysteine residue at their active site and cleave their substrates at specific aspartic acid residues. Caspases are synthesized as inactive proenzymes that are themselves activated by proteolytic cleavage. When

activated, they effect proteolysis of specific targets, such as proteins that support the nuclear membrane and the cytoskeleton, leading rapidly to the dismantling of the cell and targeting it for phagocytosis. Egl-1 and ced-9 are distantly related to one another; both are members of the Bcl2 family of proteins. These proteins are associated with membranes (plasma, mitochondrial, and ER) and are represented by multiple members in vertebrates. Their precise function remains unclear, other than that they seem to act as regulators of the activity of caspases like ced-3.

One plausible model for the activation of the cell-death pathway suggests that ced-9—probably localized to cell membranes—keeps ced-3 in an inactive state by binding to and sequestering ced-4. In response to apoptotic signals, egl-1 is either synthesized or activated. Egl-1 binds to ced-9, causing it to dissociate from ced-4 and freeing ced-4 to promote ced-3 activation and initiate the proteolytic pathway.

In *Drosophila* and vertebrates, the simple core cell-death pathway of *C. elegans* has been elaborated to a more complex network of components (Fig. 7.24). For example, mammals have about 15 members of the Bcl2 family, some of which, like *C. elegans* ced-9, suppress apoptosis (examples include Bcl2 and BclX) while others, like egl-1, enhance it (examples are Bax and Bad). Only one vertebrate homologue of *ced-4*, named *Apaf1*, has so far been found but there are at least 13 caspases which often act in specific cascades. It appears that different cell-death proteins and pathways may operate in different cell types. In addition, the mitochondrial electron transfer chain component cytochrome *c* functions as a cofactor in the vertebrate cell-death pathway, being essential for the activation of ced-4/Apaf1.

In vertebrates, there are three independent ways of activating programmed cell death (Fig. 7.24). One responds to the withdrawal of growth factors from the medium or to the absence of growth factor signalling. Under these conditions ced-4/Apaf1 is activated by cytochrome *c* released from the mitochondria. Cytochrome *c* release is also stimulated by situations of stress such as osmotic shock. This pathway for the activation of cell death is negatively regulated by members of the Bcl2 family.

A second activation pathway targets the caspases directly. For example, in vertebrates, the immune system has elaborated cell-death pathways that enable the efficient elimination of infected or foreign cells, or the termination of an immune response by the removal of the participating cells. Killer lymphocytes produce a ligand that binds to a

A

Mutant phenotype and protein product of genes
involved in programmed cell death (PCD) in *C. elegans*

Mutant	Viable	PCD	gene product
ced-9	No	excess	Bcl2 family member
ced-4	Yes	reduced	adaptor
ced-3	Yes	reduced	caspase/protease
egl-1	Yes	reduced	Bcl2 family member

B   Analysis of double mutants in genes involved in PCD

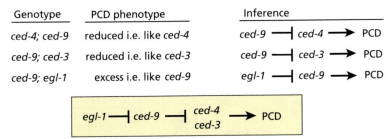

Genotype	PCD phenotype
ced-4; ced-9	reduced i.e. like *ced-4*
ced-9; ced-3	reduced i.e. like *ced-3*
ced-9; egl-1	excess i.e. like *ced-9*

Inference

ced-9 ⊣ *ced-4* → PCD

ced-9 ⊣ *ced-3* → PCD

egl-1 ⊣ *ced-9* → PCD

egl-1 ⊣ *ced-9* ⊣ ced-4 / ced-3 → PCD

C

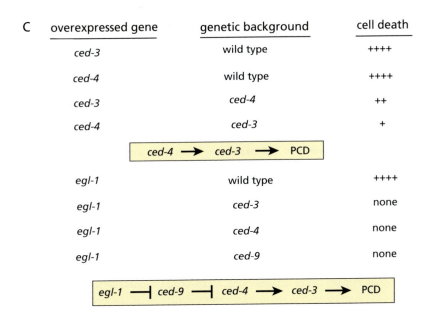

overexpressed gene	genetic background	cell death
ced-3	wild type	++++
ced-4	wild type	++++
ced-3	*ced-4*	++
ced-4	*ced-3*	+

ced-4 → *ced-3* → PCD

egl-1	wild type	++++
egl-1	*ced-3*	none
egl-1	*ced-4*	none
egl-1	*ced-9*	none

egl-1 ⊣ *ced-9* ⊣ *ced-4* → *ced-3* → PCD

Fig. 7.23. Genetic analysis of cell death in *Caenorhabditis elegans*. **(A)** Genes involved in cell death, showing the phenotypes associated with mutations in these genes (in terms of viability and programmed cell death), and the biochemical nature of their wild-type gene products. **(B)** Analysis of epistasis (see Chapter 5, Fig. 5.2) was used to order the action of the products of the genes involved in programmed cell death. Analysis of loss of function mutations establishes a linear pathway with an ambiguity at the level of *ced-3*. **(C)** Epistasis analysis using gain of function of one gene (produced by overexpression) in a background of loss of function of the other resolves the ambiguity about the functional relationship between *ced-3* and *ced-4*. If, as observed, gain of *ced-4* is blocked by loss of function in *ced-3*, but not the other way around, this means that *ced-3* is downstream of *ced-4*. The same reasoning can then be applied to order *ced-9* and *ced-4* and, as shown, to the ordering of *egl-1* and all the others. As a result, the linear pathway shown can be inferred without ambiguities. (Adapted from Metzstein, M.M., Stanfield, G.M. and Horvitz, H.R. (1998) Genetics of programmed cell death in *C. elegans*: past, present and future. *Trends Genet.* **14**, 410–16.)

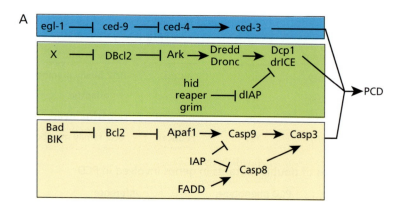

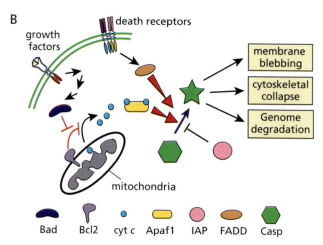

Fig. 7.24. Pathways controlling programmed cell death.
(A) Correspondence between the *Caenorhabditis elegans* (blue), *Drosophila* (green), and vertebrate (yellow) genes involved in programmed cell death (PCD). Note that, in *Drosophila* and vertebrates, there is more than one caspase and that caspases can be activated by routes other than those mediated by ced-4. (Based on a review by Meier, P. and Evan, G. (1998) *Cell* **95**, 295–8.) **(B)** Diagram indicating the subcellular location of different members of the cell death pathway, their interactions and modes of activation. As shown, there are various ways in which caspases can be activated. One responds to extracellular signals and leads to interactions between Bcl2 family members which result in the inactivation of Bcl2 in the mitochondrial membrane. As a result of this inactivation, the mitochondria release cytochrome *c* which acts as a cofactor for the caspase activator Apaf1/Ark. A second route involves dedicated cell death receptors and adaptors (such as FADD) which target the caspases directly. In many cells a family of proteins, inhibitors of apoptosis (IAP) inhibit caspase activation and are regulated by a variety of stimuli. In *Drosophila* the proteins hid, grim, and reaper act by inhibiting a *Drosophila* IAP family member.

'death receptor' on the surface of the target cells (Fig. 7.24). Clustering of the receptor leads to clustering of intracellular adaptor molecules and in turn to clustering of specific caspases; it appears that this clustering activates the proteolytic function of the caspases and so initiates apoptosis.

The third pathway (not illustrated) activates apoptosis in response to DNA damage. How this pathway operates is not yet understood, but it appears to involve a protein called p53, an important cell–cycle regulator that acts via p21 and Rb.

The classical genetic model organism *Drosophila* has naturally also been a prime subject for research into programmed cell death (Fig. 7.24). The system is very similar to that of vertebrates but in addition there are three proteins specific to *Drosophila*—hid, reaper, and grim—that are required for the extensive cell deaths that occur during

Drosophila embryogenesis. These factors contribute to the regulation of caspase activity by blocking the activity of members of the IAP family of proteins.

Death signals during development

Developmentally regulated signals can also activate cell death in target cells, for example in the separation of digits in the vertebrate limb. In this case, locally produced signals in the interdigital region cause the death of neighbouring cells (Fig. 7.25). The key signalling molecules are the bone morphogenetic proteins (BMPs), members of the TGFβ/BMP family of signalling molecules (see Chapters 4, Fig. 4.17 and 5, Fig. 5.13). If BMP signalling is blocked in the developing mouse or chick limb by infection with a dominant-negative mutant form of the BMP receptor, the animals develop webbed limbs and the digits are truncated. Caspase inhibitors also cause webbing, supporting the idea that BMP signalling in the interdigital regions activates the apoptotic pathway.

Local signalling by BMPs has also been implicated in the process of cavitation: the conversion of a solid ball of tissue into a tube (Fig. 7.25) (this should be distinguished from tube formation by the rolling up of an epithelial sheet of cells). Cavitation occurs in a number of developmental situations (e.g. in the formation of exocrine glands). During the very early development of the mouse embryo, just after implantation of the blastocyst, the inner solid core of the blastocyst (composed of embryonic ectodermal cells) undergoes cavitation, giving rise to an epithelial cell layer surrounding a central cavity. The process of cavitation has been shown to result from the opposing effects of apoptotic signals emanating from the endodermal cells that surround the ectodermal core, and survival signals from the basement membrane that lines the developing cavity. Once again, BMPs have been shown to be involved in the apoptotic signalling from the endoderm.

In contrast to the local effects of BMPs, more global cues can also initiate apoptosis during development. In tadpoles, for example, a surge of thyroid hormone during metamorphosis initiates the apoptosis that causes the tail to be shed. This is triggered by a rise in the titre of thyroid hormone and must be mediated by the transcriptional activation or repression of cell-death regulators.

In other developmental situations, it seems that cell-intrinsic cues activate the cell-death pathway. This appears to be the case in *C. elegans*, where proliferation is tightly regulated and the survival or death of cells is dictated by internal programs determined by cell lineage. However, there are also some situations in vertebrates where intrinsic controls seem to operate. For example, the cell-death pathway is known to be implicated in the elimination of internal organelles from specialized cell types such as red blood cells, skin keratinocytes, and lens epithelial cells. Why the pathway is only partially deployed, so that the whole cell is not destroyed, is not yet known.

Regulation of cell survival during development

It has been suggested that—at least in vertebrates—the default pathway for cells is to die, and that cells survive only because they receive survival signals (e.g. growth factors, or signals indicating functional attachment to other cells) that negatively regulate the initiation of apoptosis. Perhaps the most extreme example to illustrate this idea comes from the vertebrate nervous system. Vastly more neurons are produced than are needed, and development involves the mass suicide of all those that have not developed the necessary connections with target cells producing limiting amounts of survival-mediating growth factors (Fig. 7.25). This constant tension between survival and death is probably an essential property of populations of growing cells unless, as in *C. elegans*, there are rigid checks on proliferation, and the survival or death of cells is dictated by internal programs determined by cell lineage.

There is evidence that activation of the Ras GTPase by growth factors influences cell survival by regulating the machinery that in turn regulates programmed cell death. Both the MAPK and PI3K signalling pathways have been implicated in transmission of the growth factor signal. In *Drosophila*, in the developing eye, for example, signalling by the boss ligand on the R8 photoreceptor cell activates the sevenless receptor on the R7 cell and in turn the Ras-MAPK pathway. In addition to activating the transcription of genes that will confer on R7 its phenotypic and physiological characteristics, signalling through this pathway suppresses apoptosis of the R7 receptor cell by activating the pointed transcription factor and suppressing the activity of the transcription factor yan (Fig. 7.26). Mutation of the sevenless receptor stops MAPK signalling, leading instead to activation of yan, which induces expression of the death effector gene hid and kills the R7 cell (hence the 'sevenless' phenotype). Hid also appears to be subject to more direct regulation by Ras, via phosphorylation by MAPK.

In vertebrates, the PI3K pathway appears to be the main route through which Ras signalling influences cell survival (Fig. 7.26). In some cell types, the molecular basis for its action has begun to be worked out: PKB (Akt) has been shown to phosphorylate the Bcl2 family protein Bad, thereby blocking its ability to promote cell death, and it is also able to down-regulate caspase activity. This provides a way in which extracellular signals can maintain cell survival in a cell-type-specific manner, as in the case of sevenless in *Drosophila*.

Cell numbers and overall size

What regulates the total size of an organ, or indeed an organism? The answer to this question is not completely clear, but it appears that the parameter that is regulated is total cell mass rather than cell numbers. Perhaps the clearest evidence supporting this view is that diploid and polyploid variants of animals such as the salamander reach the same final size, but the polyploids have both larger cells and a smaller total number of cells.

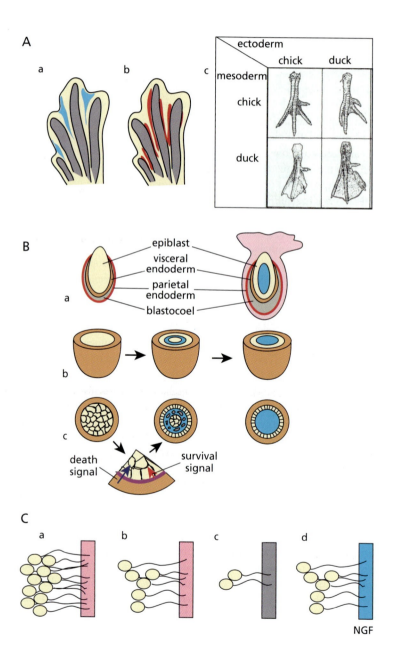

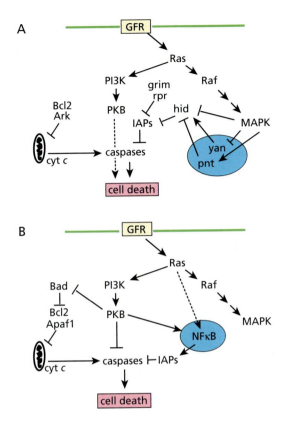

A

B

Fig. 7.26. Extracellular regulation of programmed cell death. **(A)** In *Drosophila*, the Ras signalling pathway (see Fig. 5.15) is directly involved in the regulation of programmed cell death, at least during the development of the R7 photoreceptor. Activation of the pathway leads to inhibition of hid expression (by suppressing the activity of the transcription factor yan and activating the transcription factor pnt) and the suppression of cell death. In the absence of Ras signalling, yan activity leads to the expression of hid and this results in the death of R7, probably via inhibition of the IAP. In other tissues, activation of PI3K, which can be mediated by Ras, also suppresses cell death but in this case by activation of the universal cell death pathway. **(B)** In vertebrates, there is no evidence for a direct involvement of MAPK activity in cell survival. However, Ras-dependent activation of PKB (Akt) does lead to the suppression of cell death by coordinately regulating several effectors of the pathway. In addition, PKB inhibits the activity of caspase9 and promotes the activity of NFκB by promoting the degradation of IκB; this results in the synthesis of antiapoptotic proteins.

The control of tissue and organ growth during development involves both cell-intrinsic mechanisms that limit the size of an individual cell and the number of times it can divide, and mechanisms that rely on signalling and cell interactions to coordinate the survival, growth, and division of groups of cells. Although it appears that the parameter that is 'sensed' by the growing organ is its total cell mass, in practice this usually means the total number of cells, though as the similar size of diploid and polyploid

Fig. 7.25. Programmed cell death in development and pattern formation. Examples of processes in which programmed cell death is used to sculpt structures, regulate cell numbers or generate differentiated cells in developing embryos. **(A)** During the final stages of limb development, digits become sculpted in a species-specific manner through the death of the cells in the interdigital space. Cell death is associated with the expression of various BMPs (blue) in the interdigital space (a) and of BMP receptors (red) over the boundaries of the digits (b). (c) The development of the limb depends on interactions between the mesoderm and the ectoderm (see Chapter 12). Experiments involving combination of these tissues from different species indicate that the patterns of cell death are determined by the mesoderm. The feet of ducks and chicks differ in that ducks, but not chicks, have an epidermal membrane that forms a web. Towards the end of development both chicks and ducks have a web but whereas in the chick these cells die, in the duck they remain. Chimaeras between ectoderm and mesoderm show that the webbing only develops when the duck mesoderm is a component of the chimaera, suggesting that the mesoderm is the source of the death signals that kill these cells in the chick. **(B)** During the early stages of mouse development, the epiblast is sculpted

by a process of cell death that results in it acquiring a central cavity. Diagrammatic longitudinal (a) and transverse (b) sections through a mouse embryo, showing development of the cavity (blue). (c) The cavity develops as a result of a combination of cell death signals released from the visceral endoderm, and survival signals from the basement membrane that separates the epiblast from the visceral endoderm. Only epiblast cells that are in direct contact with the basement membrane survive. (After Coucouvanis, E. and Martin, G. (1995) Signals for death and survival: a two step mechanism for cavitation in the vertebrate embryo. *Cell* **83**, 279–87.) **(C)** During the development of the nervous system, more neurons are produced than are needed to establish functional connections. (a) After arrival at a target (pink), not all connections survive and (b) many of the neurons die. This trimming effects suggests that the neurons are competing for some factor that is present in the target in limiting amounts. Experiments in tissue culture demonstrate the existence of substances, neurotrophins, that maintain the survival of neurons. (c) An artifical target (grey) that attracts growing axons will be more effective in promoting the survival of neurons if (d) it contains a neurotrophin such as NGF (blue).

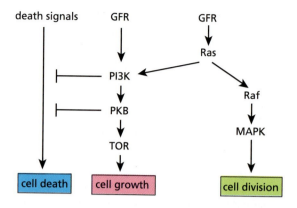

Fig. 7.27. Pathways of interaction between growth, cell division, and differentiation. Outline of the major interactions that link proliferation, cell growth, and apoptosis and ensure that cells survive as long as they are growing.

salamanders shows, cell size can also contribute to the reckoning. Cell numbers are largely determined by the balance between cell proliferation and cell death. Earlier in this chapter we have suggested that in organisms with indeterminate numbers of cells, cell proliferation and apoptosis go hand in hand: the fact that the rapidly proliferating cells of a developing organ are primed for cell death helps to ensure that the organ does not grow out of control. As soon as the organ reaches a size where survival factors become limiting, cell death is called into play. At the molecular level, the protein Myc, which is expressed only in proliferating cells, has also been shown to promote apoptosis (Fig. 7.18). The balance between the proliferative and apoptotic functions of proteins such as Myc may be one of the factors involved in regulating organ and tissue growth.

As well as programmed cell death, however, cell-cycle arrest also appears to play an important role in size control. Mice that are mutant for the Cki p27 grow much larger than normal mice (Fig. 7.13) despite the fact that the programmed cell-death pathway is still active in the mutants, suggesting that in normal mice, cell proliferation is restrained by p27. Other Ckis, such as p21, probably act similarly, as exemplified by the irreversible cell cycle arrest that occurs in differentiating muscle cells (Fig. 7.20). Cell death has also been shown to play a part in this system, however: individual cells that do not activate p21 expression undergo apoptosis, suggesting that the total amount of muscle mass that develops is regulated both by cell-cycle arrest in the differentiated cells and by apoptosis of all precursors that do not effect this arrest.

How do cell interactions and systemic signals coordinate and ultimately limit cell proliferation to achieve size control during development? Growth factor signalling through PI3K, which as we have seen influences both proliferation and programmed cell death, is likely to play a central role in coordinating these processes and, together with other regulatory inputs, in tissue-dependent size regulation. Regulatory pathways linking cell-cycle progression to cell differentiation probably also play a role, through the action of factors such as E2F and Rb that regulate the G1–S transition (Fig. 7.8) and are in turn affected by the activity of cell-specific differentiation factors. The activity of E2F and Rb is also known to be modulated by signalling pathways, suggesting the possibility that proliferation may be linked to developmental requirements by connecting the regulation of E2F expression or activity to cell interactions. This can only be achieved by the integration of information from linked signalling networks through the sorts of mechanisms discussed in Chapters 5 and 6; examples of some of the signalling pathways that are likely to participate in this regulation are illustrated in Fig. 7.27.

SUMMARY

1. Cell division and cell death are fundamental cellular 'routines' that make important contributions to development because their balance determines the total number of cells of an organism at any given time of its development.

2. The numbers and timing of cell divisions are modulated through the cell cycle, a succession of distinct molecular phases that are common to all cells and organisms.

3. The basic cell cycle is a universal sequence of alternating periods of mitosis (M) and DNA synthesis (S), separated by two gaps of variable lengths (G1 and G2). Different developmental stages are characterized by variations of this basic pattern, generated by extending or eliminating the G phases.

4. The molecular activities of each of the phases depend on phase-specific complexes between a regulatory subunit (cyclin) whose concentration oscillates throughout the cycle, and an associated kinase, cyclin dependent kinase (Cdk) which modifies the targets of the complex.

5. The signalling events that mediate cell interactions impinge on the cell cycle and regulate its tempo and its activities, ensuring that cells are produced when and where they are needed and in the necessary numbers.

6. Little is known about how the size of organs or tissues is determined in a developing organism. However, size regulation is likely to depend on a balance between external cellular variables (cell interactions mediated by signalling molecules) and internal ones (transcription factors and cell-autonomous programs of gene expression).

7. Programmed cell death, or apoptosis, is essential for normal development in all multicellular organisms. It is an active process that is regulated by extracellular stimuli. The basic molecular pathways of cell death and its regulation are conserved.

8. Many of the signals that regulate cell death also regulate the cell cycle, enabling the two processes to be balanced during development. This balance is used to determine the size and shape of organs and tissues.

COMPLEMENTARY READING

Conlon, I. and Raff, M. (1999) Size control in animal development. *Cell* **96**, 235–44.

Hengartner, M. (2000) The biochemistry of apoptosis. *Nature* **407**, 770–6.

Jacobson, M., Weil, M., and Raff, M. (1997) Programmed cell death in animal development. *Cell* **88**, 347–54.

Murray, A. and Hunt, T. (1993) *The cell cycle. An introduction.* Oxford University Press.

Neufeld, T. and Edgar, B. (1998) Connections between growth and the cell cycle. *Curr. Opin. Cell Biol.* **10**, 784–90.

Chapter 8

The generation of lineages: A developmental routine

In developmental biology, a lineage refers to the pattern of descent through which a cell of an organism can be traced back to the zygote. Therefore, an account of the lineages of all the cells of an organism provides an important basis for understanding the development of that organism. Ideally, such an account should include the number of divisions that gave rise to every cell, the state of their progenitors before and after each division in terms of gene expression and protein activity, and the molecular influences that governed the transitions between these states.

In this chapter we shall see that there are different types of lineages; we explore how they are generated and how they relate to the assignment of cell fates and the progressive emergence of an organism. We begin by describing how lineages allow different parts of the organism to be traced back to their origins in the blastoderm, which appears to be strategically organized in a way that outlines the different parts of the animal. We then search for some common molecular mechanisms that underlie the generation of these lineages, and which might be universal. This will be followed by an introduction to the generation of more complex lineages and their influence on the construction of large-scale patterns like those which give rise to particular structures and whole organisms.

Cell lineages and cell interactions

The rates of cell division in different lineages and the spatial arrangements of their component cells are variable during the development of an organism. These variations in the dynamics of individual lineages set up transient and terminal relationships between cells that are used, through specific cell interactions, to create cell diversity and pattern. In this way, lineages serve as a way of linking an increase in total cell mass to an expansion of the repertoire of different cell types.

The importance of transient cell interactions is evident during the development of the nematode *Caenorhabditis elegans*, which develops according to a stereotyped pattern of cell divisions. The number and pattern of divisions is fixed and, at first sight, this rigidity suggests that in this organism the fate of any particular cell depends, primarily, on that of its parents. Therefore, in order for diversity to emerge during cell division amidst the elements of the lineage, each mother cell must find a way of allocating different programs of gene expression to its two daughters. As we shall see, there are several molecular devices for doing this and sometimes they are tightly associated with the autonomous development of a particular lineage. However, several experiments have shown that, even when patterns of cell division are rigidly determined, interactions between cells are equally important in the assignment of cell fates. Furthermore, the programs of gene expression that underlie the development of specific lineages are determined by cell interactions.

One example of the importance of cell interactions in the context of fixed lineages is provided by the early development of the nematode *C. elegans*. At the four-cell stage, the cells of a *C. elegans* embryo have a defined and reproducible arrangement. The ABp cell lies adjacent to P2 and gives rise, through a stereotyped lineage to (amongst other cells) some body wall muscles while its sister, ABa, gives rise to some pharyngeal muscles (Fig. 8.1). If, by a gentle experimental manipulation, ABa is placed next to P2 instead of ABp, it develops like ABp (Fig. 8.1), and ABp develops like ABa. This indicates that the fate of the two daughters of A does not depend on some differential partitioning of information associated with the division

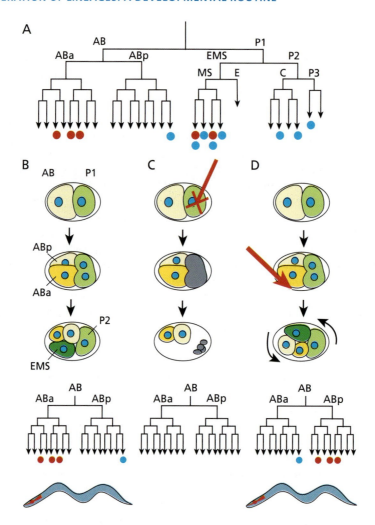

Fig. 8.1. Lineages and cell interactions in *Caenorhabditis elegans*. (A) In *C. elegans*, the first divisions generate a set of founder cells (ABa, ABp, EMS, C, and P3) from which the different lineages that generate the organism will emerge. Pharyngeal (red) and body wall (blue) muscle cells are derived from the indicated precursors through stereotyped lineages. In the diagram, each bifurcation represents a cell division; the spatial arrangement of the daughters is indicated with anteriorly positioned ones on the left and posteriorly positioned ones to the right. The vertical dimension represents time. The diagram is not to scale. **(B)** During early development, the posterior daughter of AB (ABp) comes to lie adjacent to P2 and EMS, whereas ABa does not have both of these contacts. The normal contribution of ABa and ABp to the muscles is indicated in the diagrammatic lineage shown below and corresponds to that

shown in A. **(C)** Elimination of P1 early in development abolishes the ability of ABa and ABp to produce muscle, suggesting that an interaction between P1 and AB or between their daughters is necessary for the development of the muscle fate from the AB daughters. **(D)** By use of a fine pipette, the relative positions of ABa and ABp can be exchanged so that ABa comes to be positioned next to P2. This manipulation has no effect on the development of the embryo other than a change in left/right asymmetry. Thus, after the manipulation ABp behaves as ABa would have done in the wild type. In other words, it is the position of the two descendants of AB relative to P2 that determines their fate, rather than their lineage. (Adapted from Priess, J. and Thomson, N. (1987) Cellular interactions in early *C. elegans* embryos. *Cell* **48**, 241–50.)

of their mother, but on an interaction of one of the daughters with P2. Consistent with this, in the absence of P2 and EMS, ABp does not develop normally. This example shows how, even when lineages are invariant, the fates of their component cells can be defined primarily by cell interactions.

This is not an isolated example and highlights the point that the function of very rigid lineages may not be so much to create rigid programs of cell fate assignment that rely simply on descent, but to generate a series of spatial relationships between particular cells that will be the same from one embryo to another.

Lineages and fate maps

In embryos with defined numbers of cells and stereotyped patterns of division like *C. elegans*, the descent relationships, and therefore the lineages, can be discerned by simple observation (Fig. 8.1). In other embryos, for example *Drosophila* or amphibia and other vertebrates, only the very early divisions can be followed in this way. At later stages, the large number of cells and their general inaccessibility to marking and experimental manipulation make it difficult to follow the lineages of specific cells by inspection. These problems can be overcome, at least partly, by the use of lineage tracers (Fig. 8.2). In this procedure, specific cells are labelled early in development with a dye coupled to a fluorescent reagent or an enzyme, and allowed to divide. The progeny of the original cell, and the parts of the embryo that those progeny contribute to, can be examined at any time by visualizing the tracer (Fig. 8.2). A wide variety of lineage tracers are available but the recent development of green fluorescent protein (GFP) and its many variants (see Chapter 6, Fig. 6.2) promises to allow lineage tracing to be done *in vivo*, following the dynamics not only of particular cells, but also of individual proteins whose changes of location within and across cells over time make essential contributions to the emergence of pattern.

Studies of lineages during early stages of development can be used to draw up a fate map, showing the contribution of cells or regions of the early embryo to the final structures of the organism (Figs. 8.2, 8.3). Fate maps are particularly useful because they can reveal early spatial relationships between cell populations that are later pulled apart by cell division and cell movement. Such dispersals are very common in vertebrate embryos and might obscure early spatial relationships that are responsible for causal cell interactions. Fate maps can also reveal groups of cells that are initially separated but are brought together by cell movements.

Comparing available fate maps of early embryos reveals that there are some organisms, for example, *C. elegans*, whose development can be described at a single-cell level of resolution (Fig. 8.3). In these organisms every cell matters, and its descent can be traced precisely through the invariant spatial and temporal pattern of cell divisions that gives rise to the adult animal. Organisms of this type are said to have 'determinate lineages'. However, in many other organisms the total numbers of cells and divisions are not so precise, and lineages have to be traced against a background of large masses of cells dividing in an apparently unordered manner. In these organisms, which are described as having 'indeterminate lineages', multicellular territories matter more than individual cells. These territories often behave as units, thus taking on the role played by individual cells in determinate organisms. When cells divide, their progeny are allocated particular fates according to their position; the role of the cell lineage is to amplify a population of cells and place it in the right geographical position (Fig. 8.3).

The classical terms for embryos with determinate and indeterminate lineages are 'mosaic' and 'regulative', respectively. The rationale behind these names is that mosaic embryos are hard-wired and unable to repair the loss of any of their elements, while regulative embryos can compensate for such losses. However, closer examination of many embryos originally described as mosaic suggests that there may be no such thing as totally mosaic development, and that, as we have seen in one example in the nematode (Fig. 8.1), a rigid lineage might simply be a way of creating particular cell interactions reproducibly. In general, the fate of a cell in an embryo is always conditional on its interactions with other cells, even when the lineages are fixed.

Stem cell lineages

The simplest possible lineage is one in which a cell, upon division, produces two cells like itself and this process is re-iterated so that there is constant self-renewal. A variation on this theme is the type of lineage in which a cell produces one daughter like the mother and one that develops or differentiates in a different way. If, in either of these patterns, the mother cell always remains undifferentiated and retains an unlimited (or at least prolonged) potential to generate differentiating cells, the mother cell is called a stem cell, and the lineage is a stem cell lineage (Fig. 8.4).

Stem cells fulfil two important functions. First, they act

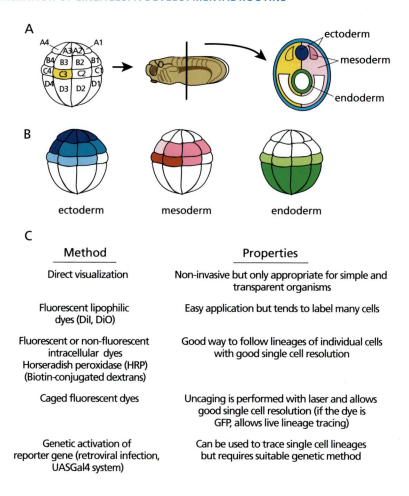

Fig. 8.2. **Tracing lineages. (A)** In an embryo, the descendants of a particular cell can be traced by injecting a cell marker that has several properties: it does not affect development, it remains restricted to the cell where it is expressed (i.e. it is cell-autonomous), it is not diluted after too many divisions, and it can be visualized in a sensitive manner. In the example shown, a blastomere of a frog embryo (C3) is injected with a lineage tracer (yellow). After injection, the embryo is allowed to develop and the distribution of the tracer is monitored at a later stage to show the descendants of C3 (yellow ectodermal and mesodermal cells in the section through the embryo). **(B)** By labelling different cells at a particular stage of development, a fate map can be created for that stage. Fate maps show that specific cells in the embryo contribute reproducibly to particular regions later in development. When the cells whose destination is being followed are marked at the blastoderm stage, a blastoderm fate map—a map of the animal on to the blastoderm—can be drawn up. In this example, experiments like those described in A allow the three major germ layers of a *Xenopus* embryo to be mapped onto the early blastomeres. The intensity of each colour reflects the probability that a given blastomere will contribute to a particular germ layer: the stronger the intensity the higher the probability. Note that cells in certain tiers, particularly C, can contribute to more than one germ layer. **(C)** Examples of lineage tracers used for fate mapping, and their properties. (For further details see Clarke, J. and Tickle, C. (1999) Fate maps old and new. *Nature Cell Biol.* **1**, E103–E109.)

as a reservoir of developmental potential that can be called upon if regeneration is required, and second, they provide an economic way of generating large repertoires of cell fates. In both cases, through successive divisions a stem cell generates progeny that will differentiate into specific cell types, while itself remaining undifferentiated. The cells of the germ line are the 'ultimate' stem cells: they generate cells that will give rise to a whole organism, including new germ cells that retain this potency. Stem cells have also

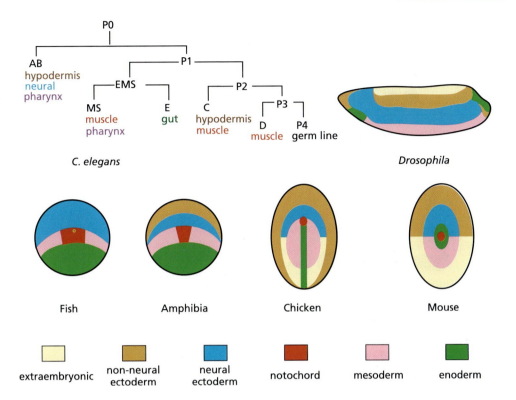

Fig. 8.3. Fate maps. The techniques described in Fig. 8.2 can be used to generate blastoderm fate maps of different organisms. The fate maps identify discrete and continuous areas of the blastoderm that contribute to specific germ layers. The fate maps shown belong to *Caenorhabditis elegans*, *Drosophila*, zebrafish (fish), *Xenopus* (amphibia), chicken, and mouse. The fate map of *C. elegans* was determined by observation while the others represent varying mixtures of labelling and observation. Note the similarity between the fate maps of the four vertebrate embryos, which represent variations on a common plan.

been identified in the haematopoietic system, epidermis, neural crest, and intestine of higher vertebrates. Recently, they have also been found in certain regions of the nervous system in vertebrates and they probably also exist in other tissues and cell types. Stem cells that give rise to all the different cell types within a given tissue are said to be 'pluripotent' in that they give rise to 'many' different cells (blood in the case of the haematopoietic system, neurons in the nervous system) but they are not 'totipotent'—this term is reserved for cells (e.g. the germ cells) that can give rise to 'all' cell types of an organism.

Cells within the epithelium of the epidermis, or of the lining of the small intestine, display the basic properties of stem cells. Both of these structures are constantly abraded and damaged by their contact with the environment and must have a mechanism for regeneration. In both cases, the epithelium has an underlying repository of cells that act as stem cells (Fig. 8.4). The epidermis is a multilayered structure and the stem cells are found in the basal layer. Every division of a stem cell generates one daughter that remains a stem cell, and one 'blast' or 'transit amplifying cell' that divides a small number of times to amplify the population before leaving the stem cell repository and migrating towards the surface as it begins to differentiate. In the intes-

tine, the stem cells are located in a ring near the bottom of small pockets, or 'crypts', at the base of the intestinal villi, and as cells differentiate they migrate upwards towards the tip of the villus. In both the intestinal epithelium and the epidermis, the stem cell population occupies a specific position adjacent to a layer of cells that appear to provide it with some of the factors necessary to maintain its function as a stem cell (Fig. 8.4).

In the two cases we have just described, the progeny of the stem cells differentiate into one or two different cell types. In contrast, the stem cells of the haematopoietic system (haematopoietic stem cells, HSC) are able to give rise to virtually all of the many cell types—at least nine—

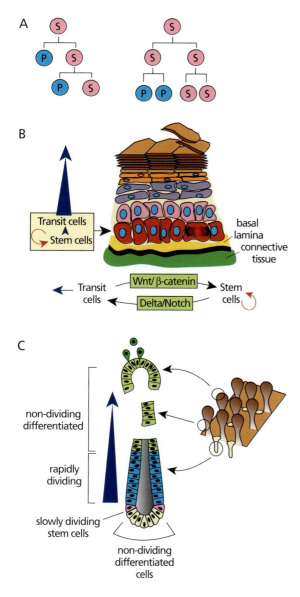

Fig. 8.4. Stem cell lineages and stem cells. (A) Two examples of stem cell lineages. In the lineage on the left, a stem cell (S) divides to generate a cell like itself and another cell (P for precursor) that will differentiate. In the lineage on the right, the first division of S generates two stem cells. One of them self-renews (S) while the other divides once to generate two differentiating cells (P). **(B)** The epidermis of vertebrates contains several layers of cells at different stages of differentiation. At the innermost side there is a layer of dividing cells: the basal cell layer. Some of these cells are stem cells and it is upon these that the renewal of the layer depends. The interaction between these cells and the underlying basal lamina is essential for maintaining the stem cell state. Another cell population in this layer is the transit amplifying cells; these are daughters of the stem cells that are destined to differentiate. When cells leave the basal cell layer, they stop dividing and begin to differentiate towards the enucleated squamous cell type in the outermost layer of the epithelium. The differentiated cells in the outer layer are full of keratin. Wnt (Fig. 5.25) and Delta/Notch signalling (Fig. 5.9) play opposite roles in the maintenance of the stem cell population. Wnt signalling via β-catenin is required for the maintenance of the stem cell fate, whereas Delta/Notch signalling is required for the transit cell fate. **(C)** In the intestine, the epithelial cells that are in contact with the lumen suffer considerable wear and tear, and are constantly renewed from a basally located stem cell population. This population is in close contact with a population of non-dividing differentiated cells which probably play a role similar to that of the basal lamina in the epidermis. The cells at the bottom push the overlying cells upwards, maintaining a continuing flux of cell renewal. Wnt signalling appears to play a role in the maintenance of the stem cell population.

that make up the system (Fig. 8.5). It is not easy to see how the capacity to generate such a large number of different cell types from a single cell is maintained, but that this is indeed the case has been demonstrated in experiments in which an animal is depleted of its blood cells by irradiation. The animal is then seeded with serially diluted suspensions of cells from the bone marrow of a donor animal. The dilutions are such that in some cases only a very small number of cells is injected into the animal. After 1–2 weeks, colonies of developing blood cells, each derived from a single stem cell, are visible in the spleen of the host animal or can be detected in the circulating blood, suggest-

ing that the complete repertoire of haematopoietic cell types can be generated from a single cell (Fig. 8.5). Quantitative analysis of these experiments shows that the new blood cells must be derived from a population of pluripotent stem cells that are present at a frequency of 1 in 1000 donor cells. A further inference is that, because there are many more cell types than could arise separately from divisions of a single stem cell lineage, the stem cell must give rise to several precursor cells that themselves initiate different sublineages. Throughout all the cell divisions that generate the system, the original cell or some of the precursors it gives rise to must maintain pluripotency.

Stem cell factors

Very little is known in molecular terms about what makes a stem cell a stem cell. There might be cell-intrinsic 'stem cell factors' that are common to all stem cells and maintain the 'open' undifferentiated state that characterizes them. In

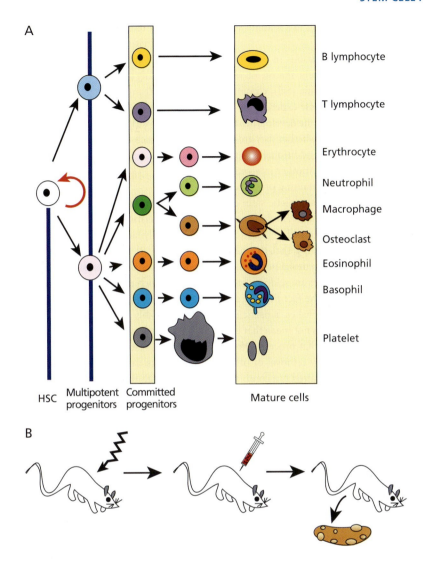

Fig. 8.5. Haematopoietic lineages and haematopoietic stem cells (HSCs). (A) Outline of lineages and sublineages in the haematopoietic system. From a stem cell, a common lymphoid progenitor will give rise to the T and B cells, and a second progenitor will give rise to all myeloid derivatives, which include erythrocytes, neutrophils, macrophages, eosinophils and several other cell types as indicated. **(B)** Experiments that reveal the existence of stem cells in the haematopoietic system. The criterion that is used for the definition of a stem cell in the haematopioetic system is that it can reconstitute haematopoiesis in mice that have received an almost lethal dose of irradiation which has impaired the activity of the bone marrow. These mice are then injected with different dilutions of cells, marked genetically so that they can be identified against those of the host, and the degree of reconstitution is measured after several days. This is done in a variety of ways: a common way is by determining the number of colony-forming units (CFUs) that appear on the spleen. Each colony (visualized as a clear spots in the figure) is derived from a single cell.

general terms, these factors should be capable of maintaining the characteristics of a particular cell type and at the same time preventing the development of new characteristics that would take it along the path of differentiation. To date, little is known about such factors that might be common to all stem cells. Some factors are known, however, that affect the development of stem cells in a manner specific to particular systems.

In the haematopoietic system, mutations in two transcription factors: the zinc finger protein Ikaros, or in a bHLH protein called stem cell leukaemia (Scl), eliminate all cell types from the blood. While these experiments do show that these factors are essential for the development of the various cell types, it is not yet clear whether they are required exclusively in the stem cells, or additionally in the precursors of the various sublineages. Another interesting observation is that homeodomain-containing proteins of the Hoxb4 class are enriched in haematopoietic stem cell populations. Overexpression of the human *Hoxb4* gene in a culture of mouse bone marrow cells caused a 50-fold enrichment of the HSC content, suggesting that this gene may be involved in regulating the self-renewal potential of this type of stem cell. However, cell fates are seldom determined by a single factor. In general, the fact that the absence of one transcription factor abolishes a particular developmental process or cell type does not mean that this is the only factor that can determine that process. Often, the same process is equally affected by removal of other factors, thus revealing that what is required is a combination of factors and that each component of this combination is equally necessary for the functioning of the ensemble (see e.g. Chapter 5, Fig. 5.37).

In addition to cell-intrinsic factors, such as Ikaros, Scl, or Hoxb4, maintenance of the pluripotent state might also require extrinsic factors supplied by other cells. Less is known about these factors, but the expression of specific cell-surface receptors in some stem cell types, and the influence of particular cytokines on their development, does suggest that environmental factors are important in the maintenance and function of stem cell populations. In mammals, for example, mutations in a receptor tyrosine kinase encoded by the *White* (*W*) gene, and in the *Steel* gene which encodes the ligand for White, cause anaemia. Steel factor has been shown to play a role in the survival and/or proliferation of several different stem cell types including HSCs, primordial germ cells and intestinal crypt cells, and might be a factor involved in the maintenance both of adult stem cell populations and of stem cell properties during embryonic development.

In epithelia such as the epidermis and the intestinal epithelium, the precise spatial location of the stem cells, within specialized groups of cells that secrete large numbers of signalling molecules and growth factors, suggests that their environment may be a source of factors necessary for their survival, self-renewal or other properties. It appears that one factor important in maintaining the stem cell fate, both in the intestine and the epidermis, is Wnt signalling (see Chapter 5, Fig. 5.25). Increased activity of the Wnt signalling pathway results in an increase in the size of the stem cell precursor population, sometimes at the expense of differentiating cells. In the case of the epidermis this might work by antagonizing signalling by the Notch receptor, which favours the differentiated state (Fig. 8.4).

Stem cell-like lineages

The nervous system rivals the vertebrate haematopoietic system in terms of its large number of component cells and certainly surpasses the haematopoietic system in its enormous cell diversity and functional complexity. In mammals, the nervous system is estimated to contain 10^{14} neurons which establish on the order of 10^{25} different connections. The number of 'different' neurons is not known but is probably greater than the number of different cell types in the haematopoietic system. The development of the nervous system is very similar in both vertebrates and invertebrates, and relies on the generation of neural precursors which divide a number of times to generate neurons. These precursors divide in a stem cell-like manner, that is, when they divide they generate one cell that differentiates, and one that remains undifferentiated and will divide in the same way as the precursor cell. However, while in a small number of cases the precursors remain pluripotent, in most cases they lose some of this potential as they divide, and so should be considered 'stem cell-like' rather than 'true' stem cells. Although the generation of neural precursors is a basic theme in the development of the nervous system of invertebrates and vertebrates, there are some strategic differences between these two types of organisms (Fig. 8.6 and 8.7).

In insects (Fig. 8.6), the central nervous system is derived from a set of neuroblasts which delaminate from an epithelial sheet of undifferentiated cells (see details of the development of the nervous system in Chapter 10). Neuroblasts divide a number of times along the apico-basal axis, to generate, with each division, a ganglion mother cell (gmc) and another neuroblast. The gmc itself divides once to generate two neurons with specific identities. These neurons then emit axonal projections and generate the three-dimensional mesh that constitutes the nervous system. In vertebrates, the neural tube, which gives rise to all the neurons of the central nervous system, is a single-cell-layered proliferative epithelium (Fig. 8.7). The apical surface of the epithelium faces the lumen of the tube. The

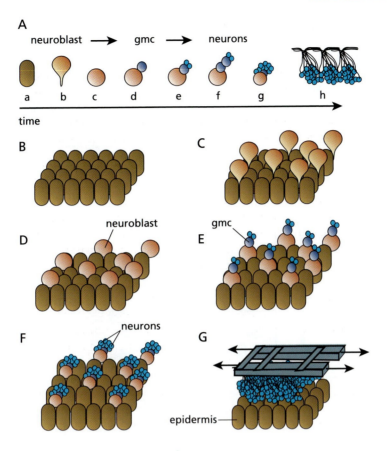

Fig. 8.6. Neural lineages and sublineages in insects. (A) The generation of neural precursors in insects has been studied in most detail in grasshoppers and *Drosophila*. It involves a sequence of segregation (a, b), asymmetric division (c–g) and differentiation by the extension of axonal projections (h) of specialized cells, neuroblasts. **(B, C)** In the first step, some cells within the epithelial lager of ectodermal cells grow in volume and delaminate towards the interior of the embryo, which is on the basal side of the ectodermal layer. These cells are the neural precursors or neuroblasts. **(D, E)** Each of these cells divides several times in a stem-cell mode to generate a number of daughter cells, ganglion mother cells (gmcs) (c, d in A). Each gmc divides once to give rise to two neurons (e in A). Ectodermal cells that do not become neural precursors early in this process continue to divide and then later will become epidermal cells. An important difference between ectodermal cells that have not yet adopted a fate, and neural cells, is the plane in which they divide: the ectodermal cells divide in the plane of the epithelium, whereas neural precursors generate gmcs by divisions perpendicular to this plane. **(F, G)** After dividing a number of times, which can be over one hundred in some cases, the neuroblasts disappear and the neurons emit their axonal projections to form the nerves.

plane of division of the cells that make up the epithelium seems to be crucial for the specification of their fate. When the cells divide in the plane of the epithelium they produce two identical proliferating cells, but when the division is perpendicular to this plane it gives rise to an outer (more basal) cell that migrates away from the apical side and becomes a neuron, and an inner (or more apical) cell that remains connected to the apical side of the epithelium and continues to proliferate to generate further precursors and neurons.

For many years it was thought that the nervous system did not contain a stem cell population of the kind that exists in the haematopoietic system. While this is probably true in the sense that there is no neural precursor cell that can give rise *de novo* to the entire nervous system, there are certain areas of the vertebrate brain that continue to produce neurons throughout the life of an individual. This observation led to the discovery in vertebrates of a population of neural stem cells in the deep layers of the brain that persist into adult life and that are capable of generating the basic cell types of the nervous system.

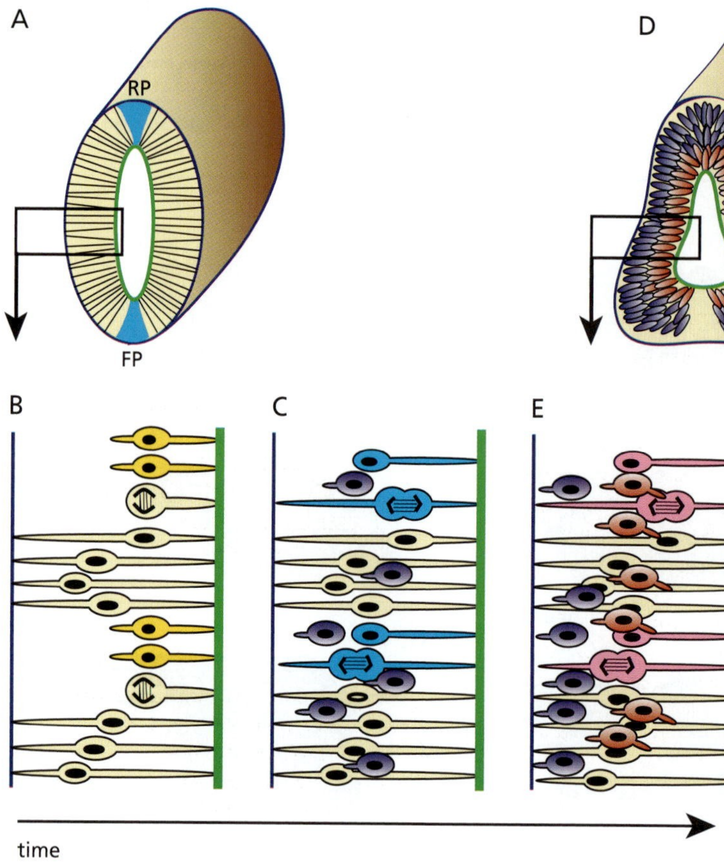

time

Fig. 8.7. Neural lineages and sublineages. (A) In vertebrates, the central nervous system is also derived from an epithelium, the neural tube. In contrast to the situation in insects, where epidermal and neural precursors are intermingled, all the cells of the neural tube participate in the generation of the nervous system. Two regions of the tube, the floor plate (FP) and the roof plate (RP), do not differentiate into neural elements but act as sources of signals for the differentiation of neural elements (see Chapter 10 for details). The neural tube has a lumen, coinciding with the apical side of the cells (green line), which contains a population of precursor cells that divide in a stem cell-like mode. A pattern of sequential division and proliferation produces, over time, the dense array of neurons that characterizes the nervous system. **(B)** The cells of the neuroepithelium span the width of the tube and divide symmetrically in the plane of the epithelium to generate two precursor cells (yellow), whereas **(C)** divisions perpendicular to the plane of the epithelium are asymmetric and generate a neural cell (purple), which is located on the basal side of the tube and differentiates into a neuron, and another precursor (blue). **(D, E)** Regulated iterations of this pattern generate the collection of neurons. In both insects and vertebrates, the same lineages give rise, in addition to neurons, to other neural elements such as glia.

With the possible exception of these recently discovered cell populations in the brain, neural progenitor cells do not self-renew with complete fidelity but rather appear to change subtly during development. Experiments on grasshopper neuroblasts illustrate this point (Fig. 8.8). These experiments take advantage of the fact that in this organism the neurons that emerge early have defined identities determined by their lineage. If the gmc in a particular lineage is killed, although the neuroblast continues to divide and to give rise to new gmcs, it never regenerates the neurons that would have been derived from the dead cell. This is true even if the gmc that is eliminated is the first one. It must therefore be the case that as the neuroblast transmits a particular identity to the gmc, it simultaneously loses the potential to do this.

The divisions of neural precursors are not only sequential but also asymmetric in character and sometimes in size, that is, the mother and daughter cells have different sizes and molecular constitutions and the mother cell never differentiates into a fully fledged neuron. It is likely

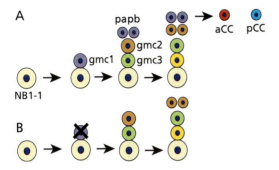

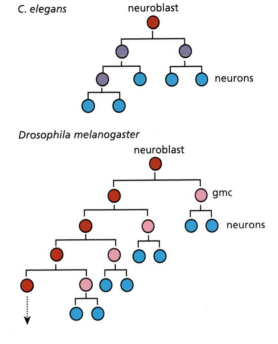

Fig. 8.8. Intrinsic factors in neuronal specification in insects. Development and experimental manipulation of the lineage of neuroblast 1-1 (NB1-1) in the grasshopper. **(A)** NB1-1 divides to produce the ganglion mother cell gmc1 and a cell that retains the properties of a neuroblast. Further divisions give rise to gmc2, gmc3 . . . gmcn; the gmcs divide once to generate two neurons. In the case of gmc1 the neurons, aCC and pCC, each have a specific pattern of connections that allow them to be identified as individuals. **(B)** The fate of aCC and pCC is determined intrinsically by the lineage, since abolition of gmc1, after it has been born but before it divides, eliminates both neurons. (From Doe, C., Kuwada, J., and Goodman, C. (1986) From epithelium to neuroblasts: the role of cell interactions and cell lineage during insect neurogenesis. *Proc. Roy. Soc. B* **312**, 67–81.)

Fig. 8.9. Pattern of neural lineages in the embryos of *Caenorhabditis elegans* **and** *Drosophila melanogaster.* In *C. elegans* each neuroblast divides in a stereotyped manner to generate a set of five neurons. In *Drosophila*, the neuroblast follows a stem cell like pattern of proliferation in which, at every division, it generates a ganglion mother cell (gmc) that will give rise to two neurons and a mother cell that can repeat the process.

that these precursors are specialized mother cells that have evolved from less versatile neural precursors like those found in other invertebrates such as *C. elegans*, where the nervous system arises from contributions of several lineages each of which produces specific cells. In this nematode, most but not all of the neural cells are descendants of the AB precursor which generates a row of cells, each of which gives rise to a neural precursor cell or neuroblast that contributes motoneurons to the nervous system through a stereotyped lineage (Fig. 8.9). These precursors are not very different from the insect neuroblasts; the main difference is that upon division, instead of generating a self renewing cell and a neuron, they undergo three divisions to generate five neurons and none of these divisions can be said to be stem cell-like (Fig. 8.9).

Complex determinate lineages

In many organisms there are both stem cells for particular cell types, and stem cell reservoirs with strict associated lineages that generate (e.g. in the haematopoietic and nervous systems) or regenerate (e.g. the epidermis and the intestinal epithelium) tissues and organs. In some cases, these

cells and their patterns of proliferation determine the three-dimensional organization of the resulting tissue. For example, during the development of the nervous system, a two-dimensional sheet of cells (neural precursors), rolled up into a tube in the case of the vertebrates, is transformed into the organized three-dimensional structure of the mature nervous system. The essential properties underlying this transformation are the polarization of the original sheet of cells, and the orientations of the cell divisions with respect to this plane (Fig. 8.6 and 8.7).

However, there are cases, particularly in invertebrate embryos, in which not just a tissue or an organ, but a whole organism, is generated from one cell through a strict determinate lineage similar to those of the nervous system or the haematopoietic system. In these organisms, of which the most familiar examples are the leech and the nematode *C. elegans*, the early cleavages give rise to a set of precursor cells that act as founder cells for the lineages which

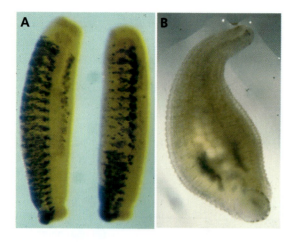

Fig. 8.10. The leech. (A) Late stage leech embryos showing the P (left) and O (right) lineages revealed after staining their progenitor teloblasts with horseradish peroxidase. **(B)** Adult leech. (Images courtesy of D. Weisblat.)

generate the cells that give rise to the animal. Seen in this context, lineages provide the basis not only for the generation of different cells, but also for their spatial arrangement.

The leech *Helobdella triserialis*

Helobdella triserialis (Fig. 8.10) is a segmented animal whose zygote divides in a stereotyped manner to generate three precursors for the endoderm and a fourth cell, the D cell, that inherits a specialized cytoplasm, the teloplasm. This cell gives rise to two sets of five stem cells called teloblasts, one set on each side of the midline (Fig. 8.11). In each set of teloblasts, one (M) gives rise to the mesoderm and the rest (NOPQ) to the epidermis and the nervous system. Each teloblast undergoes a sequence of asymmetric stem cell-like divisions that give rise to a ribbon or 'bandlet' of primary blast cells. Blast cells remain associated with the teloblast in an ordered manner and divide to generate the segments of the leech. The blast cells from each teloblast generate the elements of each segment by specialized divisions, and mix in a complex but reproducible way with the cells from other teloblasts. Labelling experiments show that a segment is composed of the progeny of more than one blast cell and that the manner of this contribution is different for the blast cells from each teloblast.

Segmentation in the leech is thus organized along the anteroposterior axis by the pattern of divisions of the teloblasts, and the degree to which a blast cell will con-

tribute to a segment is strictly determined by its birth rank, early-born blast cells contributing to more anterior segments than late ones. Evidence from a variety of experiments, in which cells are ablated or their positions altered, suggests that although every segment is fine-tuned by later cell interactions, lineage (i.e. the order of birth of the blast cells), is the driving force of the segmentation process.

In the leech, spatial constraints on the development of particular lineages form a basis for the morphogenesis of the animal: the teloblasts remain together and the bandlets they generate are parallel, thereby giving rise to a tube-like structure that prefigures the shape of the organism.

The nematode *Caenorhabditis elegans*

In addition to their potential for generating specific three-dimensional assemblies of cells, as shown in the case of the nervous system or the leech, lineages also determine cell interactions. The nematode *C. elegans* provides multiple examples of this strategy and we have already discussed one at the beginning of this chapter (see Fig. 8.1). In *C. elegans*, as in the leech, early cleavages generate several precursors that then give rise to the different germ layers and their components. The major precursors established by the early cleavages are six founder cells which together give rise to 558 different cells through a sequence of stereotyped and well characterized cleavages (Fig. 8.12).

The specification of the six founder cells relies on a mixture of intrinsic cues and cell interactions. The first division generates two cells or blastomeres: AB and P1; the fates of these cells and their immediate descendants are specified in very different ways. While the fate of the descendants of AB is determined by their interactions with surrounding cells (see Fig. 8.1), the fates of the descendants of P1 are strictly determined by lineage: when isolated, the P1 blastomere produces the correct pattern of descent and cell fates. The different requirements for AB and P1 underscore the fact that there is no rigid rule specifying how organisms will develop and how cells within them will acquire fates. Within the framework of a few well-defined molecular mechanisms, different organisms have evolved individual strategies based on the differential regulation of those mechanisms. Thus most organisms develop using a mixture of cell-autonomous, lineage-driven mechanisms, supplemented and modified by influences derived from cell interactions.

A detailed study of *C. elegans* also shows that the most strict lineages can hide important variations. Despite the apparent invariance of its lineages, the timing of some of

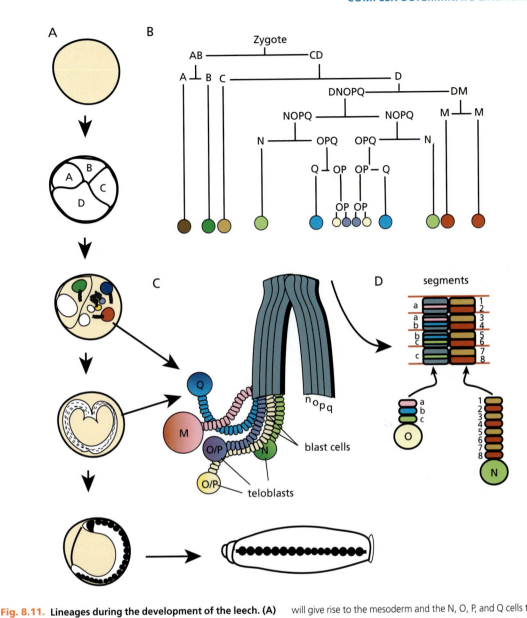

Fig. 8.11. Lineages during the development of the leech. (A) The zygote undergoes a number of stereotyped cell divisions (shown in detail in **B**) which give rise to a set of precursor cells, teloblasts (coloured), that will divide a number of times to generate the whole organism. The teloblasts behave as stem cells, each giving rise to a 'bandlet' of cells. The bandlets coalesce to generate a thick band, and later on the bands from the two sides fuse to form the main body of the leech. By the late stages of development, the shape of the animal can be discerned, with the developing nervous system (black) lying on the ventral side. The integument wraps around the mesoderm and the nervous system to create the tube-like structure. The black spots indicate the ganglia of the nervous system. **(C)** Detailed view of the role of the teloblasts. The teloblasts (M, N, O, P, Q) generate chains of blast cells, which coalesce into bandlets (m, n, o, p, q). The m bandlet lies underneath the other four. For simplicity, only one of the two sets of bandlets is shown in detail. The M cells will give rise to the mesoderm and the N, O, P, and Q cells to the ectoderm. The spatial order of the chains of blast cells in the bandlets reflects the order of their birth: the earlier born cells are located further from the teloblast than later born ones. **(D)** The blast cells divide a number of times to generate the segments. Labelling experiments show that the progeny of each blast cell contribute to a specific segment or segments in a precisely ordered pattern. There are two types of pattern. Each blast cell arising from the M, O, and P teloblasts contributes to two adjacent segments, as shown here for the O teloblast. Each blast cell arising from the N and Q teloblasts contributes only to one segment, with adjacent blast cells contributing alternately to the anterior or posterior region of every segment. The registration between the progeny of the different teloblasts within a segment generates its structure and provides the basis for the generation of different cell types within it.

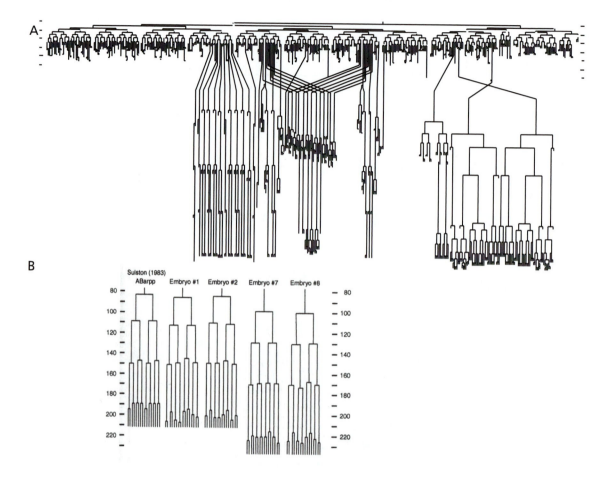

B

the divisions can vary in different individual worms (Fig. 8.12). This means that, in certain lineages, the same cells are liable to have different neighbours in different individuals. We have already seen that transient interactions between neighbours from different lineages can affect how the lineages develop, so the individual-to-individual differences must mean that the organism has ways of adapting to these variations.

Cell fate determination within lineages

As lineages develop, whether in the leech, *C. elegans*, the haematopoietic or central nervous systems, they generate different cell types and this is associated with the expression of fate-determining genes in the individual daughters of the stem cell. Many of these genes code for transcription factors whose function is to initiate the sequence of

Fig. 8.12. Lineage variability in *Caenorhabditis elegans*. (A) The complete lineage tree of *C. elegans* (Courtesy of J. Ahringer). **(B)** Computerized lineage analysis of specific lineages reveals a certain degree of variability from one embryo to another. Cell divisions are indicated, in standard fashion, by bifurcations of the vertical lines. The vertical axis represents time. Here, the lineage of the ABarpp cell is shown in different individuals. Although the pattern is identical in all of them, the timing of the divisions varies. (After Schnabel, R. *et al.* (1997) Assessing normal embryogenesis in *Caenorhabditis elegans* using a 4D microscope: variability of development and regional specification. *Dev. Biol.* **184**, 234–65.)

events that leads to the acquisition of a specific fate.

The early developmental stages of *C. elegans* provide some examples of how the cells that generate particular lineages are specified. During the early divisions that

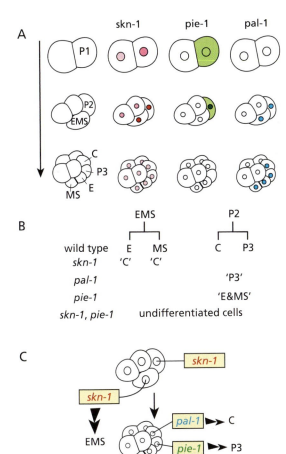

Fig. 8.13. **Assignment of cell fates to the blastomeres of the early *Caenorhabditis elegans* embryo. (A)** The development of the descendants of the P blastomere (see Fig. 8.1) relies on the allocation of transcription factor determinants to specific cells. The arrow on the left indicates time. The figure shows the distribution of the bZip transcription factor skn-1, the zinc-finger protein pie-1, and the homeodomain protein pal-1 in the blastomeres of the embryo at the two-, four- and eight-cell stages. Genetic analysis indicates that these factors act at the top of a hierarchy of cell fate assignment. **(B)** Cell fate changes in different mutant embryos. The mutants are indicated on the left. Only cells that change fate in particular mutants are indicated. **(C)** Genetic analysis of the phenotypes and product distribution in single and double mutants shows that every blastomere in the P and EMS lineages acquires its identity through a unique combination of skn-1, pal-1, and pie-1, as shown here for C and P3.

is achieved by a combination of inheritance of specific factors, which create a combinatorial code of transcriptional activity, and regulatory networks that ensure the activity of particular combinations.

Once founder cells for particular lineages have been specified, the next step requires the allocation of fates to the descendants of those cells. The founder cell may generate two different daughter cells that are subject to different intrinsic and environmental influences that determine their gene expression profile and thereby their fate. Alternatively, the original cell may act as a stem cell that generates a lineage of different kinds of cells while itself retaining its original identity (see Fig. 8.4). In this case, at the time of cell division the mother cell must be able to allocate, to the differentiating daughter, a protein that will determine the fate of the daughter. At the same time, the mother cell must ensure either that it does not itself inherit this determinant or that it can suppress its effects. The developmental program inherited by a daughter cell will then determine the expression of a specific repertoire of transcription factors that direct the fate of that cell and its descendants.

The fate of the descendants in stem cell types of lineages depends on a balance of intrinsic and extrinsic cues. This interplay is revealed by a number of experiments, particularly in systems in which individual elements of a lineage can be readily identified. For example, as we have seen earlier (Fig. 8.14), during the development of the nervous system in the grasshopper, the first ganglion mother cell (gmc1) from neuroblast 1 (NB1-1) gives rise to two sibling cells that develop as the aCC and pCC neurons. These cells can be distinguished by their characteristic axonal trajectories and cell-surface molecules. Ablation of gmc1 results in the absence of both aCC and pCC. No other gmc can

specify the main precursor cells (Fig. 8.1), particular proteins that act as determinants of cell fates are segregated to specific cells. This is illustrated by the segregation of three transcription factors (skn-1, pal-1 and pie-1) to each of the first four cells of the embryo (Fig. 8.13). *Skn-1* is absolutely necessary to specify the fate of EMS, which gives rise to the E and MS lineages. The expression of *pie-1* is always associated with the blastomere (P2, P3 . . . Pn) that will give rise to the germ-line precursors and at every cell division it segregates with this blastomere. In the absence of *pie-1*, the P2 blastomere develops EMS features. This observation, together with the fact that *skn-1* is expressed in P2, suggests that a function of pie-1 is to repress the activity of *skn-1* in P and its descendants. *Pal-1* is necessary for the proper development of P3 and C. These three factors—pie-1, skn-1 and pal-1—must play an important role in initiating developmental programs of gene expression, since in *skn-1 pie-1* double mutants, P1 fails to produce any differentiated cell of any kind. Thus, the fate of each blastomere

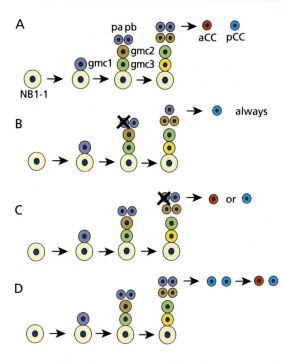

Fig. 8.14. Extrinsic factors in neuronal specification in insects. (A) Development of the wild-type neuroblast 1-1 (NB1-1) in the grasshopper. The first gmc, gmc1, gives rise to the aCC and pCC neurons. **(B)** If either of the two neurons born from gmc1 is eliminated shortly after division of gmc1 and before they differentiate, the remaining one always develops as pCC. **(C)** Eliminating either of the two daughters of gmc1 at a later stage results in the remaining neuron developing as aCC or pCC at random. **(D)** Summary of the events that lead to the specification of the aCC and pCC neurons from the lineage of the neuroblast 1-1. The pCC fate is imposed by the lineage while aCC emerges from cell interactions between the two daughters of gmc1. (From Doe, C., Kuwada, J., and Goodman, C. (1986) From epithelium to neuroblasts: the role of cell interactions and cell lineage during insect neurogenesis. *Proc. Roy. Soc. B* **312**, 67–81.)

The nervous system of *Drosophila* is very similar to that of the grasshopper and allows a genetic analysis of the mechanisms that are involved in the establishment of different fates in particular neuroblast lineages. This analysis has identified the homeodomain protein prospero as a general cell fate determinant in the central nervous system (CNS). During development prospero is made in the neuroblast but segregates to the gmc where it is required to generate the different neuronal cell fates. Vertebrates also have homologues of prospero, whose pattern of expression in the differentiating neurons suggests that they play a similar role in the assignment of cell fates.

Prospero appears to be a general determinant for the fates of most ganglion mother cells. Within individual lineages, the expression of specific transcription factors is essential for the identity of the emerging cells. Thus, for example, the first gmc to arise from NB1-1 inherits the expression of the hunchback transcription factor from the neuroblast. NB1-1 then switches off hunchback expression and activates that of another transcription factor, cas, which becomes incorporated into the second gmc. These observations correlate well with the results of the ablations and provide a molecular basis for the unique identity of every gmc.

The intrinsic patterns of gene expression characteristic of many lineages can be readily modified by external cues. In fact, these intrinsic programs include the expression of specific receptors and signalling molecules that will modulate the cell's transcriptional program by enabling it to receive signals from other cells. Thus, cytokines and growth factors in the haematopoietic system, and growth factors and other signalling molecules in the nervous system, create networks of external influences that contribute to the cell diversity associated with the lineage.

Mechanisms of segregation of determinants of lineage identity in *S. cerevisiae* and *C. elegans*

Whether a division within a lineage gives rise to a stem cell and a cell that will differentiate, or to two cells that will both differentiate, there must be mechanisms for ensuring that cell fate determinants are allocated differentially to the cells arising from the division. In the first case (the stem cell lineage) either the loss or the retention of a determinant might maintain the stem cell fate, while in the second case two (or more) different determinants might segregate into the two daughters. In either case, these proteins will

substitute for this function of gmc1, that is, only the first daughter of NB1-1 can give rise to aCC and pCC and therefore the fate of these cells has a lineage component intrinsic to gmc1. However, if one of the progeny of gmc1 is killed at random shortly after the cell's birth, the remaining cell always develops as pCC (Fig. 8.14). If the ablation is performed later, the remaining cell develops as aCC or pCC with equal probability. This experiment indicates that the pCC fate or identity has a strong lineage component and can be considered as a ground-state or default fate. It can be modified by other factors which act over time and contribute to promoting the aCC fate.

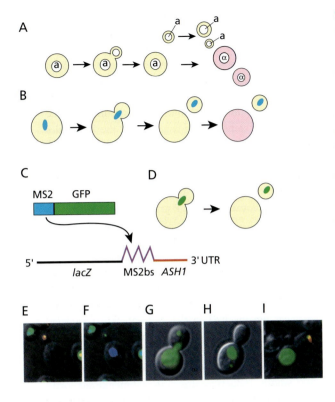

Fig. 8.15. Asymmetric localization of determinants in Saccharomyces cerevisiae. (A) The yeast *S. cerevisiae* has two different cell types, known as mating types **a** and α, which are generated after each cell division according to a patterned lineage. Divisions are asymmetric (also known as budding). The lineage pattern is such that only mother cells (i.e. cells that have previously budded) can switch, which they do at their next division. **(B)** Daughter cells are prevented from switching by the localization to the bud (and hence to the daughter after cell division) of the mRNA (blue) for the transcription factor Ash1 (Chapter 3, Fig. 3.14). Deletion/substitution experiments of different regions of the *ASH1* mRNA indicate that this process requires a 3′ UTR of this mRNA. **(C)** The process of *ASH1* mRNA segregation can be followed *in vivo* using an adaptation of GFP technology (see Chapter 6, Fig. 6.2) to follow mRNAs by making use of a protein from the MS2 bacteriophage that recognizes a specific RNA sequence (MS2-binding site, MS2bs; purple). If an MS2bs is incorporated into the *ASH1* mRNA, a fusion protein containing MS2 protein and GFP will bind to the mRNA so that its movement can be followed *in vivo* through GFP fluorescence. In the experiment shown, the coding region of *ASH1* is replaced by that of *lacZ*, to provide an independent way of monitoring the segregation of the hybrid RNA. **(D)** The *lacZMS2bsASH1UTR* mRNA behaves like the *ASH1* RNA and is localized to the daughter cell. **(E–I)** Images of the localization of the *ASH1* reporter mRNA in wild type and mutant dividing yeast cells. The images are taken from live videos. **(E)** Colocalization of the MS2:GFP protein (green) and the reporter RNA revealed with a fluorescently labelled probe against the *lacZ* RNA sequence (red). Yellow indicates colocalization of the two signals. **(F)** DNA in E revealed with a DAPI stain (blue). **(G)** Wild-type yeast carrying the *ASH1* reporter mRNA and MS2:GFP fusion protein, showing its translocation into the daughter cell. **(H)** In a *she1* mutant the *ASH1* reporter does not translocate to the daughter cell. The *SHE1* gene encodes a specialized myosin that is required for the translocation process. **(I)** Wild type: co-localization of the reporter mRNA (green) with the She1 protein (red) (yellow indicates colocalization) in the daughter cell. (Images E–I from Bertrand, E. *et al.* (1998) Localization of *ASH1* mRNA particles in living yeast *Mol. Cell* **2**, 437–45.)

need to be asymmetrically distributed within the cell, and this asymmetric localization will need to be coupled to the process of cell division.

The budding yeast *S. cerevisiae* offers a simple example of how cell polarity and asymmetric cell division can be linked to cell fate allocation. As we have seen already (see Chapter 2 , Fig. 2.9), when a haploid yeast spore germinates, it divides by budding to produce a larger mother cell that is able to switch its mating type at the next division and a smaller daughter cell that is unable to switch (Fig. 8.15). The ability to switch mating type is determined by transcription of the *HO* gene. Analysis of mutants defective in restricting *HO* expression in daughter cells led to the identification of a gene, *ASH1*, which encodes a zinc-finger protein that represses *HO* transcription. The Ash1 protein is only expressed in the daughter cell and this is achieved by a differential segregation of the *ASH1* mRNA during mitosis (see Chapter 3, Fig. 3.14). After transcription, the *ASH1* mRNA accumulates preferentially in the bud area that is destined to become the daughter cell (Fig. 8.15). Genetic studies have identified a set of genes, the *SHE* genes, that are required to prevent the accumulation of *ASH1* mRNA in mother cells. One of these genes, *SHE1*, encodes a myosin protein that is part of the machinery that transports *ASH1* RNA to the bud. Other genes involved in the regulation of this process also encode components of the cytoskeleton or regulators of the activities of these components, suggesting that the segregation of *ASH1* mRNA is tightly linked to the activity of the cytoskelton during the process of process of cell division (Fig. 8.15).

Similar examples of determinant segregation can be observed in multicellular embryos. Such patterns are particularly clear in embryos with defined lineages, such as *C. elegans* or many marine invertebrates. For example, in *C. elegans*, the P granules, ribonucleoprotein particles associated with the germ line, are segregated to cells that will be

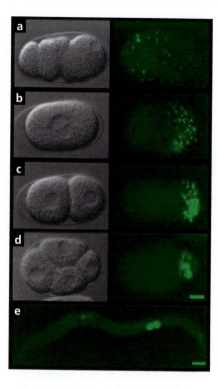

Fig. 8.16. Asymmetric localization of determinants in early embryos. In the *Caenorhabditis elegans* embryo, cytoplasmic granules known as P granules segregate to the posterior region of the embryo after fertilization and at every subsequent cleavage. This segregation allocates them to the germ cells of the nematode. The left-hand column shows Nomarski images of a *C. elegans* embryo during the first two divisions (a–d; note that the zygote, panel a, has a characteristically twisted appearance). The column on the right shows the P granules as revealed by a fluorescent probe, in the same embryos (green). The bottom picture (e) shows a worm with the germ line identified by the green stain. (Images courtesy of S. Strome.)

precursors for the germ line (Fig. 8.16). This is achieved, in the lineage that gives rises to the germ-line precursors, by the localization of the P granules to the posterior pole of the cells before division, so that the granules are preferentially allocated to the posterior daughter cell arising from the division. The iteration of this process leads through several rounds of cell division to the segregation of the granules to the cells of the germ line, which are clustered in the posterior region of the animal. As in yeast, genetics can provide useful insights into the proteins and processes that are involved in the segregation of determinants in complex organisms. For this approach to be successful it is important to focus on a defined process and to have suitable markers that can be used to analyse it.

For example, the unfertilized egg of *C. elegans* has no apparent polarity apart from a non-centrally located nucleus. Polarity is established during fertilization by the point of sperm penetration, which becomes the posterior pole of the zygote. At this stage, the polarity of the zygote is evident from visible cytoplasmic movements, from the orientation and positioning of the first mitotic spindle, and from the segregation of some cellular constituents: the P granules segregate to the posterior region of the embryo (Fig. 8.16 and 8.17), and certain determinants segregate to specific blastomeres (Fig. 8.17). Some of these processes are reiterated during successive divisions, as can be seen in the progressive allocation of the P granules to the germ-line blastomere.

Genetic analysis of this process has revealed a number of maternal-effect mutants that are defective in various aspects of the polarity of these early divisions. The genes identified through these mutants are called *par* genes, and six have been found so far. The wild-type products of the *par* genes are involved in the segregation of cell fate-associated determinants and are themselves asymmetrically distributed during cell division (Fig. 8.17). Three of the *par* genes have been studied in some detail: *par-1* and *par-2*, whose products localize to the submembrane cortical region of the posterior half of the zygote, and *par-3* whose product localizes to the anterior cortex. Each of these genes has a distinct function, that is a particular set of molecular interactions and effects as revealed by loss of function mutations, but they are all involved in the asymmetric localization of P granules and cell fate determinants. However, it is not possible to construct a simple pathway that links the activities of the par proteins. Although the functions of the proteins are clearly related, no pattern can be discerned from analysing which determinants require which par proteins for their localization; rather, each determinant seems to have a distinct relationship with the different par products (Fig. 8.17).

The defects observed in *par* mutants are very similar to those that arise from alterations in the actin cytoskeleton. This suggests that the par proteins might be involved in bringing about or coordinating the changes in the cytoskeleton that ensure determinants are correctly localized. This idea is supported by the observation that correct localization of the par proteins is accompanied by cytoplasmic streaming that requires the activity of the actin cytoskeleton and of the non-muscle myosin protein nmy-2. Thus, just as in the yeast, there is a direct link between the proteins that have to be partitioned to different cells, the cytoskeleton, and (probably) the activity of motor

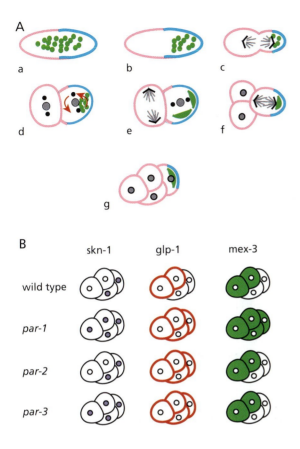

Fig. 8.17. The mechanics of asymmetric localization of determinants in the early embryo of *Caenorhabditis elegans*. **(A)** In the *C. elegans* embryo a reorganization of the cytoskeleton during the first cleavage creates an asymmetry in the mitotic spindle and in the distribution of the P granules (green) that leads to the generation of the AB cell and the smaller P cell, which contains the P granules (a–c). The second division takes place first in the AB cell (left in d and e) and later in the P cell (f, g), which divides at right angles to the AB cell after a rotation of the centrosomes (black dots). The *par* gene products are associated with the cortex and follow the patterns shown. Par-1 and par-2 (blue) are always associated with the posterior part of the P cell and its descendants, whereas par-3 (pink) is enriched cortically in all cells. Par-1 encodes a putative serine/threonine kinase that can interact with the non-muscle myosin protein nmy-2; par-2 is a protein of unknown function with a zinc-binding domain of the Ring-finger class, while par-3 contains a PDZ domain. The function of the par proteins is required not only for the cytoplasmic reorganization that precedes the first cleavage, but also for the correct orientation and positioning of the mitotic apparatus and the segregation of complexes such as the P granules. In *par* mutants, many of these processes go wrong (e.g. the first division gives rise to two equally sized cells and an abnormal distribution of the P granules). **(B)** Another important function of the par proteins is in the asymmetric localization of a variety of gene products that are involved in determining the fates of the early blastomeres. These determinants include the transcriptional regulator skn-1, the Notch family receptor glp-1, and the RNA-binding protein mex-3. The distributions of these proteins are shown in wild-type embryos and *par* mutant embryos at the four-cell stage. In the wild type, each of these products is allocated to a particular set of blastomeres at the four-cell stage: skn-1 to P2 and EMS; glp-1 to ABa and ABp; and mex-3 to ABa and ABp. Note that each different *par* gene mutation leads to a different spatial distribution of the determinants. This suggests that, although the par proteins are all part of a coordinated molecular system, each has a specific relationship with the different determinants.

proteins. In *C. elegans*, and probably also in other multicellular systems, the activity of the cytoskeleton and associated motors is likely to be regulated by an interaction between intrinsic and extrinsic cues.

Mechanisms of segregation of determinants of lineage identity in the development of the nervous system

The demonstration that segregation of determinants during cell division plays an important role in the assignment of cell fates in simple, determinate lineages, such as those of *S. cerevisiae* and *C. elegans*, suggests that similar mechanisms may operate in more complex lineages, such as those associated with stem cells or those that characterize the development of the nervous system and perhaps also haematopoiesis. Detailed genetic analysis of the development of the nervous system in *Drosophila* has identified some molecules that may mediate such processes.

The external sensory organs in the peripheral nervous system (PNS) of the fruit fly are made up of five different cells derived from a sensory mother cell (SMC or pI) through three sequential divisions (Fig. 8.18). Each of these divisions generates two cells with different developmental potential as revealed by mutations in the *numb* gene, which result in an alteration of the fates of the two sibling cells that arise from the first division, pIIa and pIIb, both of which now follow the pIIa fate. Numb is a cell-cortex-associated protein which segregates asymmetrically in each of the divisions that generate the sensory organ (Fig. 8.19). During the first division numb segregates to the pIIb cell, where mutations indicate that it is required, and forced expression of numb in the pIIa cell makes it develop as pIIb (Fig. 8.18). These experiments

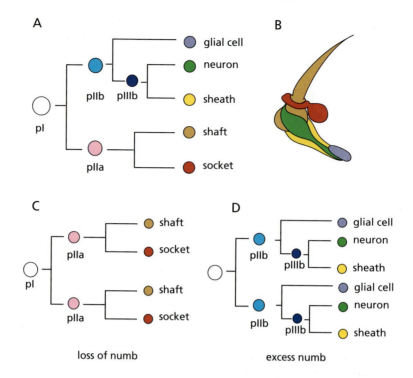

loss of numb

excess numb

suggest that after every division in the lineage of the SMC, intrinsic influences assign fates to the two daughters and that the function of numb, which is segregated differentially to the daughters, is associated with the establishment of this fate. In particular, these results suggest that the pIIb fate results from the suppression of that of pIIa.

The numb protein is also expressed during the development of more complex lineages such as those generated by the neuroblasts of the central nervous system (CNS) (see Fig. 8.20). During division of the neuroblast, numb segregates differentially between the neuroblasts and the gmcs. However, the cues that lead to the differential distribution of numb are probably different from those that operate in the PNS. In the PNS, the segregation of numb follows cues, as yet unidentified but which must lie in the plane of the epithelium, that allocate it differentially between two daughter cells. In the CNS, the numb protein is differentially distributed along the apico-basal axis. Before the division of the neuroblast, numb is distributed throughout the cortex of the cell but it then undergoes a cycle of localization and segregation which is repeated at every cell division (Fig. 8.20). As the neuroblast gives rise to the gmc, numb forms a tight crescent on the basal side of the cell just where the gmc will segregate, becomes incorporated into the gmc and is then degraded. The cycle begins again in the neuroblast in its next division. Despite this tantalizing ob-

Fig. 8.18. Neural lineages and numb in *Drosophila*. (A) The external mechanosensory organs of *Drosophila* are part of the peripheral nervous system (PNS) and each is composed of an external element, the bristle or shaft, which is associated at its base with a socket, a neuron, a sheath cell, and a glial cell. These elements are clonally derived from a precursor, pI, which divides three times to generate the five cells through the stereotyped lineage shown. **(B)** Assembly of the cells in a sensory organ (colours as in A). **(C)** In loss of function *numb* mutants, both the daughters of pI follow the pIIa pathway. **(D)** If numb is overexpressed, both daughters develop as pIIb. This suggests that both pI progeny have the same intrinsic fate, pIIa, and that pIIb develops as a result of the action of numb.

servation, genetic experiments provide no support for the idea that, in the CNS, numb acts alone in distinguishing cell fates; for example, loss of *numb* alone does not alter cell fates.

Some aspects of this cycle of numb movement and activity are closely mirrored by the prospero protein and RNA (Fig. 8.20). Prospero, as mentioned briefly earlier in this chapter, is a homeodomain protein that acts as a transcription factor involved in directing the program of gene expression in the gmcs. Before neuroblast division, prospero is tightly associated with the apical cortex of the neuroblast, but when the neuroblast divides prospero follows numb to the basal side and segregates into the gmc, where it enters the nucleus. This scenario, and the nature of the

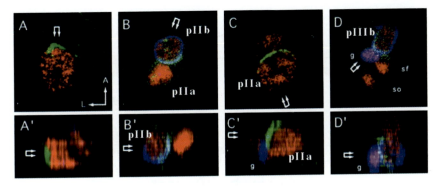

prospero protein, suggest that prospero can be viewed as a determinant of cell fate, and numb as an element of the cellular machinery that determines its spatial localization. However, the spatial localization of numb and prospero appears to be independent, as numb localization is normal in *prospero* mutants, and prospero localization is normal in *numb* mutants.

There is an increasing number of proteins known to be involved in the segregation of determinants or that are asymmetrically segregated during the generation of gmcs by the neuroblasts. This suggests that the process might require a large multiprotein complex whose activity is subject to several layers of regulation. Components of the molecular machinery involved in localizing and segregating proteins such as numb and prospero are emerging from classical genetic screens and from more modern molecular screens such as the yeast two hybrid screen (see Chapter 5, Fig. 5.6), performed mainly in *Drosophila*. Already several additional proteins have been found whose interactions with numb and prospero, as well as their distributions in the cells, indicate that they play important roles in the process of segregation (Fig. 8.20). This is also the case for the par proteins in *C. elegans*, and indeed some of the *Drosophila* components that drive the differential divisions during neuroblast division have turned out to be homologues of the par proteins (Fig. 8.20).

The intracellular movements of proteins such as numb during the division of the neuroblast suggest that their segregation, like that of the *ASH1* mRNA in yeast or the products of the *par* genes in *C. elegans*, is linked with the cell cycle and the pattern of cell division. Therefore one might predict that their movements would involve interactions with the cytoskeleton, which undergoes very dramatic changes during the mitotic process. This idea is supported by the observation that, in *Drosophila* and *C. elegans*, disrupting microfilament formation alters the segregation of determinants.

During the division of the neuroblast, the protein in-

Fig. 8.19. Asymmetric distribution of numb during the development of the adult PNS in *Drosophila*. Asymmetric segregation of prospero (blue) and numb (green) in the lineage of the adult mechanosensory organ. The red is a nuclear marker that highlights the position of the cells. The pictures represent confocal images of cells in the lineage during division (see Fig. 8.18): **(A)** pI, **(B)** pIIb, **(C)** pIIa and **(D)** pIIIb. The top panels show horizontal sections through the cells with anterior (A) up and lateral (L) to the left. The bottom panels show reconstructions along the vertical axis of the different sections shown above, with apical up. In dividing pI **(A)** and pIIa **(C)** cells, numb is localized in an anterior crescent and prospero is not detected. On the other hand, in dividing pIIb **(B)** and pIIIb cells **(D)**, prospero protein is localized tightly at the basal pole of the cell, while numb accumulates both basally and at the site of contact with pIIa and the glial cell (g). (From Gho, M., Bellaiche, Y., and Schweisguth, F. (1999) Revisiting the *Drosophila* microchaete lineage: a novel intrinsically asymmetric cell division generates a glial cell. *Development* **126**, 3573–84.)

scuteable plays an important pant in the segregation of determinants and is a good candidate for a link with the cytoskeleton. Inscuteable contains five ankyrin repeats, an SH3 domain and several alpha helices—all potential protein–protein interaction motifs. In the absence of inscuteable, none of the proteins involved in segregation localize properly, and the mitotic spindle, which in wild-type neuroblasts always becomes oriented along the apico-basal axis, is instead oriented at random. In wild-type flies, inscuteable protein remains localized in a tight apical crescent throughout the segregation of the gmc, and this localization requires an interaction between inscuteable and the par-3 homologue bazooka. Bazooka is necessary for maintaining the apico-basal polarity of the cell and can interact physically with inscuteable (Fig. 8.20). This suggests that the bazooka–inscuteable complex provides a link between the apico-basal polarity of the cell and the segregation of determinants.

The molecular machinery involved in the localization and segregation of determinants in cell lineages may be conserved more widely in the animal kingdom. Both

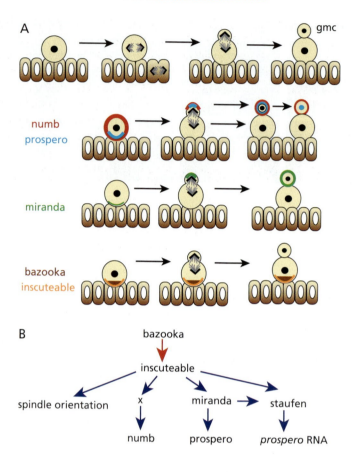

Fig. 8.20. The mechanics of determinant segregation during early neurogenesis in the *Drosophila* CNS. (A) Spatial localization of some of the proteins involved in the asymmetric segregation of determinants during the division of the neuroblasts in the *Drosophila* embryo. A neuroblast segregates from amongst the ectodermal cells to the basal side of the ectodermal epithelium (see also Fig. 8.6), where it will divide. The spindle of the first division initially assembles as if the neuroblast were about to divide in a plane parallel to the epithelium, which is the plane of division of the ectodermal cells. However, the spindle then rotates 90° and positions itself perpendicular to the epithelium, with the ganglion mother cell (gmc) appearing on the side distant from the ectoderm. Numb (red) is first present throughout the cortex of the neuroblast but as the gmc appears it becomes located in the emerging daughter cell. Prospero (blue) is localized to the apical side of the neuroblast but is translocated basally just before the cell divides and becomes incorporated into the gmc, where it then goes into the nucleus. Miranda (green), a protein with four coiled-coil domains and two leucine zipper motifs, is first localized as a crescent within the neuroblast on the side that contacts the ectoderm but as the gmc appears it becomes integrated with numb in the cortex of the gmc. In contrast, bazooka (brown), a *Drosophila* homologue of the

Caenorhabditis elegans protein par-3 with three PDZ domains, and inscuteable (orange), an SH3-containing protein, are located on the ectodermal side of the neuroblast and remain in this location throughout the cell division. **(B)** Summary of the interactions that drive the segregation of determinants during the development of the CNS in *Drosophila*. Arrows indicate that a protein is required for the localization of the one the arrow points to. The red arrow indicates a direct, physical interaction and blue arrows indicate functional interactions. Most of the functional relationships have been established by a mixture of genetics and cell biology to follow the localizations *in situ*, and yeast two-hybrid assays (see Fig. 5.6) to establish the possible physical interactions.

prospero and numb have been found in vertebrates where their expression is also tightly linked to the generation of neural cell types. Whereas the expression of numb seems to follow similar rules to those operating in the *Drosophila* CNS, expression of the vertebrate prospero homologue *Prox1* in different elements of the lineage appears to result not from differential segregation of the protein but from transcriptional control of expression of the *Prox1* gene.

A similar situation seems to apply in the PNS in *Drosophila*. It has been suggested that these differences in the regulation of prospero might arise because the segregation of determinants is tightly linked to the cell cycle, and the cell cycle has different lengths in different systems. In the *Drosophila* embryo, at the time when the CNS is beginning to form, the cell cycle is very fast. In this situation, transcriptional regulation of the *prospero* gene might be too slow to ensure that prospero is expressed in the gmc where it is needed, so it might be an advantage to regulate the segregation of the prospero protein itself. At the later developmental stage when the sensory organs of the PNS develop (during the development of the adult from the larva), or in vertebrates, the cell cycle is longer and so it may be preferable to regulate gene expression rather than localization of the protein.

It remains to be seen whether the mechanisms that regulate the spatial distribution of determinants in the nervous system of various animals follow a universal molecular theme with some variations, or whether there are more fundamental differences. However, examples such as the differential regulation of prospero segregation suggest that as more information accumulates we will again, as in so many other situations, see cell- and organism-specific strategies applied to the same basic developmental process.

Notch signalling: A recurring element in the lineage-mediated assignment of cell fates

In molecular terms, a cell fate represents a commitment to a specific program of gene expression and cell differentiation, mediated by interactions between intrinsic and extrinsic factors. In some cases, the assignment of a cell to a specific fate is a determinate and irreversible event. In *C. elegans*, for example, there are many cell fate decisions in which a cell A gives rise to cells B and C, each of which then gives rise to particular cells, all in a direct and irreversible manner. In other situations, like that of stem cells, cells have a choice and can either adopt or forfeit the fates they are offered; cell A divides to give rise to two cells, each of which has the potential to become B or remain as A. A mechanism must exist to ensure that the division of A gives rise to another cell A and a cell B. This applies, for example, to the daughters of the first ganglion mother cell from NB1-1 in the grasshopper. In this case, each cell is initially capable of adopting either the aCC or the pCC fate. The

adoption of one of these fates requires cell interactions and the suppression of the other fate. Similarly, in the *Drosophila* PNS lineage, both of the daughters of the SMC/pI are initially capable of adopting either the pIIa or the pIIb fate, but the pIIb fate develops in one of the daughters because the presence of numb suppresses the pIIa fate.

Caenorhabditis elegans also provides examples of this more plastic mode of cell fate assignment, for example during the specification of the anchor cell, a specialized inductive element crucial for the development of the vulva (see Chapters 9 and 12). During development, each of two precursor cells (Z1.ppp and Z4.aaa) has an equal chance of becoming the anchor cell while the remaining one develops as a ventral uterine blast cell. This is shown by experiments in which either of the two cells is ablated and the remaining one always develops as the anchor cell. This is a situation similar to the one we have described for the aCC and pCC neurons during the development of the nervous system and suggests that cell interactions play a role in the emergence of these cell fates (Fig. 8.21).

Genetic analysis of the specification of the anchor cell identified the product of the *lin-12* gene as a key element in this cell fate decision (Fig. 8.21). In embryos mutant for *lin-12* the two precursor cells adopt the same fate, confirming that they are equivalent as suggested by the cell ablation experiments. In *lin-12* loss of function mutants, both cells develop as anchor cells, whereas in dominant gain-of-function mutants (*lin-12d*) both develop as blast cells. This indicates that the two daughter cells share the potential to develop the anchor fate cell and that lin-12 suppresses this potential in one of the cells. The *lin-12* gene encodes a member of the Notch family of receptors (see Chapter 4, Fig. 4.22) and a combination of experimental ablations and genetic analysis suggests that, rather than defining the fate adopted by each of the cells, Notch is involved in ensuring the outcome. The lineage leads to a primary fate, the anchor cell fate, in one of the cells and the activity of lin-12, triggered by a signal from the anchor cell, suppresses this fate in the other cell and uncovers the secondary, ventral uterine blast cell fate.

Further genetic analysis of this process revealed other genes that when mutated produced phenotypes similar to *lin-12* mutants and that might therefore encode proteins that work together with lin-12. These genes encode components of the Notch signalling pathway: a DSL ligand (*lag-2*), and a CSL transcription factor (*lag-1*) (Fig. 8.21).

The nuclear targets of the lin-12–lag-1 complex are probably the genes that confer the anchor cell fate, many of which have not yet been identified. Two targets, however,

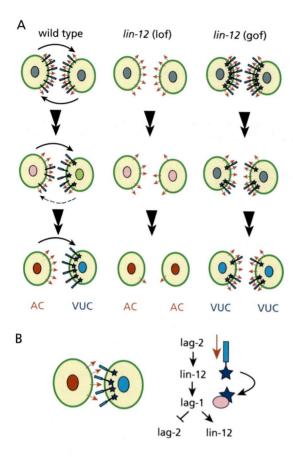

Fig. 8.21. Specification of the anchor cell through *lin-12* activity in *Caenorhabditis elegans*. **(A)** *Left*: In wild-type *C. elegans*, the anchor cell is selected stochastically from two equivalent cells (Z1.ppp and Z4.aaa), one of which develops as the anchor cell (AC) and the other as a ventral uterine precursor cell (VUC). This process is mediated by a Notch family receptor encoded by the *lin-12* gene (blue box on the cell surface) and a DSL ligand encoded by *lag-2* (red arrow). Initially, both cells express similar levels of lin-12 and lag-2 but as development proceeds, an asymmetry develops: one cell accumulates high levels of the receptor and simultaneously decreases the level of the signal, while the reciprocal happens in the other cell. As a result, the cell that accumulates high levels of the receptor develops as a VUC and the other as the AC. *Centre*: In the absence of *lin-12* both cells develop as AC. *Right*: In animals with gain-of-function mutations in *lin-12*, which create receptors that signal constitutively, both cells adopt the alternative fate, VUC. **(B)** Genetic analysis indicates that during the specification of the AC fate, small differences in the expression of *lin-12* and *lag-2* between Z1.ppp and Z4.aaa are amplified through a regulatory mechanism whereby lin-12 signalling increases its own expression and suppresses the expression of *lag-2*. This creates a stable situation in which the cell that receives the signal does not itself signal, and the cell that is signalling is unable to receive a signal. These effects of signalling between lag-2 and lin-12 are implemented by the CSL transcription factor lag-1.

are *lag-2* and *lin-12* themselves. Thus, lin-12 signalling represses expression of its ligand (lag-2) and promotes its own expression. This creates a situation in which the more a cell activates the receptor, the more receptor and less ligand it will make, with the reciprocal being true for the cell that makes the ligand. This helps in understanding the stochastic selection of the anchor cell from the two equivalent cells: once a difference has arisen in the concentration of lag-2 and lin-12 between the two cells, a self-reinforcing loop will amplify this difference and lead to one cell signalling and the other receiving and interpreting the signal (Fig. 8.21).

The specification of the anchor cell in *C. elegans* represents an example of a cell fate decision between two equivalent cells and its mechanism applies to other situations like aCC and pCC in *Drosophila* and the grasshopper. There are situations, however, in which a cell is specified not from two but from a small group of several cells which share the potential to adopt this fate. Such a group of cells is called an 'equivalence group'. One example that has been particularly well studied is the segregation of precursors

in the central and peripheral nervous system of insects (see also Chapter 10 for details of this process). Each single neural precursor develops from a cluster of cells, all of which express a number of bHLH transcription factors encoded by genes of the *achaete-scute* complex (AS-C). With time, one of the cells in the cluster accumulates high levels of these bHLH proteins and becomes the precursor. Simultaneously, the other cells of the cluster switch off the expression of bHLH proteins and become ectodermal. This inhibition is dependent on Notch signalling: in the absence of Notch, all the cells of the cluster accumulate high levels of AS-C-encoded bHLH proteins and become neural precursors. Genetic analysis of the role of the different elements of the Notch signalling pathway in this process indicates that it is driven by expression of the Notch ligand Delta in the precursor cell, and by the activation of the Notch receptor in the cells that will not develop as neural (Fig. 8.22). Because Delta signals to neighbouring cells, the process is called 'lateral inhibition'. A similar situation occurs in vertebrate neurogenesis, where Delta is expressed in the emerging neurons and signals to suppress this fate in the cells of the neuroepithelium, which therefore continue to proliferate without differentiating.

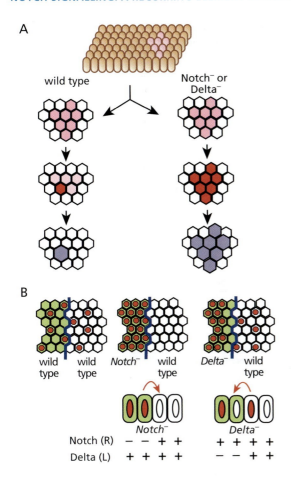

Fig. 8.22. Notch and cell fate assignment in the *Drosophila* nervous system. (A) In *Drosophila*, neurogenesis in the central and the peripheral nervous system (see Fig. 8.6, 8.18 and 8.20) is initiated by the localized expression, in defined clusters of cells, of genes encoding bHLH proteins of the achaete scute family (see Chapter 10, Fig. 10.29 for details of these processes). *Left*: In wild-type animals, the expression of these proteins (pink) is initially low and uniform in clusters of ectodermal (CNS) or epidermal (PNS) cells. As time proceeds, the levels of bHLH proteins increase in one of the cells of the cluster (red), which becomes a neural or myogenic precursor and begins to express differentiation genes (purple). At the same time, the expression of *achaete-scute* genes decays in the surrounding cells. *Right*: In mutants for components of the Notch signalling pathway (either Notch, Delta or Su(H)), the expression of *achaete-scute* genes remains uniform and all cells within the cluster develop as neural precursors. **(B)** Mosaic analysis (see Figs. 4.6, 7.14) can be used to find out where the different elements of Notch signalling are required. Most of these analyses have been performed during the development of the PNS of the adult. Cells developing as neural precursors are indicated with a red spot. In mosaics in which all the cells are wild type for Notch pathway components but some patches of cells are mutant for a cellular marker (green), neural precursors develop in a regularly spaced pattern. If the cells carrying the marker are also mutant for *Notch*, all the mutant cells develop as neural precursors, and no neural precursors develop from wild-type cells near the border with the mutant patch. That is, *Notch* mutations behave cell-autonomously. The same result is seen if the mutant cells are mutant for the nuclear effector of the pathway, Su(H), instead of for *Notch* (not shown). However, if the mutant cells are mutant for *Delta*, many of them develop as neural precursors: *Delta* behaves in a non-autonomous way. Note that in this experiment a group of mutant cells just adjacent to the boundary with the wild-type cells do not develop as neural and, in contrast to the *Notch* and *Su(H)* mosaics, some wild-type cells near the boundary do develop as neural. One explanation for these observations (illustrated diagrammatically in the bottom panel) is that there is a regulatory relationship between the receptor and the ligand. Cells that become precursors emit a strong inhibitory signal (red arrow), hence no precursors develop from among wild-type cells near *Notch* mutant patches. On the other hand, cells that lack the signal but still have the receptor can respond to the inhibitory signal from the wild-type cells—hence the row of cells that do not become precursors despite being mutant for *Delta*.

Just like the anchor cell fate in *C. elegans*, the neural fate of the cells within the equivalence group is not specified by Notch signalling, but by combinations of intrinsic and extrinsic factors that are able to instruct those cells to adopt the fate. The function of Notch activity is simply to suppress a particular fate in a subset of the cells, ensuring that only some of them develop that fate at a given time. What is the mechanism that links Notch activity to the suppression of cell fates? During neurogenesis in *Drosophila*, the activation of the CSL family protein Su(H) by the intracellular domain of Notch results in the transcription of genes encoding members of the Enhancer of split (E(spl)) family of bHLH transcription factors, which repress the transcription of the genes of the AS-C. The signalling network mediated by Delta, Notch, Su(H), and E(spl) operates in other cell fate decisions in *Drosophila* and has been shown to be conserved in vertebrates (Fig. 8.23).

Notch signalling functions in the assignment of cell fate in many different systems (Fig. 8.23). Although the components of the signalling pathway have been given different names in different organisms, the mechanism is always the same: cell fate decisions are binary and Notch signalling plays a role in ensuring that one of the two options is implemented. In the simplest situation, a cell decides between two fates, one of which (fate A, say) is ancestral (i.e. the fate of the precursor cell) while the other (fate B) is a new fate that will initiate a new

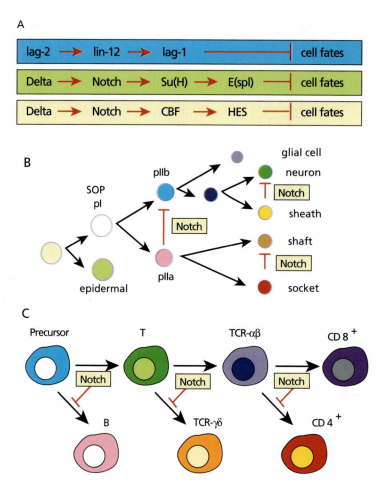

Fig. 8.23. **The Notch signalling pathway. (A)** The components of the Notch pathway in *Caenorhabditis elegans* (blue), *Drosophila* (green), and vertebrates (yellow). In many cases where cell fate assignment is associated with a specific lineage, Notch is used reiteratively to determine the outcome at each level of the lineage, as shown here for the PNS of *Drosophila* **(B)** and for the lymphocyte lineages in higher vertebrates **(C)**. The blunt-ended arrow indicates that a particular fate is achieved by suppression of the alternative one. (C adapted from Rothenberg, E. (2000) Stepwise specification of lymphocyte developmental lineages. *Curr. Opin. Genet. Dev.* **10,** 370–9.)

regulatory program of gene expression. Notch signalling acts at this point to bring about the decision, sometimes in a stochastic manner, between A and B. The cell that signals adopts the new fate and the cell that receives the signal adopts the alternative fate by repression of the new fate. In neurogenesis in *Drosophila*, fate A is epidermal and B is neural precursor. In neurogenesis in vertebrates, A is dividing neural precursor and B is non-dividing neuronal precursor. In other systems, such as the immune system in higher vertebrates, A and B are alternative differentiated fates (Fig. 8.23).

Because in most cases in which Notch signalling has been analysed in detail, Notch suppresses a particular fate, it is sometimes suggested that Notch signalling has a general function of 'preventing differentiation'. This is not strictly correct. Although the outcome of Notch signalling might indeed be an undifferentiated cell, the function of Notch is to suppress one of the alternative fates that a

cell is offered and in some cases this might not be the less differentiated state. We have already seen an example of this in the mammalian epidermis, where Notch signalling is not required to maintain the stem cell fate but favours the first differentiated state (Fig. 8.4).

A

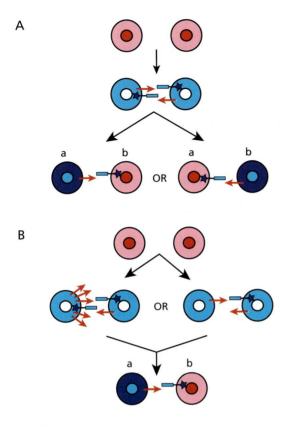

B

Fig. 8.24. Spatial regulation of Notch signalling. (A) The specification of the anchor cell in *Caenorhabditis elegans* results from stochastic regulation of the expression of *lin-12* and *lag-2* (see Fig. 8.15). Initially, the levels of expression of both genes are similar in the two precursors. A small fluctuation in the level of either lin-12 or lag-2 then results in the development of an asymmetry that is amplified, with the result that one cell has low levels of lin-12 and high levels of lag-2 while the other has high levels of lin-12 and low levels of lag-2. **(B)** In some situations, the outcome of Notch signalling is biased in space, that is, one of the equivalents always signals and the other always responds and their relative positions are fixed. This can be achieved if extrinsic or intrinsic influences define the signalling or the responding cell. The signalling cell can be defined by increasing the expression or activity of Delta (left) or by decreasing the concentration or activity of Notch (right). These asymmetries can contribute to the spatial patterning of Notch signalling.

The spatial and temporal regulation of Notch signalling

How is Notch signalling regulated in space and time so that the correct cell fate choices are specified in the right places? Sometimes, as we have seen for the specification of the anchor cell in the nematode, an initially symmetric distribution of ligand and receptor in both interacting cells activates an unstable feedback loop that amplifies small differences between them, creating a signalling cell and a receiving cell that stabilizes itself. In other situations, specific cells within the equivalence group express high levels of the ligand, either as a result of their lineage or in response to extrinsic signalling, and this creates a spatial bias in the Notch signalling event.

In either case, mosaic experiments indicate that while high Notch activity driven by its ligand suppresses a particular cell fate, low Notch activity promotes and is required for the alternative fate. In other words, the lowering of Notch activity may be as important during the assignment of cell fates as increasing it. Although, in some cases, low Notch activity is accompanied by a decrease in the concentration of the receptor at the cell surface, as in the anchor cell of the nematode, in others it is not, for example, during the segregation of many neural precursors in *Drosophila*. In these cases, the important thing may be the suppression of Notch *activity*, rather than its *expression*, and this might be mediated by specific factors (Fig. 8.24). One possibility is that there may be factors that affect the interaction between Notch and Delta. For example, the glycosyltransferase fringe acts on the extracellular domain of Notch to modulate its interactions with its ligands Delta and Serrate.

The mechanism of cell fate assignment during the generation of the PNS in *Drosophila* provides an example of how the activity of Notch can be spatially biased. As we have seen, in the cell divisions that generate the PNS, fates are always correlated with the segregation of numb, and this segregation is spatially identical in every cluster, so that fates always appear in equivalent positions. Gain- and loss-of-function mutants for *Notch* and *numb* show reciprocal phenotypes for the development of the cells of the PNS lineage (see Fig. 8.18). In the context of the requirement for Notch in these lineages (see Fig. 8.23), this can be interpreted as indicating that one activity of numb is to interfere with Notch activity: cells that have high Notch, have low numb, and cells that inherit numb show low Notch signalling. Numb inhibits Notch signalling by binding directly to the intracellular domain of Notch (Fig. 8.25). This function is conserved in vertebrates, although here it is not clear what creates the spatially asymmetrical distribution of numb. Once this distribution is established, however, numb operates by blocking Notch signalling (Fig. 8.25). So in this case, the probability of adopting a cell fate that is governed by Notch

A

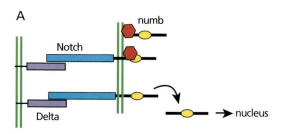

B

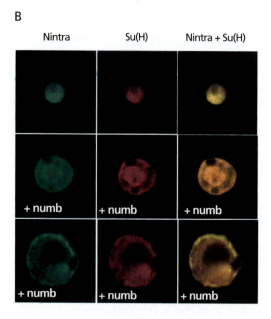

Fig. 8.25. Numb interacts with Notch and interferes with Notch signalling. (A) As a result of Delta–Notch signalling, the intracellular domain of Notch (Nintra, yellow) is released and translocated to the nucleus (see Chapter 5, Fig. 5.9). Numb (red) inhibits Notch signalling by binding to Nintra and this prevents its translocation. Numb can block the nuclear translocation of a cytoplasmic form of the intracellular domain of Notch by tethering it to the membrane. **(B)** Experiments in tissue culture provide evidence for the interactions shown in A. The left column shows the localization of Nintra (green) and the middle column the localization of Su(H) (red), while the column on the right shows the two images merged. In tissue culture cells Nintra is found in the nucleus, where it localizes the Su(H) protein (top panel). However, if numb is added to the cells expressing Nintra, Nintra is retained in the cytoplasm and sometimes in the cell cortex (middle panel) and is excluded from the nucleus (bottom panel). Su(H) also associates with Nintra in these situations. (From Frise, E. *et al.* (1996) The *Drosophila* Numb protein inhibits signalling of the Notch receptor during cell-cell interactions in sensory organ lineage. *Proc. Natl Acad. Sci.* **93**, 11925–32.)

Notch signalling is used to forfeit cell fate decisions and, in each situation, to allow a cell to remain in one of two possible states.

Analysis of mutations in two *C. elegans* genes, *pop-1* and *lit-1*, suggests that, at least in this organism, there is a second mechanism associated with many cell fate decisions. As in the case of Notch, this mechanism is not involved in the specification of individual fates but rather in ensuring that the fates, specified by other molecular mechanisms, are adopted (Fig. 8.26).

The *pop-1* gene encodes a member of the Tcf family of HMGbox-containing proteins that mediate Wnt signalling (see Chapter 5, Fig. 5.25), and *lit-1* encodes a kinase that can phosphorylate pop-1. The products of both genes are expressed throughout development but, whereas *lit-1* is expressed in all cells, the product of *pop-1* is expressed asymmetrically after each cell division along the AP axis: anterior cells have high levels of pop-1 and posterior cells low levels. In *pop-1* mutants, cells always adopt the posterior fate after division along the AP axis. The absence of lit-1 affects the same cell fate assignments but in the opposite direction: in *lit-1* mutants both cells develop the anterior fate and express high levels of pop-1. The double mutant *lit-1, pop-1* is like *pop-1* alone which suggests that a function of lit-1 is to regulate the levels and activity of pop-1 in posterior cells. One possible mechanism for this regulation might be phosphorylation of pop-1 by lit-1.

Genetic screens to identify genes that are involved in the same fate decisions and that interact with both *lit-1* and *pop-1* reveal that the activity of pop-1 is under the control

is biased by the influence of the cellular processes that govern the segregation of determinants.

Iterative molecular mechanisms for cell fate assignment

The use of DNA microarrays (see Chapter 2, Fig. 2.15) to investigate the genetic input to cell fate decisions suggests that each one might involve more than 100 genes. Thus, the information processing associated with the development of a lineage is likely to be very complex. Does this mean that each cell fate decision within the lineage will involve a specific mechanism that is unique to that particular fate, or could it be that the molecular information that defines the fate varies, but the mechanism by which the decision is made is more general? We have seen in Notch signalling an example of a mechanism that is used reiteratively in many cell fate decisions to ensure that some cells do and others do not develop along certain pathways.

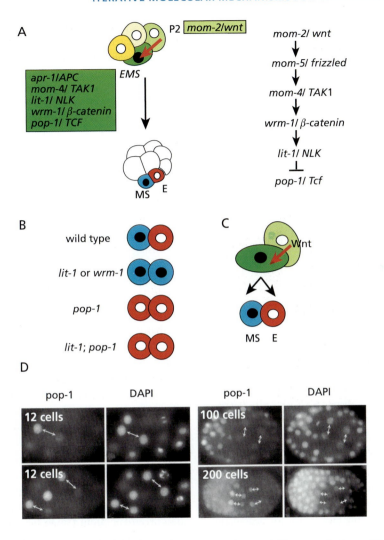

Fig. 8.26. Wnt signalling is involved in the iterative assignment of cell fates in *Caenorhabditis elegans*. (A) During the early stages of development of *C. elegans*, the P2 cell signals to EMS to establish the fates of its daughters, E and MS. Genetic analysis indicates that this signalling event is mediated by Wnt signalling from the P2 cell. The figure shows the different elements of the Wnt pathway that are involved in this event and where they are required. On the right is shown the pathway as deduced from genetic analysis. In the absence of Wnt signalling the embryo develops an excess of MS descendants (and more mesoderm), while if there is an excess of Wnt signalling MS is transformed into E, resulting in more endoderm. Antibodies against the Tcf family member pop-1 detect high levels of pop-1 (blue) in the MS cell and low levels (red) in the E cell. **(B)** Loss of function of Wnt signalling (as in *wrm-1* or *lit-1* mutants) results in the two daughters of EMS developing like MS and expressing pop-1. The reverse is observed in a *pop-1* mutant: both cells develop like E. The double mutant *lit-1*, *pop-1* is like *pop-1*. The genetic and developmental relationships suggest that the function of *lit-1* is to suppress *pop-1*. **(C)** Wnt signalling from P2 to EMS polarizes the cell in such a way that it will generate the two different daughters, E and MS, after division. **(D)** Photographs of the expression of pop-1 in *C. elegans* embryos at different developmental stages; nuclei are identified with DAPI and sister cells are indicated with arrows. The upper and lower panels at the 12-cell stage show two different planes of focus. Note that pop-1 is always more abundant in one of the sister cells and that this pattern persists until very advanced developmental stages (100–200 cells). (From Lin, R. *et al.* (1998) POP-1 and anterior-posterior fate decisions in *C. elegans* embryos. *Cell* **92**, 229–39.)

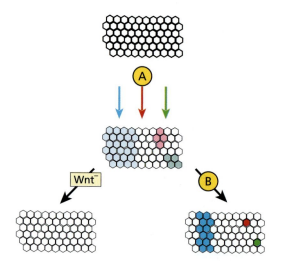

Fig. 8.27. A general role for Wnt signalling during the assignment of cell fates. Studies in *Drosophila* suggest that the assignment of cell fates relies on two different events that can be separated by genetic analysis. In a first step **(A)**, cells respond to different signals (coloured arrows) by activating the expression of particular genes in groups of cells. A second step **(B)** leads to the stable expression of these genes, usually in a subset of the cells in which the expression was initiated. In *wingless* mutants, the initial stages of induced expression are not affected but they fail to be stabilized. This suggests that Wnt is an important element in a mechanism involved in the stabilization of the fates induced by different signalling systems.

volved in a positive step that stabilizes the expression of genes associated with a particular fate. There is evidence that in *Drosophila* Wnt signalling plays a similar role and this might also be the case in vertebrates (Fig. 8.27).

Indeterminate lineages and polyclonal lineage compartments

Lineage tracing studies in the wing of *Drosophila* revealed that in some situations, the cells generated by a particular lineage may be restricted to a defined territory. In these experiments, individual cells were labelled early in development using the technique of mitotic recombination (see Chapter 7, Fig. 7.14) to assess their contribution to different structures of the adult fly. When the clones resulting from these experiments were examined in the wing it was observed that although they formed compact groups of cells that, in principle, could map to any region of the wing, the clones never crossed a line half-way through the wing (see also Chapter 7, Fig. 7.15). This line acted as a lineage restriction boundary separating two subpopulations of cells: one anterior and the other posterior. Proliferating cells always respected this boundary—even cells that had a growth advantage, such as clones of wild-type cells in a heterozygous *Minute* background (Fig. 7.15). The lineage-restricted territories were called compartments and the boundary that limited their growth the compartment boundary. These compartments are established very early in development, probably at the blastoderm stage. Each one is derived not from a single cell but rather from a small number of cells; that is, the compartments are not clonal structures but polyclonal ones (Fig. 8.28).

Compartments have been observed in other parts of the fly such as the legs, the head and the thorax, and this has led to the idea that each segment of the adult fly is made up of a collage of anterior (A) and posterior (P) lineage compartments (Fig. 8.28). In some cases, compartment boundaries were shown to be boundaries for patterns of gene expression. This led to the idea that compartments are used to restrict the developmental potential of cells by linking the commitment of groups of cells to a particular lineage with the expression of a particular gene that maintains a common fate in the cells of the lineage.

Studies on the expression and function of one gene, in particular, have fuelled the concept of the compartment as a link between lineage and gene expression: the homeobox gene *engrailed*. *Engrailed* is expressed in the cells of each posterior compartment of *Drosophila*. In the absence of

of Wnt signalling. In this case, the effect of Wnt signalling is to phosphorylate a transcription factor (pop-1), thereby inactivating it and possibly targeting it for degradation. This contrasts with the Wnt signalling pathway we have described previously (Fig. 5.25), in which signalling leads to the formation of a transcriptionally active complex between a Tcf family member and β-catenin.

At a functional level the effects of lit-1 and pop-1 on cell fates indicate that, in *C. elegans*, Wnt signalling is used as a permissive factor that regulates cell fate assignment along the AP axis (Fig. 8.26). Wnt signalling is not involved in the definition of the fates themselves but in whether or not a cell adopts the fate that is assigned to it. A plausible model is that pop-1 is involved in the repression of these fates and that Wnt signalling is permissive for the fates by blocking the activity of pop-1.

The permissive role of Wnt signalling in *C. elegans* is very similar to the role we have described for Notch signalling during the assignment of cell fates. However, while Notch signalling is involved in 'suppressing' cell fates during development, in *C. elegans* Wnt signalling is in-

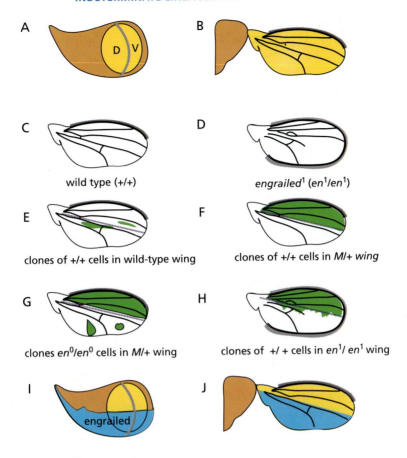

Fig. 8.28. Developmental compartments in the _Drosophila_ wing. (A) A wing disc of _Drosophila_. The territory that will give rise to the wing proper is indicated (yellow) as is the wing margin (grey arc). D and V will become the dorsal and ventral surfaces of the wing. **(B)** The disc gives rise to the adult wing, attached to the notum (brown). Only one of the surfaces is shown. **(C)** A wild-type wing has anterior and posterior regions marked by particular structures: a triple row of mechanosensory organs (indicated as a thick arc) at the anterior margin and epidermal thickenings, veins, which mark different positions along the anteroposterior axis. **(D)** In some mutants for the _engrailed_ gene (en^1), cells in the posterior region adopt characteristics of the anterior region. **(E)** Mitotic recombination clones of wild-type cells in wild-type wings distribute themselves at random, but never cross a line across the middle of the wing (dashed line). **(F)** Mitotic recombination clones of wild-type cells in a _Minute_ background out-compete surrounding cells and occupy large territories of the wing. However, they respect the same line of lineage restriction: the compartment boundary (see also Fig. 7.15). **(G)** Mitotic recombination clones of _en_ mutant cells (en^0) in a _Minute_ wing are indistinguishable from wild type in the anterior compartment. However, in the posterior compartment they tend to sort out from the surrounding cells and display growth alterations. **(H)** Clones of cells in an en^1 mutant wing behave as if there were no compartment boundary and cross from the anterior to the posterior.

(I) Wing disc showing the expression of engrailed throughout the posterior compartment. **(J)** A wild-type wing attached to the notum showing that the expression of engrailed in the adult wing corresponds with the line of lineage restriction: the AP compartment boundary.

engrailed, cells lose their 'posterior' characteristics, often develop anterior characteristics, and mix freely across the compartment boundary (Fig. 8.28). These observations suggest that engrailed endows posterior cells with a common fate which manifests itself in a number of common cellular characteristics upon which a fine-grained pattern is later laid down. The expression of several other genes, most notably the _Hox_ genes (see Chapter 2, Fig. 2.8), is also restricted by compartment boundaries at several stages of development and this has lent support to the idea that compartments are fundamental units in the developmental architecture of _Drosophila_.

These observations raised the possibility that, in the same way that there is mesoderm, endoderm or ectoderm, there are also more abstract groups of cells in the embryo related to its global architecture, and that these form the

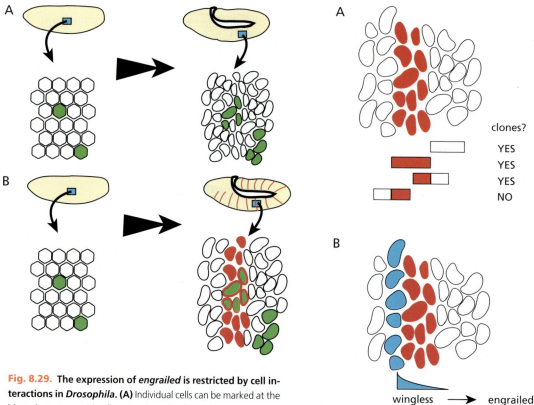

Fig. 8.29. The expression of *engrailed* is restricted by cell interactions in *Drosophila*. **(A)** Individual cells can be marked at the blastoderm stage with a fluorescent marker (green). Two hours later, after two rounds of division, their progeny can be identified as small clones of cells. **(B)** The expression of the homeodomain protein engrailed can be monitored within the clones of cells marked at blastoderm by observing the correspondence between the expression of engrailed (red) and the distribution of the fluorescent label (green). The interpretation of the observed expression patterns is discussed in Fig. 8.30. (From Vincent, J. P. and O' Farrell, P. (1992) The state of engrailed expression is not clonally transmitted during early *Drosophila* development. *Cell* **68**, 923–31.)

Fig. 8.30. Wingless signalling maintains the expression of *engrailed*. **(A)** Summary of the results from the experiment shown in Fig. 8.29. Anterior is to the left and posterior to the right. During the phase of proliferation, engrailed expression (red) is not strictly associated with a lineage. In all experiments three classes of clones were observed: clones in which none of the cells expressed engrailed, clones in which all of the cells expressed engrailed and clones in which only some of the cells expressed engrailed. However, engrailed is expressed in well-defined stripes along the AP axis of the embryo and therefore there are two possible types of clones of the last kind: those that overlap the anterior border of engrailed expression and those that overlap the posterior border. Only clones that overlapped the posterior border of engrailed expression were found. **(B)** Anterior and adjacent to the stripe of cells expressing engrailed, there is a stripe of cells expressing the Wnt family signalling molecule, wingless (blue). In the absence of wingless, engrailed expression ceases, suggesting that Wingless is responsible for maintaining the pattern of engrailed expression.

basis for the patterning of the organism. Although clonal analysis has suggested that some lineages tend to be spatially biased towards particular territories, or that transient lineage restrictions do exist in many organisms, absolute restrictions like those seen in the *Drosophila* wing have not been found elsewhere. Even in *Drosophila*, compartments have only been found in the epidermis, and the expression of *engrailed*, a gene clearly associated with the compartments in the wing, is only restricted by lineage in the adult. For example, in the embryo, where the expression of *engrailed* also demarcates differences between anterior and posterior cells within every segment (Fig. 8.29), *engrailed* expression is not associated with an ab-

solute lineage restriction. Marking single cells in the blastoderm and following their fate during the first rounds of division shows that, at this stage, *engrailed* expression is not clonal; that is, a clone of cells contains both *engrailed*-expressing and non-expressing cells (Fig. 8.29 and 8.30).

Thus, lineage restrictions of the kind revealed by the experiments in the wing of *Drosophila* appear to be a specialized developmental strategy rather than a general principle of development.

However, the experiments on *engrailed* expression in the early *Drosophila* embryo indicate that one of the boundaries of the cells expressing *engrailed*, the anterior one, acts as a transient lineage boundary (Fig. 8.30). This boundary represents an interface with cells that express the Wnt family member wingless, which is required to maintain the expression of *engrailed*. In the wing disc, and in other discs, similar boundaries can be found; the function of these boundaries appears to be to ensure, through cell interactions, the correct registration of patterns of gene expression. It appears that these boundaries, which may be transient and need not be associated with lineage restrictions, might be a more general feature of pattern formation during development (see also Chapters 11 and 12).

SUMMARY

1. Lineages describe the ancestry of cells or groups of cells and provide a basis for the coupling of cell division and its associated molecular machinery to the molecular processes that assign cell fates. They configure important routines executed by cells in developing embryos.

2. During development, in most cases, the fate of a cell within a particular lineage is conditional on cell interactions. In organisms with rigid cell lineages (e.g. many marine invertebrates, *C. elegans*), the lineages have the effect of generating reproducible interactions between individual cells.

3. There are two classes of lineages: determinate and indeterminate. In determinate lineages, every cell always generates the same daughters. This type of lineage is usually associated with mosaic organisms in which the animal cannot compensate for the loss of a particular cell. In indeterminate lineages, the fate of a cell is less dependent on that of its ancestor, and the organism can compensate for its loss. These organisms are called regulative.

4. In stem cell lineages, division of an undifferentiated precursor generates a cell that remains undifferentiated (stem cell) and another that begins to differentiate. If the stem cell can give rise to all the cells of an organism, it is said to be totipotent. If it can only give rise to a subset of the cells of an organism, for example a haematopoietic stem cell, it is said to be pluripotent.

5. The generation of cell fates in specific lineages relies on interactions between intrinsic and extrinsic factors. Intrinsic factors are usually associated with the transcriptional state of the individual cells of the lineage. Extrinsic factors are signalling molecules that mediate cell interactions.

6. Molecules that are instructive in the specification of fates in specific lineages are known as determinants.

7. The segregation of determinants to specific cells of the lineage is usually associated with regulation of the polarity of the cells and the orientation of the cell-division apparatus.

8. Notch signalling is used iteratively in binary decisions to ensure that cells adopt one fate over another.

COMPLEMENTARY READING

Artavanis-Tsakonas, S., Rand, M. D., and Lake, R. J. (1999) Notch signaling: cell fate control and signal integration in development. *Science* **284**, 770–6.

Bowerman, B. (1998) Maternal control of pattern formation in early *C. elegans* embryos. *Curr. Top. Dev. Biol.* **39**, 73–117.

Clarke, J. and Tickle, C. (1999) Fate maps old and new. *Nature Cell Biol.* **1**, E103–E109.

Jan, Y.-N. and Jan, L. (1998) Asymmetric cell division. *Nature* **392**, 775–8.

Labouesse, M. and Mango, S. (1999) Patterning the *C. elegans* embryo: moving beyond the cell lineage. *Trends Genet.* **15**, 307–13.

Nishida, H., Morokuma, J., and Nishikata, T. (1999) Maternal cytoplasmic factors for generation of unique cleavage patterns in animal embryos. *Curr. Top. Dev. Biol.* **46**, 1–37.

Schnabel, R. (1996) Pattern formation: regional specification in the early *C. elegans* embryo. *BioEssays* **18**, 591–4.

Shen, Q. *et al.* (1998) Stem cells in the embryonic cerebral cortex: their role in histogenesis and patterning. *J. Neurobiol.* **36**, 162–74.

Chapter 9

Long- and short-range influences in the generation of cell diversity

Lineages provide a 'historical' component to the assignment of cell fates; in other words, what a cell becomes depends to some extent on what its ancestors were. However, as we have seen in Chapter 8, the specific fates of cells within particular lineages also rely heavily on external influences. The initial steps in the development of the vulva of *Caenorhabditis elegans* illustrate the importance of this interplay in the assignment of cell fates (Fig. 9.1). A row of six vulval precursor cells (VPCs) are endowed by their lineage with the potential to contribute to the vulva (i.e. these cells and no others have the potential to develop as vulval cells). Which of these cells actually do follow the vulval development pathway depends on the influence of another cell, the anchor cell (AC). If the anchor cell is ablated, all the cells follow an alternative developmental program, and no vulva forms. In the presence of the AC, only three of the VPCs, those nearest to the AC, follow the vulval development pathway, giving rise to lineages that adopt primary (1°) or secondary (2°) vulval cell fates. The three cells that are furthest away give rise to lineages that adopt tertiary (3°) epidermal fates. If one or more of the cells that would normally adopt a vulval fate is eliminated, the adjacent 3° cell comes under the influence of the AC and will adopt the vulval fate. This demonstrates that all six VPCs have the potential to develop as vulval cells—a potential endowed by their lineage—but that only those in a position to receive information from the anchor cell will actually do so (Fig. 9.1).

This example shows that the position of a cell relative to other cells is an important parameter in determining the external influences it will receive; however, it also shows that distance from the instructive influence plays a role. In the example of the vulva, only the cells nearest to the anchor cells receive the signal that instructs them to adopt the vulval fate. However, as we shall see later in this chapter, there are other developmental situations in which cells appear to respond to influences whose source is much further away. This observation of instruction at a distance raises several questions: what do cells in this field actually 'see' in molecular terms, how are such influences generated, how are they transmitted across fields of cells, and how do cells interpret and respond to these influences? In this chapter, we shall try to answer these questions and in the process will explore the principles and molecular mechanisms that mediate both short- and long-range instructive interactions.

Induction and positional information

The process whereby a cell or a cell population instructs a program of gene expression or protein activity in other cells is called 'induction'. In this process, one cell or a cell population (the inducer), alters the developmental state of another (the responders). The interaction between the anchor cell and the vulval precursors in *C. elegans* is an example of induction at a single-cell level of resolution, one in which the ability to respond to the inductive signal is narrowly restricted by lineage.

The emergence of mesoderm in amphibia and other vertebrate embryos during the blastula stage—usually presented as a 'classical' example of induction—shows inductive processes in action between larger populations of cells.

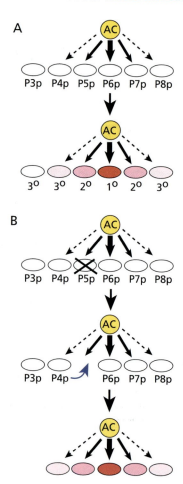

Fig. 9.1. **Induction of vulval cell fates in *Caenorhabditis elegans*.** **(A)** The vulva is generated from three vulval precursor cells (VPCs), P5p, P6p, and P7p, each of which follows a specific program of cell division and gene expression that is determined by inductive interactions from the anchor cell (AC) (for details see Chapter 11). The three cells that contribute to the vulva are selected by the adjacent anchor cell from among six vulval precursor cells (P3p–P8p) that are generated by specific lineages. The three VPCs that will adopt the vulval program are those nearest to the AC. The closest one develops a specialized developmental fate called the primary (1°) fate, and its neighbours develop a related but different fate, the secondary (2°) fate. The cells furthest from the AC do not contribute to the vulva and acquire the tertiary (3°) fate. **(B)** If either of the cells closest to the AC is deleted, one of the tertiary cells, which normally would not contribute to the vulva, will now contribute. The reason is that as a result of the space left by the ablation, one of these cells now comes under the instructive influence of the AC.

responder). If a permeable barrier is placed between the two cell populations, preventing direct cell contact but allowing molecular diffusion, mesoderm still develops from the animal cells, showing that the instruction must be contained in a diffusible substance(s) moving between the cell populations. Other experiments show that the inductive interaction is specific for a particular stage in development: vegetal cells from early embryos cannot induce mesoderm from tissue taken from embryos at later developmental stages. Although the vegetal cells are producing the inducer, the other cell population is not capable of responding.

In the two examples we have just described (mesoderm and vulva), the interacting cell populations are closely apposed to each other. However, there are other situations in which the inducing and responding cell populations are physically separated. Two examples illustrate this point. Early in the development of the vertebrate limb, a piece of posterior mesoderm within the limb bud known as the zone of polarizing activity (ZPA) has the property of instructing cells to develop particular digits in a pattern that is related to their distance from the ZPA: digit 2 develops from the most distant location whereas digit 4 develops from the cells nearest to the ZPA (Fig. 9.3; see also Chapter 12). Removal of the ZPA abolishes the patterning of the limb and if the ZPA is transplanted to an ectopic location within the limb bud, ectopic digits are induced in a pattern that is also related to the distance of the cells from the ZPA. If the ZPA is transplanted to a position in the middle of the limb bud, it affects tissue on both sides of the transplant (Fig. 9.3). This shows that the polarization of the response results from the asymmetric location of the

The early *Xenopus* embryo has two visibly different regions: the 'animal' region, from which the epidermis and the neural tissue will largely develop, and a 'vegetal' region, characterized by large yolky cells, that gives rise to most of the endoderm. During the early stages of development, cells at the interface between these two populations develop as mesoderm (Fig. 9.2) as can be seen in a variety of fate mapping experiments (see Chapter 8, Fig. 8.2). The characteristics of each set of cells can be followed with molecular markers for the different cell types. If the animal and vegetal parts of the early embryo are separated and allowed to develop on their own, neither develops mesodermal characteristics. However if, after separation, the two populations are rejoined, mesoderm develops from some of the animal cells that are close to the vegetal tissue (Fig. 9.2). A number of variations of this experiment show that the appearance of mesoderm depends on an instruction from the vegetal cells (the inducer) to the animal cells (the

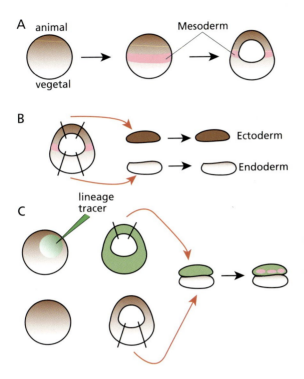

Fig. 9.2. Mesoderm induction in amphibia. (A) At fertilization, the *Xenopus* embryo has two visibly different parts: the 'animal' region and the 'vegetal' region. As the embryo cleaves, these regions give rise to animal and vegetal cells that develop as ectodermal and endodermal derivatives, respectively. Cells at the interface of these cell populations develop mesodermal characteristics (pink). **(B)** If sections from the animal and vegetal regions are cut out of the embryo and allowed to develop separately, they develop into ectoderm and endoderm, respectively. **(C)** If sections from the animal and vegetal regions are separated from each other and then rejoined, some cells in the aggregate develop as mesoderm (pink). These experiments can be done in such a way that the animal and the vegetal cells can be identified with a tracer. This allows the the source of mesodermal cells to be identified. In the experiment shown, mesoderm develops from the animal cap cells which are labelled with a green dye (see Chapter 8, Fig. 8.2).

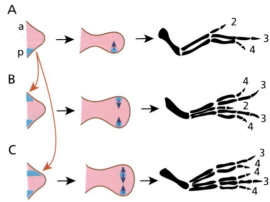

Fig. 9.3. Examples of long-range instructive processes. (A) The development of the chick limb from the limb bud. In a normal embryo, the bud grows outwards, away from the body, and the mesoderm (pink) develops the pattern of cartilage (which will later develop into bone) shown on the right. The chick limb has three digits: digit 2 develops from the more anterior region (a), and 3 and 4 follow in a posterior (p) direction. **(B)** A group of mesodermal cells at the posterior of the limb bud (shown as a blue patch) appear to play an important role in the development of the limb. If this region is transplanted to the anterior region of another limb bud, the bud develops into a limb with a mirror image pattern of digits, indicating that the transplanted region, known as the zone of polarizing activity (ZPA), can instruct the fates of cells around it. **(C)** Transplanting the ZPA to other positions within the limb generates different patterns of digits, for example transplantation to the middle of the bud leads to a limb with extra digits 4 and no digit 2. These experiments indicate that digit 4 always forms in the neighbourhood of the ZPA, and digit 2 farther away from it.

signalling source (ZPA) with respect to the responding tissue (the limb bud) rather than from some polar effect of the source of inducer. In the developing limb bud, the importance of lineage can be seen from the fact that only the cells from around the limb bud, and not from other parts of the embryo, can respond to the ZPA.

The observation that some tissues exert inductive effects at a distance and, in particular, that there often appears to be a correlation between the effects and the distance of the responding cells from the inductive source, has led to the concept of 'positional information' (Fig. 9.4). This idea creates a link between the process of cell fate assignment and pattern formation: cells develop according to where they lie in a field relative to sources of inductive molecules that have strong instructive properties, and this results in a specific pattern. A crucial feature of this idea is that the concentration of the inducer determines the cell fate and, as a corollary, that different concentrations of inducer elicit different fates. Within this framework, the patterning of the limb bud can be understood by postulating that there are diffusible substances with localized sources, which determine cell fates in a concentration-dependent manner (Fig. 9.5).

The concept of positional information depends on there being a way of creating and maintaining gradients

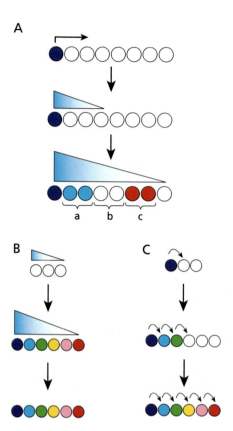

Fig. 9.4. Positional information. (A) The concept of positional information states that the developmental pathway followed by the elements of a cellular field is determined by their position relative to the source of an instructive substance (dark blue). This substance diffuses through the field and creates a concentration gradient. Cells within the field respond to the different concentrations by following different developmental pathways a, b, or c (light blue, white or red). **(B, C)** Different ways of creating cellular diversity within a field. In these examples, the increases in cell number that are normal during embryonic development are taken into account. **(B)** Gradient of positional information, as indicated in A. As the field grows, the gradient develops and this contributes to the generation of different types of cells as they become exposed to different concentrations of the inducing substance. **(C)** Short range or nearest neighbour inductive events in which cell diversity is generated, progressively, through nearest neighbour interactions.

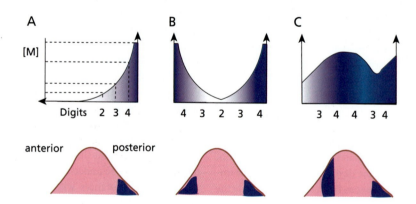

Fig. 9.5. Positional information and pattern formation. (A) In the chick limb, the ZPA is the source of a substance (M) that diffuses towards the anterior of the bud. Different concentrations of this substance promote the development of different digits, with the highest promoting digit 4, the lowest digit 2, and intermediate concentrations digit 3 (Fig. 9.3). **(B)** Transplantation of a ZPA to the an-terior region of a different limb bud would create symmetric gradients of M with a low point in the centre of the bud, accounting for the mirror image pattern of digits. **(C)** Transplantation of the ZPA to a more medial position within the limb bud results in a new land-scape of gradients which, again, can account for the observed pattern of digits on the basis of linear, dose-dependent responses.

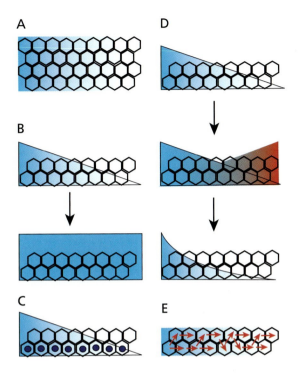

Fig. 9.6. Strategies for the generation of gradients of diffusible substances. (A) Molecules that diffuse from a source through a field of cells can form concentration gradients (blue). **(B)** On theoretical grounds, simple diffusion is not sufficient to establish a stable gradient because, over time, all cells will accumulate the same amount of the diffusing molecule. **(C)** One way to generate a gradient of a diffusible substance from a stable source is for the cells that receive the diffusible molecule to degrade it continuously so it does not accumulate. The ability to degrade the signal is indicated by a dark blue dot and if it is tailored to the rates of production and diffusion, it will generate a gradient. **(D)** An opposing gradient of a neutralizing or antagonizing molecule (red) will usually have the effect of maintaining and sharpening the original gradient. **(E)** An effective gradient may be established if the diffusible molecule does not rely on simple diffusion but on aided transport (or diffusion) across the field. This can be useful in situations in which the diffusible molecule is strongly attached to the cell surface, as in the case of hedgehog or Wnts. The cells may produce a factor that aids the extracellular movement of the molecule or, sometimes, may actively transport the diffusible substance from one cell to another.

of diffusible substances across fields of cells during development. Diffusion alone, over time, would lead to a homogeneous distribution of the substance and so there must be mechanisms for regulating diffusion in such a way that the gradient is maintained (Fig. 9.6). One such mechanism is for the substance to have a very short half life in the extracellular space, preventing it from accumulating evenly across the field. An active sink at a distance from the source, or an opposing gradient of an antagonistic molecule, can also provide the molecular conditions that enable a gradient to be maintained over time (Fig. 9.6). Alternatively, the diffusible molecule may have properties that trap it in the extracellular space, impeding its diffusion and so creating a gradient that will be slow to equalize. The signalling molecule hedgehog behaves in this way (see Chapter 5, Fig. 5.28).

Although efforts have been made to distinguish positional information from induction, it could be argued that there is little difference in principle between these concepts. Ultimately, both refer to the ability of cells to influence other cells. If there is a difference between the two terms it is an operational one: induction tends to be used for instructive cell interactions between adjacent cell populations, often in contact and usually involving a small repertoire of responses. Positional information is used mainly for effects over a long spatial range, the inducer and the responder need not be in contact, and the inducer can trigger a range of effects depending on its concentration. Cell diversity can be generated by either process or a mixture of both, depending on the circumstances, for example, Fig. 9.4 shows a relay system, in which short-range signals propagated across a field can lead to long-range effects.

Long- and short-range inductive molecules

It is generally accepted that all inductive processes are mediated by signalling molecules that surf the extracellular space in search of receptors through which to implement an instruction. In Chapter 4, we explored some of the many known families of signalling molecules and receptors, and discussed their properties; here, we want to look at them from the perspective of the patterning mechanisms in which they participate.

Signalling molecules can be classified as long- or short-range depending on the distance over which they exert their effects (see Chapter 4). This range can be determined experimentally by applying the signalling molecule to a group of cells or a tissue and measuring a response over a spatial range. The response may be, for example, the expression of a gene or a change in a cellular activity such as adhesion or movement (Fig. 9.7). The definition of 'long' range and 'short' range is obviously arbitrary, but effects at a distance of more than 5–10 cell diameters are usually considered long range. However, when a signal appears to

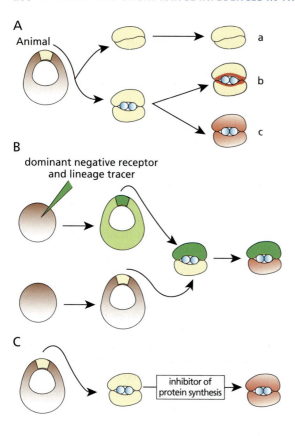

Fig. 9.7. **Determining the range of cell signalling during development. (A)** In *Xenopus*, the properties of an inductive process can be assessed in an assay in which cells that are being tested for a response are wrapped around beads coated with specific signalling molecules (blue). In the example shown, the cells are derived from the animal region and the response (red) to the signal is measured, for example by using antibodies against the products of responsive genes or by probes to measure the transcription of these genes *in situ* in the aggregates. The range of the signalling event can be measured as the distance from the beads over which the response can be observed. In this type of assay, the nature and numbers of the responding cells can be controlled. However, while the assay provides a good measurement of the response, in theory it cannot distinguish if the response is direct or dependent on relay systems (see also Fig. 9.4). a, no beads, no response. b, beads with a low concentration of a signalling molecule elicit a response only in cells neighbouring the input. c, beads loaded with a higher concentration of a signalling molecule elicit a response over a wider range. **(B, C)** Two types of experiments can be used to determine whether long-range effects are due to direct effects of a specific signalling molecule. **(B)** Some of the cells in the aggregate are injected with a dominant negative receptor for the signal. These cells can be identified in the aggregates independently because they are simultaneously injected with a lineage tracer (green). If, upon exposure to beads with the signal, only the cells that do not contain the tracer respond, this indicates that the response is dependent on the reception of the signal by all cells. **(C)** Protein synthesis is blocked in all responding cells by exposure to a protein synthesis inhibitor. If the cells respond under these conditions it means that the response does not involve the synthesis of inducible intermediaries that mediate secondary signalling events. Together, the experiments shown in B and C can provide evidence for a direct effect of a signalling molecule on the responding cell population.

be long range it is important to rule out the possibility that the effects are due to the transformation of a short-range into a long-range effect by a relay-type mechanism as shown in Fig. 9.4. This can only be done by carefully designed experiments (Fig. 9.7). The range of action of a signal can only be determined precisely in situations in which: (1) there are good markers for the response, (2) the source of the signalling molecule is localized, and (3) the field of cells upon which it acts is sufficiently large and does not change significantly in dimensions during the signalling process.

The range of a signalling molecule depends first on its intrinsic biochemical properties, and second on its interactions with other molecules that can enhance or diminish its activity. For example, hormones dissolve in the bloodstream and have systemic effects far away from their site of synthesis (Chapter 4). Retinoic acid (RA) is an example of a small molecule that can diffuse easily over long distances, either passively or associated with RA-binding proteins, and elicit an effect over a broad range of cell

diameters (Fig. 9.8). Some proteins are also capable of signalling over more than 10 cell diameters. For example, members of the TGFβ/BMP family are readily soluble in the extracellular medium; experiments in cultured tissue from *Xenopus* embryos, in which beads soaked in activin are used as a signalling source, have shown that activin will diffuse through a cell population and elicit effects several cell diameters away from its source (Fig. 9.8).

Members of the Wnt family of signalling molecules also appear to have long-range effects. However, in contrast to RA or members of the TGFβ/BMP family, Wnts interact strongly with the cell surface and other surface-associated molecules. In principle, these interactions could prevent the passive diffusion of Wnts through the extracellular space. This possibility has led to the suggestion that the

A

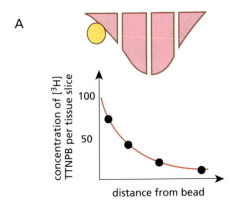

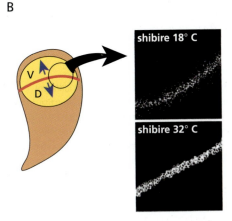

Fig. 9.8. Long-range signalling. (A) Molecules such as RA, BMPs, and activin diffuse through the extracellular space and can generate concentration gradients. This can be demonstrated by providing a source of labelled molecules and analysing their distribution at a later time. In the chick limb bud, a bead soaked in RA shows the same properties as the ZPA in the induction of new digits (see Fig. 9.3), suggesting that some of the effects of the ZPA could be mediated by RA. To test if a gradient of RA could be formed across the developing limb, beads (yellow) containing a radioactively labelled chemical analogue of RA (TTNPB), which is functional, were implanted at the site of the ZPA. After a time, the tissue was sectioned into thin strips along the anteroposterior axis, and radioactivity measured in each of the strips. The result was a gradient of radioactivity which demonstrated that RA can diffuse through the cells of the limb bud. (After Eichele, G. (1989) Retinoids and vertebrate limb pattern formation. *Trends Genet.* **5**, 246–51.) **(B)** Members of the Wnt family associate tightly with the cell surface but nevertheless can diffuse several cell diameters away from their source. In the example shown, during the development of the *Drosophila* wing, the Wnt family member wingless is expressed in a narrow stripe of cells (red) between the dorsal (D) and ventral (V) regions of the future wing (yellow). An antibody against the wingless protein reveals that in addition to a stripe of wingless protein coinciding with the region where the gene is expressed (the protein is in the extracellular space), there is a pattern of stained dots distributed on either side of the stripe. In mutants for the gene *shibire* (which encodes a dynein) the secretion and uptake of wingless protein are suppressed, and there is a very tight pattern of protein expression in the cells that make the protein. This indicates that, in the wild-type animal, the pattern of dots represents protein that is diffusing through the extracellular space and being taken up by other cells. (After Strigini, M. and Cohen, S. (2000) Wingless gradient formation in the *Drosophila* wing. *Curr. Biol.* **10**, 293–300.)

long-range effect of Wnts could be mediated by secondary mechanisms that may involve either the modification of Wnt molecules or their cell-to-cell passage by a system of vesicular transport (Fig. 9.8).

In contrast to TGFβ/BMP and Wnts, members of the hedgehog family of signalling molecules are usually considered to be short-range signalling molecules, often working only on the nearest neighbouring cells (Chapter 4 and Fig. 5.28). However, there are some situations in which hedgehogs act as long-range signals, for example, in the patterning of the vertebrate neural tube, where the the vertebrate homologue of hedgehog, known as Sonic hedgehog (Shh), can be observed to have direct effects more than ten cell diameters away from its source (Fig. 9.9). The variation in the range of hedgehog signalling in different systems seems to depend on the balance between several factors: modifications of hedgehog itself, the activity of molecules that modulate its interaction with patched, and the participation of factors in the extracellular matrix (Fig. 9.10). For

example, hedgehog is modified by the addition of cholesterol (Chapter 4, Fig. 4.14). This modification may be situation-specific, with the result that in different places the protein has different degrees of association with the cell surface, affecting its mobility through the extracellular space. Mutations in the *tout velu* gene, which is involved in the manufacture of proteoglycans, affect the range of hedgehog diffusion (Fig. 9.10), illustrating the influence of extracellular molecules on the range of signalling (see also Chapter 4, Fig. 4. 31).

With such strong constraints on hedgehog activity, the time over which a tissue can respond to hedgehog becomes an important variable. The apparent range of signalling will increase as this time increases. If the cells can respond only for a short time, signalling will appear to be

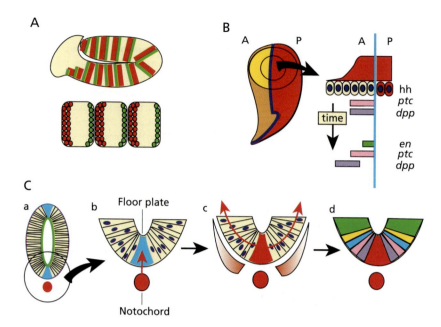

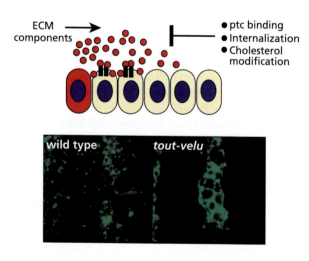

Fig. 9.9. Gradients of hedgehog. (A) In the *Drosophila* embryo, hedgehog (red) is expressed in a reiterated pattern of stripes, two to three cells wide, which define the anterior region of metameric units called parasegments (see Chapter 11, Fig. 11.16 for further explanation). Anterior to these cells and under the influence of hedgehog, a stripe of single cells (green) express the Wnt family member wingless. In the absence of hedgehog, these cells do not express wingless. **(B)** In the wing imaginal disc, hedgehog (red) is also expressed in the posterior cells (P) and diffuses towards anterior ones (A) where it induces the expression of the genes *patched* (*ptc*) (which encodes the receptor for hedgehog (see Fig. 5.23)) and *dpp* (which encodes a BMP family member), in a band that is initially three to four cells wide. With time, the response is modified; new targets appear (e.g. *engrailed* (*en*)) and different thresholds of response emerge for *dpp* and *ptc*. **(C)** (a) During the development of the vertebrate nervous system (see also Chapter 8, Fig. 8.7) two specialized cell populations act as sources of Sonic hedgehog (Shh): the notochord (a population of mesodermal cells located on the ventral side of the neural tube) and the floor plate (a population of nonneural cells in the ventral part of the neural tube). (b) Shh (red) is initially secreted from the notochord, and induces its own expression in the floor plate. (c) Shh diffuses from these cells through the walls of the neural tube, generating a gradient. (d) The gradient induces the expression of up to five different cell fates (colours), each associated with different genes, in a concentration-dependent manner (for further details see Chapter 10, Fig. 10.28).

Fig. 9.10. Modulation of the diffusion of hedgehog. Hh/Shh is indicated in red and different parameters that modulate its range of diffusion are indicated. Movement of hedgehog (detected by an antibody against the protein) in the ventral ectoderm of *Drosophila* embryos. In the wild type, hedgehog diffuses away from well-defined stripes of expression and can be detected as dots of antibody staining. In mutants for *tout velu*, which encodes an enzyme that modifies the ECM, hedgehog does not diffuse and is restricted to the cells that synthesize it. Note that, although in wild-type embryos hedgehog clearly diffuses further than one cell diameter from its source, a response to hedgehog signalling can only be detected in cells adjacent to those expressing hedgehog; the concentration of hedgehog at greater distances may be below the threshold for a response. Other evidence suggests that, as well as aiding the movement of hh from cell to cell, *tout velu* is also required in responding cells to enable them to receive the hh signal. For this reason, *tout velu* mutant embryos show a *hedgehog* mutant phenotype. (Image from The, I., Bellaiche, Y., and Perrimon, N. (1999) Hedgehog movement is regulated through tout velu dependent synthesis of a heparan sulfate proteoglycan. *Mol. Cell* **4**, 633–9.)

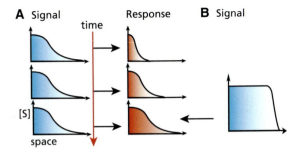

Fig. 9.11. Parameters affecting cellular responses to diffusible signalling molecule. The response of a cell to a diffusible signalling molecule can depend not only on the concentration but also in some cases on how long the responding cells are exposed to the signal. **(A)** In this example, the same concentration gradient is maintained over a period of time; this can enable cells in positions that expose them to low concentrations to increase their response over time. This type of response might underlie the changes observed in the response to hh in the *Drosophila* wing (see Fig. 9.9). **(B)** A similar response to that achieved by low concentrations over time, can be achieved by high concentrations over a shorter periods of time.

short range because it will be difficult for hedgehog to reach a sufficient concentration far away from the source. However, if the cells can respond for longer, the signalling might appear to have a longer range. Hedgehog signalling in *Drosophila* provides a good example of changes in the absolute range of signalling that may be associated with a difference in the period over which the cells can respond. In the embryo, where the time course of events is relatively short (between minutes and one or two hours at the most) the response to hedgehog is seen only in the nearest neighbours of hedgehog-expressing cells. Later on in development, when the signalling events spread over more than five hours, the range of response to hedgehog signalling increases to over five cell diameters (Fig. 9.9).

The hedgehog example highlights a principle that is likely to hold for all diffusible signalling molecules: the response is a complex function of the strength of the source, the diffusion parameters of the signalling molecule and the length of the period over which cells can respond. A low concentration of inducer acting over a long period of time—thus reaching cells far away from the source—can bring about the same response in these cells as is observed in cells closer to the source when a high concentration of signal acts over a short period (Fig. 9.11).

Other 'short-range' signalling molecules that are affected by the same parameters that affect hedgehog action include members of the EGF and FGF families of growth factors. These molecules are diffusible, and the short range over which responses are observed is likely to result from their binding to factors of the extracellular matrix

that restrict both their active concentration and their effectiveness.

Very short-range signalling molecules signal only to neighbouring cells and are usually membrane-bound proteins. Delta, Serrate, bride of sevenless, and ephrins are examples of such molecules (see Chapter 4, Fig. 4.14).

Cellular responses to inductive signals: Competence

The outcome of an inductive interaction depends not only on the characteristics of the inductive signal but, most importantly, on the ability of the responding cells to receive and interpret the signal. For example, the responding tissues can, over time, become inert or produce different responses to the same stimulus (Fig. 9.12 and 9.13). The state of responsiveness is often referred to as the 'competence' of the cells or tissue.

The competence of a cell exposed to an inductive signal depends on its molecular constitution (Fig. 9.14), which is a consequence of its developmental history. In molecular terms this means, first, whether a cell has expressed a receptor(s) that can bind the signalling molecule. If it has, the ligand–receptor interaction may be a stimulatory or inhibitory one, thereby activating or inhibiting an associated signal transduction pathway (see the examples of hh and Wnt signalling in Chapter 5). The ligand–receptor interaction may also be affected by what other information the cell has received or is receiving. For example, other extracellular molecules may affect the affinity of the ligand–

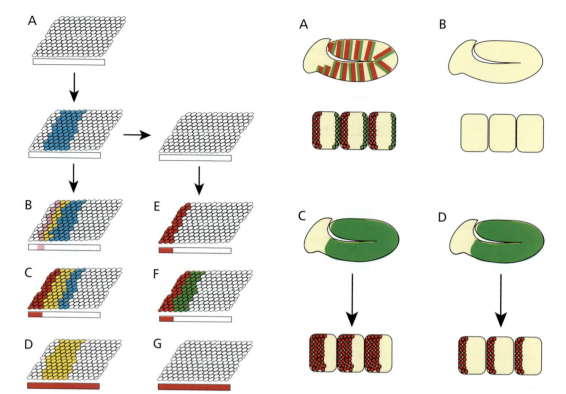

Fig. 9.12. Competence. The competence of a cell to respond to a signal can be restricted either spatially, temporally or both. **(A)** In the course of time, some of the cells within a group acquire the ability to respond to a signal in a particular way. The competence to respond is represented by the blue shading. Cells acquire this state through their lineage or through influences from their environment. The competence is lost or changes over time. **(B)** As a consequence of this competence, cells can respond to a specific signal (pink; the location of the signal is also indicated in a bar that defines the extent of the field). The response to the signal is indicated by a change to yellow within the blue area. In the example shown it is presumed that the signal can diffuse into the field from the cells that synthesize it and influence their fate. **(C)** A more intense signal (red) can result in an extension of the response to other cells within the competence domain. **(D)** However, the domain of competence is fixed and spatial extension of the signalling domain, even to the whole field, does not alter the domain of the response **(E)** After a certain amount of time, the cells lose their competence and then will not respond to the signal. **(F)** Sometimes, the loss of competence can result in a change in the response of the cells (green). **(G)** The loss of competence is absolute and means that the cells do not respond at all to high levels of the signal across the whole field.

Fig. 9.13. Spatial and temporal parameters in the response to wingless signalling in the *Drosophila* embryo. The response of the ectoderm of the *Drosophila* embryo to the Wnt family member, wingless, illustrates the principles of spatial and temporal restrictions of competence. **(A)** In the wild-type embryo, wingless is expressed in each parasegment in a single one-cell-wide stripe of cells (green). Adjacent and posterior to each stripe of wingless expression there are two to three rows of cells (red) that express two genes: one encoding the homeodomain protein engrailed and the other the signalling molecule hedgehog. **(B)** The patterns of expression of *engrailed* and *hedgehog* depend on wingless, since in a *wingless* mutant these cells cease to express *en* and *hh*. **(C)** If *wingless* expression is engineered to be ubiquitous throughout the embryo, more cells begin to express *engrailed* and *hedgehog* than in the wild type, but expression is only seen in about half of the cells in each segment. This shows that only a limited number of cells in the segment are competent to respond to wingless. **(D)** This competence is not only spatially but also temporally restricted because if wingless is expressed ubiquitously at a later stage of development (e.g. 2 hours after the time shown in C), it will not elicit any changes in the patterns of *engrailed* and *hedgehog* expression.

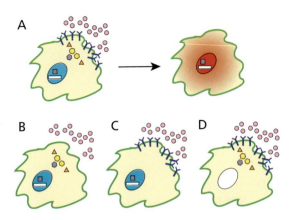

Fig. 9.14. Molecular parameters of competence. (A) A cell responds to a signal by effecting a change in state, indicated by darker shading. This response depends on the presence of a number of molecules that allow a cell to detect, receive, and interpret the signal. If any element involved in this molecular information-processing system is missing, for example **(B)** the receptor for the signal, **(C)** some of the molecules that transduce the signal or **(D)** some of the nuclear factors required for interpreting the signal, the cell may lose competence to respond.

receptor interaction or of the interaction between the receptor and some of the transducers (see, for example, JAK/STAT or BMP signalling in Chapter 5), and thereby raise or lower the threshold for a response.

As discussed before (Chapters 5 and 6) modulation of the activity of a signalling pathway can result in a variety of responses including transcriptional regulation and changes in cell shape, adhesiveness or movement. The competence of a cell also refers to its ability to effect a change in one of these activities as a result of the inductive input (Fig. 9.14).

Signalling-induced modifications of the inductive event

Once a cell has received a signal and produced a response, it may then itself become an active player in determining the outcome of the inductive process. For example, activation of a signalling pathway by the inductive signal may lead to the transcriptional activation (or repression) of genes encoding molecules that can modulate the initial signalling event. Three examples will highlight some of the possibilities.

The activity of the EGF receptor over the ventral surface of the *Drosophila* embryo provides an example of how a feedback mechanism dependent on the initial signalling event turns a short-range and coarse-grained signal into a fine-grained patterning device (Fig. 9.15). The gene *spitz* encodes a ligand for the *Drosophila* EGF receptor. Spitz is initially membrane-bound and expressed in a single row of cells in the ventral midline of the *Drosophila* embryo. Processing of spitz generates a short-range gradient of diffusible spitz. The signalling ability of this short-range gradient is rapidly modulated by the expression within the responding cells of the *argos* gene, which is a target of the EGF signalling pathway and encodes an extracellular competitive inhibitor of the spitz–EGFR interaction. In this way, the concentration gradient of spitz expression is steepened by argos, enabling sharp response thresholds to be generated (Fig. 9.15). This strategy not only contributes to the patterning of the epidermis, where argos and spitz are produced, but also to the patterning of the underlying nervous system. It seems likely that a similar process may occur during the dorsoventral patterning of the neural tube in vertebrates.

In the hedgehog signalling pathway (see Chapter 5), one of the main targets of signalling is patched, which is a component of the hh receptor complex and can act as a sink for hedgehog (see Fig. 5.28). Hedgehog signalling depends on the stoichiometry between patched and the smoothened receptor: at a constant concentration of smoothened, excess patched can act as a hedgehog-binding protein which can titrate hedgehog. The higher the levels of signalling the higher the levels of patched and the higher the concentration of hedgehog required to activate the signalling pathway—in other words, the higher the threshold for the response. This increased threshold is reflected as a reduction in the range of hedgehog signalling since less active hedgehog will reach distant cells (Fig. 5.28).

A third way in which the response of the cells can modify their competence was first demonstrated during the process of mesoderm induction in *Xenopus*. In this case, cells responding to an inductive signal themselves become the source of a signal that can affect surrounding cells and may modulate their response to the initial signal. During mesoderm induction, this process may create a global sensitivity in the responding tissue and ensure that the response of the tissue is homogeneous. This has been termed a 'community effect' (Fig. 9.16).

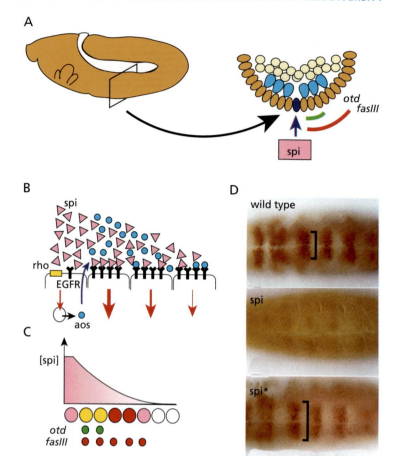

Fig. 9.15. Response modifications induced by the signalling event. (A) *Drosophila* embryo half way though embryogenesis and a transverse section through it. The EGF receptor ligand spitz is produced by a single row of cells (dark blue) along the midline of the ventral ectoderm of the *Drosophila* embryo. Spitz is made as a membrane associated form. When cleaved by the seven transmembrane domain protein rhomboid, it diffuses away from its source and activates the EGF receptor. The *orthodenticle* (*otd*) gene (green) responds to high levels of EGFR activity and is expressed in one or two cells around the source of spitz; in contrast the *fasciclin III* (*fasIII*) gene (red) responds to lower levels of spitz and is expressed over a domain of four to five cells. **(B)** Diagram of the formation of a gradient of spitz (pink triangles) over the ectodermal cells. Spitz is processed by rhomboid (rho). One of the targets of EGFR activity is the expression of *argos* (*aos*) (light blue dots) which encodes an extracellular protein that competes with spitz for binding to the EGFR.

This competition raises the threshold of the response to spitz and together with sequestration of spitz by the receptors, creates a steep gradient of response, as indicated by the thickness of the red arrows. **(C)** Diagram of the gradient of spitz and the response by the epidermal cells, which express both otd and fasIII, or fasIII alone. **(D)** Ventral region of *Drosophila* embryos at the stage shown in A, displaying the expression of the *fasIII* gene. In the wild type, *fasIII* is expressed in a wedge across the midline (square bracket). In *spi* mutant embryos, there is no *fasIII* expression. Engineered expression of a hyperactive form of spitz (spi*) from the single cell in the middle, increases the range of the signalling event as shown by the wider bracket. (Golembo, M., Raz, E., and Shilo, B. (1996) The *Drosophila* embryonic midline is the site of spitz processing and induces activation of the EGF receptor in the ventral ectoderm. *Development* **122**, 3363–70.)

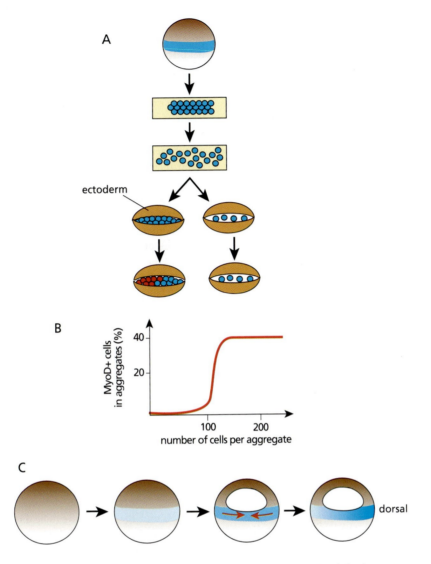

Fig. 9.16. The community effect. (A) To test their ability to develop as mesoderm, cells from the dorsolateral region (blue) of an early *Xenopus* embryo are dissected and dissociated. They are sandwiched between two layers of ectoderm, either as isolated cells, or as reaggregates. After a certain period they are assayed for expression of the muscle-specific gene *MyoD*. Cells in aggregates express MyoD (red), but isolated cells never do. **(B)** Plotting the percentage of MyoD-positive cells in the aggregates against the number of cells per aggregate shows that the aggregates need to be a certain minimum size before some of their cells begin to express MyoD. This size is about 100 cells and suggests the need for a 'community effect' in the expression of MyoD. **(C)** These experiments suggest that during normal development, a gene might be expressed initially at low levels or in small numbers of cells, and then cell interactions reinforce expression. During *Xenopus* development, it may be that, following an inductive event, MyoD is activated at low levels (light blue) and then a 'community effect' (represented by the red arrows) leads to an increase in its expression during embryogenesis. Factors regulating this effect might also regulate the spatial activity of MyoD, biasing it towards the dorsal side of the embryo. Such a mechanism could play an important role in the spatial patterning of the mesoderm. (After Gurdon, J. *et al.* (1993) A community effect in muscle development. *Curr. Biol.* **3**, 1–11.)

The ability of cells to modulate their response to a signal in a manner that is dependent on the signal is closely associated with the concept of competence and becomes particularly important when considering the effects of long-range inductive signals. Here, the changing competence and responses of cells that are first exposed to the signal may play a part in modulating the responses of more distant cells, that is, the response of a cell can lead to the production of a protein that will alter the response of the neighbouring cells to the same signal.

The morphogen concept: Graded responses to diffusible signals

As we have mentioned earlier in this chapter, the concept of positional information proposes that diffusion of an inducer across a field of cells from a localized source will generate a concentration gradient across the field, and that the position of a cell relative to the source of the information will determine its fate because different concentrations of the inducer will activate different developmental pathways (Fig. 9.4 and 9.17). The idea that a diffusible inducer can instruct developmental fates in a concentration-dependent manner has been extremely influential in developmental biology. Because the ultimate outcome of this process is a particular morphological state or form of the responding cells, substances with this property have been called 'morphogens'. This term was first used by Alan Turing in a seminal paper in which he investigated the physicochemical properties of gradients of chemical substances and their potential impact on the generation of form by biological processes. The spatial distribution of the substance is commonly referred to as a 'morphogen gradient'.

A morphogen is commonly defined as a molecule that can diffuse over a distance to form a concentration gradient and can elicit more than two different cellular responses in a concentration-dependent manner (Fig. 9.17). To qualify as a morphogen a substance must exert its effects directly on target cells. There should also be evidence that a putative morphogen functions in the required way *in vivo*.

Morphogen gradients are often envisaged as having three variables, each associated with a biological property (Fig. 9.17). First, the gradient defines a series of discrete concentrations, each of which corresponds to a threshold of response to the morphogen. Second, the changes in con- centration define a direction for the responses that can impose a local polarity on the cells, orienting them within the field. Third, the gradient has a slope that reflects the size of the field the gradient acts upon. Some experiments suggest that these variables can be used by cells as a reference for size. For example, if a fixed source acts on a growing population of cells, the extent of the growth might be determined by the slope of the gradient. Thus, theoretically, a morphogen gradient could provide cells with information about their individual identities and polarity, as well as about the size of the group they are part of.

Associated with both the morphogen concept and the closely allied concept of positional information is the idea that the same instructive molecule can elicit a different repertoire of responses in different tissues or organisms: that is, the instructive nature of the concentration-dependent molecular events is universal, while its interpretation is species- or tissue-specific. FGF or Shh, for example, may induce the development of different organs in different regions of the embryo (see Chapter 1, Fig. 1.5) and perhaps also different responses in different organisms.

Candidate morphogens

For a molecule to act as a morphogen it has to fulfil a number of experimental criteria. First, it has to be able to direct more than two different 'states' or responses. This is a fairly easy requirement to satisfy and many molecules have been designated 'morphogens' on the basis of this criterion alone. Second, the effects have to be direct and not mediated by secondary events. Third, it has to be shown to function in the required way *in vivo* and finally, the cells upon which it acts have to be initially equivalent so that the differential responses can be related solely to the differences in concentration of the putative morphogen.

Many molecules have been tested as candidate morphogens, and many have been found, under controlled experimental conditions, to form concentration gradients and to elicit different responses at different concentrations. For example, beads soaked in different concentrations of retinoic acid (RA) and implanted at strategic positions within developing vertebrate limbs mimic the activity of the ZPA and induce novel patterns of digits that could be related to the concentration of RA; as a result of these experiments RA was hailed as the first identified morphogen (Fig. 9.8). Since then, several other candidates have

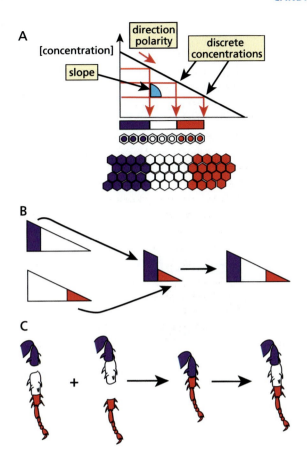

Fig. 9.17. Morphogen gradients. The 'French flag' analogy has provided a useful model to illustrate the concept of the morphogen gradient **(A)**. It is suggested that the French flag represents a pattern generated by a field of cells. To achieve this, cells have to acquire fates that allow them to become red, white, or blue. The instructions for this pattern are laid down by a diffusible signal from one side of the field. Cells respond to different concentrations of this signalling molecule and become blue (at high concentration), white (medium concentration), and red (low concentration). A cell is instructed to become one of these colours solely on the basis of its position. **(B)** An important corollary of the morphogen concept is that if the original pattern is interrupted, the morphogen will re-establish a gradient and restore the pattern. Using the French flag analogy once again, if the white cells are excised, and the blue and red cells apposed, the original French flag pattern will be regenerated. This is explained by proposing that the morphogen diffuses from the high-concentration (blue) region to the low concentration (red) region and that the cells measure the slope of the resulting gradient, re-establishing white fates at intermediate concentrations. **(C)** Experiments on the regeneration of cockroach legs illustrate the ideas shown diagramatically in B. If cockroach legs are cut at two different levels and pieces from different levels are rejoined, growth ensues which tries to repair the section that has been lost.

emerged, in particular the transcription factors bicoid and dorsal and the signalling molecules dpp and hedgehog in *Drosophila*, and members of the TGFβ/BMP and hedgehog families of signalling molecules in vertebrates.

In each case, it can be shown that different concentrations of these molecules can elicit qualitatively different responses and that these responses have important developmental consequences. However, it has been more difficult to demonstrate that some of these candidate morphogens fulfil the other criteria: that different concentrations directly elicit different responses from equivalent target cells, and that they function in this way *in vivo*. In the following sections we shall discuss what is known about the roles and mechanisms of action of several candidate morphogens. We shall then examine how closely these candidates fulfil the required criteria.

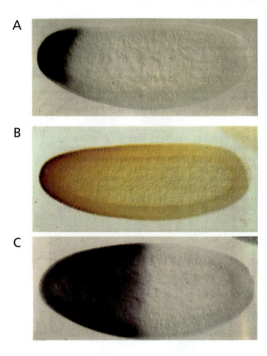

Fig. 9.18. Bicoid activates *hunchback* expression in the *Drosophila* blastoderm. (A) Distribution of the mRNA for bicoid at the anterior end of a Drosophila blastoderm. **(B)** Distribution of the gradient of bicoid protein over the anterior half of the *Drosophila* blastoderm, revealed with an antibody against bicoid. **(C)** Expression of the *hunchback* gene in response to the bicoid protein. Note the correlation of this expression profile with that of bicoid. (From St. Johnston, D. *et al.* (1989) Multiple steps in the localization of *bicoid* RNA to the anterior pole of the *Drosophila* oocyte. *Development* Suppl. 13–19.)

Candidate morphogens in *Drosophila*

Bicoid

Bicoid is a homeodomain-containing transcription factor expressed during early embryogenesis in *Drosophila* (see Chapter 3, Fig. 3.24). At this time the embryo is a syncytium and the *bicoid* mRNA is tightly bound to the anterior pole, where it is translated. The protein diffuses through the syncytial embryo towards the posterior pole and forms a steep anterior-to-posterior concentration gradient spanning two or three orders of magnitude (Fig. 9.18). Specific molecular and morphological landmarks in the embryo are associated with particular concentration thresholds of bicoid and the positions of these markers can be shifted by changes in local concentration. For example, the domain of expression of *hunchback*, a target of bicoid expression, can be shifted posteriorly or anteriorly according to the profile of the bicoid gradient (Fig. 9.18, 9.19). This effect has been shown to be mediated by direct binding of bicoid to the promoter of *hunchback*, and to rely on

an interplay between high- and low-affinity binding sites for bicoid that are present in the *hunchback* regulatory region. Experiments in which engineered *hunchback* promoters with different complements of bicoid binding sites were linked to a reporter gene showed that binding to the high-affinity sites leads to a broader pattern of *hunchback* expression than binding to the low-affinity sites, which restricts *hunchback* expression to the anterior end of the embryo (Fig. 9.19).

Although the differential affinity of bicoid for its binding sites can explain its role in establishing the pattern of *hunchback* expression, this is not enough on its own to explain how bicoid regulates the setting up of fine-grained patterns of morphology and gene expression. Bicoid has some effects that are clearly not dependent on its concentration: for example, at the posterior end of the embryo, very low concentrations of bicoid are required to activate some genes (such as *even skipped*, Fig. 9.20 and see also Fig. 3.16 and 3.17) that are also activated in more anterior regions where the concentration of bicoid is high. In other words, for these genes it is the simple presence of bicoid

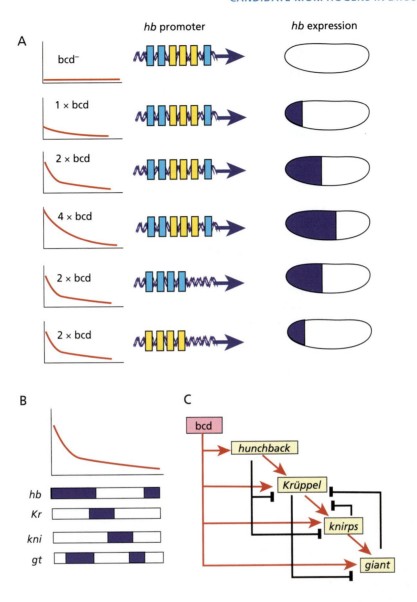

Fig. 9.19. Responses to bicoid are determined by affinity for its binding sites and by combinatorial interactions with other transcription factors. (A) Changes in the dose of bicoid (bcd) and/or of the complement of its high- and low-affinity binding sites on the *hunchback* promoter change the position of the gradient of *hb* expression. The column on the left shows the profiles of the bicoid gradient generated by different doses of bicoid. Increasing the dose of bicoid in the context of a wild-type *hb* promoter broadens the zone of *hb* expression. The response to bcd results from an integration of its binding and transcriptional initiation activity at high- and low-affinity sites on the *hb* promoter. The activity of individual sites can be demonstrated in experiments in which they are placed in front of a basal promoter and a reporter gene and then assayed in *Drosophila* embryos. Low-affinity binding sites (yellow) drive expression of the reporter in regions where the concentration of bicoid is high. High-affinity binding sites (blue) drive expression over a broader region which includes domains of low bicoid concentration. **(B)** Distribution of the expression of bicoid-responsive genes in response to the gradient of bicoid (top) in the *Drosophila* blastoderm. **(C)** Whereas bicoid is the sole input for *hb* expression, it is only one element of the combinatorial regulatory system for the spatial distribution of *Krüppel* (*Kr*), *knirps* (*kni*), and *giant* (*gt*) expression. Both positive and negative interactions are involved in establishing the expression patterns. (After St. Johnston, D. and Nüsslein Volhard, C. (1992) The origin of pattern and polarity in the *Drosophila* embryo. *Cell* **68**, 201–19.)

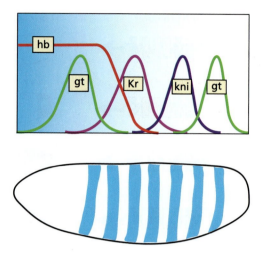

that is required, not a specific concentration. At the promoters of these targets, bicoid acts in combination with other transcription factors to activate transcription. The arrangement of the bicoid binding sites in the promoters of genes such as *eve* is also important because it can generate cooperative interactions between bicoid molecules that will affect the rate of transcription. Observations such as these show that, in addition to its concentration-dependent effects on the *hunchback* promoter, two other mechanisms contribute to bicoid's role in the establishment of fine-grained pattern in the embryo: combinatorial interactions with other transcription factors, and cooperativity.

Dorsal

The three mechanisms that mediate bicoid's action—differential affinities, combinatorial interactions with other factors, and cooperativity—are also seen in the action of dorsal, a transcription factor of the NFκB/Rel family (see Chapter 5, Fig. 5.11) that can trigger a range of responses along the dorsoventral axis of the *Drosophila* embryo in a concentration-dependent manner (Fig. 9.21). In contrast to bicoid, however, different concentrations of dorsal have been shown to activate different genes directly on the basis of differential affinity of dorsal-binding sites in their promoters; in this respect, dorsal comes closer than bicoid to fulfilling all the requirements of a morphogen.

In the early *Drosophila* embryo, graded activation of the Toll receptor by its ligand spätzle (see Chapter 4, Fig. 4.26 and Fig. 9.21) leads to a gradient of dorsal protein in the nuclei, with low levels dorsally and high levels ventrally. This gradient results in several responses: high concentrations of dorsal activate the genes *twist* and *snail* in the ven-

Fig. 9.20. Distribution of bicoid-responsive gene expression in the *Drosophila* blastoderm and the resulting pattern of *even skipped* expression. *Top:* As a result of interactions outlined in Fig. 9.18 each target gene is expressed as a gradient over a discrete region of the blastoderm. The anteroposterior gradient of bicoid is indicated in blue. The pattern of *even skipped* expression in seven stripes results from stripe-specific activities of bicoid and the products of its target genes (see Fig. 3.18 for details).

tral region from low-affinity binding sites, and lower concentrations in the ventrolateral region activate *rhomboid* from high-affinity binding sites. In addition, dorsal represses the expression of *dpp* and *zerknüllt* (*zen*) throughout the ventral and lateral regions (Fig. 9.21). Several other genes are expressed in domains that have the same ventral boundary as *rhomboid* but extend to different levels dorsally. These patterns are reminiscent of the response of engineered *hunchback* promoters to bicoid (Fig. 9.19) and it is likely that they reflect different arrangements and complements of high-affinity dorsal-binding sites in the promoters of the genes. The initial responses to dorsal are sharpened by interactions between some of its downstream targets and, in some cases, by interactions of these targets with dorsal itself (Fig. 9.21).

Both bicoid and, in particular, dorsal, clearly have some of the key properties expected of morphogens. In one important respect, however, they are not good examples to support the classical concept of the morphogen: both act intracellularly on nuclei within a syncytial cytoplasm. In its classical formulation, the morphogen concept refers to patterning within a field of cells, requiring that the morphogen be transported (or diffuse) within the field and elicit its effects by acting directly on target cells.

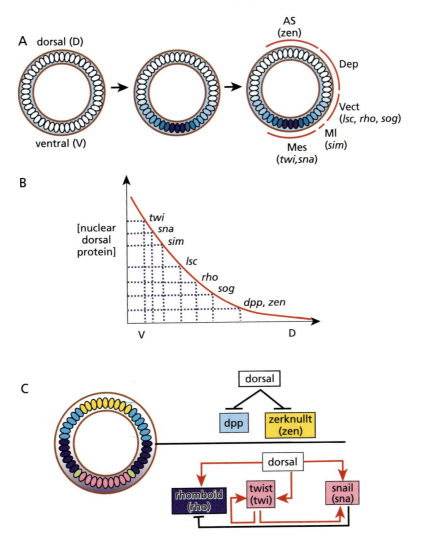

Fig. 9.21. Patterning of the dorsoventral axis of *Drosophila* by a gradient of dorsal. (A) The figure represents cross-sections of blastoderm stage *Drosophila* embryos, with dorsal at the top and ventral at the bottom. The dorsal transcription factor (blue) remains uniformly distributed throughout the syncytial cytoplasm. During the early stages of development, differential activation of the Toll receptor along the dorsoventral axis (see Fig. 4.26, 5.11) results in differential translocation of dorsal into the nucleus: high Toll activity promotes nuclear translocation and therefore creates a gradient of nuclear dorsal with a high point ventrally. As a result of this gradient of nuclear dorsal, a number of genes are activated along the DV axis in a concentration-dependent manner. These genes exert an instructive influence on the development of different cells: *twist (twi)* and *snail (sna)* promoting mesoderm (Mes) development, *single minded (sim)* a specialized group of cells referred to as mesectoderm or midline (Ml), *lethal of scute (lsc)*, *rhomboid (rho)*, and *short gastrulation (sog)* ventral ectoderm (Vect) and nervous system, *decapentaplegic (dpp)*, dorsal ectoderm (Dep) and *zerknüllt (zen)* extraembryonic tissue, amnioserosa (AS). **(B)** The expression of each of these genes is activated at a precise threshold of dorsal. **(C)** The pattern of response shown in B results from both repressive and activating effects of dorsal, acting on dorsal binding sites of various affinities in different targets. For example, dorsal activates with different affinities the transcription of *twist* and *snail* (pink), *single minded* (green) and *rhomboid* (dark blue), and suppresses the expression of *zen* (yellow) and *dpp* (light blue). This results in a pattern that is strictly dependent on the nuclear concentration of dorsal. Removal of dorsal reveals this interaction by eliminating the expression of the activated genes and extending the expression of those that dorsal represses. Regulatory interactions between some of the genes activated by dorsal refine the initial pattern, for example, *rhomboid* would be expressed in the most ventral region of the embryo but for the fact that it is repressed by snail. Positive regulatory interactions, like those of twist with snail and on itself, help to refine and sharpen the patterns. The outcome of these interactions is the subdivision of the blastoderm into different territories with defined boundaries along the DV axis.

Dpp

Both the signalling protein hedgehog, and dpp, a member of the TGFβ/BMP family of signalling molecules, can diffuse through the extracellular space and form concentration gradients that participate in patterning fields of cells. Hedgehog has been shown to act in a concentration-dependent manner and to elicit the expression of a number of target genes during the development and patterning of the wing, one of which is *dpp* itself (Fig. 9.9). We shall discuss hedgehog's qualifications to be considered a morphogen more fully later, in the context of its role in the development of the vertebrate nervous system (Chapter 10, and see also Fig. 9.9); here, we shall concentrate on dpp.

Dpp determines cell fates in a concentration-dependent manner both in the embryo and during the development of the adult from the larval stage. In the embryo, it is expressed over the dorsal third of the circumference of the embryo, where the concentration of dorsal is low (see Fig. 9.21 and 9.22). Cellularization of the syncytial embryo occurs soon after the dorsal protein gradient has been set up, so dpp acts in a cellular context. Genetic analysis has shown that *dpp* is required for patterning along the dorsoventral (DV) axis. The wild-type embryo has a number of landmarks that identify different cell fates along the DV axis. In the absence of *dpp* these differences disappear and the embryos develop with a uniform fate of ventral ectoderm all round. The influence of different concentrations of dpp can be demonstrated in experiments in which embryos genetically engineered to have a homogeneous lateral pattern along their DV axis, are injected with defined amounts of dpp. As a result, these embryos develop new patterns that are directly correlated with the amount of dpp they receive (Fig. 9.22). Under these conditions it is possible to establish a correlation between dpp signalling and particular DV cell fates: the highest levels of dpp signalling correlate with most dorsal development and lowest levels with more ventral fates. Interestingly, the wild-type embryo has no graded distribution of dpp at this stage, suggesting that there must be mechanisms that can generate a gradient of dpp activity from a uniform distribution of the molecule. One candidate to play this role is the short gastrulation (sog) protein which is involved in antagonizing the activity of dpp (Chapter 4, Fig. 4.27). Sog is expressed ventrally to dpp and may generate an opposing gradient that will sharpen the effects of dpp.

During the development of the adult fly from the larval stage, dpp displays properties more similar to those of a classical morphogen. The adult fly develops from groups of cells, the imaginal discs, that contain spatially defined sources of dpp. The cells of the imaginal discs respond to dpp in a distance-related manner that can be correlated with different amounts of dpp. In the developing wing primordium, in particular, the *dpp* gene is expressed in a narrow stripe, on either side of which target genes are expressed in nested patterns (Fig. 9.23 and see also Chapter 7, Fig. 7.17). Loss of function of *dpp* results in the loss of expression of these genes, and if *dpp* is expressed ectopically elsewhere in the wing the typical nested pattern of target gene expression develops relative to the new dpp source. These responses are not due to relays because constitutive activation of the Type I dpp receptor thickvein leads to cell-autonomous expression of the target genes, and conversely, loss of function of the receptors results in a cell-autonomous loss of expression of the target genes (Fig. 9.23). Further studies on the response to dpp during wing development have shown that dpp can elicit the expression of two targets at two different thresholds: *spalt* (*sal*) at high concentrations and *optomotor blind* (*omb*) at lower concentrations.

However, some experiments suggest that the pattern of response to dpp might not be a simple response to a linear gradient of dpp. The establishment of *omb* and *sal* expression has not been observed directly during normal development, but it has in experiments in which small ectopic sources of dpp are established. If the concentration of dpp is low, *sal* and *omb* are initially expressed in similar domains, and only after 50 hours does *omb* expression extend further into regions with low levels of dpp. This is likely to be the situation in the early development of the disc and suggests that the gradient of dpp is not the only variable that determines the response of the cells. The extension of *omb* expression requires the activity of a second BMP family member, the glass bottom boat (gbb) protein. Gbb is ubiquitously expressed in the developing wing, and in its absence, the patterns of *sal* and *omb* are very similar (Fig. 9.24). Gbb acts through its own Type I and Type II receptors and can activate the expression of *omb*. Therefore, it is likely that during normal development, the receptors for gbb synergize with thickvein and the Type II dpp receptor punt, to boost the response to low levels of dpp (Fig. 9.24).

There is evidence that other variables also come into play in establishing the final pattern of response to dpp, as we discuss later in this chapter.

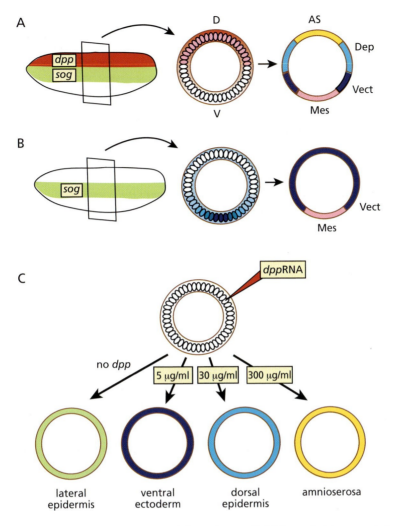

Fig. 9.22. Patterning of the ectoderm of the *Drosophila* embryo by graded activity of dpp. (A) As a consequence of the gradient of dorsal, the ectoderm of the early *Drosophila* embryo is subdivided into two large domains: a ventral domain where *sog* is expressed and a dorsal domain where *dpp* is expressed. The products of these two genes interact to pattern the ectoderm into three broad regions: a very dorsal region, the amnioserosa (AS), which forms an extraembryonic membrane that will not contribute any tissue to the fly, a dorsal epidermal region (Dep), which will form mostly epidermis, and a ventral ectodermal region (Vect), which will give rise to epidermis and central nervous system (see details in Chapter 10). **(B)** Removal of dpp expands the ventral ectoderm all around the embryo, whereas removal of sog results in an expansion of the dorsal epidermis (not shown). **(C)** The role of dpp can be demonstrated by injecting different concentrations of dpp into embryos that have been manipulated genetically to create a homogeneous lateral pattern throughout. In these experiments, increasing the concentration of dpp induces increasingly more dorsal levels of development (After Ferguson, E. and Anderson, K. (1992) Decapentaplegic acts as a morphogen to organize dorsal-ventral pattern in the *Drosophila* embryo. *Cell* **71**, 451–61.)

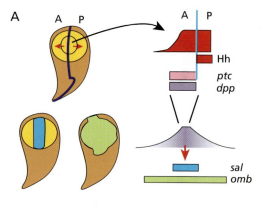

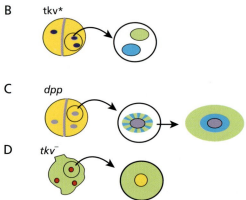

Fig. 9.23. Morphogen gradients in the imaginal discs. (A) The wing imaginal disc of *Drosophila* will give rise to the wing and the thorax of the adult fly (see Chapter 12 for details). In response to hedgehog, cells along the AP compartment boundary express *dpp* (purple) and create a gradient of dpp that spans the developing wing. Dpp induces in a concentration dependent manner the expression of two genes: *spalt* (*sal*) (light blue) in a domain adjacent to the source of dpp, and *optomotor blind* (*omb*) (green) over a broader domain. **(B)** Clones of cells expressing an activated form of tkv (tkv*), a Type I dpp receptor, express *sal* and *omb* in a cell-autonomous manner. This shows that the expression of these genes is not due to a secondary effect triggered by dpp. **(C)** Clones of cells that express *dpp* in ectopic positions elicit new patterns of *sal* and *omb* expression that mimic the wild-type pattern, but in the wrong region of the disc. **(D)** Removal of the dpp receptor tkv in clones of cells (red) results in the loss of *sal* and *omb* expression in those clones (yellow) and shows that the responses observed in the experiments shown in B and C are due to dpp acting through its receptor. (After Nellen, D. *et al.* (1996) Direct and long range action of a dpp morphogen gradient. *Cell* **85**, 357–68; and Lecuit, T., Brook, W., Ng, M., Calleja, M., and Cohen, S. (1996) Two distinct mechanisms for long range patterning by decapentaplegic in the *Drosophila* wing. *Nature* **381**, 387–93.)

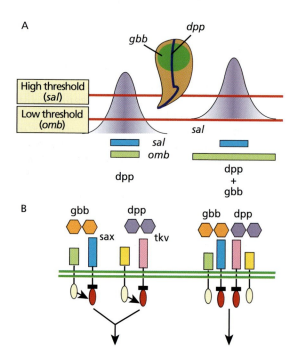

Fig. 9.24. Establishing thresholds of responses to dpp. (A) Glass bottom boat (gbb) is a *Drosophila* member of the TGFβ/BMP family of signalling molecules that is involved in changing the threshold of the response to dpp during wing development. In contrast to the spatially restricted expression of the *dpp* gene (dark purple line), the *gbb* gene is expressed (green) throughout the wing primordium. In the absence of gbb, the cells at the edges of the dpp concentration gradient, where the concentration of dpp protein is low, do not express *omb*. However, in the presence of gbb the thresholds change and now *omb* is expressed at the lowest concentrations of dpp. **(B)** How gbb works and achieves its effects on the response to dpp is not clear. Gbb has its own Type I (saxophone) and Type II receptors, which might act synergistically with the dpp receptors (thickvein and punt) to allow expression of *omb* at low dpp concentrations. At the molecular level, this might be achieved by convergence downstream of the receptor (*left*), or by the formation of complex receptors with different signalling activities (*right*). (After Haerry, T. *et al.* (1998) Synergistic signalling by two BMP ligands through the SAX and TKV receptors control wing growth and patterning in *Drosophila*. *Development* **125**, 3977–87; and Podos, S. and Ferguson, E. (1999) Morphogen gradients, new insights from DPP. *Trends Genet.* **15**, 396–402.)

Candidate morphogens in vertebrates

In vertebrates, the best available evidence for the existence of molecules with the properties of morphogens comes from studies on the function of members of the TGFβ/BMP family of signalling molecules, in particular activins.

Activins, members of the TGFβ/BMP family of signalling proteins, play an important role in the induction and patterning of the mesoderm in vertebrates (Fig. 9.2 and 9.16). In *Xenopus*, several experiments indicate that activin can induce concentration-dependent responses that could be used to pattern the mesoderm (though it is possible that the endogenous inducer may actually be a related family member rather than activin itself). Activin can be synthesized *in vitro*, and applying different amounts of activin to cells derived from the animal region of *Xenopus* embryos can direct their development into different mesodermal derivatives: low concentrations induce the development of mesenchyme and blood, whereas high concentrations induce the development of muscle and heart. These patterns of differentiation correlate with the expression of different sets of genes, which respond to different activin concentrations with well defined thresholds; for example the gene encoding the T box transcription factor *Xbrachyury* (*Xbra*) is activated at low activin concentrations and the homeobox gene *Xgoosecoid* (*Xgsc*) at higher concentrations (Fig. 9.25).

Although these experiments show that different concentrations of activin can induce different cell fates, they do not address the question of whether activin can form a gradient in a field of cells or whether the proposed long-range signalling mechanism actually operates *in vivo*. The first of these questions can be answered by placing some radioactively labelled activin in the middle of a group of responding cells and measuring its distribution over the surrounding tissue at the time of the response. The result of this experiment suggested a diffusion range of about 120 μm (about seven cell diameters), and a very low effective concentration (Fig. 9.26).

Details of the range and properties of the response to an activin gradient were obtained by sandwiching animal cap cells around beads loaded with either a high or a low concentration of activin and testing the responding cells for the expression of various mesodermal genes as activin was released from the source and spread out over the field of cells (Fig. 9.26). In response to a source containing a high concentration of activin, cells close to the source of activin expressed *Xgsc* while the expression of *Xbra* was detected further away from the source. The *Xgsc*- and *Xbra*-expressing cell populations did not overlap. It was also possible to measure the response of the cells over time, as activin was released from the source. When this was done, cells close to the source first expressed *Xbra* and then switched to expressing *Xgsc*—presumably as the concentration of activin built up—in agreement with the observation that *Xbra* responds to low and *Xgsc* to higher concentrations of activins. The possibility of a relay system was eliminated by placing a non-responding population between the activin-loaded beads and the responding cells; under these conditions, the responding cells still activated the expression of genes as predicted from their distance from the source. It was also possible to demonstrate, by showing that isolated cells responded in the same way as cells that were allowed to have varying degrees of contact with other cells, that each individual cell can assess the concentration of activin and express specific genes in a concentration-dependent fashion.

Altogether these experiments show that activin can form a diffusion gradient over cells from early *Xenopus* embryos and can trigger concentration-dependent responses. Whether activin acts in this way during normal development is not clear. However, members of the Nodal family of signalling molecules, which are structurally related to activin, have been shown to mediate many of the effects of activin during the development of *Xenopus* and other vertebrates, and would probably act in a similar manner to activin in the experiments we have described.

Another molecule that has been shown to display some of the properties of morphogens is Sonic hedgehog. In vertebrates, during the early stages of neural tube development, Shh initiates the patterning of the neural tube by determining the fate of different classes of motor neurons in a concentration-dependent manner from a localized source (see Fig. 9.9). However, as in the case of *Drosophila* hedgehog, most of the experiments describe a correlation between Shh concentration and the cellular response, rather than demonstrating a direct instructive function of Shh alone, and it is not yet clear what role other factors play in the generation of these responses (see also Chapter 10).

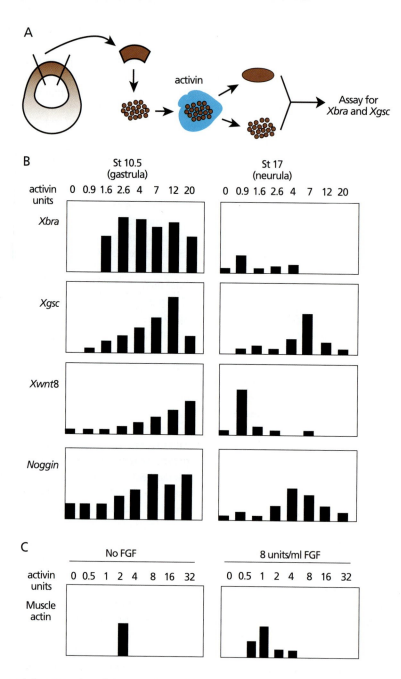

Fig. 9.25. Activins and the patterning of the mesoderm. (A) To test the effects of activins on the patterning of the mesoderm, cells from the animal cap of an early *Xenopus* embryo are dissociated and then treated with different concentrations of activins for a given period of time. After this, the cells are either reaggregated or left dissociated, then assayed for the expression of various genes, each of which is characteristic of a particular domain of the mesoderm. **(B)** Cells from embryos at different developmental stages respond differently to a range of activin concentrations. While at the neurula stage the target genes respond with very sharp thresholds, earlier on (at the gastrula stage) they show very broad dose response profiles. (Adapted from Green, J., Smith, J., and Gearhart, J. (1994)

Slow emergence of a multithreshold response to activin requires cell contact dependent sharpening but not prepattern. *Development* **120**, 2271–8. **(C)** The response to activin can be modified by other signalling molecules (e.g. FGF). In this example, a general mesodermal marker, a gene encoding muscle actin, is used to follow the response of the animal caps. In the absence of FGF, cells from early embryos show a sharp response profile to activin but addition of some FGF to the medium changes and broadens the response. (Adapted from Green J. *et al.* (1992) Responses of embryonic *Xenopus* cells to activin and FGF are separated by multiple dose thresholds and correspond to distinct axes of the mesoderm. *Cell* **71**, 731–9.)

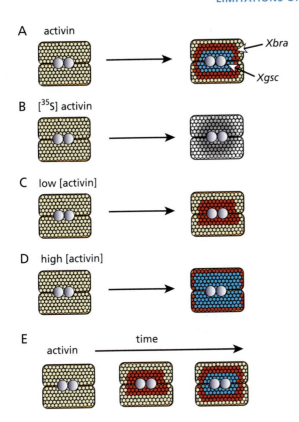

Fig. 9.26. **Parameters of the cellular response to activin.** Experiments in which beads are coated with different concentrations of activin illustrate some of the properties of the responses to a gradient of activin in vertebrate cells. The experiments are similar to those described in Fig. 9.2 and 9.7). **(A)** Ectodermal cells from *Xenopus* embryos, which can respond to mesoderm inducers and express mesodermal genes, are used to sandwich activin-coated beads. An intermediate dose of activin leads to a mixed response in which cells near to the beads (exposed to high concentrations of activin) express genes such as *Xgoosecoid* (*Xgsc*), whereas cells located at a distance from the beads express genes such as *Xbrachyury* (*Xbra*). **(B)** If the beads are coated with radioactively labelled (e.g. [35S]) activin, it is possible to observe the generation of a gradient of activin over the ectodermal cells by exposing an X-ray film to produce an autoradiogram. **(C)** Coating the beads with a low concentration of activin leads to the expression of *Xbra* throughout the ectodermal cells, whereas **(D)** loading the beads with a very high concentration of activin leads to broad expression of *Xgsc*. **(E)** The response of the cells can be observed at different times; in this case, it is possible to observe how, initially, cells located near the beads express *Xbra* and then switch to express *Xgsc*, as the cells distant from the bead begin to express *Xbra*.

Limitations of the classical concept of the morphogen

It is clear from the examples we have discussed that graded concentrations of some signalling molecules (and, in the special context of the *Drosophila* syncytium, transcription factors) can elicit graded responses, and that in some cases these responses are not just quantitatively but also qualitatively different. However, detailed analysis of these examples also makes it clear that the classical concept of the morphogen, although it has provided a very important conceptual framework for thinking about the relationship between the assignment of cell fates and pattern formation, is less helpful when the analysis moves to the level of molecular mechanisms.

The main difficulty with the concept lies in the requirement that the information embodied by different concentrations of the morphogen should be transmitted *directly* to the nucleus of the responding cell, where it should be transformed into qualitatively different responses and act in an instructive manner. In almost every case that has been studied, more detailed investigation of the mecha-

nisms by which putative morphogens act has revealed a more complicated situation in which the thresholds for different responses are set up, not by a linear gradient of the morphogen alone, but by combinatorial interactions between the morphogen and other factors, some pre-existing and some induced by the morphogen. These interactions develop and change with time, as the initially 'naïve' cell is altered by the information it has received.

For example, in studies on the role of dpp on the patterning of the wing in *Drosophila*, it is clear that the dpp concentration gradient alone is insufficient to direct the observed responses, which appear to involve complex regulatory interactions with other molecules at the cell surface (see Fig. 9.24). We have already discussed the role of the uniformly distributed TGFβ/BMP family member gbb in modulating the effective range of dpp signalling. However, there are additional elements involved in the response to dpp. In particular, the Type I dpp receptor thickvein (tkv) and a nuclear protein, brinker, act as negative regulators of the activity of dpp. At high concentrations tkv can bind dpp and reduce its effective signalling concentration, as patched does with hedgehog (Fig. 5.28). Thus, an effective response to dpp requires a reduction in the concentration of tkv, and this is observed in the developing wing. It is possible that the reduction in the concentration of tkv may help to explain why gbb is required for the expression of

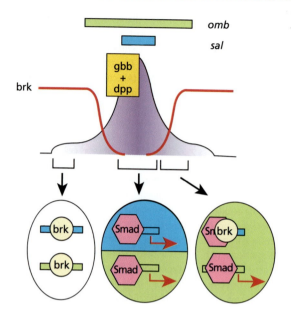

Fig. 9.27. **Nuclear factors affecting cellular responses to gradients of signals.** In the wing imaginal disc of *Drosophila*, the response to the combined activity of gbb and dpp (see Fig. 9.24) is further refined by the activity of the nuclear protein brinker (brk), which acts as a repressor of targets of dpp activity. The expression of the *brk* gene is under negative control by dpp and therefore is low where dpp is high. This results in two opposing gradients across the AP axis of the disc. High concentrations of brk repress the expression of *sal* and *omb*. As the concentration of brk decays, thresholds are established in which the response is defined by the relative levels of gbb/dpp and brinker activity. (After Podos, S. and Ferguson, E. (1999) Morphogen gradients, new insights from DPP. *Trends Genet.* **15**, 396–402.)

omb at low dpp concentrations (Fig. 9.24): in this region, the combination of low dpp and low receptor might not be sufficient to elicit a response.

The nuclear protein brinker represses genes that respond to dpp. The expression of both *tkv* and *brinker* is repressed by dpp, creating a regulatory loop that sensitively controls the response of the cells to dpp: the more dpp there is, the better it will signal, because it will reduce the amount of tkv and brinker in the cells. If the concentration of dpp drops there will be a steep decay in the response because the concentration of the negative regulators will increase concomitantly. This means that the graded response to dpp is not a simple measure of different concentrations but a complex function of its interactions with other extracellular proteins, such as gbb, and of the balanced effects of transcriptional activators and repressors (Fig. 9.27). In addition, the gradient of hedgehog, on which dpp depends, overlaps with that of dpp and might contribute to the qualitatively different responses by interactions between the effectors of these signalling pathways on targets in the nucleus. The complexity of this system is likely to be generally representative of many other situations involving graded concentrations of inducers.

Recent experiments on the initial responses of cells from *Xenopus* embryos to different concentrations of activin also suggest a more complicated situation than was at first apparent. It appears that the thresholds established for different responses may not be simply a direct function

of morphogen signalling alone (Fig. 9.28). These experiments measured responses at earlier time points after the exposure to activin began than had been used in previous studies. In one set of experiments, dispersed cells from animal caps were suspended in medium containing different concentrations of activin. The first response of the cells, both at high and low concentrations of activin, was the expression of both *Xbra* and *Xgsc*, with *Xbra* more strongly expressed than *Xgsc*. However, *Xbra* expression was rapidly down-regulated at high activin concentrations, in a process that required protein synthesis (but not the participation of *Xgsc*). The fact that these responses were observed in dispersed cells shows that they do not depend on cell–cell interactions. Similar results were obtained with activin-loaded beads sandwiched between animal caps: the first cells to 'see' the activin diffusing from the source activated both *Xgsc* and *Xbra* expression, but by three hours later *Xbra* expression had disappeared from the cells nearest the source, which now only expressed *Xgsc*, in agreement with previous studies. The inference from both sets of experiments is that, in some situations, different concentrations of activin might not elicit the expression of different target genes directly. Rather, secondary responses come into play—perhaps like those observed in response to bicoid or dorsal in the *Drosophila* syncytial blastoderm—that sharpen and resolve initially broad and overlapping domains of target gene expression.

It must also be kept in mind that *in vivo*, one signalling system probably never acts in isolation. In experiments where different concentrations of activin are combined with other signalling molecules that are present in the embryo at similar stages, the classical linear response observed with activin alone quickly becomes non-linear (Fig. 9.25). Given the complex interactions between signalling path-

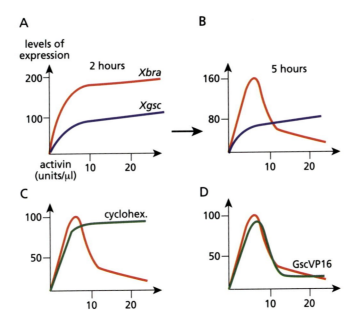

ways and the combinatorial effects that operate at promoters (Chapters 3 and 5), the question arises whether the linear response observed to activin in experimental situations also occurs *in vivo*.

It is clear from this discussion that cells can respond to different concentrations of signalling molecules and create different responses and that this is an important variable during development. However, it also appears that the concept of a morphogen in its original formulation might not provide quite the right framework for understanding what happens at the level of molecular mechanisms. Perhaps only spätzle/dorsal in the early *Drosophila* embryo, as well as activin and related molecules very early in vertebrate development, actually fit all the requirements at this level. Although, in all the cases we have discussed, and probably in many others, graded concentrations of putative morphogens contribute to a fine-grained pattern, it is also clear that the concentration gradient of the morphogen does not establish this pattern directly. More often than not, these fine patterns result from an interplay between combinatorial interactions at promoters, and changes in competence that can develop over time. Within this framework, morphogens are leading elements in a sequence of pre-patterns in which at each stage the number of inputs increases and the contribution of the morphogen is diluted.

It is perhaps interesting that the few cases in which a

Fig. 9.28. Activin as a ruling element in hierarchical prepatterns. In experiments similar to those of Fig. 9.25, animal cap cells are disaggregated, cells are exposed to different concentrations of activin, then reaggregated. The response, assayed as expression of *Xgsc* and *Xbra*, is measured 2 or 5 hours after treatment. **(A)** At 2 hours there is a clear induction of both genes although *Xbra* appears to be expressed more strongly than *Xgsc*. **(B)** After 5 hours, there is a clear separation of responses, with *Xbra* expressed at low doses of activin and *Xgsc* at high. An early broad response of *Xbra* is sharpened with time to produce a much more precise pattern of gene expression. The sharpening of the response is not dependent on interactions between cells because it will occur whether the cells are reaggregated after treatment or not. **(C)** The sharpening of the *Xbra* response requires protein synthesis since it is blocked by cycloheximide (green). **(D)** Repression mediated by Xgsc is not responsible for sharpening the *Xbra* response. Xgsc does act as a repressor, but turning it into an activator by overexpressing a construct in which it is fused to the activator domain of the viral VP16 transcriptional activator has no effect on the profile of *Xbra* expression. (After Papin, C. and Smith, J. (2000) Gradual refinement of Activin induced thresholds requires protein synthesis. *Dev. Biol.* **217**, 166–72.)

direct dose response curve can be drawn come from very early developmental stages, when cells are naïve and their competence is very wide. This might mean that morphogens—in the spirit of the original formulation—might only be found in these situations.

The limitations of the historical concept of the morphogen in no way undermine the vital importance of graded information in developing systems; they merely require a readjustment in how these processes are understood. In the following sections we shall discuss the available information about how graded molecular information is generated, and how it is measured by cells.

The generation of graded molecular information

The basic mechanism that underlies the generation of graded information *in vivo* is the free or assisted diffusion of molecules in syncytia or fields of cells (Fig. 9.6). The medium through which the molecules must diffuse can vary greatly in different developmental situations. For example, the syncytial nature of the *Drosophila* embryo allows the bicoid protein to diffuse from a local source over a very large distance, about 500 μm, which spans hundreds of nuclei (Fig. 9.18). In the case of dorsal, the nuclear gradient that is generated depends on a gradient of spätzle-mediated Toll activation (see Chapter 4, Fig. 4.26, and Fig. 9.21). At first sight, spätzle might appear to reflect a more general situation, in which molecules involved in generating gradients have to interact with the extracellular space. However, the medium through which spätzle moves—the perivitelline space—is not a field of cells, and is probably quite permissive to simple diffusion. In most developmental situations, in contrast, signalling molecules have to diffuse through a thick extracellular matrix where they encounter complex interactions that regulate their movement (see Chapter 4, Fig. 4.30, and Chapter 6). These interactions will play a role in the generation of graded molecular information.

While in most cases a gradient is formed by passive or active diffusion from a source, in some cases a uniform distribution of a signalling molecule can be transformed into a graded signal. The case of dpp in the early *Drosophila* embryo provides a good example of this (Fig. 9.29). As we have seen above, different concentrations of dpp elicit different cell fates, but dpp is uniformly distributed within the dorsal third of the embryo. This suggests that there is a mechanism that creates a gradient of activity from this uniform distribution of the signalling molecule. Two molecules appear to contribute to this mechanism: the chordin homologue short gastrulation (sog) and the sog protease tolloid (see Chapter 4, Fig. 4.27). Tolloid is expressed in the same pattern as dpp, and sog is expressed in a complemen-

tary pattern. The negative effect of sog on dpp thus creates a gradient of dpp activity with its highest point in the most dorsal region of the embryo. The gradient is steepened because tolloid, a protease that cleaves and inactivates sog, is coexpressed with dpp (Fig. 9.29). Exactly how sog antagonizes dpp signalling in the early embryo is not clear because although sog can interact with and inactivate some BMP family members, it does not do so directly with dpp.

A similar set of interactions modulate the activity of activin and BMPs in *Xenopus* and other vertebrates and these interactions play a role in establishing and refining graded molecular information early in development (Fig. 9.29).

The interpretation of graded molecular information

Although much is known about how gradients are established, very little is known about how cells read and interpret graded information. Two exceptions are, perhaps, the cases of bicoid and dorsal which illustrate how, at the transcriptional level, an interplay between differential affinity of binding sites and combinatorial interactions between different factors can generate a series of graded responses (Fig. 9.19). Much less is known about how graded information at the cell surface is translated into graded information in the nucleus. Even in the case of dorsal, whose gradient of nuclear concentration is generated by a gradient of Toll activation, this is not entirely clear at a qualitative level and still less in quantitative terms.

One set of experiments has addressed the question of how a cell senses different concentrations of a signal (Figs. 9.30, 9.31). There are two possibilities. First, a cell could sense the total number of receptors that are occupied at any given time. Alternatively, the cell might be measuring a ratio of occupied to unoccupied receptors. Experiments designed to answer this question for the activin receptor (Fig. 9.30) support the first possibility, that is, the cell can compute the absolute number of occupied receptors. Thus, under the experimental conditions, occupancy of 2% of the receptors led to activin-dependent expression of *Xbra* whereas a 6% occupancy led to expression of *Xgsc*. Overexpression of the number of receptors at the cell surface did not change the number of receptors at which the different responses were elicited (Fig. 9.30 and 9.31). How receptor occupancy is transformed into the different responses of target cells remains to be discovered.

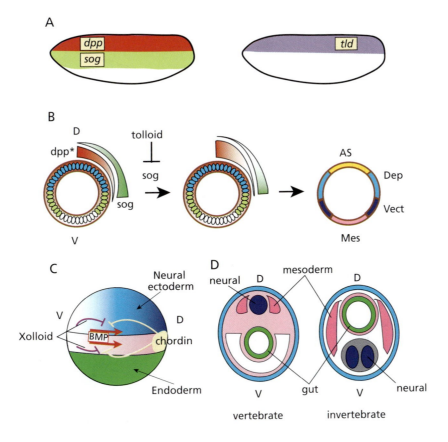

Fig. 9.29. The generation of BMP activity gradients. (A) Lateral view of early *Drosophila* embryos showing the patterns of expression of *dpp*, *short gastrulation* (*sog*) and *tolloid* (*tld*). While *sog* is expressed in the ventral ectoderm, *dpp* and *tld* are coexpressed over the dorsal quadrant of the embryo. **(B)** Transverse sections through embryos shown in A. Interactions between sog and dpp, diffusing through the early embryo from opposite sources, create a gradient of active dpp (dpp*) from dorsal (D) to ventral (V). This gradient is steepened by the effects of tolloid, probably acting on a complex that includes dpp and sog, and results in the different fates along the DV axis: amnioserosa (AS), dorsal epidermis (Dep), and ventral ectoderm (Vect). **(C)** In the early *Xenopus* embryo, BMPs are secreted from the ventral (V) region of the embryo and pattern the mesoderm and the ectoderm. The organizer (yellow) (see also Figs. 1.12, 4.29) secretes the sog homologue chordin, which antagonizes the activity of BMPs. The gradient of active BMP resulting from this interaction is further steepened by Xolloid, acting like tolloid in *Drosophila*. **(D)** The distributions and activities of dpp/BMP, sog/chordin and tolloid/Xolloid, and the outcome of their interactions, are inverted between vertebrates and invertebrates with respect to the DV axis. In both groups of animals these interactions define the nervous system within the ectoderm but, whereas in insects they operate over the ventral (V) region, in vertebrates they act in the dorsal (D) region.

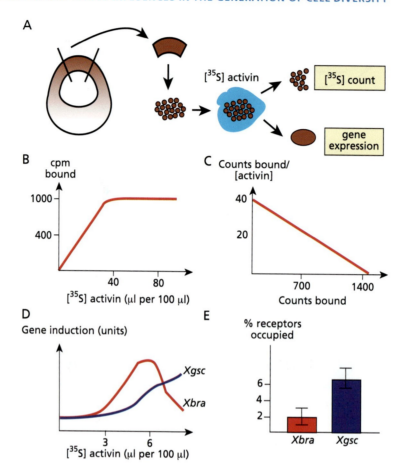

Fig. 9.30. The interpretation of graded information. Measurements of activin receptor occupancy in *Xenopus* animal cells. (A) Experimental design for the measurement of receptor occupancy. Animal caps are isolated and after dissociation, the cells are incubated with [^{35}S]-labelled activin for a short period of time. This allows the activin to bind its receptor. The cells are washed to remove unbound activin, then separated into two pools: one is used to count the [^{35}S] activin on the surface of the cells and the other is used to measure the expression of activin targets. **(B)** Amount of [^{35}S] activin bound to cells, expressed in radioactive counts per minute (cpm bound), when they are exposed to different concentrations of ac- tivin. Increasing the concentration of [^{35}S] activin increases the concentration of [^{35}S] activin bound to cells until saturation is reached. **(C)** Scatchard analysis of the experiment shown in B (see Chapter 4, Fig. 4.2) shows that activin binds to a single receptor species. **(D)** Concentration-dependent induction of *Xgsc* and *Xbra* in the experiment described in A. **(E)** The results shown in B and D allow a calculation of the percentage of activin receptors occupied for the induction of *Xgsc* and *Xbra* above background. (After Dyson, S. and Gurdon, J. (1998) The interpretation of position in a morphogen gradient as revealed by occupancy of activin receptors. *Cell* **93**, 557–68.)

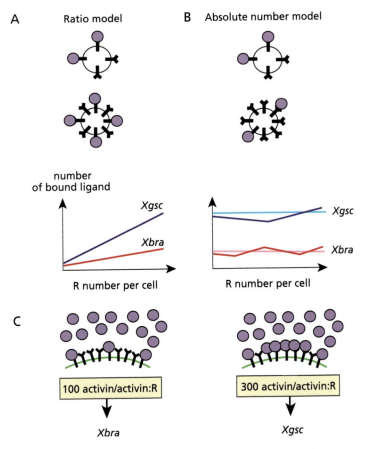

A   Ratio model   **B**   Absolute number model

number of bound ligand

Xgsc

Xbra

R number per cell

Xgsc

Xbra

R number per cell

C

100 activin/activin:R

Xbra

300 activin/activin:R

Xgsc

Fig. 9.31. Cellular responses to signalling molecules as a function of the ratios or absolute number of receptors occupied. (A, B) Two possible models to explain the differential responses to different concentrations of activin. **(A)** If cells were to respond on the basis of the ratio of bound to unbound receptors (ratio model) the number of bound ligands to generate the same response would increase if the number of receptors is increased experimentally. The plot at the bottom represents this situation. **(B)** If on the other hand, the response is dependent on the absolute number of receptors occupied (absolute number model), the number of bound ligands required to generate a response would not have to change when the number of receptors is increased experimentally. The plot at the bottom compares a theoretical prediction for this model (light colours) with the experimental data (dark colours). **(C)** The data summarized in Fig. 9.30 indicate that *Xbra* is induced when 100 receptor molecules are occupied, while *Xgsc* is induced when about 300 receptor molecules are occupied. (After Dyson, S. and Gurdon, J. (1998) The interpretation of position in a morphogen gradient as revealed by occupancy of activin receptors. *Cell* **93**, 557–68.)

Long- versus short-range patterning mechanisms in pattern formation

Although it is possible to set up experiments which demonstrate that diffusible molecules can elicit effects at a distance, and that their different concentrations can play a role in the assignment of cell fates, it is not straightforward to find out whether *in vivo* pattern formation is mediated by long- or short-range influences.

A series of classical experiments using the epidermis of hemimetabolous insects illustrate these difficulties. As these insects develop, they periodically have to slough off their outer rigid cuticle so that they can continue to grow. This process is called moulting. After each moult the epidermis expands by cell division and secretes a new cuticle, in such a way that the pattern of structures characteristic of each segment is maintained. These controlled adjustments to the epidermis offer an ideal experimental system for investigating how fields of cells are patterned. Just before a moult, the epidermis can be cut, sections removed, and the

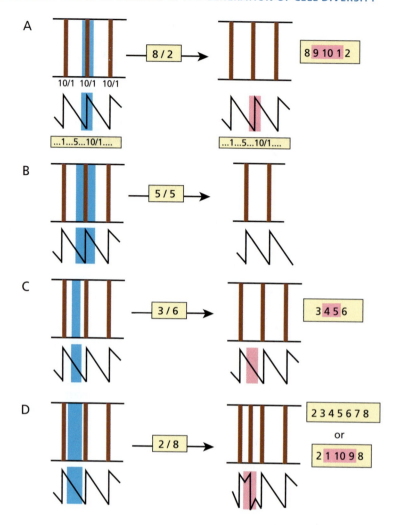

Fig. 9.32. Long- and short-range cell interactions and pattern formation. Diagrams of the abdominal segments of *Oncopeltus* before (*left*) and after (*right*) experimental manipulations in which different levels along the anteroposterior axis are brought together, and the response of the tissue is observed after a moult. Brown thick lines represent the segment boundary. Below each cartoon there is a representation of the segment from anterior to posterior as a gradient of positional values (from 1 to 10) with a sharp cut off (10/1) at the segment boundary. Excised regions are indicated in blue and regenerated ones in pink. **(A)** If the segment boundary is eliminated (represented here as an apposition of level 8 from one segment and level 2 of the adjacent one), it is regenerated without alterations of the pattern (represented by the filled in values 8 **9 10 1** 2). **(B)** Elimination of a larger piece of tissue spanning the segment boundary but which apposes similar levels of the segment (5 and 5), does not trigger any regeneration, As a result of this mani-

pulation, the insect loses one segment. **(C)** Deletion of a small region within the segment spanning two different positional values (e.g. 3 and 6), results in a regeneration of the missing pattern (represented as filled in values 3 **4 5** 6). **(D)** Deletion of a large region within the segment spanning positional values 2 and 8 results in a novel pattern that always includes a new segment boundary. Formally, this is similar to filling in the missing positional values in the order 2 **1 10 9** 8, a sequence which includes the 10/1 discontinuity and therefore a segment boundary. There is another way of filling in the gap created in the experiment: 2 **3 4 5 6 7** 8. The difference between the two is that the second one involves a longer sequence of positional values. In a situation like this, cells always choose the shorter route. (Adapted from Lawrence, P. and Wright, D. (1981) The regeneration of segment boundaries. *Phil. Trans. Roy. Soc. Lond. B.* **295**, 595–9.)

cut ends rejoined (Fig. 9.32). If this microsurgery is done in such a way that cells from similar positions within adjacent segments are apposed to one another, there is no pattern regulation after the moult. However, when cells from different positions within segments are apposed, new patterns of cuticle appear after the moult. These patterns are very specific: interactions between, say, positions A and B always produce pattern X, whereas interactions between positions A and C always produce pattern Y (Fig. 9.32). One way of interpreting these experiments is to assume that there is a graded series of 'positional values' across the segment. When cells are apposed in ways that do not alter their relative positions within one segment, there is no response, but when their relative positions are changed, they react.

Two different explanations can be proposed for how these positional values are set up. One invokes the concepts of morphogens and positional information and suggests that segment boundaries are the sources of morphogens whose graded distribution across the segment generates the pattern as a response (Fig. 9.17). The microsurgical operations in experiments on moulting insects alter the landscape in a way that forces the cells to regulate and reorganize to fill in the pattern. A different model suggests that cells can sense the differences in positional values and react by triggering a sequence of nearest-neighbour cell interactions that fills in the missing values. The second model can account for some bizarre but reproducible patterns that are difficult to explain on the basis of morphogen gradients (e.g. Fig. 9.32D).

As is often the case, the truth may well lie in a mixture of both explanations: short-range cell and molecular interactions drive much of the response, while a long-range influence that emanates from the segment boundary regulates some global properties of the system. The reasoning behind this suggestion is that, as we have seen throughout this chapter, although long-range influences can certainly trigger responses, changes in concentration of such inducers alone cannot produce a very fine-grained pattern. This requires both the participation of further molecular elements and the contribution of parameters such as time and changing competence.

SUMMARY

1. Cells can respond to different concentrations of the same signalling molecule by activating different genes.

2. If the signalling molecule is secreted from a localized source in a field of cells and can diffuse, it will form a concentration gradient that can be used to pattern the field. Cells will be exposed to different concentrations depending on their distance from the source and will respond by activating different genes.

3. The term 'positional information' refers to the ability of cells to acquire fates depending on their position relative to the source of a diffusible signalling molecule.

4. Morphogens are secreted signalling molecules that can elicit at least three different cellular responses in a direct and concentration-dependent manner.

5. The ability of a cell to respond to a signal is called its competence. This can mean the presence of receptors for the signal, the availability of a signal transduction pathway or an open transcriptional state for that signal.

6. A signalling molecule can alter the pattern of response of the target cells by altering their competence (i.e. changing the expression of receptors, state of nuclear activity or modulating the interactions between the signal and the receptor).

7. Gradients of signalling molecules are often used to establish coarse-grained patterns of responses that are refined through interactions between the original signalling molecule and its targets.

8. Short-range signals also play a role in pattern formation and, *in vivo*, patterns must result from an interplay between long- and short-range signals.

COMPLEMENTARY READING

Gurdon, J. B., Dyson, S., and St Johnston, D. (1998) Cells' perception of position in a concentration gradient. *Cell* **95**, 159–62.

Gurdon, J. B., and Bourillot, P. Y. (2001) Morphogen gradient interpretation. *Nature* **413**, 797–803

Lawrence, P. (1981) The cellular basis of segmentation in insects. *Cell* **26**, 3–10.

Lawrence, P. A. and Struhl, G. (1996) Morphogens, compartments, and pattern: lessons from *Drosophila*? *Cell* **85**, 951–61.

Neumann, C. and Cohen, S. (1997) Morphogens and pattern formation. *BioEssays* **19**, 721–9.

Pages F. and Kerridge S. (2000) Morphogen gradients. A question of time or concentration? *Trends Genet.* **16**, 40–4.

Rivera Pomar, R. and Jaeckle, H. (1996) From gradients to stripes in *Drosophila* embryogenesis: filling the gap. *Trends Genet.* **12**, 478–83.

Wolpert, L. (1989) Positional information revisited. *Development* **107**(Suppl.) 3–12.

Wolpert, L. (1996) One hundred years of positional information. *Trends Genet.* **12**, 359–64.

Chapter 10

Cell-type specification:
A developmental operation

The generation of cell diversity provides the raw material for the development and patterning of embryos. Cell diversity is achieved by the allocation of specific programs of gene expression to particular lineages, such that they become different from each other and generate different types of cells. In Chapter 8 we considered the molecular mechanisms that come into play during the establishment of some lineages, and described ways in which the patterns of cell division seen in Chapter 7 can be coupled to the assignment of cell fates through lineage-intrinsic mechanisms. In Chapter 9 we explored the ways in which cell interactions can influence cell fates. Here and in the following chapters we discuss how the molecular and cellular operations that we have described previously are used to specify different cell types and organize them in space to generate the component elements and structures of different organisms.

In this chapter we explore how the basic cellular routines seen in earlier chapters are used to generate specific cell types, focusing our attention on two cases: the generation of muscle, and of the neurons of the central nervous system. Using these examples, we shall see some of the different strategies through which particular lineages are endowed with specific programs of gene expression, and how these programs are developed and modulated by influences from the cells' environment to generate unique patterns of gene expression that confer specific identities on individual cells within these lineages.

The specification of particular cell types is a progressive process that requires the accumulation of molecular information. It will become clear that the development of a specific cell type does not simply involve the acquisition of a series of 'molecular options' for patterns of gene expression, but often also involves the forfeiting of others; that is, the expression of genes associated with commitment to a particular lineage usually carries with it the suppression of gene expression associated with fates not taken. In some cases, this suppression is an active molecular process and a cell fate results from the interplay between the activation and suppression of the activity of specific genes.

As we have seen in Chapter 8, the information that defines cell-type-specific profiles of gene expression stems from two sources: intrinsic 'running' programs of gene expression and their modification by external influences. The latter relies on the relative spatial arrangement of different cell types or lineages which act as the sources and receivers of information. For this reason, in the examples we discuss we shall describe certain aspects of the embryological context within which the particular lineages develop.

Methods for the spatial and temporal analysis of gene activity

The 'mining' of genomes by developmental biologists during the last twenty years has uncovered an extraordinarily large number and variety of genes that are expressed during embryogenesis, and revealed the dynamic nature of their expression patterns (Fig. 10.1). Genes are often expressed transiently in particular cells in the embryo and, sometimes, the same gene is expressed in very different lineages or cells. Questions as to the function of these genes can only be partially answered by classical genetic loss-of-function analysis. Mutation of a particular gene will eliminate the activity of the gene in the whole organism; this can generate defects very early in development that will preclude analysis of the function of the gene in particular lineages and cells that would normally appear at later stages. For this reason, techniques have been developed

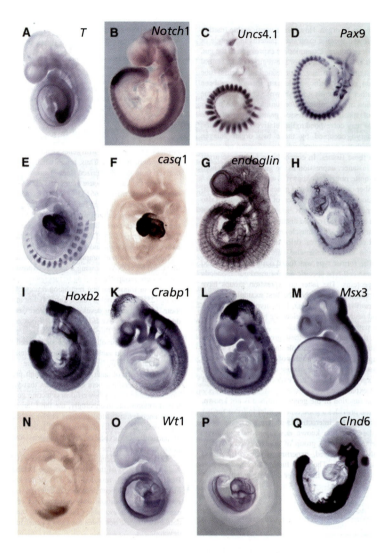

Fig. 10.1. Patterns of expression of diverse genes in mouse embryos. The genes were identified by screening a cDNA library. The gene names are indicated in each panel; if a panel has no name it is because the gene is so far known only by a cDNA number. The structures highlighted by their expression patterns are: **(A)** tail bud and notochord; **(B)** presomitic mesoderm; **(C)** caudal half of each somite and mesonephros; **(D)** sclerotome and pharyngeal endoderm; **(E)** heart and somite/myotome; **(F)** heart; **(G)** blood vessel endothelium; **(H)** blood vessels; **(I)** hindbrain and trunk; **(K)** neural crest; **(L)** neuroblasts; **(M)** dorsal neural tube excluding fore- and midbrain; **(N)** forelimb buds; **(O)** urogenital ridge; **(P)** mesothelium; **(Q)** endoderm and otic vessicle. (From Neidhardt, L. *et al.* (2000) Large scale screen for genes controlling mammalian embryogenesis, using high throughput gene expression analysis in mouse embryos. *Mech. Dev.* **98**, 77–93.)

that allow gain- and loss-of-function mutations to be targeted to particular cells.

A valuable experimental tool for analysing the function of a specific gene in a lineage is the ability to alter the spatial and temporal pattern of expression of the gene by altering its regulatory region. In this way, the influence of the gene product on the development of particular cells or groups of cells can be tested (Fig. 10.2). Different experimental techniques for doing this have been devised for different organisms. Sometimes it is useful to express the gene ubiquitously, at higher levels than normal, and without spatial constraints. This can be done by using heat shock promoters and other molecular techniques that enable the timing and level of gene expression (but not its spatial extent) to be controlled (Fig. 10.2).

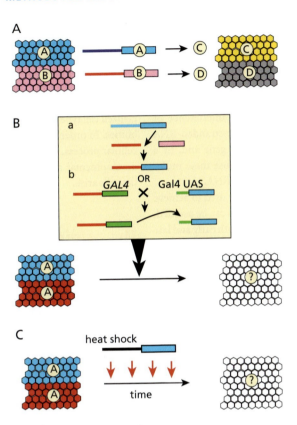

Fig. 10.2. Some methods for the altering the spatial and temporal expression of genes. (A) Two genes, A (blue) and B (pink), are expressed in two different sets of cells and are associated with a developmental transition of these cells. The cells expressing gene A develop into cell population C (yellow), while those expressing gene B develop into population D (grey). The expression of A and B may determine the development of those cells or may simply be passively associated with the transition. One way to distinguish these possibilities is to express gene A in population B or gene B in population A. This can be done in a variety of ways, some of which are shown here. **(B)** One way relies on exchanging the regulatory regions of A and B and assaying the effects of the chimaeric genes in embryos. The exchanges can be done in a direct or an indirect way. **(a)** In the direct exchange, the regulatory region of gene A is simply substituted by that of gene B and this will result in the gene A being expressed in the pink cell population. **(b)** Another way of achieving the exchange is to use a 'two-part system' with the help of the yeast Gal4 transcriptional activator, which recognizes a specific DNA sequence, the Gal4-binding UAS (abbreviated to Gal4UAS in the figure). The Gal4-binding UAS sequence can be placed upstream of gene A, and the regulatory region of gene B upstream of the coding region of *GAL4*. Gal4 will then drive the expression of A in the domain of B. This technique creates a conditional expression of A and is very useful in situations when mis-expression of gene A can kill the embryo or have secondary effects which preclude the analysis of its effects in the desired place and at the desired time. It has been used very effectively in the analysis of developmental events in *Drosophila*, where a large number of strains have been constructed in which *GAL4* is expressed in specific domains or even individual cells. In this way, the mis-expression of genes of interest can be very finely controlled. Flies containing *GAL4* can be kept separately from those containing the Gal4-binding UAS, so if the expression of A in the domain of B is lethal, the lethality will only be produced when the two strains are crossed. Useful as this technique is in *Drosophila*, there are not many reports of its successful use in vertebrates. **(C)** Ubiquitous expression can be achieved by placing the gene of interest under the control of a ubiquitously active promoter, or under the control of a promoter that is responsive to heat shock and will be activated at high temperature. The use of heat shock promoters can create conditional mis-expression and also allows the timing of expression to be controlled as the promoter will not be switched on until a heat shock is given.

The ability to target gene expression also provides a way of eliminating the activity of particular genes in specific contexts by expressing a dominant negative molecule that competes with the normal one. In this way, the requirement for a particular molecule can be assessed. Another technique to create loss of function relies on the use of double-stranded RNA (also called RNA interference or RNAi). Cells will readily take up double-stranded RNA molecules corresponding to a particular gene and, by a mechanism that is not well understood, use these molecules as guides to destroy the endogenous RNA.

Finally, targeted gene expression can be used to achieve limited rescue of a mutant phenotype. This can be useful if, as is often the case, a gene is expressed at both early and late developmental stages. Rescuing the early mutant phenotype by targeted gene expression will make it possible to study the effects of the mutation at later stages.

Altogether, these techniques allow a detailed dissection of the activities of individual genes in the specification of particular cells and lineages, and their influence on pattern.

Muscle development: Myogenesis

Muscle cells are derived from the mesoderm, which is one of the primary cell types generated early in development (see Chapter 1). In vertebrates, mesoderm arises as a result of an inductive event mediated by members of the BMP family of signalling molecules (see Chapter 9, Fig. 9.2). As the mesoderm is specified, signalling through members of the Wnt, BMP, and FGF families of signalling molecules combines to outline different developmental potentials in these cells (see Fig. 4.29). As a consequence of these secondary signalling events, a population of mesodermal cells is selected to become muscle. In *Drosophila*, the mesoderm results from the activity of the transcription factor dorsal as induced by the activity of the Toll receptor (see Chapter 4, Fig. 4.26 and Chapter 9, Fig. 9.21).

The formation of muscle, known as myogenesis, is initiated by the commitment of a subset of mesodermal cells to express a number of transcription factors which in turn regulate the expression of genes encoding muscle-specific cytoskeletal elements (e.g. muscle myosin) and cytoplasmic enzymes (e.g. creatine kinase) as well as proteins required for the electrophysiological properties of the muscle. Depending on the particular constellation of these proteins, different kinds of muscle cells become fine-tuned for specific functions (e.g. cardiac, smooth or striated in vertebrates, somatic and visceral in flies). Here, we focus on the development of vertebrate skeletal muscle and the somatic muscle of the fruit fly to illustrate the molecular steps that lead to the definition and development of a specialized cell type from one with a broad spectrum of possibilities. In order to understand the molecular basis of this cellular process, it is helpful to have an appreciation of the developmental and embryological context within which the cells arise.

The origins of muscle cells

In vertebrate embryos, during gastrulation the mesoderm comes to lie underneath the ectoderm. Shortly afterwards, a second morphogenetic movement, the invagination and folding of the ectoderm to form the nervous system, places a mass of cells that forms the neural tube in the middle of the mesodermal cell population. This splits the mesoderm into two masses of unpatterned cells which arrange themselves on either side of the neural tube. Some mesodermal cells are left underneath the ventral section of the neural tube and develop into a specialized structure: the notochord (Fig. 10.3). As development proceeds, the masses of mesodermal cells adjacent to and on either side of the neural tube become organized along the anteroposterior axis into paired spheres of epithelial cells: the somites (Fig. 10.3).

Different regions within the somite differentiate different kinds of cells (Fig. 10.4). The most ventral cells become mesenchymal and form the sclerotome, which gives rise to ribs and vertebrae. The more dorsal regions form the dermomyotome, which gives rise to muscles and the dermis of the skin. The positioning of the dermomyotome within the global coordinate system of the embryo provides these cells with a set of spatial references which lead to their natural subdivision into medial (closer to the neural tube, also called epaxial) and lateral (furthest from the neural tube, also called hypaxial) regions. Cells from the medial region turn inwards and form the myotome, from which myoblasts (muscle progenitor cells) will arise for the axial muscles of the body. Most cells from the lateral region migrate away from the position of the somite, and give rise to body wall and limb muscles (Fig. 10.4). Cells that lie between these two domains contribute to other tissues including the dermis and elements of the vascular system.

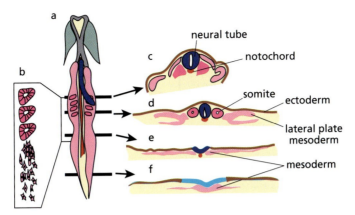

Fig. 10.3. The development of the mesoderm in a vertebrate (chick) embryo. Longitudinal view of an early chicken embryo (a) in which the progress of gastrulation, from anterior (top) to posterior (bottom), reveals different stages in the arrangement and positioning of the mesodermal cells (pink). At the anterior end, on either side of the neural tube (blue), masses of mesodermally derived epithelial cells, somites, form by aggregation of mesenchymal cells (b). These cells are the source of muscle precursors. At the posterior end of the embryo (f), gastrulation is less advanced; in this region, the mesoderm lies underneath the ectoderm (brown). As one proceeds anteriorly one can observe more advanced stages of development:

after the invagination of the mesoderm, the invagination of the neural plate (e) gives rise to the neural tube (dark blue). Concomitantly with this, the mesoderm splits to form the somites (dark pink) on either side of the neural tube, and the notochord (red) underneath the most ventral region of the neural tube (d). Mesoderm that does not get incorporated into the somites forms what is known as 'lateral plate' mesoderm. Later on (c), the cells in the somite begin to proliferate and become divided into two major populations: the sclerotome (dark pink), which will give rise to the ribs, and the dermomyotome (lighter pink), which will give rise to muscle and dermis.

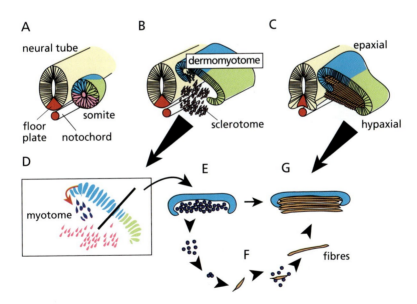

Fig. 10.4. Myogenesis in a vertebrate embryo. (A–F) Different stages in the development of the somites. **(A)** The somite is an aggregate of epithelial cells. Shortly after being formed each somite is divided along its dorsoventral axis into a ventral region (pink) which will give rise to the sclerotome, and a dorsal cell population (green and blue) which will give rise to the dermomyotome. **(B)** Cells in the somite proliferate and begin to differentiate into different cell populations. Cells in the sclerotome become loose and migrate from the somitic domain to give rise to the ribs. The dermomyotome becomes subdivided into three major domains: two of them, known as epaxial (blue) and hypaxial (green), will give rise to muscle, while

cells in between these domains (not illustrated in this figure) will contribute to the dermis. **(C)** Muscle fibres begin to assemble in the ventral region of the dermomyotome as it proliferates and grows. **(D, E)** Detail of the cellular dynamics in the dermomyotome. In a way that is not clearly understood, cells from the epaxial region come to lie underneath the dermomyotome where they become myoblasts (dark blue). Cells from the hypaxial region (green) migrate away and give rise to body wall and limb muscles. The sclerotome is indicated in pink. **(F)** When myoblasts stop proliferating they fuse with one another and give rise to myotubes. **(G)** Myotubes bundle together to form the myofibre.

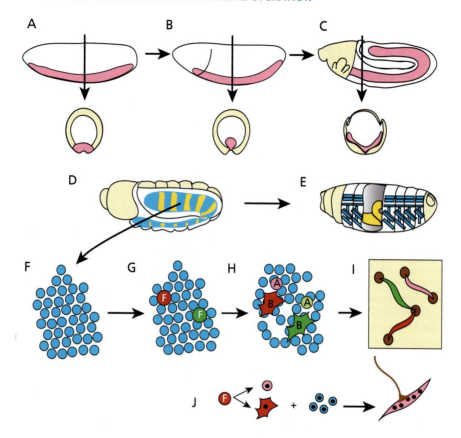

Fig. 10.5. Myogenesis in an invertebrate embryo: *Drosophila*. (A–C) Embryos are drawn with anterior to the left and posterior to the right. A transverse section through each embryo at the position marked by an arrow is shown underneath. **(A)** In *Drosophila*, cells with high levels of the dorsal transcription factor become defined as mesoderm (pink and see Chapter 9, Fig. 9.21). **(B)** These cells change shape, invaginate during the process of gastrulation **(C)** and spread over the underlying ectoderm. These processes coincide with germ band extension, a process whereby the cells of the embryo rearrange and the embryo curls over itself. **(D)** Shortly afterwards the mesoderm becomes segmented in register with the ectoderm and, within every segment, is subdivided into four cell populations. The two larger populations are shown here: one, located at the anterior region of every segmental unit (yellow) that will give rise to the visceral mesoderm, and another, located at the posterior end (blue) that will give rise to the somatic mesoderm cells, which will differentiate into the corresponding muscle fibres **(E)**. The process of muscle development during this transition is summarized in **F–I. (F, G)** Within the cells of the somatic mesoderm, single cells called founder cells (labelled F) emerge from clusters of cells located at specific locations, enlarge, and move towards the epidermis. Each founder cell acts as a seed for a particular muscle. **(H)** Each founder divides once to generate two precursors (A and B) for two different muscles. **(I)** These cells now draw surrounding mesodermal cells to fuse with them to generate multinucleate muscle fibres with a particular size and attachment (indicated by brown circles) in the epidermis. The individuality of each muscle is emphasized here by the different colours. **(J)** Detail of the differentiation of the founder cell. First, it divides to generate two precursors, each with particular characteristics. The precursors will fuse with surrounding mesodermal cells which have not been specified as founders and generate multinucleate muscles which attach to the epidermis and are innervated in specific ways. The information specifying the attachments and innervation is likely to be encoded in the founder.

Myoblasts proliferate until a defined time of development, when they exit the cell cycle, begin to express differentiation proteins (see Chapter 7, Fig. 7.20), and fuse with other myoblasts to form multinucleate syncytial myotubes. Further recruitment of more myoblasts to these syncytia converts them into fibres which attach to specific places and where they are innervated by neurons (Fig. 10.4).

In insects (Fig. 10.5), the somatic muscle is also derived from the mesoderm and has a segmental basis. However, in this case there are no somites, and muscles arise from segmental units derived from the overall segmental subdivision of the embryo. Their development has been best characterized in *Drosophila*, whose muscle cells, like those of vertebrates, are syncytial and follow a broadly similar pattern of differentiation.

After gastrulation, the mesoderm underlies the ectoderm and becomes subdivided into segmental units along the anteroposterior axis (Fig. 10.5). After cell proliferation, within every segment two populations of cells can be discerned: a posterior population that will give rise to the somatic muscle (the muscle of the body wall), and an anterior population that will give rise to the visceral (gut) muscle. Within the posterior region, two different sorts of myogenic cells (cells that have the potential to become muscle precursors) arise: single cells that act as seeds for individual muscles and are therefore called 'muscle founders', and other cells that contribute to the myotubes by fusing with the founders but cannot themselves act as founders. Each founder divides once to generate the precursors of two different muscles (Fig. 10.5), which initiate and drive the process of fusion. After fusion, individual syncytia attach to the epidermis and become innervated in specific patterns.

The existence of the two sorts of myogenic cells in *Drosophila* embryos is most clearly demonstrated by the effects of mutations that prevent the fusion of myoblasts to form myotubes. In these mutants, muscles form but each contains a single nucleus and is derived only from the population of muscle founders. At present there is no evidence for such muscle founders in vertebrates. They might exist, but at a lower frequency than in *Drosophila*, where their larger numbers might reflect the need to define the many different kinds of muscles needed for the complex movement of the *Drosophila* larva. It appears that although a vertebrate might have more muscle fibres than *Drosophila*, the *Drosophila* larva has many more different *kinds* of muscles and might therefore require a higher degree of individual cell definition.

Establishment of different mesodermal fates

The fates of mesodermal cells within the somites of vertebrates or the segments of *Drosophila* are not hard-wired but depend on interactions between the mesodermal cells and their surroundings. This is demonstrated by experiments in which the relative positions of the cells within the mesoderm are altered. For example, if a chick somite is rotated 180° around its own axis early in embryogenesis, its subsequent development and differentiation are entirely normal. This indicates that cells that were initially ventral are developing as dorsal, and *vice versa* (Fig. 10.6). Similarly, in *Drosophila*, if mesodermal cells are mixed early in development by transplantation, they develop according to the position where they find themselves and can give rise to any mesodermal derivative of any segment. These experiments show that their development is not hard-wired by their lineage (ancestry) but rather is determined by their position in the events that follow gastrulation. As development proceeds this potential becomes restricted: if the rotation of the chick somite is carried out later in development, cells will develop according to their origin at the time of the experiment rather than being influenced by their new location (Fig. 10.6). This indicates that, at some point in development, cells become irreversibly committed to a particular fate and that they will express that fate in an autonomous manner.

These observations, in particular the regulation of fates after early rotation, suggest that the fates of the different mesodermal cells within the somite might be instructed by surrounding cells, in particular by the cells in tissues close to the somite: the notochord and the neural tube. This idea can be tested by surgical experiments. Placing extra notochord tissue, or a piece of the floor plate from the neural tube, in ectopic positions induces the tissue closest to the transplant to develop as sclerotome (Fig. 10.7). However, removing the notochord altogether abolishes muscle development, suggesting that the notochord is also required for the correct development of other regions of the somite. In *Drosophila*, mesodermal fates are also induced by cells surrounding the mesoderm but in this case, the developing nervous system does not play a role, and the inducing population is the ectoderm, against and under which the mesoderm develops.

Various experimental approaches (and serendipitous observations) have led to the identification of factors mediating the inductive interactions. For example, transplantation experiments, usually in the chick embryo, have been used to locate the source of the inducing activities. Then, signalling molecules expressed in the inducing tissues have been tested for their ability to induce muscle development *in vivo* and *in vitro*. In this way, for example, it was shown that Sonic hedgehog (Shh), which is expressed in the notochord and the floor plate (see Chapter 9, Fig. 9.9), can substitute for the notochord in the assays outlined in Fig. 10.7. Other approaches to the identification of signals involved

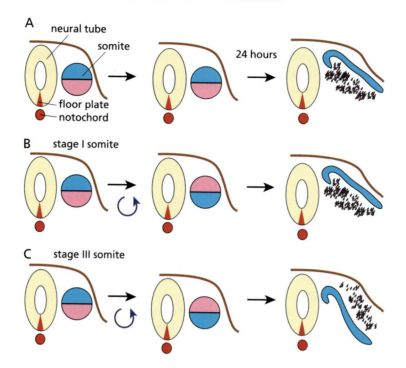

in the assignment of fates in the mesoderm have made use of mutants in *Drosophila* and mouse, and of molecular biology *in vivo* and in tissue culture, to identify inducing agents.

Altogether, these experimental approaches have revealed that the molecules that mediate the assignment of cell fates within the mesoderm are components of major signalling pathways: Wnt, hh, BMP, Notch/Delta, and RTK/Ras. Perhaps not surprisingly, no single molecule is sufficient for the assignment of the different fates within the mesoderm, which depend on the cooperation of a number of signals from spatially defined sources. Because many of these signalling molecules can diffuse (see Chapter 9), the positioning of their sources is crucial in determining the specific functional concentrations to which the development of different mesodermal derivatives responds. In this regard, it is interesting that in vertebrates, *Shh* mutants have severe defects in the development of the ribs. This suggests that in regions near to its source (i.e. at high concentration), Shh promotes the formation of the sclerotome, probably at the expense of the muscle tissue. However, in tissue culture experiments, lower concentrations of Shh have been found to promote myogenesis, an effect that is enhanced by the activity of some members of the Wnt family, Wnt1 and Wnt3, which *in vivo* are ex-

Fig. 10.6. Plasticity of myogenic precursors in the early embryo. Transverse sections through chick embryos showing the neural tube (yellow) and the notochord (red) underneath. A somite is shown to the right with the dermomyotome (blue) and the sclerotome (pink). The epidermis (brown) overlies both structures. The floor plate, a specialized region of the neural tube with inductive properties, is also indicated. **(A)** During normal development, the dorsal region of each somite (the dermomyotome, blue) gives rise to muscle and the ventral region (the sclerotome, pink) gives rise to the ribs and the vertebrae (see also Fig. 10.3). **(B)** Rotating the somite by 180° early in development (indicated as stage I) does not alter the final pattern of cells contributed by each region (i.e. cells that were initially ventral and that would have given rise to the sclerotome behave according to their new location and give rise to myogenic precursors). **(C)** If the rotation is performed at later stages (e.g. stage III somite), cells behave according to their origin (i.e. cells that were initially ventral still give rise to sclerotome despite their new position). This results in an inversion of the relative position of the muscle precursors and those of the ribs and vertebrae.

pressed in the dorsal section of the neural tube (known as the roof plate) and the epidermis (Fig. 10.8). This action of Shh is consistent with the observation that, although transplanting the notochord (a source of Shh) induces the tissue nearest to it to develop as sclerotome, the notochord is also

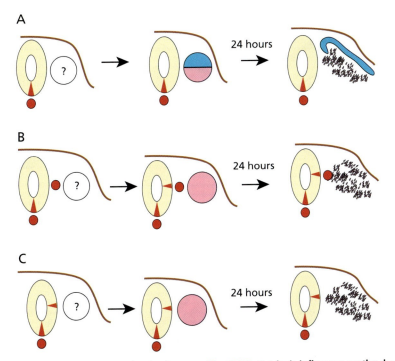

essential for muscle development (Fig. 10.7). The interaction between Shh and Wnts in promoting myogenesis is likely to play a role *in vivo* and even though single mutants in any of these genes do not show major defects in muscle development, some double mutants do. For example, double mutants for some of the Wnts, notably Wnt1 and Wnt3, show reduced myogenesis.

The importance of concentration gradients, and of the balance between levels of functionally interacting signalling molecules, is further illustrated by the role of BMP family members in muscle development. In tissue culture, BMPs have the opposite effect to that of a combination of Shh and Wnts, in that they suppress the myogenic fate. *In vivo*, BMPs are expressed in the dorsal neural tube and the lateral plate mesoderm where, the tissue culture observations suggest, they create an inhibitory field (Fig. 10.8). However, the BMP inhibitors follistatin and noggin (see Chapter 4, Fig. 4.27 and later in this chapter) are expressed by specific subsets of cells in the dermomyotome and the notochord, where they are likely to play a role in the development of muscle by locally antagonizing the action of BMPs and creating concentration gradients (Fig. 10.8). Variations in the concentration of BMPs are likely to be functionally significant. For example, high levels of BMP4, expressed in the lateral plate mesoderm, have been shown to promote the development of the hypaxial muscles.

Fig. 10.7. Extrinsic influences on the development and patterning of the mesoderm. The experiments shown in Fig. 10.6 indicate that cells within the somite acquire their fate from the surrounding tissues. Further experiments identify the inducing cell populations. (For labels, see Fig. 10.6.) **(A)** The neural tube and the notochord are possible sources of information for the unpatterned mesoderm (indicated by a question mark) because of their position relative to the somite. The experiments in B and C test this possibility. **(B)** Transplanting notochord tissue to a position between the somite and the neural tube induces the dermomyotome to develop like the sclerotome. This suggests that the notochord contains information for the development of the sclerotome and that the proximity of cells in the somite to this structure determines their developmental fate. At the same time, it suggests that the notochord can suppress the potential of cells close to it to develop as muscle. **(C)** The notochord induces floor plate in the neural tube, so in the experiments shown in B, the generation of extra sclerotome at the expense of myotome could be due to the new floor plate formed. To test this, the experiment in B is repeated with floor plate tissue, with similar results. Thus, the floor plate and the notochord have similar inductive abilities. Removal of the notochord or the floor plate, however, abolishes muscle development, suggesting that these structures are not only required for the development of the sclerotome but modulate the development of other cells in the somite.

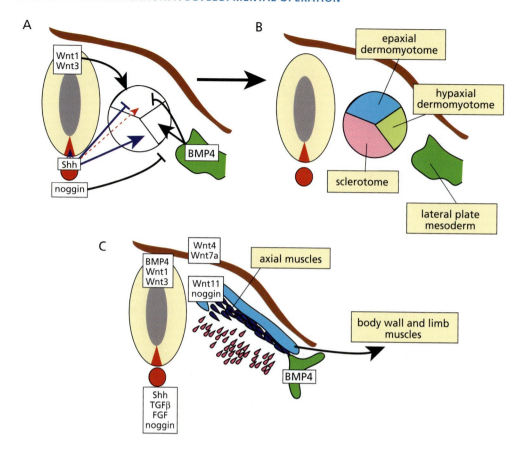

Fig. 10.8. Spatial organization and influences of different signalling molecules during muscle development in vertebrate embryos. (A, B) Early sources and effects of different signalling molecules on the subdivision of the somite into different cell populations, as inferred from studies on mutants and tissue culture experiments. Some molecules (e.g. BMP4 and Shh) have dual effects: while promoting the development of a fate in a specific cell population, they suppress other fates. Many of these molecules can diffuse over a long range and this might explain the observation that they can promote two different fates at different distances. For example, high concentrations of Shh promote the development of sclerotome in the ventral cells of the somite and suppress the development of muscle. However, in the cells of the dorsal region, which are located further away and therefore must receive a lower concentration, Shh can promote muscle development. The presence of other signals, and the local concentration of specific antagonists (e.g. noggin in the case of BMP4) also serve to modulate signalling in space and time. **(C)** As the development of the cells within the somite proceeds, the signals act on different cell populations to promote further differentiation. The signalling landscape also changes, becoming more complicated as can be gauged by comparing A and C.

Some evidence from the phenotypes of transgenic mice that are mutant for TGFβ/BMP family members or their antagonists supports the suggested roles of these signalling molecules and their antagonists in muscle development. Mice mutant for follistatin, for example, have a reduced muscle mass (although the early stages of muscle development do not seem to be impaired).

The same signalling systems play a role in the subdivision of the mesoderm of *Drosophila* (Fig. 10.9). Dpp and hedgehog are involved in the specification of visceral mesoderm, and the somatic muscle is promoted by the Wnt family member wingless. Combinations of these signalling systems also define other mesodermal derivatives like the heart, from the same population of precursor cells. In this case, however, the nervous system does not play a major role role in these signalling events and the source of the inducing elements is the ectoderm that overlies the mesodermal cells.

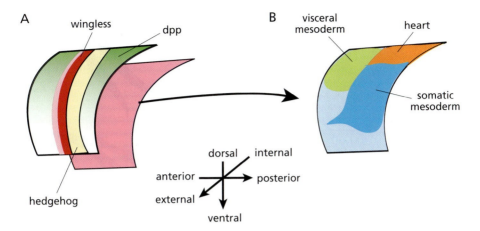

Fig. 10.9. Spatial organization and influences of different signalling molecules during mesoderm development in *Drosophila* embryos. In *Drosophila* the source of signals for the patterning of the mesoderm (pink on the left) is the ectoderm. Early in embryogenesis, populations of cells located at specific positions in the ectoderm act as sources of BMP (dpp), Wnt (wingless), and hedgehog, to pattern the underlying mesodermal cells. Wingless promotes the development of somatic mesoderm, wingless together with dpp promotes development of the heart, and dpp with hedgehog promotes development of visceral mesoderm. The spatial coordinates of the system are indicated.

In principle, the assignment of cell fates in the mesoderm could be implemented exclusively through positive inductive influences mediated by activators. However, as we have seen, a more likely situation is that fates are assigned through a balance of negative and positive effects of different signalling systems. Alternatively, it could be that a cell fate is a default state associated with a particular lineage, and is actively repressed until removal of this repression allows it to emerge. The role that signalling molecules play in each case is slightly different. In vertebrate myogenesis, there are clearly positive inductive influences at work but there is also evidence that de-repression is involved. For example, disaggregated late blastomeres from several different species can differentiate *in vitro* into muscle. The implication of this observation is that many cells of the early embryo have a latent myogenic potential that is actively suppressed. The function of the cooperation between the Wnt and Shh pathways would then be to antagonize this repression, and that of BMPs to collaborate with it.

Myogenic regulatory factors and the commitment to the myogenic pathway

In vertebrates, the regulatory hierarchies that specify cells to become muscle cells have been unravelled by a combination of tissue culture experiments and genetic approaches, mainly in mice. So strong is the conservation of both molecules and mechanisms operating in muscle development across the vertebrate group that, despite some strategic differences in the deployment of some of the proteins, conclusions drawn from studies in one system have usually proved generally applicable with only a few minor variations.

Factors capable of promoting myogenesis in vertebrates were first identified by experiments designed to find genes that could trigger muscle development in cultured cells that by themselves would not activate a myogenic program. The existence of these molecules had been inferred from heterokaryon experiments (see Chapter 3, Fig. 3.1 and Fig. 10.10); in these experiments, when myoblasts were fused with other cells they imposed a myogenic developmental program on these cells. Subsequently, several genes were isolated that when transfected into cells in tissue culture could induce muscle developmental programs (Fig. 10.10). Several of these genes—*MyoD*, *Myf5*, *MRF4*, and *myogenin*—encode transcription factors of the bHLH family. Similar experiments also identified the PRD homeodomain transcription factor Pax3 as capable of activating myogenesis in some cell types in tissue culture. *MyoD*, *Myf5*, *MRF4*, and *myogenin* are all expressed specifically in the dermomyotome region of the somite (Fig. 10.11). *Pax3*

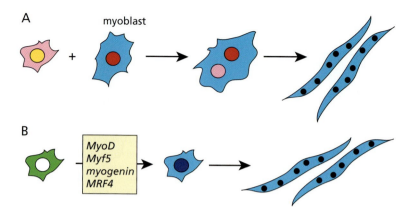

Fig. 10.10. Search for myogenic factors. (A) Heterokaryons between myoblasts (blue cytoplasm) and other types of cells (pink cytoplasm) induce the expression of muscle specific proteins from the nuclei of the non-muscle cells (indicated as a change in the nucleus from yellow to pink). This suggests the presence of proteins in the myoblast that act in a dominant manner to impose muscle programs of gene expression. **(B)** Transformation of the fibroblast cell line 10T1/2 with cDNAs extracted from myoblasts identified a number of genes that were able to induce the differentiation of these cells into myotubes. These genes, *MyoD*, *Myf5*, *myogenin*, and *MRF4*, all encode bHLH proteins and induce the formation of myotubes in cells and in conditions where this would not normally occur.

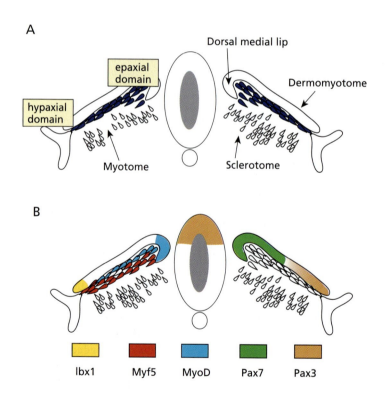

Fig. 10.11. Expression of myogenic regulatory factors in the embryo. (A) Subdivision of the myogenic domain within a chicken somite at an advanced stage of development. **(B)** Patterns of expression of different myogenic regulatory factors at this stage, identifying different cell populations already commited to give rise to particular sets of muscles. (Adapted from Borycki, A. and Emerson, C. P. (2000) Multiple tissue interactions and signal transduction pathways control somite myogenesis. *Curr. Topics Dev. Biol.* **48**, 165–224.)

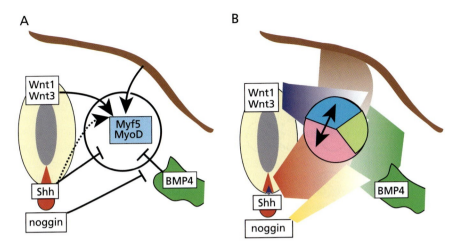

Fig. 10.12. Signalling networks act on the expression of myogenic regulatory factors. (A) Effects of the different signalling molecules on the expression of Myf5 and MyoD in the early stages of somite development (see also Fig. 10.8). The identity of the signal from the ectoderm (brown) is not known for certain but may be Wnt7a. **(B)** The signalling events depicted in A result in a complex landscape of signalling molecules over the somite that leads to specification of different cell fates: sclerotome (pink), epaxial myotome (blue), and hypaxial myotome (green). Each of these results from a combination of signalling molecules and is maintained by the myogenic regulatory factors. Genetic experiments indicate that in addition to signalling from the surrounding tissues, there are also interactions (arrows) between the sclerotome and the dermomyotome that are dependent on myogenic regulatory factors, but whose direct mediators have not been identified.

is expressed very early in the dermomyotome but, in contrast to the bHLH proteins which are expressed exclusively in myogenic lineages, *Pax3* is also expressed in the neural tube (Fig. 10.11), suggesting that its function in myogenesis depends on its interaction with other factors. Although, in tissue culture, forced expression of any of these genes on its own can induce myogenesis, the temporal profile of their patterns of expression together with regulatory interactions revealed by genetic experiments (see below) suggest that in the embryo they assemble a combinatorial network over time.

The expression of all these myogenic transcription factors is under the control of the cocktail of signalling molecules provided by the cells surrounding the somite (Fig. 10.12). *Pax3*, *MyoD*, and *Myf5* have all been shown to be targets of Wnt and Shh signalling, strengthening the idea that the myogenic program is activated by a combination of these signalling pathways, acting on genes encoding transcription factors that confer lineage specific properties. In contrast, Notch and BMP signalling repress *MyoD* expression, consistent with their suggested roles as inhibitors of myogenesis. *MRF4* and *myogenin* are expressed later and are associated with the differentiation of the muscle cells.

As we have seen (Fig. 10.6), manipulation of chick somites at different developmental stages shows that the myogenic fate is acquired progressively, and this is reflected in the temporal pattern of the expression of the myogenic genes. In the mouse, *Myf5* and *Pax3* are expressed early, in overlapping and dynamic patterns within the somite;

MyoD is switched on shortly afterwards and *MRF4* and *myogenin* are expressed much later (Fig. 10.13). Studies on the effects of loss of these genes in transgenic mice, together with tissue culture experiments, have led to a model for how interactions between the myogenic proteins mediate myogenesis in the mouse. In this model, Myf5 cooperates with Pax3 to activate expression of *MyoD* in the myotome; MyoD then initiates a series of stimulatory feedback loops that maintain the expression of all three genes. This model can account for the observation that while individual mutants for *Myf5*, *MyoD* or *Pax3* show few problems in muscle development, double mutants *Myf5 MyoD* or *Myf5 Pax3* lack muscles and muscle precursors

A   B

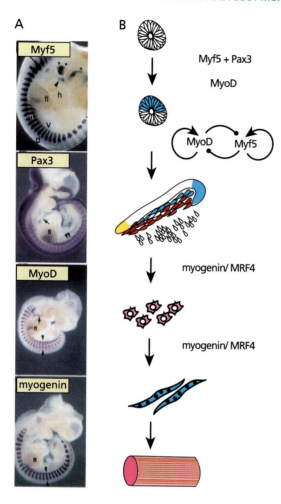

Myf5 + Pax3

MyoD

MyoD   Myf5

myogenin/ MRF4

myogenin/ MRF4

Fig. 10.13. Expression and activity of myogenic regulatory proteins in vertebrates. (A) Patterns of expression of genes encoding the different myogenic regulatory factors during mouse development. (From Tajbakhsh, S. *et al*. (1997) Redefining the genetic hierarchy controlling skeletal myogenesis: Pax3 and Myf5 act upstream of MyoD. *Cell* **89**, 127–38.) **(B)** Sequence of expression of different myogenic regulatory factors and associated events during muscle development.

also heterodimerize with myogenic bHLH factors. Id proteins are bHLH proteins that have an intact HLH interaction domain but lack crucial residues in the basic domain so that they do not bind DNA. These proteins can therefore form inactive heterodimers with myogenic factors, thereby titrating their function. The combinatorial interactions of myogenic factors with each other and with other bHLH partners provide a large number of different regulatory possibilities, some of which activate muscle gene expression while others repress it, contributing to a finely balanced regulatory network.

A key group of proteins that collaborate with the myogenic bHLH proteins in the development of skeletal muscle are members of the MEF2 family of MADS domain transcription factors (Fig. 10.15). These proteins are expressed in several mesodermal and non-mesodermal cell types and, although they cannot promote myogenic development on their own, they do so in cooperation with myogenic bHLH proteins. During the specification of skeletal muscle, MEF2 family members interact with heterodimers between E (E12, E47) proteins and bHLH myogenic factors, but do not interact with E protein homodimers. *In vivo*, myogenic bHLH factors initiate the expression of *MEF2*, and thus create a positive regulatory control that helps to reinforce and stabilize the myogenic potential of the cells. This has two important effects: first it contributes to the onset of expression of muscle specific genes, whose promoters and enhancers recognize combinations of MEF2 family members and bHLH proteins and, in addition, it creates a feedback loop to maintain the expression of myogenic genes (Fig. 10.15).

In *Drosophila*, too, there are bHLH and MADS box proteins associated with the development of the muscles (Fig. 10.16). Genetic experiments suggest that the functional equivalent of the myogenic factors is the bHLH protein twist and that the homologue of MyoD, nautilus, is involved in some aspects of muscle differentiation. Shortly after gastrulation, *twist* expression becomes modulated along the anteroposterior axis in such a way that the

(Fig. 10.14). The cooperation between Myf5 and Pax3 is underscored by the observation that in the *Myf5 Pax3* double mutant, *MyoD* is not expressed. These experiments also revealed interactions between the different cell populations of the somite: *Myf5* mutants lack ribs even though *Myf5* is not expressed in the sclerotome. This suggests that *Myf5* expression in the muscle lineages must activate a series of cell interactions between the two cell populations.

Experiments *in vitro* show that the myogenic bHLH proteins, as is characteristic of members of this family (see Chapter 3), act by forming heterodimers with ubiquitous members of the E class of bHLH transcription factors, particularly E12 and E47. The heterodimers bind a consensus DNA sequence, GANNCT, which is present in many muscle specific genes. In addition to the E proteins, bHLH members of the Id (for 'inhibitors of differentiation') class

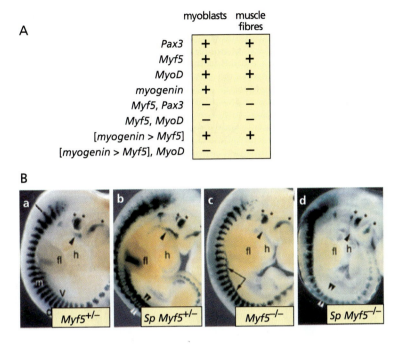

	myoblasts	muscle fibres
Pax3	+	+
Myf5	+	+
MyoD	+	+
myogenin	+	−
Myf5, Pax3	−	−
Myf5, MyoD	−	−
[myogenin > Myf5]	+	+
[myogenin > Myf5], MyoD	−	−

Fig. 10.14. Effects of loss of function of different myogenic regulatory factors. (A) Phenotypes of single and double mutants for different myogenic factors. The mutant listed as [*myogenin > Myf5*] has the coding region of the *Myf5* gene replaced by *myogenin*, while the last mutant in the table is, in addition, mutant for *MyoD*. **(B)** Muscle development in *Myf5* and *Pax3* (*Splotch* (*Sp*)) mutants, highlighted by an insertion of β-galactosidase in one copy of the *Myf5* gene, thereby eliminating its activity. (a) *Myf5* heterozygous embryos show no defects in the development of muscle fibres. *Splotch* (b) and *Myf5* (c) mutants show some deviations from normality but these become much more pronounced in *Sp Myf5* double mutants, in which muscle is scarce and abnormally organized. (From Tajbakhsh, S. *et al.* (1997) Redefining the genetic hierarchy controlling skeletal myogenesis: Pax3 and Myf5 act upstream of MyoD. *Cell* **89**, 127–38.)

posterior region of every segment has high levels of twist, while anterior regions show low expression (Fig. 10.17). Loss or reduction of *twist* expression results in aberrant muscle development, whereas overexpression leads to an overproduction of muscle at the expense of other mesodermal derivatives. As development proceeds, the expression of *twist* decays.

The *Drosophila* Wnt, wingless, has been shown to be required for the maintenance of high levels of twist and for the development of a great deal of the somatic musculature. It acts in two ways: directly, through its effector, the complex of β-catenin and the Tcf transcription factor (see Chapter 5, Fig. 5.25), and indirectly, by activating the expression of the forkhead domain transcription factor encoded by the *sloppy paired* (*slp*) locus. *Drosophila* also has MEF2 homologues which, as might be predicted from their function in vertebrates, are under the control of twist and provide a link between the specification of myoblasts and their differentiation. In contrast to *twist*, the expression of *DMEF2* does not decay during development (Fig. 10.16). The homologue of *MyoD*, *nautilus*, is expressed in a small subset of muscle precursors where it is required for some aspects of their differentiation. *Drosophila* also has E-like proteins (for example, the homologue of E12 encoded by the *daughterless* gene) and Id family members (for example, encoded by the gene *extra macrochaetae*). As

might be expected, mutations in the genes encoding these factors affect muscle development to varying degrees, probably because of their interactions with twist, but their exact roles remain to be explored.

Together, the information from vertebrates and *Drosophila* suggests that myogenic bHLH proteins and members of the MEF2 family of MADS proteins are key elements of a transcriptional regulatory network which is involved in defining the myogenic cell state and activating

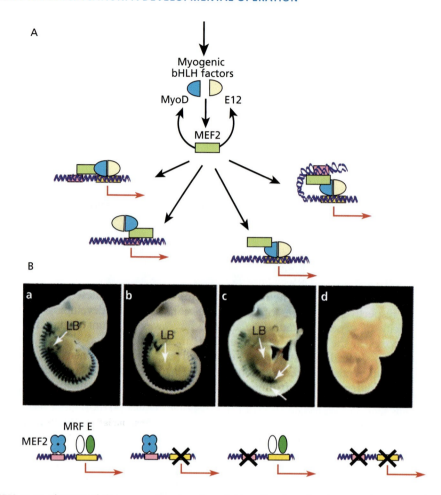

Fig. 10.15. MEF2 as an element of the myogenic network.
(A) The expression of *MEF2* is induced in the myogenic lineages by the myogenic regulatory factors. MEF2 can then associate with several of these regulatory factors in a promoter-specific manner and thus contribute to the regulation of muscle-specific gene expression. The different possible interactions between MEF2, MyoD and E12 are shown. MEF2 can contribute to the combinatorial interactions that select genes for expression in the myogenic lineages either by directly binding to their promoters or indirectly, by mediating protein–protein interactions. **(B)** Example of the combinatorial control of the *myogenin* promoter during mouse development. *Top*: Expression of *myogenin-lacZ* in 11.5-day-old mouse embryos.

Bottom: Structure of each *myogenin* promoter driving the expression of *lacZ* in the panel above. (a) In the wild type, the promoter has a binding site (pink) for MEF2 and an E box (yellow) which is bound by MRF and an E protein. *Myogenin* is expressed in the somites and the limb buds (LB). (b) A mutation in the E box does not alter the expression of *myogenin* in the somites. (c) Mutation of the MEF2-binding site, however, significantly reduces the expression of *myogenin*. (d) Mutation of both the E box and the MEF2 binding site eliminates the expression of the reporter. (From Firulli, A. and Olson, E. (1997) Modular regulation of muscle gene transcription: a mechanism for muscle cell diversity. *Trends Genet.* **13**, 364–9.)

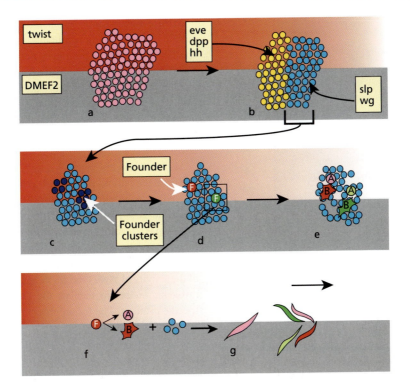

downstream genes that mediate muscle differentiation. The target genes of this network have complex regulatory regions that appear to consist of multiple modules containing binding sites for ubiquitous, myogenic, and gene-specific factors; the integrated output from these regulatory modules achieves fine control of muscle-specific gene expression in different muscle subtypes.

Within this broad framework, however, many variations are possible, with the result that it is not possible to extrapolate from the function of a particular protein in one animal to the function of its homologue in a different animal. For example, the vertebrate homologue of twist acts as a repressor of skeletal muscle development, whereas in *Drosophila* it promotes the development of a similar type of muscle. MyoD plays a central role in myogenesis in mice, whereas it is only essential in a small subset of muscles in *Drosophila*. And in birds, the temporal patterns of expression of *Myf5* and *MyoD* are different from those in mice (for example, in the chick, *MyoD* is expressed before *Myf5*). What this means is that the network is more important than any of its individual members, which appear to have a latent versatility that can emerge if conditions alter. This concept might explain why, for example, myogenin can substitute for most of Myf5 function in *Myf5* mutants

Fig. 10.16. The molecular biology of myogenesis in *Drosophila*. (A) The development of muscle in *Drosophila* takes place against a background of mesodermal cells which express the bHLH protein twist (red) and the MADS domain protein, DMEF2 (grey). Time runs from left to right in each consecutive panel. The expression of twist decays in the mesodermal cells as development proceeds whereas that of DMEF2, although modulated, remains in all mesodermal cells. (a) After gastrulation, twist and DMEF2 are expressed in all mesodermal cells. (b) Shortly afterwards, an interplay between transcription factors and external signals defines two cell populations within every segmental unit: one of low twist expression (yellow) defines a competence to become visceral mesoderm, and one of high twist (blue) a competence to become somatic muscle. The low twist domain is defined by the homeodomain transcription factor even skipped and is reinforced by dpp and hedgehog (hh). The high twist domain is defined by an interaction between the transcription factor sloppy paired (slp) and signalling by wingless. (c, d) Within the group of cells that gives rise to the somatic muscle, clusters of cells emerge from which founders (F) for specific muscles will be defined (see Fig. 10.18 for details of this process). (e, f) The founders divide once to give rise to precursors (A and B) which will seed the muscles and (g) fuse with other myoblasts. By this stage, the expression of twist is very low but DMEF2 is still expressed and is needed for the proper differentiation of muscles.

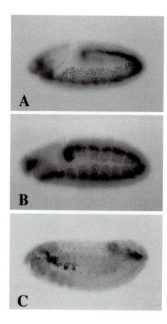

Fig. 10.17. Expression of twist during *Drosophila* embryogenesis. (A) Shortly after gastrulation, the twist protein is expressed uniformly by all mesodermal cells. **(B)** As cells proliferate, the expression of twist is modulated, with high levels in the posterior region of every segment. **(C)** After proliferation, and coinciding with the differentiation of the muscles of the larva, twist expression decays in most mesodermal cells, except in some which become the precursors for the muscles of the adult fly. (Images courtesy of M. Baylies.)

even though it plays a very different role during normal muscle development.

Muscle differentiation

Muscle cell differentiation does not begin until myoblasts have completed a proliferative phase and exited from the cell cycle. During the proliferative phase, the network of myogenic factors maintains stable expression of the myogenic genes and gradually builds up large promoter complexes at the regulatory regions of genes encoding muscle structural proteins. However, the fact that these proteins do not appear in the cells until myoblasts have left the cell cycle suggests there must be a mechanism that inhibits their expression during the proliferative phase.

In mice mutant for *myogenin*, muscle cells do not differentiate normally. In these mutants, muscle precursors are specified properly and proliferate normally, but are unable to initiate the synthesis of muscle structural proteins, such as muscle myosin, or to achieve the appropriate innervation. Observations such as this suggest that there is a boundary that separates specification and proliferation from differentiation, and that myogenin is essential for crossing this boundary.

As mentioned in Chapter 7 (Fig. 7.20), the link between exit from the cell cycle and muscle cell differentiation is via the cell cycle inhibitor p21, whose expression in the myogenic lineage appears to be under the control of myogenin. This step is irreversible and involves parallel activation of a number of cell cycle regulators, in particular Rb. p21 inhibits the activity of cyclin D1, which is in turn an inhibitor of the activity of MyoD and Myf5, and thereby allows these myogenic factors to regulate the expression of muscle-specific proteins. The fusion program must also be activated simultaneously.

In *Drosophila*, the bHLH myogenic factors have not been shown to act like vertebrate myogenin in promoting cell-cycle exit, but it is known that a decrease in the expression of *twist* is correlated with cell-cycle exit and muscle differentiation. Differentiation appears to depend, as in vertebrates, on a *MEF2* family member, *DMEF2*. Flies mutant for *DMEF2* show no deviations from normal development until the beginning of differentiation, when muscle cells fail to fuse and show other defects in the expression of differentiation proteins, supporting the suggestion that DMEF2 is an essential element in the differentiation of muscle.

Different muscles have characteristic sizes, sites of attachment and innervation, which determine their function. Very little is known about how these properties are specified in vertebrates. In *Drosophila*, however, it is clear that they are encoded in the muscle founders: cells that segregate from the mesoderm and act as seeds for individual muscles (Fig. 10.16 and 10.18). The segregation of founders involves the iterative integration of a number of signalling pathways and signalling events. Different muscles arise from the expression of different combinations of transcription factors, which endow these precursor cells and the muscles they give rise to with different properties (Fig. 10.18). The transcription factors are expressed in response to specific combinations of intrinsic and extrinsic influences (Fig. 10.19), similar to those discussed in Chapter 8.

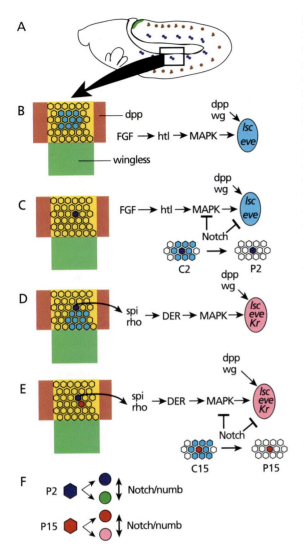

Fig. 10.18. Specification of muscle founder cells in the *Drosophila* embryo. (A) Half way through embryogenesis, the *even skipped* gene is expressed in ventral clusters of neurons in the nervous system (brown) and dorsal mesodermal cells (blue). These cells represent precursors of specific muscles and heart elements which arise as a result of mesodermal and ectodermal influences. **(B)** At the same time, the BMP family member dpp (red) and wingless (green) are expressed in the overlying ectoderm in groups of cells perpendicular to each other and their intersection defines groups of cells in the ectoderm and the mesoderm. In the mesoderm (yellow), a group of these cells (blue) receives an additional influence through FGF signalling via its receptor heartless (htl), and expresses the genes *lethal of scute* (*lsc*), which encodes a bHLH protein, and *even skipped* (*eve*), encoding a homeodomain protein. Lethal of scute defines a general myogenic fate and even skipped confers a specific identity on the cells that express *lsc*. There are other clusters in the *Drosophila* embryo that express *lsc* and which give rise to different muscles; this particular one is called C2. **(C)** Lateral inhibition mediated by Delta/Notch signalling, together with an effect of Notch on MAPK signalling, restrict the expression of *lsc* and *eve* and the activation of MAPK to one cell, which becomes the P2 precursor (dark blue). **(D)** The P2 cell begins to express spitz (spi), which is a ligand for the EGF receptor, and the positive modifier of spitz, rhomboid (rho), which contribute to the activation of *lsc*, *eve*, and *Krüppel* (*Kr*) in a second adjacent cluster of cells, C15. **(E)** Events similar to those that lead to the emergence of P2, lead to the emergence of a P15 from C15. The expression of *Kr* in P15 creates a new combinatorial code that distinguishes it uniquely from P2. **(F)** After the two precursors have arisen, they divide once to give rise to two founders, which are separated by the activity of the Notch regulator numb: P2, which gives rise to two *eve*-expressing pericardial cells and one muscle founder cell, and P15, which gives rise to one muscle founder and another cell of unknown fate. (After Frasch, M. (1999) Controls in patterning and diversification of somatic muscles during *Drosophila* embryogenesis. *Curr. Opin. Genet. Dev.* **9**, 522–9; Carmena, A. *et al.* (1998) Combinatorial signalling codes for the progressive determination of cell fates in the *Drosophila* embryonic mesoderm. *Genes Dev.* **12**, 3910–22.)

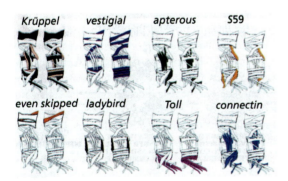

Fig. 10.19. Different muscles in *Drosophila* are identified by individual combinations of transcription factors. In the abdomen of the larva each half-segment has 30 muscles and each muscle is characterized by the expression of a unique combination of genes, many of them encoding transcription factors. There are both internal and external sets of muscles. Each panel shows a view of the external (*left*) and internal (*right*) set, with muscles expressing a particular gene highlighted in colour. Krüppel, vestigial, apterous, S59, even skipped, and ladybird are transcription factors. Toll and connectin are cell surface molecules. (From Baylies, M., Bate, M. and Ruiz Gomez, M. (1998) Myogenesis: a view from *Drosophila*. *Cell* **93**, 921–7.)

Neurogenesis: The central nervous system

Neurogenesis is the process through which nerve cells are generated and arranged in space to configure the nervous system of an animal. During neurogenesis, just as during the development and patterning of the mesoderm, a basic generic cell fate (neural in this case) is amplified and diversified into a number of different cell types, in particular neurons and glia. A unique property of these specialized cells is their organization in linked networks whose properties and patterns of connectivity govern physiological and behavioural aspects of the organism (Fig. 10.20 and 10.21).

The nervous system consists of the peripheral (PNS) and central (CNS) nervous system, which interact to produce and process information. The PNS functions mainly to receive sensory information from the outside and channel it to the CNS, which contains the elements that process that information and generate outputs (e.g. movement in response to visual or auditory stimuli). In some situations, the CNS can also generate information in the absence of sensory input from the periphery. Some aspects of the development of the PNS in *Drosophila* have been discussed in Chapter 8 in the context of the generation of specific lineages (Fig. 8.18). Here, we shall discuss how the neurons of the CNS are generated in vertebrates and insects.

The origins of the central nervous system

In all organisms, the central nervous system is derived from the ectoderm, a primary cell type that also gives rise to the epidermis (see Chapter 1). The cellular mechanisms through which the central nervous system arises from the ectoderm vary slightly from one organism to another but follow the same general outline: neural precursors segregate from ectodermal cells and those ectodermal cells that are not selected to be neural develop as epidermis. As in the case of muscle development, the differences in neurogenesis between different organisms lie mainly in the strategies within which shared molecular mechanisms are deployed; the elements that mediate those mechanisms are by and large conserved. The main difference between vertebrates and insects is that while in vertebrates a group of cells that will give rise only to neural tissues is segregated very early from the ectoderm (Fig. 10.20), in insects neural precursors segregate from intermingled precursors for both neural and epidermal tissues (Fig. 10.21).

Many of the classical studies on the early development of the nervous system in vertebrates have been carried out in amphibian and chicken embryos, largely because they are favourable experimental systems for cell and molecular biology. However, information emerging from complementary genetic studies on zebrafish and mice suggests that the essential features of the process are common to all members of the vertebrate group.

In vertebrates, the primordium for the nervous system arises within the ectoderm when this tissue associates with the mesoderm. During gastrulation, as mesodermal cells invaginate they sweep underneath the ectodermal region. Transplantation of the mesodermal organizer results in induction of a secondary axis and an associated nervous system (Fig. 1.12), suggesting that the mesoderm has an instructive role in the formation of the neural fate. As a result of their contact with the mesoderm, ectodermal cells form a conspicuous plate, the neural plate, which buckles and invaginates to form a tube: the neural tube (Fig. 10.3 and 10.20). Within the neural tube, neurons differentiate from the lateral regions while on the ventral and dorsal sides of the tube specialized strutures develop: the floor plate on the ventral side and the roof plate on the dorsal side (Fig. 10.20). Ectodermal cells that are not contacted by the mesoderm during gastrulation become flat and adopt an epidermal fate, while those at the boundary between the neural and epidermal cells are the source of the neural crest, a population of stem cells with a number of possible fates which include the precursors of the PNS (Fig. 10.20).

In *Drosophila* and grasshoppers, a neurogenic region that can give rise to both neural and epidermal cells is defined within the ectoderm by a combination of the nuclear concentration of the dorsal protein and the activity of the BMP family member dpp (see Chapter 9, Fig. 9.29) (Fig. 10.21). From this population of cells, neural precursors, called neuroblasts, arise (Fig. 10.21). The neural potential of the ectodermal sheet is demonstrated by the observation that ablation of a neural precursor as it emerges triggers the development of one of the surrounding cells, which normally would not develop as a neuroblast, to adopt the neural fate. This observation suggests that within the ectodermal sheet, cells share an equivalent neural potential but that only some of them develop this potential. It also indicates that when an ectodermal cell develops as a neuroblast, it inhibits the surrounding cells from adopting the same fate. This process is known as lateral inhibition (see also Chapter 8, Fig. 8.22). Once segregated from the ectodermal sheet, the neuroblast divides a number of times to

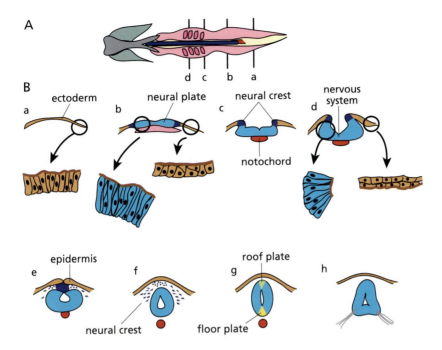

Fig. 10.20. Neurogenesis in vertebrates. (A) Chicken embryo at an advanced stage of gastrulation (similar to that shown in Fig. 10.3). Anterior is to the left and posterior to the right. Bars indicate the positions of the transverse sections shown below. **(B)** Sections through different stages of development of the nervous system indicating the fates and, in some cases (a–d), the activities of different populations of cells. (a, b) The cells that will give rise to the nervous system appear as a thickening of the ectoderm shortly after gastrulation, as the mesoderm (pink) contacts the ectoderm. The interaction with the mesoderm defines three populations of cells: those directly over the mesoderm will form the neural plate (light blue) and give rise to the nervous system; those that are not in contact with the mesoderm will give rise to the epidermis (brown); and those that lie at the border of these cell populations (dark blue) act as the source of the neural crest, which will give rise to migratory cells that include the elements of the PNS. (c, d) The formation of the neural plate induces cell shape changes that lead to the invagination of the prospective neural cells. During this process, some mesodermal cells remain associated with the neural cells, and give rise to the notochord (red). (e) As it invaginates, the neural plate closes on itself and generates a tube, the neural tube. The apical side of the cells forms the lumen of the tube. (f) After the formation of the tube, the neural crest precursors become motile and migrate towards their target sites. (g) Within the neural tube, two populations of cells are defined, the roof plate and the floor plate, that will not participate in the formation of neurons; cells in the other parts of the tube proliferate and begin to generate neurons. (h) As the cells in the tube divide, it changes its morphology to accommodate the increase in cell numbers. Axonal projections extend from the cells.

generate a sequence of ganglion mother cells, each of which divides once to generate a pair of neurons (Fig. 10.21 and Chapter 8, Fig. 8.6). In these organisms, the peripheral nervous system is derived from epidermal cells at a later stage in development, but the basic mechanism, including the involvement of lateral inhibition, is the same as in the CNS.

The emergence of neural tissue

The correlation between the movements of the mesoderm and the appearance of the neural plate in vertebrate embryos suggests that in these organisms mesodermal cells influence the development of the neural plate. A first hint at what the effect of the mesoderm on the ectoderm might be came from some observations in *Xenopus* embryos. If isolated caps of animal cells are left undisturbed they will develop as epidermis. However, if they are dissociated into individual cells they develop neural characteristics (Fig. 10.22). One explanation for this observation is that in the ectoderm, neural fates are suppressed in favour of epidermal ones by cell interactions. Dissociation counteracts this suppression either because it breaks up cell–cell contacts essential for the inhibition or because it dilutes the concentration of an inhibitor of neural fate. If the latter is the case, induction of neural tissue would

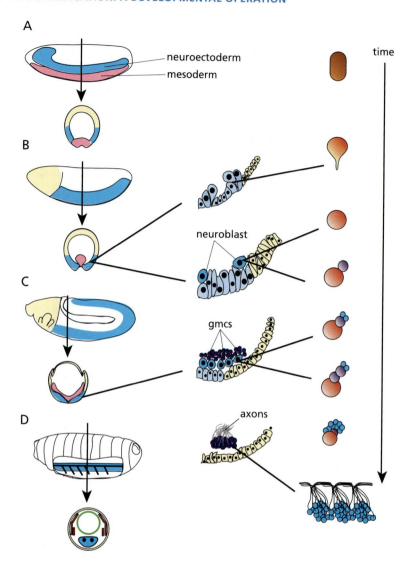

Fig. 10.21. Neurogenesis in *Drosophila*. Embryos are shown with anterior to the left and posterior to the right. Arrows through each embryo indicate the positions of the transverse sections that are shown below. The central column of diagrams shows some of the cellular events associated with neurogenesis at each stage, while the column on the right shows diagrams of the progression from ectodermal cells (top) to neurons with axonal projections (bottom). **(A)** The nervous system arises at the blastoderm stage from a population of ectodermal cells (blue), adjacent and dorsal to the mesoderm (pink) (see also Chapter 9, Fig. 9.21). **(B)** Shortly after gastrulation, the first neural precursors, neuroblasts, segregate from amidst the ectodermal cells. **(C)** Once segregated from the ectoderm, neuroblasts begin to divide to give rise to strings of ganglion mother cells (gmcs), each of which will divide once to give rise to neurons. **(D)** The nervous system separates from the epidermis, which is composed of the cells that have not become neural, and begins to differentiate. (Middle column, after Hartenstein, V. and Campos Ortega, J. (1985) *The embryonic development of* Drosophila melanogaster. Springer, Berlin.)

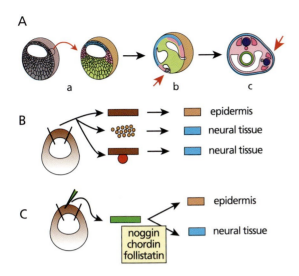

Fig. 10.22. **Experimental definition of neural induction. (A)** In amphibia, transplantation of the organizer region from one embryo to another (a) results in the induction of a secondary axis (b) and nervous system (c). If the host and donor tissues are labelled differently it can be shown that the induced tissue develops from the host and that, therefore, it is induced by the graft. **(B)** The fate of cells from the animal cap of an amphibian embryo is influenced by whether they are aggregated or dissociated. If left undisturbed they develop as epidermis, but dissociation triggers neural development. Cell fate can be monitored by the expression of genes specific for either neural or epidermal tissue. The effect of dissociation can be mimicked by placing an intact animal cap (brown) in contact with mesoderm (red). **(C)** The last experiment in B suggests that the mesoderm contains a neural-inducing activity and also suggests an assay for substances with this activity. In these experiments, proteins or cDNAs are injected into animal caps to test their ability to induce the development of neural tissue. Most have no effect on the development of the animal caps, but a few induce them to develop as neural. Analysis of the inducing factors shows that they belong to a family of BMP antagonists: noggin, chordin, and follistatin (also see Fig. 4.27).

result from the inhibition of this repression by signalling from the mesoderm. The observation that the neural plate develops from the ectoderm as the mesoderm sweeps underneath it suggests that the mesoderm could be a source of neural inducers. This explanation is supported by the observation that placing mesoderm in contact with animal caps will induce the ectoderm to develop into neural tissue. Altogether, these experiments suggest an assay for substances or genes with neurogenic potential: such substances should be able to induce neural development if applied to intact animal caps in the absence of mesoderm (Fig. 10.22).

Three proteins, noggin, chordin, and follistatin, are capable of mimicking the effects of the mesoderm or of dissociating animal caps. Because these proteins are expressed in the organizer region but not in the ectoderm early in development, they are good candidates for inducing agents (see Chapter 9 for a discussion of induction). All three proteins are antagonists of BMP signalling (also see Chapter 4, Fig. 4.27 and Chapter 9, Fig. 9.29). Further experiments have shown that in amphibia members of the BMP family (particularly BMP2 and BMP7) are expressed throughout the ectoderm before gastrulation and, when overexpressed, can repress neuralization and maintain the epidermal fate in dissociated animal cap cells.

Thus, rather than acting as a source of neural inducers, the mesoderm is a source of antagonists of BMPs, which in turn act as antagonists of neural development. As the mesoderm invaginates and comes into contact with the ectoderm, it secretes suppressors of BMP signalling that locally release the potential neural fate of the ectodermal cells (Fig. 10.23). In a further refinement of this regulatory mechanism, the activity of one of these inhibitors, chordin, is itself antagonized by the diffusible metalloprotease Xolloid. It has been proposed that Xolloid acts as a molecular 'sink' that maintains and sharpens the gradient of chordin diffusing from the mesoderm (see Chapter 9, Fig. 9.29) and that the gradient of Xolloid might play a role in the definition of different fates within the developing neural plate, with high levels directing epidermis, low levels directing neural cells, and intermediate levels neural crest cells.

Genetic analysis in zebrafish has provided some support for this view of neural induction. Mutants for a member of the BMP family, *BMP2b/swirl*, have an expanded neural plate whereas in mutants for the gene encoding chordin (*chordino*), the neural plate is reduced. These phenotypes are consistent with the suggestion that, in vertebrates, neural tissue develops by suppression of the epidermal fate that is promoted by members of the BMP family of signalling molecules.

It is now clear that despite some variations associated with their different strategies for early development, the induction of neural tissue and formation of the neural plate in amphibia, fish, and mammals involves the downregulation of BMP signalling. An example of a strategic difference in how this mechanism is deployed is provided by chicken embryos, in which BMP antagonists do not induce neural tissue, and the region of the embryo that will give rise to this tissue is devoid of BMPs by the time gastrulation begins. This suggests that, in these embryos, there is an event, before the classical neural induction by the mesoderm, that pre-patterns the neural tissue and makes it

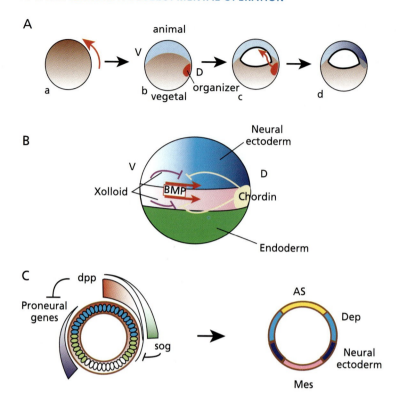

competent for the mesodermal signals. This might also be the case, though to a lesser degree, in amphibian embryos, where it has been noticed that the ability of animal cells to become neural increases with time and that by the time gastrulation is underway, the cells from the animal cap have a very strong potential to develop as neural (Fig. 10.23). The event that mediates the establishment of this competence is likely to originate at the organizer and act in the plane of the epithelium. Thus, the commitment of cells to the neural fate might be a two step event: the first step, which depends on molecular activities before gastrulation that are not well understood, imposes a neural potential on cells from the animal cap; the second event is dependent on the mesoderm and gastrulation. The balance between the two might be different in different animals.

In *Drosophila*, the neural region is outlined where the chordin homologue short gastrulation (sog) antagonizes signalling by the TGFβ/BMP family member dpp (Chapter 9, Fig. 9.29). However, in this case, the antagonism serves only to maintain and refine the cell fates that have been established through the activities of the transcription factor dorsal (see Chapter 9, Fig. 9.21). In particular, the interaction between sog and dpp determines the extent of the neurogenic region within the ventral ectoderm. Sog is

Fig. 10.23. Mechanism of neural induction. (A) Evolution of neural induction during *Xenopus* development. Each figure shows a *Xenopus* embryo during the early stages of development. (a, b) An influence of unknown identity within the early embryo (indicated by the red arrow) predisposes the animal cap for neural development (blue crescent). (c) As gastrulation proceeds, the mesoderm, moving beneath the ectoderm (red arrow) begins to reveal the neural potential in the ectoderm and (d) induces the neural precursors (dark blue). **(B)** Activities of BMP antagonists in the establishment of the nervous system in vertebrates. The maintenance of BMP expression requires BMP signalling, setting up a positive feedback loop that is broken where BMP signalling (red) is suppressed by BMP antagonists such as chordin (yellow), produced by the mesoderm. This action of the BMP antagonists makes an important contribution to establishing the low levels of BMP on the dorsal side of the embryo that are essential to permit neural development. On the ventral side of the embryo, the inhibitory action of Xolloid (purple) on chordin helps to maintain high levels of BMPs **(C)** In *Drosophila*, the activity of dorsal subdivides the embryo into different domains along the dorsoventral axis (see Chapter 9, Fig. 9.21). This process leads to the expression of the *chordin* homologue *short gastrulation* (*sog*) over a ventral sector of the ectoderm, and of *dpp* over a dorsal sector. In the ventral ectoderm, the activation of proneural genes by the gradient of dorsal, and the maintenance and refinement of these patterns of expression by dpp and sog, combine to promote the development of the nervous system.

expressed in the neurogenic region, where it can effectively inhibit the activity of any dpp that diffuses from the more dorsal regions (Fig. 10.23). Since dpp can suppress neurogenesis, the activity of sog allows the development of the central nervous system.

Overall, there is a broad similarity between vertebrates and insects in that in both there are cells with a more or less overt competence to develop as neural, and the role of the BMP antagonists is to reinforce this competence until the cells establish their neural fate.

Axial organization of the nervous system

Transplantation experiments in *Xenopus* suggest that, as the mesoderm sweeps past the ectoderm, different regions

of the organizer can trigger neural development characteristic of different anteroposterior levels of the animal. Implanted organizer tissue from an embryo at an early stage of gastrulation can induce anterior neural tissue, but organizer tissue invaginating at a later time during gastrulation induces only posterior neural tissue. These identities can be revealed both by the axial structures developed in the experimental embryos and by markers characteristic of different axial regions. These experiments indicate that the mesoderm not only acts to implement the neural potential of the ectoderm but that it also patterns it along the anteroposterior axis (Fig. 10.24). The precise mechanism for this event is not clear. Neural induction alone is not sufficient to trigger the regionalization since, for example, treatment of animal caps with noggin yields neural elements with a uniform anterior character. Several candidates for the anterior/posterior influence have been identified. Two factors can induce more posterior fates in the nervous system: FGF and, in a concentration-dependent manner, retinoic acid.

The most likely targets of anteroposterior specification in the nervous system are the *Hox* genes, whose patterns of

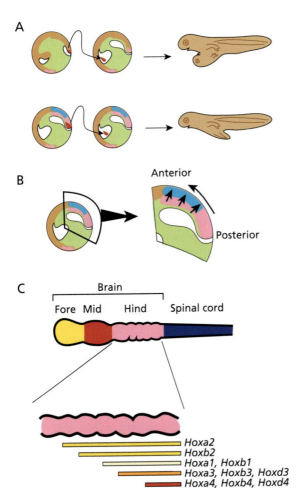

Fig. 10.24. Anteroposterior organization of the nervous system. (A) Transplantation of mesoderm from the organizer region (red) to other locations of the early *Xenopus* embryo induces an ectopic body axis. If the transplantation is carried out at an early stage, the mesoderm will induce a complete body axis (*top*), but if mesoderm from progressively older embryos is transplanted, it induces progressively more posterior structures without additional anterior ones (*bottom*). **(B)** The experiments summarized in A suggest that, in addition to inducing neural tissue within the overlying ectoderm, the mesoderm provides cells with information about their axial position that will be translated into specific patterns of differentiation. This information is indicated diagrammatically by arrows from the mesoderm (pink) to the neural ectoderm (blue). Experiments in which the mesoderm is not allowed to invaginate but is forced to remain in the same plane as the ectoderm suggest that there are also planar signals (arrow in the same plane as the neural ectoderm) involved in the specification of the anteroposterior axis. **(C)** The nervous system shows a fine-grained pattern of cell identities along the anteroposterior axis. This pattern is revealed at the morphological level within the brain region, which is subdivided into fore-, mid- and hindbrain domains. The hindbrain is further subdivided into metameric (segmental) domains, rhombomeres, which are identified by the pattern of nerves that emerge from them. Differences between rhombomeres can be related to the expression and function of genes from the different *Hox* clusters. Some examples are indicated. (From Lumsden, A. and Krumlauf, R. (1996) Patterning the vertebrate neuraxis. *Science* **274**, 11209–1115.)

expression contribute to the definition of cell identities along the AP axis (see also Chapter 2, Fig. 2.8). Consistent with this possibility, as gastrulation proceeds the genes of the different *Hox* clusters are activated in the nervous system in combinatorial patterns characteristic of different anterior/posterior levels. These patterns of expression imprint a positional code on the cells and contribute to the individual identities of neural precursors. In mutants for different *Hox* genes, cells change their anteroposterior specification.

In *Drosophila*, the emerging nervous system also arises with an anteroposterior imprint that is set up, in part, by *Hox* genes. However, in this case the information for the differential activation of *Hox* genes is not derived from the mesoderm but from the cascade of transcription factor activity that subdivides the blastoderm (see Chapter 3, Fig. 3.24 and Chapter 9, Fig. 9.19 and 9.20) and so in this organism there is a very close correlation between the layout of the body plan and the specification of different cell types.

The commitment to the neurogenic pathway

Just as for other cell types, the establishment and maintenance of a 'neural cell fate' must rely on transcription factors whose activity ensures that neural programs of gene expression are activated and stably maintained. However, there are some differences between vertebrates and insects in the way these transcription factors are regulated, as might be expected from their different modes of neurogenesis. In vertebrate embryos, there is a stable non-ectodermal population of cells with a potential neural fate, from which neural precursors are drawn. This suggests that there are factors that can endow cells with a neural fate and keep them in a committed but non-differentiated state. In insects, on the other hand, neural precursors are drawn directly from the ectoderm and factors that maintain a 'neural fate' might not be necessary. We shall look at these two strategies in turn.

In vertebrates, factors with the ability to promote neural fate were first identified in functional assays designed to search for genes that were expressed in the developing nervous system and that could transform ectoderm into neural tissue. The system used for these searches was the *Xenopus* embryo. In these embryos, a first round of segregation of neural precursors occurs at the neural plate stage, before the formation of the neural tube, and results in the establishment of three columns of neurons, called primary neurons. Because this process takes place within a flat sheet of tissue (the neural plate), it allows a simple assay for the number and spatial arrangement of neural precursors (Fig. 10.25). In this assay, genes and proteins are tested for their potential to modulate the neural potential of the

Fig. 10.25. Identification of neurogenic genes in vertebrates. (A) In *Xenopus* embryos, shortly after the formation of the neural plate but before its invagination, three parallel columns of neural precursors can be seen on either side of the dorsal midline. This pattern can be easily observed on the surface of the embryo and therefore provides a convenient landmark for the extent and pattern of the developing nervous system. The pattern can be used to search for genes whose overexpression can increase or decrease the amount of neural tissue. In these experiments, mRNAs or cDNAs for potential regulatory genes can be injected into one of the cells at the two-cell stage of the *Xenopus* embryo. The effect of the factor encoded by the gene can be observed by comparing the neural plate on the injected side of the animal with the normal pattern on the uninjected side. In most cases, there is no effect (b) but sometimes either more (a) or fewer (c) neural cells will be observed on the injected side. An increase will identify molecules that are involved in promoting neural tissue, whereas a decrease identifies repressors of neural fate. **(B)** From the normal expression patterns of the molecules identified by the assay shown in A, and their ability to affect each other's expression, a regulatory hierarchy can be worked out. Many of the genes encode transcription factors that respond to BMP signalling. BMP signalling stimulates expression of genes such as *GATA1* that promote epidermal development, while BMP antagonists stimulate expression of genes, such as the *Sox* and *Zic*-related genes, that promote neural development. In the neural plate, the activity of these genes triggers a regulatory cascade which culminates in the expression, in columns, of genes (highlighted in pink) such as *neurogenin* and *NeuroD*, which establish the neural fate. **(C)** Spatial expression of genes that establish the neural fate in the columns characteristic of the first wave of neural precursors in *Xenopus*. (a) Against a background of expression of the activators of neural differentiation in the neural plate (blue) a *neurogenin* gene from *Xenopus* (*Xngnr*) becomes activated in stripes (pink cells). (b) *Xngnr* expression leads to the activation of *Delta*, which in turn modulates the activity of *Xngnr* and the segregation of neurons. (c) As neurons are born (black cells) in a spaced pattern, they express genes including *NeuroD* and *MyT1*.

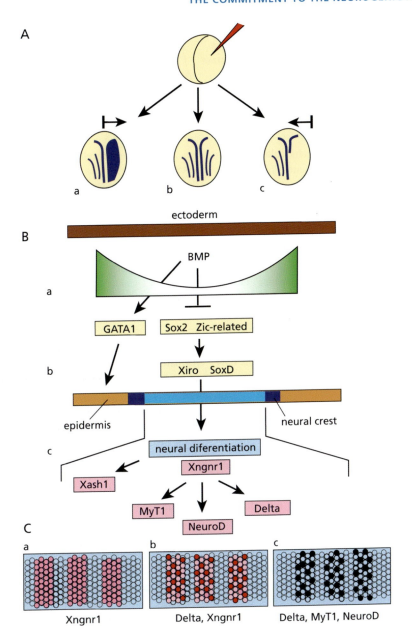

ectodermal cells. This assay is conceptually similar to the one that led to the discovery of the myogenic proteins.

A set of bHLH proteins—neurogenin, NeuroD, and Xash/Mash—can extend the neural potential of the plate in this assay. During normal development their expression is restricted to the columns of neurogenic cells in the developing nervous system. Like other bHLH proteins, these proteins work as dimers with members of the family of E proteins, suggesting that they participate in a neurogenic network comparable to the myogenic network that involves MyoD, myogenin, and Myf5. The cloning of homologues of many of these genes in other vertebrates, and analysis of their function and patterns of expression, support the suggestion that they are general mediators of neural potential during development.

In addition to genes encoding bHLH proteins, the overexpression assay has uncovered other gene families capable of affecting the extent the extent of the neural plate, and

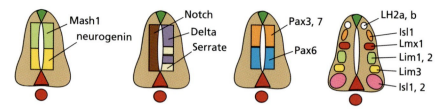

which are also expressed in the neurogenic columns or the neurogenic region early in development. These include genes encoding HMG box proteins of the Sox family (e.g. SoxD and Sox2), and proteins of the Zic-related family of zinc-finger transcription factors (e.g. Zic1, Zicr1, and Zic3). The assay has also identified factors that promote epidermal development by suppressing neural fates (e.g. GATA1) (Fig. 10.25).

The genes encoding the neural-promoting factors, called neurogenic genes, can be arranged into functional hierarchies by observing their patterns of expression, their response to overexpression of other genes of the group, their response to neural-inducing agents, and their effects on the process of neurogenesis. These experiments have led to a picture in which these genes form a regulatory cascade that progressively endows cells with a stable neural fate. The outcome of this regulatory hierarchy is the production of columns of neural precursors within the neural plate.

Not surprisingly, the onset of expression of the neurogenic genes is tightly linked to the process of neural induction. Experiments *in vivo* and in culture indicate that, in all cases, their expression is suppressed by the activity of BMPs and is promoted by BMP antagonists such as noggin, chordin, and follistatin. On the other hand, genes that promote epidermal development (e.g. *GATA1*), show the opposite behaviour, that is, their expression is stimulated by BMPs and is suppressed by BMP antagonists (Fig. 10.25). The Zic-related genes seem to provide a direct regulatory link between neural induction and the expression of genes that establish the neural fate They are directly induced by BMP antagonists, expressed throughout the neural plate and can promote neural development in overexpression experiments in animal caps by activating the expression of *neurogenin* and *NeuroD*. One member of this family, *Zic2*, can repress the appearance of neurons and is expressed in a striped pattern that alternates with the neurogenic columns. This pattern is consistent with its being a repressor of neurogenesis (Fig. 10.25) and suggests that it plays a role in the restriction of the neural potential to columns of cells within the neurogenic region.

Fig. 10.26. Neuronal cell identities in vertebrates. In higher vertebrates, the neural tube is subdivided from dorsal to ventral into domains of gene expression whose overlaps generate codes that can define populations of neurons. Some of these domains are shown here in transverse sections through a neural tube. Each population of neurons defined in this way is a column of cells that runs parallel to the anteroposterior axis; in this respect, these cell populations are similar to the columns of neural precursors in the neural plate of *Xenopus* embryos.

The expression of the *Zic* and *Sox* genes presages and regulates the expression of *neurogenin*, whose function is to establish neural fate in a manner analogous to *Myf5* and *MyoD* during myogenesis. Neurogenin activity then leads to the expression of *NeuroD*, which will commit cells to the neural fate, and of *Delta*, which mediates inhibition of the neural fate in the sister cell of the neuron, allowing it to remain as an uncommitted neural precursor (Fig. 10.25 and see also Chapter 8, Fig. 8.22 and 8.23). Expression of *neurogenin* throughout the whole neural plate leads to a 'salt and pepper' pattern of neurons that correlates with the expression of *Delta*. The position of *Xash/Mash* in the network is unclear at the moment.

Similar neurogenic genes have been identified in other vertebrates such as the mouse but many of them are not expressed until after the closure of the neural tube (Fig. 10.26), so that the stripes or columns of cells seen in the neural plate in *Xenopus* now become stripes of cells at different levels of the dorsoventral axis of the tube. Experiments in the chick involving removal and grafting of different tissues indicate that the floor plate and the roof plate of the neural tube, which do not themselves develop as neural tissue, are sources of signals that pattern the neural tube along the dorsoventral axis (Fig. 10.26).

The driving force behind the columnar arrangement of the regions with neurogenic capacity is not yet clear, but it is likely to be the basic coordinates that lay down the patterning of the early embryo, that is, positional information.

The birth of neurons: Individual cell-type specification

The nervous system is characterized by a vast array of neuronal identities that vary in their precise functions and connectivity, contributing to a system of enormous subtlety and functional complexity. For example, the neurons of the different regions of the brain differ from one another in morphology, biochemistry, and spatial arrangement, and all of these properties are in turn different in the neurons of the spinal cord. Each neuron has a specific pattern of connections and, in most cases, a unique cytoarchitecture. Studies of gene expression in vertebrates and invertebrates indicate that this diversity is associated with the activity of different combinatorial arrangements of transcription factors and it is likely that, as microarray technology is applied to this problem (see Fig. 2.15), the details of these combinatorial interactions will emerge fully. In the meantime, these studies suggest that it is very difficult to separate 'individual identity' from 'tissue identity' during neurogenesis, in other words individual neural elements appear to be born with an identity, rather than being first defined as neural and then acquiring a particular identity as happens during myogenesis.

A great deal of the work on how neuronal cell diversity arises in vertebrate embryos has been done in the chicken and in the mouse. The individual identities appear progressively in a pattern that seems to be determined by a cartesian grid of positional values along the AP and DV axes. As we have suggested already, the influences at work along the AP axis are likely to be combinations of Hox proteins and changing landscapes of retinoic acid and growth factors. The DV axis is characterized by the definition of functional differences between neurons: the most ventral region gives rise to motor neurons while the most dorsal regions generate commissural neurons, and intermediate regions give rise to interneurons (Fig. 10.27). These broad classes are correlated with the expression of certain genes induced by the activity of the roof plate and floor plate. In particular, they appear to correlate with the expression of a subset of LIM homeodomain proteins (Fig. 10.26 and 10.27). For example, the LIM domain protein Islet1 is expressed in all motor neurons early in development and is absolutely required for their development: mutants for *Islet1* have no motor neurons. Other LIM-containing proteins are similarly associated with other types of neurons.

Transplantation and ablation experiments show that the floor and roof plates play a central instructive role in establishing these patterns (Fig. 10.28). Individual cell diversity

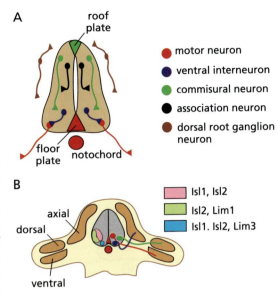

Fig. 10.27. Patterning of the vertebrate neural tube. (A) Different regions of the neural tube differentiate different types of neurons, suggesting that they are responding to spatially arranged cues. Two structures that could influence the patterning of the neural tube are the roof plate and the floor plate. **(B)** Different populations of neurons are defined by particular combinations of transcription factors (see Fig. 10.26). Many of these combinations direct the expression of particular members of the Islet family of LIM transcription factors, which endow neurons with specific properties (for example, the ability to innervate particular regions of the animal).

within a given class of neurons arises in response to interactions between long- and short-range influences. Thus, different motor neurons arise on the basis of their position with respect to the notochord and floor plate, which act as sources of a concentration gradient of Shh (Fig. 10.28). Along the dorsal region of the neural tube the signalling molecules associated with the roof plate are TGFβ/BMP family proteins; in particular, BMP4, BMP7 and dorsalin act in a similar manner to Shh to specify the dorsal sets of neurons.

Once a neuron is born it begins to differentiate by developing a polarized projection, the axon, with a specialized tip, the growth cone, which samples the environment and directs the growth of the axon in search of targets with which it can interact and form connections. These targets can be another neuron or a muscle.

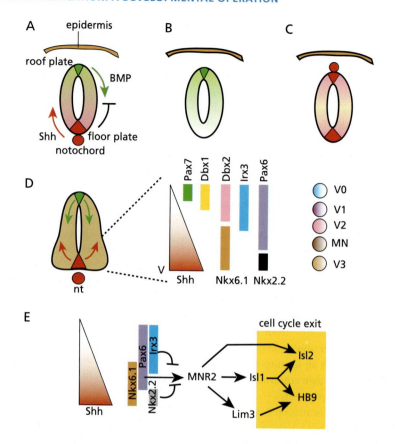

Fig. 10.28. Establishment of neuronal cell identities in verte-brates. (A) In the neural tube, the floor plate and the roof plate are sources of signals that determine the identities of the developing neurons. The floor plate, induced by the underlying notochord, is a source of Shh, whereas the roof plate is a source of BMPs. **(B)** Re-moval of the floor plate and the notochord extends the domain of dorsal fates ventrally, indicating that Shh can suppress the develop-ment of dorsal fates. **(C)** Placing a notochord on the dorsal side of the neural tube induces a new floor plate and ventral neural fates on the dorsal side of the tube. This indicates that, in addition to sup-pressing dorsal fates, Shh promotes ventral fates. **(D)** A concentra-tion gradient of Sonic hedgehog originating from the region of the floor plate and notochord plays an important role in the establish-ment of different motor neuron classes in the ventral region of the

neural tube. In this process, Shh has two types of target: some that are expressed throughout the neurogenic region but become re-pressed by Shh to different degrees along the DV axis (*Pax7*, *Dbx1*, *Dbx2*, *Irx3*, *Pax6*), and others (*Nkx6.1* and *Nkx2.2*, which encode homeodomain transcription factors) that are activated by Shh. These two sets of genes interact to specify five domains along the DV axis (V0, V1, V2, MN, V3), which correlate with the specification of different kinds of neurons. **(E)** Scheme showing how the neurons are specified by a cascade of regulatory interactions. (Adapted from Tanabe, Y. and Jessell, T. (1996) Diversity and pattern in the devel-oping spinal cord. *Science* **274**, 1115–23; and Jessell, T. (2000) Neuronal specification in the spinal cord: inductive signals and transcriptional codes. *Nat. Rev. Genet.* **1**, 20–9.)

Neurogenesis in *Drosophila*

In contrast to vertebrates, in insects the precursors of the central nervous system are intermingled in the ectoderm with the precursors of the epidermis. The generation of neural precursors follows sequential waves of cell fate decisions in which clusters of cells located at specific positions within the ectoderm begin to change shape. One cell from each cluster then emerges as a neural precursor or neuroblast while the surrounding ones revert to the ectodermal fate. After successive waves of this process, the epidermis develops from those ectodermal cells that have not become neural (Fig. 10.21 and see Chapter 8).

In *Drosophila*, the route to the identification of neurogenic genes involved genetic analysis of loss-of-function mutations. This analysis revealed a network of bHLH proteins, centred on those encoded by the *achaete-scute* complex (AS-C), which are essential for the development of the nervous system. The members of this family of transcription factors, whose expression is under the control of dorsal, delimit the neural potential of a region of the ectoderm and provide a molecular context for the action of factors involved in neural differentiation. By 'molecular context' we mean that, as in the case of the myogenic factors, these transcription factors bind to the promoters of a number of genes and, in collaboration with other factors, provide a common denominator for their expression in neural lineages.

The expression of the AS-C genes is patterned along the AP and DV axes indicating that, as in the columns of neural precursors in vertebrate embryos, it responds to positional information. From these domains of expression, individual precursors arise. In embryos mutant for *Notch* or *Delta*, the mediators of lateral inhibition, the ectodermal cells surrounding the precursors also develop neural potential, suggesting that in the wild type the precursors inhibit the cells around them from adopting the same fate (Fig. 10.29 and 10.30).

In addition to bHLH proteins, other proteins also play a role in the specification of neural precursors. In particular, three homeodomain proteins, vnd, ind, and msh, expressed in non-overlapping rows parallel to the anteroposterior axis of the embryo, are also required for adoption of the neural fate (Fig. 10.29). In the absence of these proteins, neurons from those positions either do not develop or have their fates altered, indicating that the proteins play a role in the estabishment of neural precursors. These transcription factors are neural-specific and probably act as mediators of the positional information that contributes to neural fate.

Thus, AP and DV molecular coordinate systems create a combinatorial grid that defines an ectodermal domain with neural potential and, within this domain, selects precursors. The selection is mediated by interactions with the transcriptional cascade that patterns the blastoderm. In *Drosophila*, when a cell is committed to the neural fate, its identity is also defined. This is achieved by combinatorial interactions between transcription factors that endow cells with a neural fate (the bHLH and homeodomain proteins) and other transcription factors and signalling proteins that participate in patterning the AP and DV axes (Fig. 10.29). Neural precursors then divide following precise lineage patterns to generate the neurons that will assemble into the nervous system (Fig. 10.29 and see Chapter 8, Fig. 8.6).

The three early columns of neural precursors parallel to the AP axis during early neurogenesis in *Drosophila* are very similar to the three columns of primary neurons that are visible in the neural plate in zebrafish and *Xenopus* and that, after the transformation of the neural plate into the neural tube, lie at different dorsoventral levels of the tube. In higher vertebrates, there are many more colums of neurons. In *Drosophila*, each of the columns arises within a region defined by the expression of a particular homeobox gene: *vnd* medially, *ind* intermedially, and *msh* laterally. Homologues of these genes, *Nkx2.2*, *Gsh1*, and *Msx1*, respectively, have been found in vertebrates. In the mouse these genes define three domains that might correspond to the three early neurogenic domains of fish and frogs and can be correlated, by position, with the domains of *Drosophila* (Fig. 10.31). This striking conservation perhaps reflects a primitive pattern in the development of the nervous system.

Regulatory hierarchies in the establishment of cell fates

A comparison of cell fate specification in skeletal muscle and in the neurons of the central nervous system hints at some general principles that might govern how specific lineages are established during development and how the cells within them are diversified. For example, a cell type can be established by cross-regulatory interactions between lineage specific transcription factors, and individual cell identifies within this cell type can be generated by different combinations of transcription factors.

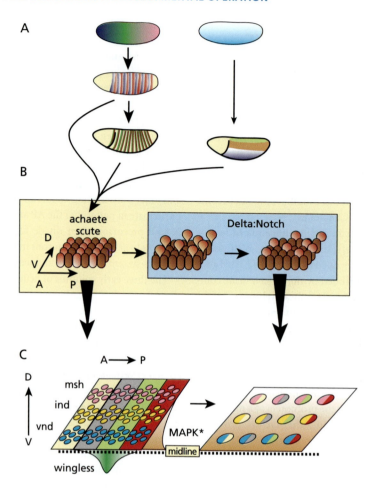

Fig. 10.29. Neurogenesis in *Drosophila*. (A) During the blastoderm stage, cascades of transcriptional regulators subdivide the *Drosophila* embryo into domains of gene expression along the anteroposterior (left) and dorsoventral (right) axes. One of the outcomes of the DV patterning system is the definition of the neurogenic region (dark-blue gradient and see also Chapter 9, Fig. 9.21). **(B)** Within the neurogenic region bHLH proteins encoded by the *achaete-scute* complex are expressed in patterned rows of cells (red) which presage the appearance of neural precursors. The pattern of expression of *achaete* and *scute* is dynamic and becomes refined to single neuroblasts through the process of lateral inhibition mediated by Delta and Notch. This process also gives rise to a set of regularly spaced neural precursors. **(C)** Other factors also contribute to the specification of neural precursors. Along the DV axis, three homeodomain transcription factors (msh, ind, and vnd) are expressed in parallel and adjacent rows of ectodermal cells, which they help to endow with both neural fate and neural identity. In addition, MAPK activity is graded along this axis, with a high point in the most ventral region, and contributes to these identities. The transcription factors and signalling molecules (e.g. wingless) that subdivide the AP axis also make a contribution. The combination of factors from the AP and DV axes creates a unique positional code for each neural precursor, endowing it with a specific identity, as represented by the different colour combinations on the right. (Image C adapted from Skeath, J. (1998) The *Drosophila* EGF receptor controls the formation and specification of neuroblasts along the dorsal-ventral axis of the *Drosophila* embryo. *Development* **125**, 3301–12.)

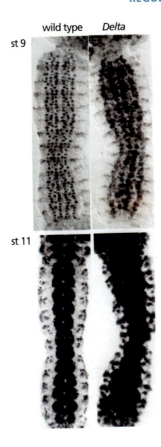

wild type   *Delta*

st 9

st 11

Fig. 10.30. Lateral inhibition control of neuroblast segregation in the *Drosophila* embryo. Pattern of neuroblasts at two stages of embryogenesis in *Drosophila* (stages 9 and 11) as revealed by the expression of a β-galactosidase reporter under the control of the *scratch* gene, in the wild type and a *Delta* mutant. In the mutant, the process of lateral inhibition is defective and the number of neuroblasts is dramatically increased. (From Seugnet, L., Simpson, P., and Haenlin, M. 1997. Transcriptional regulation of *Notch* and *Delta*: requirement for neuroblast segregation in *Drosophila*. *Development* **124**, 2015–25.)

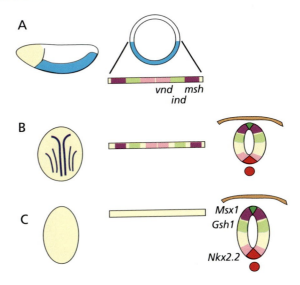

Fig. 10.31. Comparison of early neurogenesis between vertebrates and *Drosophila*. The diagrams on the left represent the neurogenic regions of a *Drosophila* **(A)**, *Xenopus* **(B)**, and mouse or chicken **(C)** embryo. In *Drosophila* anterior is to the left, while in the vertebrate diagrams anterior is at the top. The diagrams in the centre represent transverse sections across the neurogenic region, with the midline in the centre of the bar. The diagrams on the right show, for the vertebrates, cross sections through the neural tube at a later stage of development. In each organism, one of the early stages in neurogenesis is the definition of longitudinal stripes of neural precursors (brown, green, pink). Whereas in *Drosophila* and *Xenopus* these columns appear while the ectoderm lies as a flat epithelium, in mouse some of them appear after the formation of the neural tube. In *Drosophila* each of the columns is associated with the expression of a specific homeobox gene: *vnd*, *ind*, and *msh*. These genes have not been identified in *Xenopus*, but it is likely that they exist. Mouse homologues of these genes (*Nkx2.2*, *Gsh*, and *Msx1*) are expressed in positions that can be related to the expression domains of the *Drosophila* genes. Intervening regions between the stripes indicate that there are additional neurons that are specified by other genes.

The ways in which some other lineages, such as the haematopoietic system, are established and diversified suggest that these principles might indeed be general (Fig. 10.32). In the haematopoietic system, a stem cell produces at least 10 different sublineages and all of the cell types in every lineage can be replenished *de novo* from stem cell precursors (see Chapter 8, Fig. 8.5). However, despite this complexity, the sublineages—for example, lymphoid or myeloid—are established by combinations of bHLH and Pax proteins, and individual cell identities are established by hierarchies of transcription factors. At each step, too, a

binary choice applies that in some cases is regulated by Delta/Notch signalling (Fig. 10.32 and see Chapter 8). In general outline, this process resembles the specification of cell types in muscle and the nervous system.

The central feature of the processes of cell fate specification that we have discussed in this chapter is the selective activation of lineage specific genes and, within this context, establishment of the stable expression of cell-type specific genes. This is achieved in two steps. In the first, a population of cells (myoblasts, neural or haematopoietic precursors) is selected to be the source of cells destined for the

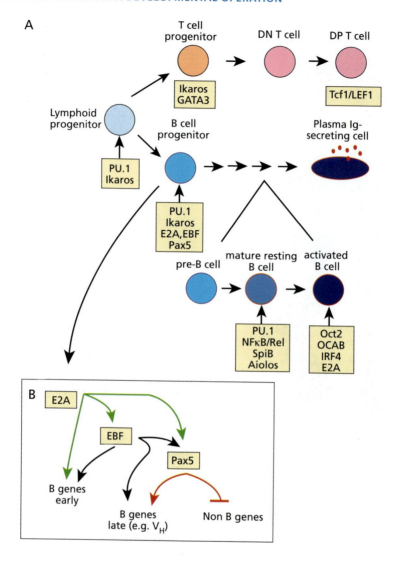

particular fate; that is, a group of cells becomes committed to a lineage characterized by a specific program of transcriptional activity. In the second step, genes are activated within the cells that confer the characteristics of the cell type, through a hierarchy of interactions between signalling pathways and transcription factors. The genes that are activated in the first step provide a molecular context within which the products of the second set of genes can act to direct a specific pathway of differentiation.

Comparison of the different cases we have discussed suggests some conserved molecular principles underlying each of these steps (Fig. 10.33):

Fig. 10.32. The assignment of cell fates in the B and T cell lineages of the haematopoietic system. (A) Progression of cell fates in the T and B cell lineages. Each cell type is associated with a specific combination of transcription factors (shown in yellow boxes) and the transition between states is associated with additions or new combinations of transcription factors. Signalling pathways govern these transitions by acting as regulatory factors for the genes that mediate the next step, and sometimes also contribute to the identity of these cells directly. **(B)** Detail of the regulatory network that modulates the transition between a lymphoid progenitor cell and a pre-B cell. There is a hierarchy of regulatory relationships. Some of its elements are concerned not only with the activation of genes characteristic of B cells but, particularly in the case of Pax5, with the repression of genes that are not expressed in B cells.

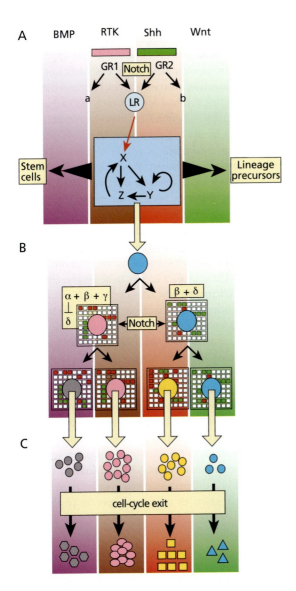

Fig. 10.33. General principles in the process of cell fate assignment. (A) Lineage-specific regulators (LR) become activated in a specific cell or groups of cells where there is an overlap in the expression and/or activity of a set of general regulatory proteins (e.g. GR1 and GR2), each of which might have regulatory targets in other cells but whose overlap defines the cell or cell population in question. LR leads to the activation of lineage- or cell-type-specific genes (X, Y, Z) which establish a regulatory network that will maintain a particular identity. This lineage identity manifests itself in the establishment of, depending on the situation, lineage precursors or stem cells. Little is known about how stem cells are established and maintained. These interactions take place in the context of spatial and temporal landscapes of signalling molecules (always BMPs, RTKs, Shh, and Wnts) which contribute to the specificity of the regulatory interactions. **(B)** After the precursor (or precursors, blue) has been established, it will proliferate to expand the lineage and, in the process, it will often diversify, that is, it will generate different classes of cells within the lineage, such as different kinds of muscles, neurons, or haematopoietic cells. This is achieved, usually, through binary processes in which different cells activate combinations of transcription factors that endow them with specific properties. In the example shown, one cell (pink) activates three transcription factors (α, β, γ) while the other (blue) activates two (β and δ). Both have one factor in common (β) but two different ones, and this generates differences in output. The squares with coloured dots represent microarrays which reveal different patterns of gene expression in different cells, mediated by their individual codes of transcription factors. Often, an important element in the generation of these codes is the suppression, in one particular cell, of genes that would confer a different identity; in the case shown, α suppresses the expression of δ and thus prevents the appearance of the $\beta\delta$ combination that would enable the blue coloured cell to appear. In a further layer of regulation, Notch signalling is used to ensure a distinction between these fates. This process can be iterated a number of times to generate several classes of cells within a given lineage. Each combination of transcription factors is dependent on the combination existing in the parental cell, resulting in the creation of cell-specific codes and the activation of different batteries of genes, as shown in the 'portraits' of microarrays. As in the earlier events, these hierarchical regulatory interactions take place within spatial and temporal contexts of signalling molecules which thus contribute to the cell-fate decisions. **(C)** A given cell type within a lineage proliferates and after a certain time, which is usually encoded in the regulatory relationships of the hierarchy, it may exit the cell cycle. One possible way that this can be achieved is through a transcriptional event that activates the expression of a cell-cycle regulator that withdraws the cell from the cycle and, at the same time, activates batteries of differentiation genes. Once again, these events take place within the context of the same signalling molecules as those that act at earlier stages.

1. Commitment to a particular lineage, understood as a specific program of gene expression, is established by the activation of cell-type-specific programs of gene expression. This is often achieved through the combined activities of transcription factors, none of which is specific for a cell type but whose overlapping expression in a multicellular domain results in the activation of new transcription factors that define a new cell population. The same outcome could result from the combination of a broadly expressed transcription factor and the activity of signalling molecules that overlap at the cell population in question. This rule applies to global histotypes, such as mesoderm or

nervous system, as well as to broad cell types within these, for example, skeletal muscle, cartilage or haematopoietic system within the mesoderm. For example, Pax3 is involved in both neural and muscle development, where it is required for the activation of specific programs of gene expression, in combination with signalling molecules and other transcription factors of varying specificity.

Commitment to a lineage leads to the definition of a population of precursor cells that have both the broad characteristics of the lineage and the ability to be multipotent within this lineage. The mechanisms that establish these populations of stem cells are not known.

2. The maintenance of lineage-specific patterns of gene expression, and thereby of the commitment of cells to a particular fate, is accompanied by the establishment of a self-reinforcing pattern of activity of lineage-specific transcriptional regulators. This is most clearly demonstrated in the case of the myogenic regulatory network but it is likely also to be true of other systems and most obviously in the case of stem cells. It is intriguing that in the three examples we have discussed these transcriptional regulators belong to the same family: the bHLH transcription factors.

3. The progressive definition of individual cell types within a lineage is accompanied by a progressive restriction in cell-fate options which depends on a functional hierarchy governed by the regulatory relationships between different transcription factors. As a result, combinatorial codes of transcription factors are established that confer identities on individual cells. Two classes of transcriptional activities seem to be associated with these steps: some that endow cells with specific characteristics, and some that suppress characteristics of other related cell types. This is clearly demonstrated by the role of Pax5 during the devel-

opment of B cells. In the absence of Pax5 pre-B cells do not develop as B cells, suggesting that it acts positively to promote B cell development. However, in the appropriate growth factor environment, these Pax5-deficient pre-B cells can differentiate into one of many haematopoietic cell types, suggesting that one way in which Pax5 achieves its effect is by suppressing non B cell-type-specific genes.

4. Growth factors and other signalling molecules contribute to and sometimes govern the transitions between different levels of the hierarchies of transcription factors. Often, the same signal operates at different levels, the response being different because the nuclear context in which it operates is different.

5. As a lineage develops and its elements progress towards a specific cell type, transitions in cell state occur at which a cell chooses between alternative programs of gene expression. These choices appear to be binary: either between two alternative new fates or between retaining an existing identity and acquiring a new one. Delta/Notch signalling is essential at these choice points and is used reiteratively during the assignment of cell fates and cell identities (see Chapter 8), to govern the choice between adoption or forfeiting of a fate.

6. In many cases, withdrawal from the cell cycle is synchronized with activation of differentiation programs that allow the genes targeted by the cell-type-specific network to be expressed (see Chapter 7). This is not invariably the case; for example, in the intestinal epithelium, and some parts of the immune system, cell proliferation is maintained within a differentiated cell population to provide a constant renewal of cells for a specific physiological purpose.

SUMMARY

1. A particular cell type results from a sequence of restrictions in developmental potential within a cell or group of cells. In molecular terms, this means a commitment to a particular program of gene expression that will lead to a specific pattern of cell differentiation.

2. The first step in the definition of a particular cell type is the activation of cell-type-specific genes through the spatial and temporal overlap of the activity of proteins, usually transcription factors, each of which is not associated with particular cells. However, these overlaps create unique transient interactions that lead to the expression of cell-type-specific genes.

3. Cell-type-specific regulators define and maintain a particular state. In the case of muscle they are members of the family of bHLH myogenic regulators, and in the case of the nervous system they are bHLH proneural factors.

4. Within the context of cell- or lineage-specific regulators, cells become different (e.g. different muscles in the muscle lineage and different neurons in the neural lineages). This is achieved

through combinations of the basic signalling system of the cell and the internal running programs of gene expression. Very often, the differences between elements of the lineages depend on the activity of particular homeodomain proteins.

5. At any stage there are binary decisions: a cell adopts or does not adopt a fate. The decision is mediated by Notch signalling.

6. At a certain point in the lineage, cells exit the proliferative phase and begin to differentiate. This transition is brought about by lineage-specific transcription factors and signalling molecules acting on the machinery that governs the cell cycle. Very little is known about the details of this transition.

COMPLEMENTARY READING

Baylies, M., Bate, M., and Ruiz Gomez, M. (1998) Myogenesis: a view from *Drosophila*. *Cell* **93**, 921–7.

Borycki, A. and Emerson, C. P. (2000) Multiple tissue interactions and signal transduction pathways control somite myogenesis. *Curr. Topics Dev. Biol.* **48**, 165–224.

Chitnis, A. (1999) Control of neurogenesis—lessons from frogs, fish and flies. *Curr. Opin. Neurobiol.* **9**, 18–25.

Sasai, Y. (1998) Identifying the missing links: genes that connect neural induction and primary neurogenesis in vertebrate embryos. *Cell* **21**, 455–8.

Frasch, M. (1999) Controls in patterning and diversification of somatic muscles during *Drosophila* embryogenesis. *Curr. Opin. Genet. Dev.* **9**, 522–9.

Jessell, T. (2000) Neuronal specification in the spinal cord: inductive signals and transcriptional codes. *Nat. Rev. Genet.* **1**, 20–9.

Lumsden, A. and Krumlauf, R. (1996) Patterning the vertebrate neuraxis. *Science* **274**, 1109–15.

Ordhal, C., Williams, B., and Denetclaw, W. (2000) Determination and morphogenesis in myogenic progenitor cells: an experimental embryological approach. *Curr. Topics Dev. Biol.* **48**, 319–67.

Rothenberg, E., Telfer, J., and Anderson, M. (1999) Transcriptional regulation of lymphocyte lineage commitment. *BioEssays* **21**, 726–42.

Streit, A. and Stern, C. (1999) Neural induction, a bird's eye view. *Trends Genet.* **15**, 20–4.

Tajbakhsh, S. and Buckingham, M. (2000) The birth of muscle progenitor cells in the mouse: spatiotemporal considerations. *Curr. Topics Dev. Biol.* **48**, 225–68.

Chapter 11

Patterns in one and two dimensions

The generation of a particular cell type, such as a neuron, muscle or blood cell, is a sequential process in which the developmental potential of the cell is progressively restricted in response to a combination of intrinsic and extrinsic cues. As the fates of cells become directed towards particular cell types, individual identities emerge, for example specific types of muscle cells, neurons or blood cells that perform specific functions. This fine-grained cell fate assignment usually relies on the integration—often at the level of individual cells—of a series of spatially organized cues. In some cases these cues are also used to organize cells in space and, as a consequence, they can generate patterns.

The realization that organisms are the result of serial and integrated processes of pattern formation has led to a search for principles that can account in a general way for these processes. In earlier chapters we have already encountered some cellular and molecular mechanisms that are used by cells and groups of cells to pattern themselves. However, we viewed these essentially as isolated events. In Chapter 10 we saw how some of these mechanisms can be used to generate cell diversity. In this chapter and the following one, we shall see how these same mechanisms and routines can be linked and integrated to generate spatial arrangements of cells that are often transformed into specific structures or patterns.

Pattern formation is analysed at the molecular level in basically the same way as the specification of cell types. Through genetics, the functions and interrelationships of genes involved in the process can be deduced by identifying and characterizing mutants in which the normal functions of the genes are impaired. Molecular biology enables the expression patterns of genes to be altered (see Chapter 10, Fig. 10.2 for a summary of ways of doing this) in order to test their influence on specific processes. Understanding the spatial regulation of gene expression is particularly important in the context of pattern formation.

In this chapter we concentrate on patterns involving cell interactions in one or two dimensions: that is, processes that take place within rows or sheets of cells (essentially one dimension), and patterning processes involving uni- or bi-directional interactions between two adjacent sheets or layers of cells (two dimensions) (Fig. 11.1). Such processes can account for many of the structures of the embryo. We have chosen three examples, each from a different model organism, which highlight features that are likely to be common to all processes of pattern formation.

Development and patterning of the *Caenorhabditis elegans* vulva

The development of the *C. elegans* vulva is a simple example of how interactions between defined cells in specific positions can generate patterns. In the *C. elegans* hermaphrodite, 22 epidermal cells from the AB lineage (see Chapter 8, Fig. 8.1 and 8.3) assemble into a structure, the vulva, that provides a connection between the gonad and the outside (Fig. 11.2). This connection is used for laying fertilized eggs, but otherwise is not required for viability. This property has been exploited in a number of genetic screens to identify genes involved in the development of this structure by searching for mutants in which either there is no vulva (vulvaless) or in which there is more than one vulva (multivulva). These studies have yielded important insights into the genetic control of pattern formation, the molecules that are involved and how these molecules interact.

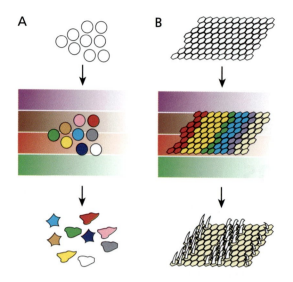

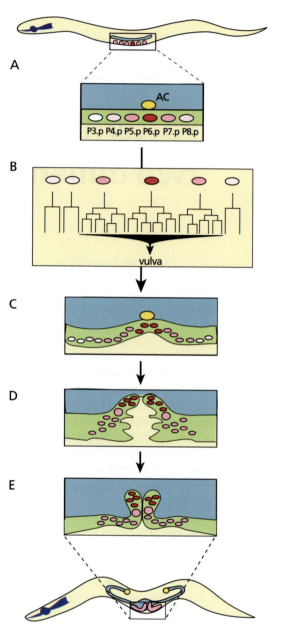

Fig. 11.1. **Fig. 11.1. Patterning in one and two dimensions.**
(A) The generation of cell diversity relies on an interplay between intrinsic factors (usually networks of gene activity, represented by different colours within the cells) and extrinsic influences (the effects of signalling molecules, represented by coloured shadings surrounding the cells) on the profiles of gene expression of individual cells. These patterns of gene expression produce precise patterns of differentiation. In some cases (e.g. the haematopoietic system), the cells that are generated are not restricted to a specific spatial arrangement.
(B) In other cases, the differences emerge within a spatially fixed cell population and cells acquire identities on the basis of where they lie relative to other cells (e.g. the patterning of two-dimensional sheets of cells). This process relies on spatially fixed sources of signalling molecules (coloured shadings surrounding cells) which generate cellular responses that are also spatially fixed (different colours within cells) and leads to a tapestry of cell diversity which results in spatially fixed patterns of differentiation.

Fig. 11.2. **The development of the vulva in the *Caenorhabditis elegans* hermaphrodite. (A)** Overview of an L1 stage larva, showing (box) the region containing the cells from which the vulval precursors will emerge, and the overlying gonadal tissue with the anchor cell (AC). There are six cells that have the potential to contribute to the vulva, the vulva precursor cells (VPCs). These cells are given the designations Pn.p with n = 3–8 (see also Fig. 9.1). The colours of the different VPCs indicate that they have different fates. **(B–D)**. After being endowed with specific fates during the third lar-

val period (L3), each of the VPCs undergoes three divisions characteristic of its fate. Only three of the six cells normally contribute to the vulva. In the other three (P3.p, P4.p, and P8.p) this fate is suppressed and they contribute to the hypodermis. **(E)** After completing their divisions, the vulval cells participate in a series of morphogenetic changes which lead to the construction of the adult vulva. (After Greenwald, I. (1997). Development of the vulva. In *C. elegans II*. Riddle, D., Blumenthal T., Meyer, B., and Priess, J. R. (Eds) Cold Spring Harbor Laboratory Press.)

The development of the vulva can be divided into three temporal phases. During the first, which lasts throughout the first two larval stages (L1 and L2), the vulva precursor cells (VPCs) are generated and specified as such. During a second phase the VPCs are assigned individual fates that will be expressed during the third larval stage (L3). This sequence of events takes the form—characteristic of *C. elegans*—of the generation of precisely defined lineages, which give rise to different kinds of cells. In the final step, the vulva is assembled through a series of morphogenetic events involving the progeny of the different VPCs (Fig. 11.2).

Generation and specification of the vulva precursor cells (VPCs)

The development of the vulva begins in the first larval stage when six cells from the right side and six cells from the left side of the larva interdigitate to form a row of 12 precursor cells in the middle region of the worm (Fig. 11.3). Each of these cells (Pn where n = 1–12) divides once to generate an anterior cell (Pn.a) which becomes a neuroblast, and a posterior cell (Pn.p) which becomes either a vulval or hypodermal cell. The decision to adopt the neural or the vulval/hypodermal fate requires the lin-26 zinc-finger-containing protein in the Pn.p cells to repress the neural fate; in *lin-26* mutants both the Pn.p cells and the Pn.a cells adopt the neural fate. In wild-type animals, the subsequent fate of each of the Pn.p cells is determined largely by its position in the array, under the influence of extrinsic signals (Fig. 11.3). The cells that lie closest to the centre, Pn.p with n = 4, 5, 6, 7, and 8, become the VPCs. Pn.p cells that lie more laterally, with n = 1, 2, 9, 10, and 11, adopt the so-called F fate and fuse with surrounding hypodermal cells. One cell, P3.p, behaves in this way in approximately 50% of wild-type animals; in other animals it behaves as a VPC. Another cell, P12.p, undergoes one division, generating an anterior cell that fuses with the hypodermal cells and a posterior cell that undergoes programmed cell death. Even though P4.p through P8.p, and often P3.p also, have the potential to adopt vulval fates, in the wild type only three of them (P5.p, P6.p, and P7.p) will actually do so.

A homeodomain protein of the Dfd class, encoded by the *lin-39* gene, is absolutely necessary for the specification of the VPCs; in its absence all 12 cells adopt the F fate and fuse with the hypodermis. The extrinsic influences that determine this fate require Ras and Wnt signalling, which have been implicated in the regulation of *lin-39* expression

(Fig. 11.3). In the absence of Ras signalling, *lin-39* expression is low in P5.p–P7.p and some of these cells acquire the F fate, while in mutants for *bar-1*, which encodes a *C. elegans* homologue of the Wnt pathway component β-catenin/armadillo, *lin-39* expression is low in all six VPCs and disappears from them prematurely. As a consequence many of these cells develop the F fate. Double mutants for both *bar-1* and *let-23* (encoding the EGF receptor that activates the Ras signalling pathway) display synergistic effects, leading to many of the prospective VPCs adopting the F fate (Fig. 11.3). This suggests that the regulation of *lin-39* expression requires interactions between these signalling systems. The spatial bias in the effects of the two signalling systems on different Pn.p cells (see table in Fig. 11.3) suggests that *bar-1* alone regulates *lin-39* expression in P3.p–P4.p, while in P5.p–P7.p Ras operates as well. It has been suggested that the variable fate of P3.p may be associated with a weaker level of Wnt signalling received by this cell which results in a stochastic cell fate decision. At the moment, however, it is not clear where the signals that activate *lin-39* lie or come from.

The signal from the anchor cell: Ras signalling and the specification of VPCs

Following the specification of six Pn.p cells (P3.p–P8.p) as VPCs, a signal from the overlying anchor cell (AC) of the gonad triggers a sequence of cell fate assignments which, during a long cell cycle that encompasses L2 and most of L3, leads to the specification of the VPCs as 1°, 2°, and 3° VPCs (see Chapter 9, Fig. 9.1). Only the lineages generated by the 1° and 2° VPCs actually adopt vulval fates while the progeny of the 3° VPCs fuse with the hypodermis. The 3° VPCs can adopt the vulval fate if there is a problem with the 1° and 2° cells, indicating that they provide a reservoir of developmental potential.

The specification of individual fates depends on the AC (Fig. 11.4 and see also Fig. 9.21). In the wild type, the VPC nearest to the AC (P6.p) adopts the 1° fate, while P5.p and P7.9 adopt the 2° fate. Screening for vulvaless and multivulva phenotypes has uncovered a large number of genes that are involved in the AC signal or its relay to and interpretation by the VPCs. These studies have shown that the Ras signalling pathway plays a prominent role in the specification of different VPC fates (Fig. 11.4). Loss of function mutations in elements of the Ras signalling pathway result in absence of the vulva, while gain of function mutants

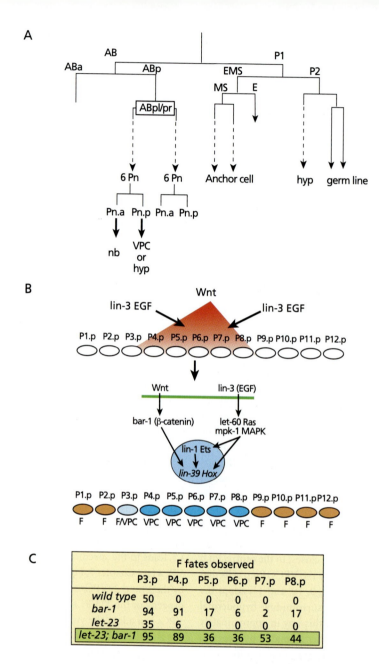

C

F fates observed						
	P3.p	P4.p	P5.p	P6.p	P7.p	P8.p
wild type	50	0	0	0	0	0
bar-1	94	91	17	6	2	17
let-23	35	6	0	0	0	0
let-23; bar-1	95	89	36	36	53	44

Fig. 11.3. Selection of VPCs from among Pn.ps. (A) The vulva precursor cells (VPCs) are selected from the posterior daughters of the ABp lineage. These cells divide once to generate an anterior cell (Pn.a) which will develop as a neuroblast and a posterior one (Pn.p) which can develop as a VPC. **(B)** The assignment of the VPC fate to the Pn.p cells requires activation of the homeobox gene *lin-39* in P4.p–P8.p, through the combined activities of Wnt and Ras signalling (the latter involving the EGF signalling protein lin-3 and the EGF receptor let-23). Of the other Pn.p cells, P3.p adopts the VPC fate in 50% of wild-type animals, probably because Wnt signalling to this cell is only partially penetrant. P1.p, P2.p, P9.p. P10.p, and P11.p adopt the F fate and fuse with the hypodermis. P12.p divides once to generate a cell that adopts the F fate and another that undergoes programmed cell death. **(C)** The fate of Pn.p (n = 3–8) is established by combined Wnt and Ras signalling, as shown by the effects of mutations in the genes encoding a β-catenin homologue (*bar-1*) and the receptor that activates the Ras pathway (*let-23*). In this experiment the fate of each cell was observed in a number of worms with mutations in different genes and scored. The results are expressed as the percentage of Pn.ps expressing the F fate. Note the spatial bias in the requirement for Wnt signalling (*bar-1*), reflected in the strong requirement for *bar-1* in P3.p and P4.p compared to the other cells, and also the synergistic effect of *bar-1* and *let-23*. (After Eisenmann, D. M. *et al.* (1998) The β-catenin homolog BAR-1 and LET-60 Ras coordinately regulate the Hox gene *lin-39* during *Caenorhabditis elegans* vulval development. *Development* **125**, 3667–80.)

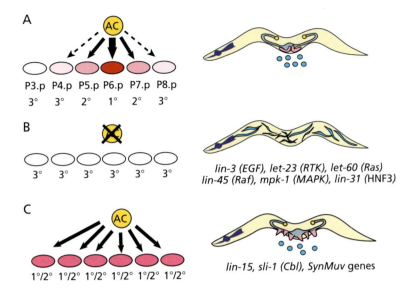

A

P3.p P4.p P5.p P6.p P7.p P8.p
 3°   3°   2°   1°   2°   3°

B

 3°   3°   3°   3°   3°   3°

lin-3 (EGF), let-23 (RTK), let-60 (Ras)
lin-45 (Raf), mpk-1 (MAPK), lin-31 (HNF3)

C

1°/2° 1°/2° 1°/2° 1°/2° 1°/2° 1°/2°

lin-15, sli-1 (Cbl), SynMuv genes

Fig. 11.4. Genetic analysis of the patterning of the VPCs. The AC inductive event. Diagram of the inductive events that generate VPCs in the wild type or in different mutants (left), and the worm that results from such events (right). **(A)** In the wild type, the anchor cell (AC) induces primary (1°) and secondary (2°) fates amongst the VPCs. These fates are ranked according to the position of each VPC relative to the AC. **(B)** A vulvaless phenotype results in an inability to lay the fertilized eggs, which hatch inside the worm. This 'bag of worms' phenotype allows an easy screen for nematodes in which all or most cells develop the 3° fate. This phenotype can be produced by ablation of the AC and also by a number of mutations, including several (e.g. *lin-3*, *let-23*, *let-60*) which identify elements of the Ras signalling pathway. **(C)** The opposite phenotype to vulvaless is one in which multiple vulvas develop inside the nematode because more than one Pn.p cell has acquired the 1° fate. Examples of mutations conferring this phenotype are shown.

result in multivulva phenotypes. Mosaic analysis of these mutations (see Chapter 4, Fig. 4.6) indicates that the AC is the source of the signal and that the Ras signalling pathway acts in the VPCs to mediate the response and effect the different fates.

One of the mutations that results in absence of the vulva is in the *lin-3* gene. This gene encodes a homologue of EGF, suggesting that lin-3 may act as a signal. This is supported by mosaic analysis, which shows that *lin-3* is required in the AC and not in the VPCs. The intensity of the lin-3 (EGF) signal that activates the Ras pathway (see Chapter 5, Fig. 5.14 and 5.15) is an important factor in the distinction between 1° and 2° fates. This is suggested by experiments in which individual VPCs, isolated from other VPCs and the AC through targeted ablations, are subject to different levels of the lin-3 signal. The VPCs adopt different fates depending on the intensity of the signal: high signal leads to the 1° fate, intermediate to the 2° fate and low to the 3° (Fig. 11.5). This hierarchy of response correlates with the hierarchy of cell fates adopted by the VPCs by virtue of their distance from the AC. P6.p is the closest and therefore develops the 1° fate because it receives the highest level of signal. However, while this experimental situation indicates that the levels of lin-3 can be instructive with regard to the fate of isolated VPCs, other factors influence the 1° versus 2° decision in the wild-type animal. This can be shown in mosaic experiments in which the consequences of making particular VPCs mutant for elements of the Ras pathway are assayed (Fig. 11.6). These experiments show that Ras-independent interactions between VPCs also play a role in the establishment of the 1° and 2° fates.

Gain of function mutations in Ras signalling lead to a multivulva phenotype because many more cells than normal adopt the 1° and 2° VPC fates. However, there are other multivulva mutants in which Ras signalling occurs normally, suggesting that the wild-type role of the genes that are mutant in these animals is to suppress the VPC fate in parallel or interacting with Ras signalling. These genes fall into two groups (A and B) and genetic analysis suggests that the two groups have redundant functions in this process: whereas single mutants in either of the groups (A or B) do not show any phenotype, double mutants that combine a mutant gene from group A and a mutant gene from group B have a strong multivulva phenotype. For this reason, the phenotype and the genes are called synthetic multivulva (*SynMuv*) (Fig. 11.7). Mosaic analysis of these

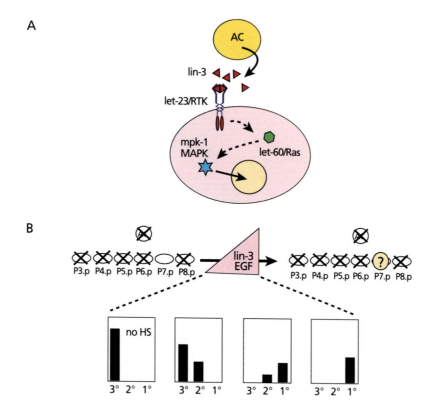

Fig. 11.5. The level of lin-3 signalling affects the specification of the VPC fates. (A) The *C. elegans* Ras pathway components acting in specification of the VPCs as a result of signalling from the anchor cell (AC). **(B)** To test whether different levels of the lin-3 signal could act to specify the 1° and 2° fates of the VPCs, a soluble form of lin-3 (lin-3 EGF) was engineered to be expressed in nematodes under the control of the heat shock promoter (HS) (see Chapter 10, Fig. 10.2). EGF is usually tethered to the membrane and its activation requires a proteolytic event (see Chapter 9, Fig. 9.15). The experimental animals were set up such that only one Pn.p was present. This was achieved by ablating all VPCs except P7.p. This cell was then exposed to different levels of lin-3 by regulating the intensity of the heat shock (indicated by the increasing gradient) which will be translated into differing amounts of lin-3. The fate of P7.p was assessed under these conditions by following the lineage that results from the experiment. In general, the fate of P7.p correlated with the levels of expression of lin-3: increasing the levels of lin-3 EGF increased the percentage of 1° fates developed from P7.p. (After Katz, W. *et al.* (1995) Different levels of *C. elegans* growth factor LIN-3 promote distinct vulval precursor fates. *Cell* **82**, 297–307.)

Fig. 11.7. The molecular basis of multivulval phenotypes. (A) The development of the vulva depends on a signalling event from the AC to the VPCs. **(B)** Removal of the AC or the signal that emanates from it abolishes the specification of different VPC fates; all cells adopt the 3° fate and fuse with the hypodermis. **(C)** Gain of function in Ras signalling leads to a multivulva phenotype because many more cells adopt the 1° and 2° VPC fates. **(D)** *Synthetic Multivulva (SynMuv)* mutants are multivulval mutants in which Ras signalling occurs normally. These phenotypes result when the worms are double mutants for a gene in each of two groups named A and B. **(E)** Single mutants, or double mutants within the same *SynMuv* group, will not produce a multivulva phenotype. **(F, G)** The SynMuv phenotype is dependent on Ras, but not on the lin-3 signal from the AC. In the absence of lin-3, the VPCs of *SynMuv* mutants still adopt the 1° or 2° fate **(F)**, but this is not the case if the *SynMuv* mutants lack Ras signalling **(G)**. This observation indicates that the function of the *SynMuv* genes is to antagonize the basal activity of the pathway, which in the mutants becomes high enough to promote specification of VPCs. **(H)** Some of the mutations underlying the SynMuv phenotypes are cell-autonomous within the VPCs and some appear to act on the VPCs from the hypodermis. This has led to the suggestion that, in addition to negative regulatory mechanisms that dampen Ras signalling in the VPCs, there is a signal from the hypodermis that represses the outcome of Ras signalling. Together, both mechanisms establish thresholds for Ras signalling.

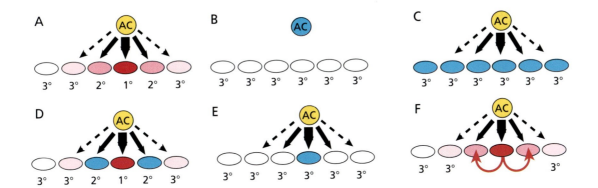

Fig. 11.6. Mosaic analysis of the requirements for Ras signalling in the specification of VPCs. (A) The experiments described in Fig. 11.5 suggest that the intensity of signalling from the AC cell correlates with the fates of the different VPCs. This can be tested further by making mosaics for mutations in either the EGF receptor encoded by *let-23* or in the EGF homologue encoded by *lin-3*. In these experiments particular cells are made mutant for either gene (blue) and their phenotype assayed. **(B)** If the AC, and only the AC cell, is made mutant for *lin-3* all VPCs adopt the 3° fate. This is consistent with the idea that lin-3 from the AC provides the initial input for the specification of the VPCs. **(C)** If all the VPCs are made mutant for elements of the let-23 signalling pathway (i.e. they cannot receive the lin-3 signal), they develop the 3° fate. **(D)** If P5.p or P7.p are mutant for *let-23*, they still develop the secondary fate. This suggests that Ras signalling to P5.p and P7.p themselves is not nec-essary to establish their 2° fate. **(E)** If P6.p is made mutant for *let-23* it develops the hypodermal 3° fate, and so do P5.p and P7.p. This indicates that the fate of P5.p and P7.p is induced by P6.p after it adopts the 1° fate. **(F)** Altogether, these experiments indicate that Ras signalling is necessary and sufficient for the specification of the 1° fate of P6.p and that a second signal from this cell to its neighbours instructs their 2° fate. The experiments described in Fig. 11.5 indicate that the 2° fate assignment occurs against a background of Ras signalling, but Ras signalling is not absolutely necessary for this event. (After Simske, J. and Kim, S. (1995) Sequential signalling during *Caenorhabditis elegans* vulval induction. *Nature* **375**, 142–6; and Koga, M. and Oshima, Y. (1995) Mosaic analysis of the *let-23* gene function in vulval induction of *Caenorhabditis elegans*. *Development* **121**, 2655–66).

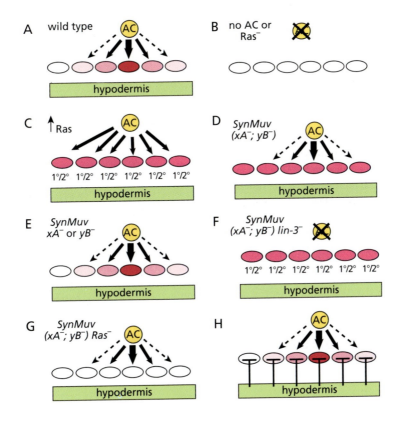

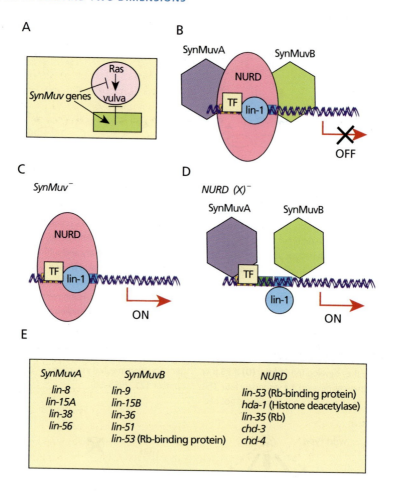

Fig. 11.8. The SynMuv complex. (A) The nature of the products of the known *SynMuv* genes suggests that they are involved in a repression system that antagonizes Ras signalling in the VPCs. This system is probably established autonomously in the VPCs but might have an additional input from the hypodermis. **(B)** At the molecular level, the nature of the *SynMuv* genes (see E) suggests that their function is likely to be to recruit large chromatin remodelling complexes (e.g. NURD) to the promoters of VPC-specific genes that are under the control of Ras (the promoter is shown as binding the Ets transcription factor lin-1 and a cell-type-specific transcription factor, TF). This complex keeps the VPC-specific genes in the off state. Mutations in some components of the SynMuv complex **(C)** or in proteins involved in chromatin remodelling **(D)** have the same effect and produce multivulva phenotypes by causing derepression of the expression of VPC-specific genes. Mutations in the SynMuv complex, and in elements of the chromatin remodelling complex, act in a synergistic manner. **(E)** List of some of the *Synmuv* and *NURD* genes and, where known, the nature of the products that they encode.

genes shows that some of them are autonomous to the VPCs, but that others are required in the hypodermal layer that lies adjacent to the VPCs. This indicates that the hypodermal layer imposes a repressive activity on the VPCs.

The development of the vulvas in *SynMuv* mutants requires the activity of Ras, indicating that the function of the *SynMuv* genes is to antagonize the activity of this signalling pathway (Fig. 11.7). Some of the *SynMuv* genes encode proteins related to the Rb (retinoblastoma) protein, which in other systems have been implicated in the regulation of the choice between cell cycle progression, and cell cycle exit and differentiation (see Chapters 7 and 10). Other *SynMuv* genes encode elements of a large chromatin remodelling complex, NURD (Fig. 11.8). The picture that emerges from these observations is one in which hypodermal cells set the activity of NURD and associated proteins in the VPCs and this establishes thresholds of activity for the effectors of Ras signalling.

VPCs: lin-12 signalling

As mentioned above, there is evidence that factors other than Ras signalling are involved in the decision between the 1° and 2° VPC fates. The activity of the SynMuv complex could set up different thresholds that impose different requirements in different cells. However, other experiments indicate additional influences. For example, as mentioned previously, analysis of VPC fate in animals mosaic for the *let-23* gene, which encodes the RTK that receives the signal from the AC, show that let-23 is not required in P5.p and P7.p for them to adopt the 2° fate (Fig. 11.6). Furthermore, if P6.p has no *let-23* activity, P7.p does not develop a 2° fate, whether or not it has *let-23* activity itself. This suggests that, after being specified as a 1° VPC, P6.p plays a role in the specification of 2° fates.

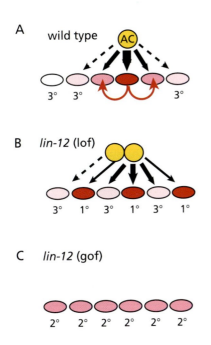

Fig. 11.9. *lin-12* and VPC specification. (A) Mosaic analysis of the requirement for Ras signalling in the VPCs (Fig. 11.6) shows that lin-3 activity alone is not sufficient to specify the 2° fate and that this fate depends on a signal from the P6.p (1°) cell. **(B)** Loss of function (lof) mutations in *lin-12* produce a duplication of the AC (see Chapter 8, Fig. 8.21) that will lead to higher levels of Ras signalling to the VPCs. However, in these mutants no 2° cell fate is established and the VPCs develop an alternating pattern of 1° and 3° fates. **(C)** In contrast to the situation of lof mutants in *lin-12*, in gain of function mutants (gof) all VPCs develop a 2° fate even though in these mutants there is no AC. This suggests that lin-12 mediates a signal that establishes the 2° fate.

Signalling through the lin-12 receptor is a candidate for the influence that specifies the 2° fate in P7.p. In *lin-12* loss-of-function mutants no VPCs adopt the 2° fate, even though there are two anchor cells and, as a result, an excess of 1° cells is specified. Conversely, in dominant gain-of-function mutants all six VPCs adopt the 2° fate, even though there is no AC and no cells adopt the 1° fate (Fig. 11.9).

Observations such as these have led to the suggestion that even though the 2° fate of the VPC probably requires Ras signalling, the function of this signalling might be to modulate the activity of lin-12, which conveys the actual signal for this fate (Fig. 11.10). In this model, Ras signalling from the AC promotes the primary fate, whereas lateral lin-12-mediated signalling antagonizes this fate and promotes the 2° fate. Recent studies suggest that lin-12 has two temporally separate roles in this process and that the cell cycle plays an important part in setting the sequence of events. In the late G1 and S phase of the cell cycle, when the decision between the 1° and 2° fates occurs, signalling through lin-12 is effectively antagonized in P6.p by high levels of Ras-mediated signalling from the AC, but is able to oppose the 1° fate in P5.p and P7.p where the Ras-mediated influence of the AC is smaller. As the cell cycle progresses and P6.p adopts the 1° fate, lin-12 expression disappears from this cell but is maintained in the other Pn.ps. The choice between the 2° and 3° fates is not made until later on, after the S phase, when lin-12 activity induces the 2° VPC fate in P5.p and P7.p in response to an unknown P6.p-derived signal that does not involve the lin-12 ligands lag-2 or apx-1. The remaining Pn.ps that are not exposed to this signal adopt the 3° fate (Fig. 11.10).

The fact that different cell-fate assignments are made at different stages of the cell cycle suggests that there must be a mechanism for coordinating cell-cycle progression with these developmental events. Mutations in a set of genes known as heterochronic genes have been found that alter the timing of the G1–S transition in the VPCs by regulating a Cki (cyclin-dependent kinase inhibitor, see Chapter 7). Some of these mutations block or delay VPC division and also block vulval development. The role of the heterochronic genes, which also affect some other developmental events in the worm, seems to be to help synchronize the cell cycle with the development of competence to respond to patterning signals.

In addition to Ras and lin-12, the specification of the different VPCs also requires the activities of *lin-39* and *bar-1*; the evidence for this is that weak mutant alleles of these genes enable all six VPCs to be specified but they all adopt

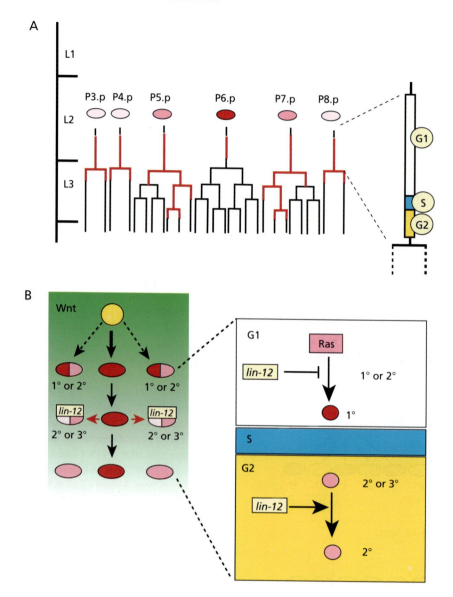

Fig. 11.10. Stepwise specification of 1° and 2° fates in the VPCs. (A) Pattern of lin-12 expression (red lines) in the VPCs and their descendants during the three larval stages L1, L2, and L3. The diagram on the right shows the relative lengths of the phases of the long cell cycle that encompasses L2 and part of L3, and during which the 1°, 2°, and 3° fates are specified. In P6.p, lin-12 is only expressed before the first cell division of the precursor, but in some of the descendants of P5.p and P7.p its expression extends into L3. **(B)**

It has been suggested that during G1 the lin-3 signal from the anchor cell acts via the Ras pathway to induce the primary cell fate in the neighbouring VPCs. In P5.p and P7.p, the expression and activity of lin-12 can effectively antagonize Ras signalling, which is weaker in these cells than in P6.p. During G2, an unknown signal from the 1° VPC activates or maintains the activity of lin-12 in P5.p and P7.p so that they can adopt the 2° fate at the expense of the 3° fate. These events all occur against a background of Wnt signalling.

the 3° fate. These observations suggest that bar-1 (Wnt pathway) signalling might be involved in the specification of the fates of individual VPCs as well as earlier in the initial establishment of the VPCs, but this time in the context of AC signalling (Fig. 11.10).

Even though the fates of the VPCs depend on their positions relative to the AC, it is likely that they are primed to adopt one or another fate by their developmental history. Thus, P3.p, P4.p, and P8.p will tend to adopt the 3° fate if left undisturbed. Such a bias is already evident from the strong dependence of P3.p and P4.p on bar-1 signalling and might even be set up by bar-1.

In summary, any influence of a graded signal from lin-3 on the specification of the 1° and 2° fates is biased by the lineage of the cells and modulated by the activity of *Syn-Muv* genes, lin-12 and Wnt signalling. The experiment showing the effects of different doses of lin-3 on the fate of the VPCs indicates that the lin-3 signal can override other influences if it is artificially strengthened, but it is clear that in the normal context all the other parameters act to modulate this underlying basic theme.

The sequence of events that leads to the specification of the VPCs follows a pattern that is common to many processes of cell-type specification. During the progressive restriction of the fate of cells, the same signalling pathways may be used sequentially, probably triggering different responses in different cellular contexts.

Execution of the fates of the VPCs

After P5.p, P6.p, and P7.p have each been specified as a 1° or 2° VPC, each initiates a three-division lineage during the late L3 larval stage to generate the components of the vulva (Fig. 11.11). The Pn.p cells that have not become 1° or 2° VPCs fuse with the hypodermis after the first division. The development of the vulval lineages is autonomous (i.e. ablation of any of the cells does not alter the lineages or fate of

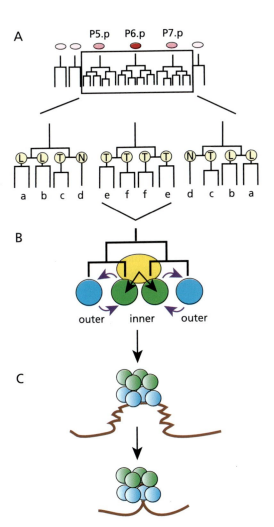

Fig. 11.11. Patterns of division and cell fate specifications in the lineages of the VPCs. (A) Each of the three VPCs that will give rise to the vulva divides a number of times to generate a total of 22 cells that will assemble the vulva. In the course of these divisions, the differences between 1° and 2° fates become evident in the patterns of division of the daughters and granddaughters of the VPCs, depending on whether these cells divide longitudinally (L) or transversely (T). In the 2° lineages one of the cells does not divide (N). Each of the final cells will be different and will contribute differently to the assembly of the vulva. These fates are labelled a–f. **(B)** Some details of the assignment of cell fates in these final lineages have been analysed for the progeny of P6.p. The four granddaughters of P6.p align in a row and develop two different fates: the inner cells (green) will develop as F cells and will lie on the inside of the structure, whereas the outer cells (blue) will develop as E cells and lie on the outside of the vulva. Each of these cells then divides once and contributes in a different way to the morphogenesis of the vulva (see Fig.11.12). The establishment of these fates depends on three processes: internal polarity within every division that predetermines some of the properties of the different cells, an induction from the AC (yellow) to the inner cells that makes them adopt the F fate, and interactions between the cells that reinforce the differences established by the AC. **(C)** During the final stages, these cells divide, rearrange, and interact with surrounding cells to form the external part of the vulva. (After Wang, M. and Sternberg, P. (2000) Patterning of the *C. elegans* 1° vulval lineage by RAS and Wnt pathways. *Development* **127**, 5047–58.)

adjacent cells). After two divisions each, P5.p, P6.p, and P7.p generate a group of 12 cells (four from each Pn.p) with a pattern of fates that is symmetrical around those generated by P6.p. The fates of the cells are characterized by the orientation of the division in the third round of the lineage, which can be either longitudinal (L) or transverse (T). Two cells that are generated by the second round of division and do not divide again are given the designation N. The fates generated by the two divisions each of P5.p, P6.p, and P7.p are, respectively, LLTN, TTTT, and NTLL, giving rise to a primordium of 22 epithelial cells (Fig. 11.11). The process of cell-fate specification within these lineages follows the same rule as earlier ones in which different cell fates are generated from a combination of cell autonomous and non-autonomous mechanisms. For example, the *lin-11* locus encodes a transcription factor with a home-odomain and a LIM domain that is expressed in the lineages of P5.p (LLTN) and P7.p (NTLL). In *lin-11* mutants, cell fates are altered in the two lineages, both becoming LLLL. The expression of *lin-11* is under the control of Wnt signalling since mutants for *lin-17*, which encodes a member of the frizzled family of receptors for Wnt, have altered patterns of *lin-11* expression which correlate with altered cell fates. In *lin-17* mutants the P7.p lineage generates an LLLL or LLTN set of fates instead of the normal NTLL.

In contrast to the lineages of P5.p and P7.p, P6.p produces a row of four cells, all of which divide transversely to generate two different differentiated cell types, E and F, depending on their position in the vulva: the inner cells give rise to the F fate and the outer cells to the E fate. A combination of genetic analysis and experimental manipulation of the lineage of the 1° VPC has identified three processes involved in bringing about this distinction. First, there are intrinsic differences between the inner and outer cells, derived from the polarity of the divisions of P6.p. Then a local signalling event from the AC determines the F (inner) fate. Finally, signalling between the inner and outer cells reinforces the effect of the AC and contributes to the development of the E and F fates. Very little is known about the molecular events that mediate these processes Genetic analysis indicates that Ras and Wnt signalling are essential elements but seem to act in a partially redundant manner (i.e. although each is necessary for the process, neither has been shown to be sufficient).

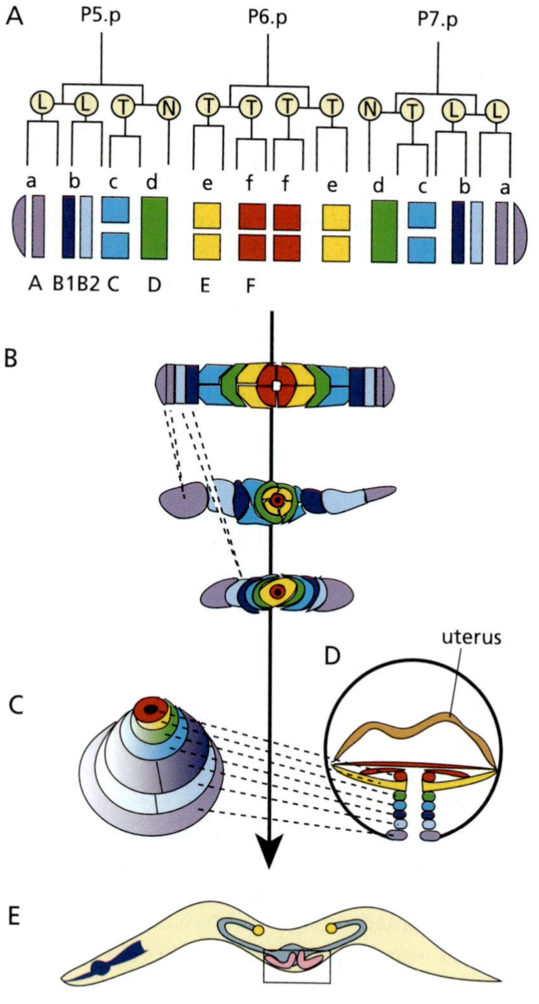

Morphogenesis of the vulva

The morphogenesis of the vulva (Fig. 11.12) from the 22-cell primordium has been studied at the cellular level by staining the cell nuclei and the adherens junctions at the

Fig. 11.12. Morphogenesis of the vulva. (A) Generation of the 22 different cells that will assemble the vulva from the three VPCs. **(B)** In the course of differentiation, a sequence of cell movement, fusions, and changes of shape assemble the toroidal structure that is the vulva. The fusions of the A and B cells are indicated by dotted lines. Vulval morphogenesis requires the concerted action and interaction of all the cells. **(C)** Three-dimensional view of the late vulva with colours representing the contributions of each of the different cell fates to the final structure. Code as in A. **(D)** Transverse view of a hermaphrodite showing the relationship between the vulva and the uterus. **(E)** *Caenorhabditis elegans* with the vulva highlighted by the box. (Adapted from Sharma-Kishore, R. *et al.* (1999) Formation of the vulva of *C. elegans*: a paradigm for organogenesis. *Development* **126**, 691–9.)

apical surface of the cells (on the ventral surface of the animal's body) with different fluorescent antibodies and observing patterns of cell migration and fusion with a confocal microscope. Reconstructions from serial sections through the developing vulva enabled the morphogenetic events to be visualized in three dimensions. Morphogenesis involves the selective migration and fusion of epithelial cells from the vulval primordium to form a tube of seven toroidal cells between the uterus and the epithelium. The toroidal cells are formed when corresponding pairs or sets of epithelial cells from the anterior and posterior sides of the primordium send out processes that travel towards the midline of the vulva and meet there. The innermost cells do this first (i.e. the progeny of P6.p), followed by cells from progressively more outer positions; as each ring of cells is formed, it displaces the ring formed by its inner neighbours towards the inside of the animal's body, so that the rings stack up. The cells making up each ring then fuse. Finally, the anchor cell penetrates the centre of the stack and fuses with a uterus cell to form a channel between the uterus and the vulva.

Little is known to date about the molecular processes that occur during morphogenesis. However, it is known that each half (anterior and posterior) of the vulva develops autonomously, and several mutants have been found that have defects in various aspects of vulval morphogenesis. Eventually, a combination of biochemical and genetic approaches will, it is hoped, elucidate the molecular basis of the cell–cell recognition events, cell shape changes, migrations and fusions that are involved.

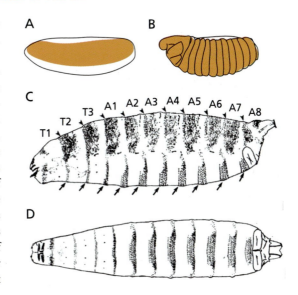

Fig. 11.13. The larval epidermis of *Drosophila*. (A) The epidermis is derived from the ectoderm, a region of the *Drosophila* blastoderm that will also give rise to the nervous system (brown, and see Chapter 10). **(B)** During epidermal development, ectodermal cells divide an average of three times, undergo a series of morphogenetic movements that allow them to cover a large surface area, and differentiate by secreting cuticle. The embryo shown here is at a late stage of embryogenesis. **(C, D)** Drawings showing lateral (C) and ventral (D) views of the cuticle of a newly hatched larva. Anterior is to the left and, in C, dorsal to the top. Segments are indicated (T, thoracic; A, abdominal) with arrows in the ventral region indicating the segment boundaries. Note the fine detail of the pattern of every segment and how it changes within and across segments.

Development and patterning of the larval epidermis of *Drosophila*

The *Drosophila* larva has a clearly segmented epidermis with a precise arrangement of denticles and hairs that form in characteristic positions within individual segments and display variations from segment to segment along the anteroposterior axis (Fig. 11.13). Although the patterning of the cells of the epidermis is obviously a two-dimensional process, much of the detailed analysis to date has addressed the question of how the intra- and intersegmental patterning along the anteroposterior axis is generated, in other words, how pattern is generated in essentially one cellular dimension. Nevertheless, it is likely that the way in which cells integrate the inputs from the two axes (anteroposterior and dorsoventral) is not very different from the way

in which they integrate multiple inputs within one axis. So even though the two axes are almost always treated separately, it is important to keep in mind that they input information into the same cellular integrator.

The patterning of the larval epidermis in *Drosophila* is in several respects an extrapolation of the situation that we have described in the nematode vulva, to a larger and less precisely determined number of cells. Over many years, studies on this system have not only yielded insights into the developmental strategies that generate this particular pattern but have also revealed many details of the workings of specific signalling pathways, for example, Wnt and hedgehog, that have since been shown to operate in a similar way in other organisms.

The larval epidermis develops from a large blastodermal primordium through a small number of cell divisions (two or three), and a process of differentiation in which cells within a sheet change their relative positions and then secrete cuticle (Fig. 11.13). The development and patterning of this sheet of cells relies on the establishment of spatial references for long- and short-range signalling, generation of cell diversity in the form of position-specific cell fates, and morphogenesis revealed as the expression of those cell fates in particular cell shapes and appearances. The differences between segments are due to the activity of the *Hox* genes (see Chapter 2, Fig. 2.8 and Chapter 3, Fig. 3.20) which act on a basic common pattern plan and modify it to create segment-specific structures. Mutations in the homeotic genes or in regulators of their activities reveal this underlying common pattern plan.

Establishment of references and polarity

The first stage in the patterning of the epidermal cells is the definition, in the blastoderm, of a large population of cells that will give rise to the ectoderm. This population of cells, which occupies essentially all but the most dorsal and ventral regions of the embryo (excluding the anterior and posterior poles), is specified by a gradient of nuclear dorsal protein activity along the dorsoventral axis, as discussed in Chapter 9 (Fig. 9.21). At the same time that the ectoderm is specified it is subdivided into specific domains of gene expression that define cell populations along the dorsoventral axis, and into segmental (also known as metameric) units along the anteroposterior axis (Fig. 11.14).

Looking at these events in more detail: along the DV axis during the early stages of development (Fig. 11.15), the patterns laid down by the nuclear dorsal protein gradient result, after cellularization, in the creation of two signalling centres at opposite ends of the epidermal sheet. On the dorsal side, a gradient of dpp activity subdivides the epidermis into dorsal and ventral regions and initiates its patterning along this axis from dorsal to ventral (see Chapter 9, Fig. 9.21, 9.29). The ventral region, characterized by low dpp activity, becomes the neurogenic region after invagination of the mesoderm during gastrulation. The expression of the transcription factor single-minded (sim) in the row of ventral-most cells leads to the expression of the EGF receptor ligand spitz and to a gradient of EGF receptor activity that patterns the epidermis from ventral to dorsal (see Chapter 9, Fig. 9.15). This is clearly reflected in the different rows of neural precursors which arise in response to

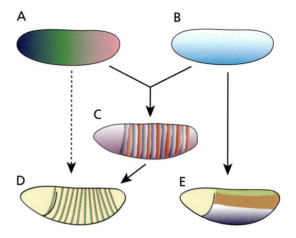

Fig. 11.14. Primary subdivisions of the ectoderm during blastoderm formation. During the formation of the blastoderm of *Drosophila*, interactions between maternal morphogens along the anteroposterior (AP) **(A)** and dorsoventral (DV) axes **(B)** trigger the subdivision (after cellularization and gastrulation) of the ectodermal primordium into periodic domains of gene expression along the AP axis **(C and D)** and dorsoventral territories along the DV axis **(E)**. The AP patterning system **(A)** activates first 'pair rule' genes **(C)**, which create an overlapping and transient pattern of transcriptional regulation with a two-segment periodicity that serves as a template for the expression of specific genes in one cell-wide stripe per segment **(D)**. These genes demarcate the basic segmental pattern of the larva. The dorsoventral patterning system **(B)** divides the ectodermal primordium into three broad regions that are evident after the mesoderm has invaginated during gastrulation **(E)**: the most dorsal region (green) gives rise to an extraembryonic tissue, the amnioserosa; the middle region (ochre) gives rise to the dorsal epidermis; and the ventral region (blue) is the neuroectoderm which contains precursors for the nervous system and the ventral epidermis. For details of the activities of these patterning systems, see Chapter 9.

different levels of MAPK activity (see Chapter 10, Fig. 10.29). The pattern of expression of dpp is rapidly refined to two parallel stripes: one most dorsally and the other most ventrally.

The pattern along the anteroposterior axis (Fig. 11.16) is triggered by a cascade of transcriptional regulation initiated by the bicoid transcription factor (see Chapter 9, Fig. 9.19, 9.20) that ultimately leads to the activation of a set of genes known as pair rule genes. Pair rule genes encode a varied collection of transcription factors with dynamic patterns of expression which, during the transition from a syncytium to a cellular blastoderm, become restricted

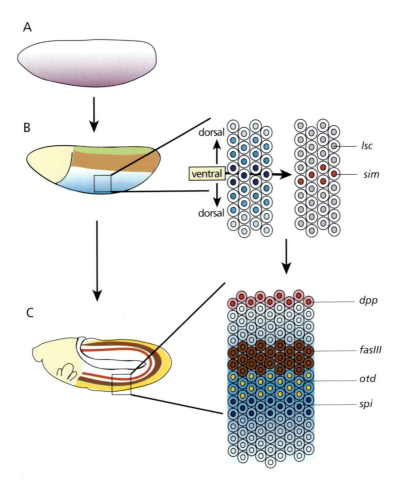

Fig. 11.15. Establishment of signalling references along the DV axis. (A) In the *Drosophila* blastoderm, a gradient of nuclear dorsal protein with a high point in the ventral region (purple) triggers a series of events that lead, after cellularization and gastrulation, to subdivision of the segmental ectoderm **(B)** into dorsoventral domains of gene expression that correspond with different fates (see Chapter 9, Fig. 9.21). The ventral ectoderm (blue) will give rise to the ventral epidermis and the central nervous system. One response to the dorsoventral patterning system is the expression of the transcription factor encoded by the *single minded* (*sim*) gene at the ventral midline (red), and of a gradient of another transcription factor, encoded by the *lethal of scute* (*lsc*) gene (grey), which outlines the primordium for the nervous system. **(C)** The patterning system that is set in motion during the blastoderm stage serves as an initial template for the appearance of spatially localized sources of signalling molecules after gastrulation; in particular, the EGF receptor ligand spitz along the ventral midline (blue) and dpp (red) midway through the dorsoventral axis. These regions act as sources of long-range patterning activities that subdivide the ectodermal primordium into different domains. In the example shown a gradient of spitz (blue) subdivides the ectodermal primordium into stripes of gene expression indicated by the expression of *fasciclin III* (*faslII*, brown) and *orthodenticle* (*otd*, yellow) (see also Chapter 9, Fig. 9.15).

to small overlapping domains within every alternate metameric unit. These domains define a transient periodic mosaic of gene expression which is transformed into cell identities through the activation of genes expressed periodically in every metameric unit.

The products of two pair rule genes, the homeodomain transcription factors even skipped (eve) and fushi tarazu (ftz), play crucial roles in the final stages of the subdivision of the ectodermal primordium and the definition of the signalling centres. In the cellular blastoderm, eve and ftz are transiently expressed in alternating graded stripes within which they regulate the expression of some of the genes required to initiate the patterning of the epidermis.

In particular, the expression of the signalling molecules hedgehog and wingless and of the transcription factor engrailed under the control of eve and ftz serves to define metameric units called parasegments which act as basic references for the patterning of the epidermis (Fig. 11.16). A parasegment is a metameric unit that is out of frame with the final segment and consists of the posterior half of one segment together with the anterior half of the adjacent segment. The expression of the pair rule genes *eve* and *ftz* each spans roughly the width of a parasegment.

The expression of *hedgehog* (*hh*) and *engrailed* (*en*) is activated by eve and ftz, whereas that of *wingless* (*wg*) is repressed by these proteins. As a consequence of these regulatory interactions, *hh* and *wg* come to be expressed in

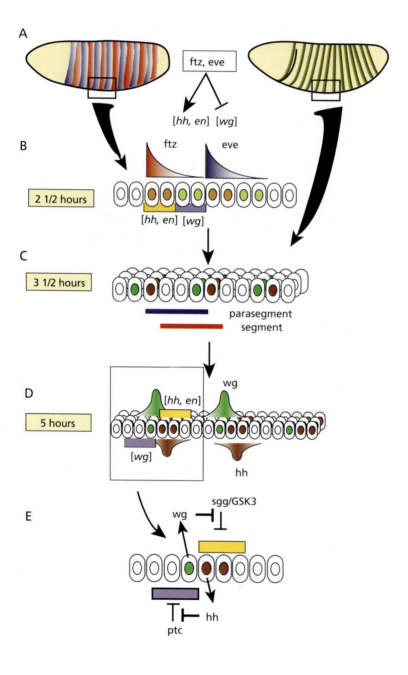

adjacent stripes of cells, with *hh* at the anterior (highest levels of eve and ftz) and *wg* at the posterior (lowest levels of eve and ftz) margin of each parasegment. This interface defines the parasegment boundary, which acts as the central reference for the generation of cell diversity and pattern in the larval epidermis. It also endows each metameric unit with an anteroposterior polarity that will be reflected in regional differences during differentiation. For this reason genes such as *hh* and *wg* have been called segment polarity genes. The expression of *hh* and *wg* becomes mutually dependent shortly after it is initiated; for example, in the absence of *wg*, the expression of *hh* is correctly initiated in adjacent cells but not maintained. The absence of *hh* has the same effect on *wg* expression (Fig. 11.16 and 11.17).

The function of eve and ftz is not only to initiate the expression of some segment polarity genes but, more importantly, to create domains of competence for their expression. For example, high levels of eve and ftz lead to the definition of a group of cells with the ability to express *hh* and *en*. However, this domain is bigger than that in which these genes are normally expressed. This is explained by two interacting effects: first, the stable expression of these genes requires high levels of wingless which,

in the wild type, are only received by the cells nearest to the source of this molecule (Fig. 11.16). Increasing the level or spatial domain of wingless reveals the domain of competence for the response to wingless (see Chapter 9, Fig. 9.13). In addition, the graded expression of eve and ftz along the AP axis is transformed into a gradient of competence, with anterior cells having a higher probability of expressing *hh* and *en* than posterior ones. This graded competence ensures a sharp response to the diffusing wingless protein.

On the other hand, since eve and ftz repress the expression of *wg*, their gradients of expression also lead to a gradient of competence for the expression of *wg* that is reinforced and maintained by the activity of hedgehog, in a similar way to that in which wingless reinforces the graded competence for the expression of *hh* (Fig. 11.16).

Expression of the pair rule genes ceases shortly after blastoderm stage and then the expression of *hh* and *wg* is maintained through reciprocal interactions which also serve to stabilize the parasegment boundary. Simultaneously with this event, the expression of genes encoding elements of the hh and wg signalling pathways is modulated in a periodic way (Fig. 11.18). These transitions from homogeneous to spatially modulated patterns of expression

---

Fig. 11.16. Establishment of signalling references along the AP axis. (A) The pair rule genes encode a family of transcription factors which represent the last layer of a cascade of transcriptional regulation that transforms a coarse-grained pattern of gradients of transcription factors into striped patterns of gene expression (see Figs. 3.18 and 9.19 for details). There are several members of the pair rule family and their interactions culminate in a pattern of seven stripes each of the homeodomain transcription factors fushi tarazu (ftz) (red) and even skipped (eve) (purple) (*left*). An important function of these transcription factors is to subdivide the embryo into segmental units along the anteroposterior axis by localizing the expression of transcription factors such as engrailed (en), and of signalling molecules, such as hedgehog (hh) and wingless (wg), to single stripes of cells per segmental unit (*right*). This pattern results from the ability of ftz and eve to activate the expression of *en* and *hh* and to repress the expression of *wg*. **(B)** The stripes of ftz and eve expression alternate and are graded along the AP axis. This results in the definition of two domains of gene expression within every stripe: an anterior domain (yellow) in which *hh* and *en* can be expressed through the activity of ftz and eve, and a posterior domain (purple) corresponding to low levels of ftz and eve, in which *wg* can be expressed because of a decay in the repressive activity of ftz and

eve. These domains are usually referred to as domains of competence for the expression of these genes. **(C)** With time, the activity of the pair rule regulators decays and the expression of *en/hh* (brown) and *wg* (green) becomes restricted to those cells within the competence domain that have a higher probability of expression. This pattern of gene expression defines a series of units along the AP axis that are called parasegments. The interface of *en/hh* and *wg* expression is called the parasegment boundary. These units are out of frame with the future segments of the larva by one cell. **(D)** As cells begin to proliferate, the expression of *en/hh* and *wg* is stabilized around the parasegment boundary, with *wg* expressed in only one cell at the posterior end and *en/hh* in about two or three cells within each parasegment. Bi-directional gradients of wingless and hedgehog proteins are generated from their spatially localized sources. **(E)** During these early stages, the expression of *en/hh* and *wg* is mutually interdependent: wingless is required for the expression of *en* and *hh*, and hh is required for the expression of *wg*, and these relationships create a positive feedback loop. The resulting patterns of expression reflect the fact that even though there are gradients of both hh and wg, the *en/hh* expressing cells have a lower threshold for the response to wingless than the *wg* expressing cells have for the response to hedgehog.

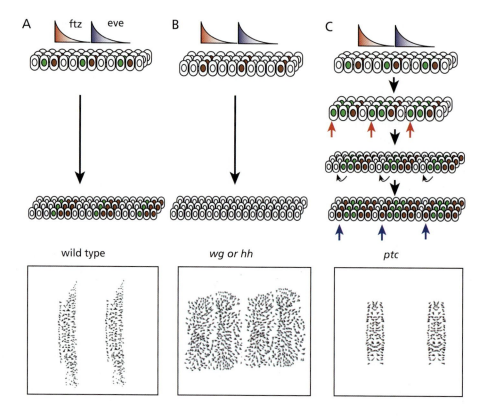

wild type   wg or hh   ptc

reflect both the activities of these signalling pathways (Chapters 4, 5, and 9) and the progressive emergence of pattern over the ectodermal cells. Thus, for example, the expression of the receptor patched (ptc) becomes restricted to the cells that receive high levels of the hh signal, while armadillo protein, the *Drosophila* homologue of β-catenin, increases in cells that receive the wingless signal (Fig. 11.18).

Similar interactions and changes in patterns of expression occur along the dorsoventral axis and two hours after the onset of gastrulation the epidermis of *Drosophila* contains stable signalling centres. Along the AP axis and centred around the parasegment boundary there is a group of cells expressing wingless and another group expressing hedgehog. Along the DV axis there are two stripes of dpp, and in the midventral region there is one stripe that acts as a source of the EGF receptor ligand spitz. In the following discussion we shall concentrate on the AP axis for it is patterning along this axis that has been studied in most detail and serves as a model for patterning of other systems.

Fig. 11.17. **Molecular basis of segment polarity mutant phenotypes. (A)** After the blastoderm stage and as the cells begin to proliferate, the expression of *hh/en* and *wg* that has been initiated by the pair rule genes, *ftz* and *eve*, becomes mutually dependent and contributes to the final pattern of the cuticle. **(B)** In the absence of *wingless* (*wg*) the expression of *hh* and *en* is initiated normally but it soon decays because there is no stabilizing input from wingless. The same is true for *wg* in the absence of *hh* and the resulting phenotype is very similar. **(C)** In the absence of *patched* (*ptc*) there is ectopic expression of *wg* within its domain of competence (red arrow). This extends the domain of wingless expression, which now reaches within the domain of competence for *hh/en* expression (black bent arrow). As a consequence of this signalling event and as the cells proliferate, a new stripe of *hh/en* expression (blue arrows) appears next to the new source of wingless. The pattern of the cuticle that emerges in a *ptc* mutant is not therefore a direct result of the loss of *ptc* function, but rather a regulated response of the cells to this primary event. (After Martinez-Arias, A., Baker, N., and Ingham, P. (1988) The role of segment polarity genes in the definition and maintenance of cell states in the *Drosophila* embryo. *Development* **103**, 157–70.)

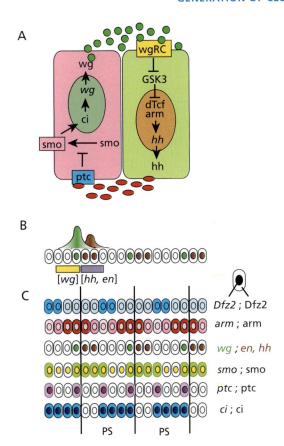

Fig. 11.18. Transcription of segment polarity genes and expression of their protein products during the establishment of the parasegment boundary. (A) Basic signalling events that stabilize the expression of *wg* and *hh/en* across the parasegment boundary during the early proliferative stages of *Drosophila* development (for details of these signalling pathways see Chapter 5, Fig. 5.22, 5.25). **(B)** Reciprocal signalling between hh and wg leads to the generation of two stable gradients of hh and wg proteins whose sources are out of phase by one cell. **(C)** Outline of cells within a few parasegmental (PS) units during the stabilization of the parasegment boundary (line). Patterns of transcription (shown as coloured shading in the nuclei and italicized symbols) and protein expression (coloured shading on cell surface and cytoplasm, and Roman symbols) of segment polarity genes during the stabilization phase. The patterns are the result of the interactions outlined in A. For example, although *armadillo* is expressed by all cells, its protein product is stabilized in cells that receive wingless signal above a certain threshold. A similar situation is observed with *smoothened*, whose protein product is stabilized in cells that receive high levels of hedgehog signalling. On the other hand, the expression of a wingless receptor, Dfrizzled2, is down-regulated by high levels of wingless signalling.

Generation of cell diversity with respect to the coordinates

The outcome of the patterning events mediated by regulated cell signalling over the epidermis of *Drosophila* is the pattern of the ventral cuticle of the *Drosophila* larva. This is a repeated mosaic of cell identities displayed by a sheet of cells about 12 cells wide along the AP axis and 36–40 cells wide along the DV axis. The pattern is a simple one: the anterior third contains six rows of denticles, contributed by six cells, and the rest is smooth in appearance. Each row of denticles is characterized by a particular shape, size, and polarity and the cells that make it up express a specific pattern of genes (Fig. 11.17). This means that there is an exquisite precision in the definition of cell fates at a single cell resolution. Mutations in the *wg* and *hh* genes have dramatic effects on this pattern, suggesting that these genes play an important role in its generation. However, while it is tempting to speculate that graded concentrations of these two proteins instruct this pattern directly, there is no direct evidence to support this idea. In fact, the loss of one function in a particular element of either of these signalling pathways does not generate a particular aberrant pattern. Instead, it triggers a series of alternative patterns of gene expression that are, eventually, translated into the pattern characteristic of that mutant (Fig. 11.17). Furthermore, mutants for *hh* and *wg* still have cell diversity, as reflected in the pattern of differentiation. Analysis of other mutants of this kind suggests a picture in which the initial pattern of wg and hh signalling outlined by the pair rule genes serves as a pre-pattern or template for new patterns of expression of signalling molecules which, both through their activity and their interactions with pre-existing molecules, generate complexity progressively. These studies suggest that pattern arises through a continuous interplay between short- and long-range influences in space and time. This idea is very similar to the proposals, discussed in Chapter 9, for how the coarse-grained information embodied in long-range 'morphogens' such as bicoid or spätzle is transformed into fine-grained information, and probably represents a universal strategy in pattern formation.

The origin of the pattern of the larval cuticle can be traced to shortly after the stabilization of the parasegmental signalling centre. At this stage, antibodies against the wingless protein reveal an asymmetric distribution of wingless expression: the protein can be easily detected anterior but not posterior to the cells expressing *wg*

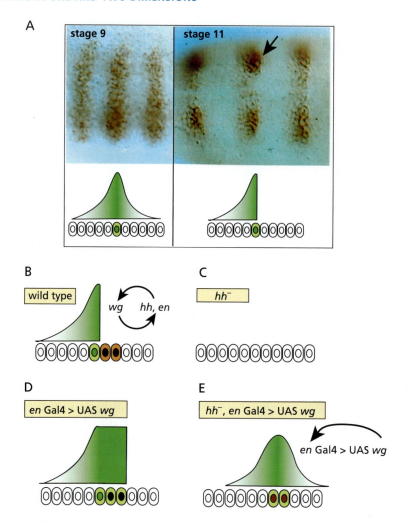

Fig. 11.19. Asymmetric signalling by wingless at the parasegment boundary. (A) About 2 hours after gastrulation, antibodies against the wingless protein reveal a change in the spatial distribution of wingless. Up to this time (*stage 9, left*) the protein can be observed on either side of the secreting cells but then (*stage 11, right*) the posterior diffusion of the protein appears to be impeded. The sharp boundary of protein expression is defined by the cells that express *en* and *hh*. **(B–E)** Experimental analysis of the influence of different molecules on the boundary of wingless protein expression indicates that the asymmetric distribution requires the activity of hedgehog. **(B)** In the wild type, the reciprocal interactions between wingless and hedgehog at the parasegment boundary maintain the expression of wingless and lead to its asymmetric distribution at stage 11. **(C)** In the absence of *hh*, no gradient of wingless is observed due to the loss of *wg* expression. **(D)** Expression of *wg* in the domain of *en/hh* does not abolish the sharp posterior boundary of wingless expression, but changes its position. The new posterior boundary coincides with that of the domain of expression of *en* and *hh*. Expression of *wg* in the domain of *en/hh* is achieved by placing the *wg* gene under the control of the Gal4-binding UAS (see Chapter 10, Fig. 10.2) which is then driven by a Gal4 protein expressed under the control of *engrailed*. **(E)** To assess the role of *hh* in the establishment of the sharp domain of wingless protein, *wg* is expressed in the domain of *hh/en* in a *hh* mutant. In this mutant, the normal expression of *wg* is lost and this should lead to the loss of *en* expression. However, the expression of wingless under the control of *en*Gal4 creates an artificial feedback loop that maintains expression of *en* in a *hh*-independent manner. This makes it possible to study whether hedgehog is required for the asymmetric distribution of the wingless protein. In these embryos, the sharp posterior boundary of wingless protein is abolished, indicating that hedgehog plays a role in its establishment. (After Sanson, B. *et al.* (1999) Engrailed and Hedgehog make the range of Wingless asymmetric in *Drosophila* embryos. *Cell* **98**, 207–16.)

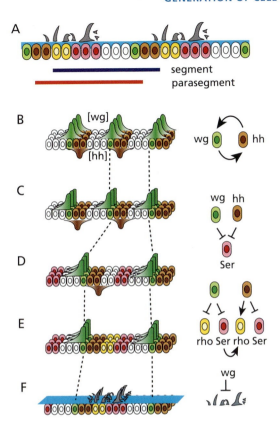

Fig. 11.20. **The generation of cell diversity along the AP axis.**
(A) The cuticle of the mature embryo is secreted by a layer of epidermal cells that display a complex pattern of gene expression which identifies almost every row of cells in a unique manner. The expression of different genes is indicated by different colours and the correlation with different denticle patterns is indicated. **(B–G)** Progressive emergence of complexity and pattern in the ventral epidermis of the *Drosophila* embryo. **(B)** Following gastrulation, interactions between hedgehog (hh) and wingless (wg) maintain the parasegment boundary as a patterning centre in every repeated unit (see Fig. 11.16 and 11.18). **(C)** At the onset of stage 11, the distribution of the wingless protein becomes asymmetric; wingless signals over a long range in the anterior direction and a short range (only to neighbouring cells) posteriorly. The gradient of hedgehog is still symmetric around its source. **(D)** Both wingless and hedgehog repress the expression of the Notch ligand Serrate (Ser) which, as a result, is expressed (pink) in the domain of low signalling by the two molecules. **(E)** Opposing effects of wingless and hedgehog on the EGF receptor ligands and regulators lead to the activation of EGF receptor signalling (represented by expression of rhomboid, rho, yellow) in between the cells that express Serrate and those that express hedgehog, and to the definition of those cells. **(F)** These interactions create a repeated pattern of wingless, hedgehog, EGF, and Notch signalling that is transformed into individual cell identities and patterns of differentiation. In addition to its effects on the expression of the different ligands, wingless suppresses the appearance of denticles over the domain in which the wg protein is expressed. (After Alexandre, C., Lecourtois, M., and Vincent, J. P. (1999) Wingless and Hedgehog pattern *Drosophila* denticle belts by regulating the production of short range signals. *Development* **126**, 5689–98 and references therein.)

(Fig. 11.19). This pattern presages the polarity of the final pattern of the cuticle, which is very different anterior and posterior to it. It is not yet clear what mediates the transition from the symmetric to the asymmetric pattern of wingless protein, however, experiments described in Fig. 11.19 suggest that it requires the activity of hedgehog.

From this stage onwards, cell diversity emerges within the metameric unit, orchestrated by interactions between wingless and hedgehog signalling. The resulting pattern of cell diversity will ultimately be used for the generation of different cell types, but can initially be detected in the spatial arrangements of the genes that will determine the final pattern (Fig. 11.20). Thus, the expression of the Notch ligand Serrate is repressed by Wingless and Hedgehog and therefore Serrate is only expressed out of the range of these signalling molecules. Serrate signals on either side of its domain of expression and, in combination with hedgehog, defines a domain of expression for rhomboid, an integral membrane protein that helps spitz activate the EGF receptor. Because spitz is ubiquitous at this stage, the pattern of rhomboid expression establishes a local pattern of EGF re-

ceptor activation as revealed by the expression of argos in a diffuse stripe (see Chapter 9, Fig. 9.15 for details of interactions (in a different context) between rhomboid, spitz and argos). This process of sequential local inductions establishes a very fine-grained pattern of receptors and ligands for the major signalling pathways which then cooperate to establish the differences between different rows of cells.

During this process, wingless, like other Wnts, stabilizes the expression of cell fates induced by a variety of short- and long-range influences. In particular, the expression of the wingless protein in a gradient anterior to the parasegment boundary establishes the 'smooth' cell fate by suppressing the formation of denticles (Fig. 11.20). Thus the asymmetric distribution of wingless during the second half of embryonic development is transformed into an asymmetric patterning of naked cuticle along the AP axis of each segment, with the range of wingless determining the domain of denticle development.

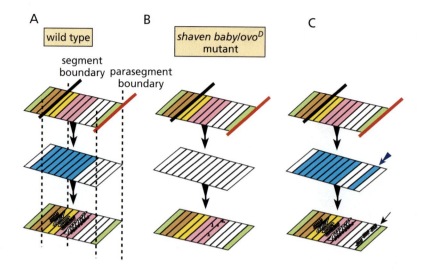

The emergence of the pattern

The final step in the generation of pattern is the translation of the different cell identities generated over time by cell interactions into patterns of cuticle secretion. The formation of denticles and hairs is associated with specific patterns of actin polymerization and a clear correlation can be established between the patterns of gene expression across the segment and the type of denticle that is formed (Fig. 11.21 and 11.22).

Genetic analysis identified a mutant, *ovo^D^/shaven baby* (*svb*), in which most denticles are abolished but there is no effect on cuticle secretion. The *ovo^D^/svb* gene encodes a zinc-finger DNA-binding protein and its expression correlates with the regions that make denticles ventrally. Ectopic expression of *ovo^D^* causes the development of ectopic denticles, leading to the conclusion that *ovo^D^* is likely to act as a final integrator of the cell interactions discussed above, defining a spatial domain of competence for the production of denticles. *ovo^D^* probably also interacts with key elements of these pathways to produce position-specific denticles (Fig. 11.22). The pattern of the cuticle also shows variations along the dorsoventral axis, suggesting that in addition to the signalling systems that operate along the AP axis, *ovo^D^* also integrates signals from the DV axis; however, these signals remain to be explored.

The cuticle of the *Drosophila* larva not only shows a position-dependent pattern within each segment, but also differences in pattern between different segments. For example, the pattern of denticles and amount of naked cuticle is very different in different segments (Fig. 11.13).

Fig. 11.21. Differentiation of the cuticle pattern. A transcription factor encoded by the *ovo^D^/shaven baby* (*ovo^D^/svb*) gene determines the competence to make denticles within the different cell identities established over time in the epidermis of the larva (Fig. 11.20). **(A)** In the wild type there is a clear correlation between the expression of *ovo^D^/svb* (blue) and the making of denticles and hairs in the cuticle. The mosaic of gene expression described in Fig. 11.20 is thus translated into a fine pattern of cell differentiation. **(B)** In the absence of *ovo^D^/svb* there are no denticles even though individual cells still acquire identities. **(C)** Expression of *ovo^D^/svb* in cells that do not normally make denticles (dark blue arrow) forces the appearance of denticles in an ectopic position.

These segmental differences suggest an interaction between the homeotic genes, which are involved in positional specification along the anteroposterior axis (Chapter 2, Fig. 2.8), and the genes that pattern the cuticle. Some of these interactions are likely to be mediated by specific molecules that act as linchpins between different molecular functions. An obvious element of the pattern that varies along the AP axis is the amount of smooth cuticle, which could be a target of this integration.

Development and patterning of the teeth

The developmental processes described in this chapter so far highlight the role cell interactions play in the generation and patterning of cell diversity. In the first example, we discussed the generation of a pattern, the nematode vulva, by a small number of cells with individual identities. In the second case, in the *Drosophila* larval epidermis, there

A

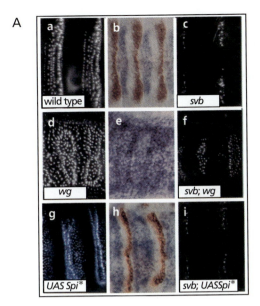

B

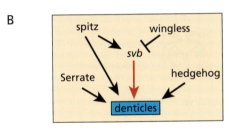

Fig. 11.22. Regulation of the spatial expression of *ovo^D/svb*.
(A) (a) Detail of the ventral cuticle of a wild-type embryo showing that the anterior region of each segment makes denticles. Two consecutive segments are shown and the denticles appear as white refringent dots in this preparation. (b) The domains of expression of *ovo^D/svb* (blue) and of *wingless* (brown) outline the pattern of denticles. The region that expresses denticles coincides with that which expresses *ovo^D/svb*. (c) In an *ovo^D/svb* mutant most of the denticles disappear and those that remain are very sparse. (d) In a *wg* mutant, the whole segment is covered with denticles. (e) This pattern correlates with the expression of *ovo^D/svb* in a *wg* mutant, which also occupies the whole segment. (f) The denticles that appear in a *wg* mutant require *ovo^D/svb* because they mostly disappear in a *wg*, *ovo^D/svb* double mutant. (g) Ubiquitous expression of an activated version of the EGF receptor ligand spitz (spi*), by use of the Gal4/UAS system, results in the appearance of ectopic denticles. (h) This ectopic expression correlates with expression of *ovo^D/svb* (blue), without changes in the expression of *wingless* (brown). (i) The denticles elicited by spi* disappear in the absence of *ovo^D/svb*. (From Payre, F., Vincent, A., and Carreno, S. (1999) *ovo^D/svb* integrates Wingless and DER pathways to control epidermis differentiation. *Nature* **400**, 271–5.) **(B)** The experiments discussed in A suggest that wingless represses and spitz/EGF receptor signalling activates the expression of *ovo^D/svb*. The pattern of denticles therefore results from the interactions shown in Fig. 11.21 and 11.22.

were a large number of cells, but they were all of the same type. During development, in many other situations, patterning is coordinated between two different cell types and requires continuous interactions between them. Interactions between epithelial and mesenchymal cells are one example, and represent a common feature in the patterning of many tissues. The development and patterning of vertebrate teeth is a good model for this type of interaction (Fig. 11.23).

Teeth, found only in vertebrates, are made of specialized sets of extracellular matrix (ECM) proteins, such as dentin and enamel, secreted by ectodermal cells. The teeth arise as a result of a series of reciprocal interactions between an epithelial cell population from the jaw and a mesenchymal cell population derived from a subpopulation of neural crest cells (see Chapter 10). Just like the segments of an insect or the different VPCs of the vulva, teeth have position-dependent characteristics: the molars (specialized for grinding) and the incisors and canines (specialized for biting and tearing) each form in distinct regions of the jaw.

Teeth are formed at periodic locations along the developing jaw. At each of these locations, the oral epithelium thickens to form the dental lamina and buds inward, into the mesenchyme, which condenses around the bud. The epithelium then folds and extends invaginations into the mesenchyme, forming the crown of the tooth and its characteristic cusp(s). At the stage when the crown of the tooth is beginning to form, it is possible to distinguish three types of cells in the developing tooth: the cells of the enamel organ, (including a specialized group of cells, the enamel knot) the dental papilla, and the dental follicle (Fig. 11.23). The enamel organ, derived from the epithelium in the region of the crown of the tooth, generates the ameloblasts which make enamel. The dental papilla, derived from the mesenchyme that underlies and contacts the enamel organ, makes the tooth pulp and the odontoblasts, which secrete dentin. The dental follicle, which is derived from the more peripheral mesenchyme that surrounds the developing tooth, gives rise to the cementoblasts and to the fibrous material that will connect the roots of the tooth to the bone.

As in the other patterning processes we have discussed, a phase of initiation and generation of these cell types is followed by one of morphogenesis and differentiation. In the case of the tooth, the phase of morphogenesis is all-

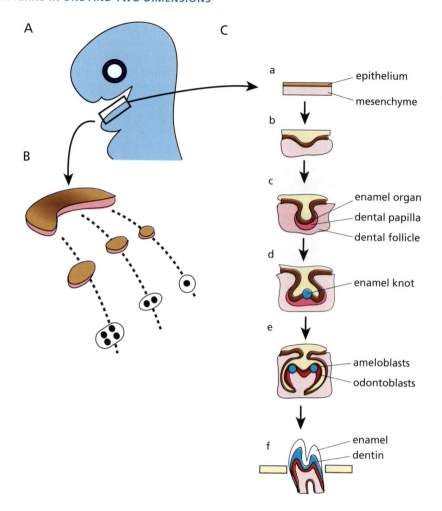

Fig. 11.23. Overview of tooth development and patterning in the mouse. (A) Teeth represent the outcome of an ordered series of cell interactions between a small set of epithelial and mesenchymal cells located in the anterior region of the head. The lower teeth arise in a structure called the first branchial arch (boxed), which gives rise to the lower jaw. **(B)** Within this region at about 10–12 days after fertilization, there is a thickening of the epithelium overlying mesenchymal tissue derived from the neural crest. The epithelium then condenses at specific periodic locations from where individual teeth will develop. **(C)** The development of teeth progresses through an ordered series of stages (a–f). (a, b) The epithelial condensations (called the dental lamina) bud inwards and give rise to the enamel organ. (c) During the 'bud stage' the mesenchymal cell population, which is derived from neural crest cells in the midbrain region, is divided into two. The population adjacent to and underlying the epithelium (the dental papilla) gives rise to the pulp and to the odontoblasts which will secrete the tooth-specific ECM molecules. The mesenchymal population that surrounds the developing tooth, the denticle follicle, gives rise to the cementoblasts and to the structures that will connect the tooth to the bone. (d) The 'cap stage' appears shortly after the establishment of these cell populations and is characterized by a specialized cell population of non-dividing cells that appears in the enamel organ: the enamel knot. This structure is important for the patterning of the tooth since there is an enamel knot associated with each of the cusps of the tooth. (e) At the 'bell stage', the cells that will differentiate the tooth begin to appear. The epithelium gives rise to the ameloblasts and the mesenchyme to the odontoblasts. (f) In the final stages of differentiation, the ameloblasts and odontoblasts produce the proteins that give strength to the tooth and the dental follicle develops the connections to the bone.

important, as it is on this that the shape of the tooth—the most important characteristic in defining its function—depends.

Initiation of tooth development

Tooth development is initiated with the determination, within the developing head, of the region where teeth will form. In the mouse, this region is marked by the expression of homeodomain transcription factors of the Lhx class (Lhx 6 and 7) in neural crest cells derived from the midbrain region which occupy a specific position within the head. Expression of these genes is initiated in the region of the neural crest mesenchyme that contacts the oral epithelium, suggesting a role for the epithelium in the initial stages of this process. The key signalling molecule in this interaction is probably FGF8, which is expressed in the oral epithelium at the right stage and can induce expression of the *Lhx* genes in culture systems derived from the developing jaw region.

Once the region has been defined, the initial stages of tooth development also depend on the epithelium overlying the neural crest-derived mesenchyme. Tissue recombination experiments in mice show that epithelium from stage 10 to 12 (i.e. from an embryo at 10–12 days after fertilization) can induce tooth formation in neural crest-derived mesenchyme that normally would not contribute to the teeth. The reverse, however, is not true, supporting the suggestion that the process is initiated in the epithelium. The first visible sign of tooth development is a thickening of the epithelium followed by a compaction at specific sites, around which mesenchymal cells also compact (Fig. 11.23). These compactions coincide with the expression of genes encoding the PRD domain transcription factor Pax9 and the homeodomain transcription factor Msx1 in small defined groups of cells within the mesenchyme. *Pax9* is essential for tooth development as mice mutant for this locus lack all teeth. The expression of *Pax9* and *Msx1*, like that of *Lhx6* and *7*, is promoted from the overlying epithelium, and members of the FGF family of signalling molecules (FGF8 and FGF9) are good candidates for this function (Fig. 11.24). BMP signalling both from the epithelium (BMP4) and the mesenchyme (BMP2) is also involved in defining the region of *Pax9* and *Msx1* expression. These BMPs activate *Msx1* expression; however, they suppress *Pax9* expression, thus locally antagonizing the inductive effect of FGFs and restricting the locations at which teeth can develop. Thus the interactions between FGF and BMPs are responsible for the spatial restriction of the tooth-making capacity (Fig. 11.24).

Other signalling systems are also known to be involved in the initiation and early stages of tooth development, though their relationships to FGF and BMP signalling in this system are unknown. Shh, for example, is expressed in the early thickening of the dental epithelium, perhaps under regulation by FGF signalling, and its receptor patched is expressed in the mesenchyme, but the function of Shh signalling at this early stage is not yet clear. The effects of mutations in the *Lef1* gene, which encodes a Tcf family transcription factor that mediates signalling through the Wnt pathway (see Chapter 5, Fig. 5.25) suggest a role for Wnt signalling as well: mice lacking *Lef1* form no teeth, arresting at the tooth bud stage.

The interactions described in the preceding paragraphs and summarized in Fig. 11.24 define the region from which some of the molars will develop. Other teeth arise later in development; their locations are probably defined by similar interactions between the epithelium and the mesenchyme, mediated by the same growth and transcription factors.

After some time, the epithelium loses its ability to influence the mesenchyme, which then itself acquires tooth-inducing abilities. This can be shown by simple tissue recombination experiments using mesenchyme and epithelium from two different species, one of which does not make teeth. For example, if frog mesenchyme is sandwiched with salamander flank epithelium, the epithelium will give rise to teeth. This shows that the reason the frog does not normally make teeth is not that its mesenchyme cannot send the necessary inductive signals, but that its epithelium cannot respond.

The rise in the expression of *Pax9* in the mesenchyme not only identifies the region from which teeth will arise, but also correlates with the acquisition of signalling properties by this group of cells, as *BMP4* is expressed under the control of Msx1 and Pax9, which now itself becomes refractory to inhibition by BMP4 (Fig. 11.24). The mesenchymal expression of *BMP4* is an important element in the acquisition of inductive ability by the mesenchyme. The switch in the hierarchy of control between Pax9 and BMP4, as first Pax9 and Msx1 are under the control of BMP4 signalling from the epithelium, and then BMP4 signalling from the mesenchyme is controlled by Pax9 and Msx1, exemplifies a fundamental mechanism underlying the interactions between different cell groups: the use of transcriptional regulation to locate and relocate sources of signalling molecules.

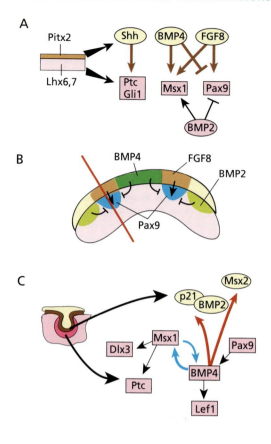

Fig. 11.24. Early stages of epithelial–mesenchymal interactions during tooth development. (A) One of the first signs of tooth development is the expression of genes encoding the Lhx6 and Lhx7 transcription factors in the mesenchyme (pink) where it contacts the oral epithelium (brown) in the first branchial arch. *Lhx6* and Lhx7 expression is probably initiated in response to FGF8 signalling from the oral epithelium, where another transcription factor Pitx2 is expressed very early in development and perhaps orchestrates the expression of various signalling molecules. This is followed by the setting up of signalling systems involving members of both the FGF and BMP families of signalling molecules. At these early stages, the inductive interactions (summarized in the figure) are from the epithelium to the mesenchyme. There is some evidence that signalling by Shh is also involved in the early stages of tooth development but its relationship to the other signalling systems is at present unknown. **(B)** The positions at which teeth will develop are highlighted by the local expression of *Pax9* in the mesenchyme. These positions are defined where BMPs (BMP2 from the mesenchyme and BMP4 from the epithelium) locally antagonize an inductive effect of FGF8 on *Pax9*. The localized expression of the BMPs is essential for the restriction of tooth development to specific sites, such as the one indicated by a red line. **(C)** At the bud stage the epithelium has lost its signalling capacity, which has now passed to the mesenchyme. Here, BMP4 is essential in establishing the expression patterns of several transcription factors, and in signalling to the epithelium to induce the expression of new genes that will promote its continued development. A positive feedback loop between Msx1 (induced in the mesenchyme from the epithelium) and BMP4 ensures the sustained expression of several genes in this group of cells.

Morphogenesis and differentiation

Once the bud of the tooth is formed, a phase of growth and morphogenesis begins that will lead to the development of individual teeth, each with a characteristic shape and size. There are two important stages in these processes, called the cap stage and bell stage because of the characteristic shape formed by the developing tooth (Fig. 11.23). The tooth bud develops into the cap as a result of patterned proliferation of the mesenchymal and epithelial cells. This is driven by signals from the mesenchyme, in particular BMP4.

A characteristic feature of the cap stage is the appearance in the epithelium of a transient population of cells called the enamel knot (Fig. 11.23 and 11.25). This is a group of cells, located centrally in the bud, which express the Cdk inhibitor p21 and the transcription factor Msx2 and which do not proliferate. The signal from the mesenchyme that induces the enamel knot has not been identified with certainty, but is likely to be BMP4. The cells of the enamel knot are the source of several signals that are likely to play a role in the morphogenesis of the tooth.

These signals include initially BMP2, 7, and Shh, and later BMP4 and FGF4. Circumstantial evidence indicates that these cells signal both to the mesenchyme and in the plane of the epithelium to direct the development and patterning of the tooth. This evidence consists of the striking correlation between the appearance of the knot and the events that pattern the tooth, and the known potential of the molecules that are expressed from the knot. In fact, it has been shown that the enamel knot can substitute for the ZPA in the patterning of the chicken wing (see Chapter 9, Fig. 9.3).

Signals from the enamel knot prepare the way for the formation of odontoblasts by the dental papilla, and of ameloblasts by the enamel epithelium. The onset of differentiation of these cell types is associated with the disappearance of the enamel knot, which undergoes apoptosis, and a change in the shape of the tooth as it enters the bell stage. At this point, the dental epithelium grows and spreads to generate the cusp(s): a single cusp in the case of incisors and canines, and more than one in the case of the molars. The formation of additional cusps in molars is as-

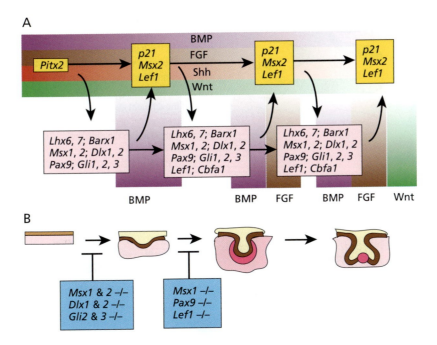

Fig. 11.25. Molecular networks during the growth and patterning of the teeth. (A) Reciprocal signalling events associated with different stages of tooth development. Epithelial sources and interactions are shown at the top (boxed yellow), and mesenchymal ones underneath (boxed pink). The coloured gradients indicate that varying concentrations of the signalling molecules shown play an important role in modulating the circuits of transcriptional regulation characteristic of each stage (i.e. that all the transitions between transcriptional states happen against a background of these signalling molecules). In the epithelium, all events require signalling by BMPs, FGFs, Shh, and Wnts. On the other hand, some specific signalling events from the mesenchyme require BMPs, FGFs or Wnt as indicated. **(B)** The stages of tooth development corresponding to the molecular events shown in A. Below the diagrams are shown the points at which mutations in the indicated genes result in an arrest in tooth development. In some cases, mutations in two genes (e.g. *Msx1* and *Msx2*) are needed to block a particular stage. As both genes are expressed in wild-type animals, this reveals a certain degree of redundancy. (After Jernvall, J. and Thesleff, I. (2000) Reiterative signalling and patterning during mammalian tooth morphogenesis. *Mech. Dev.* **92**, 19–29.)

sociated with the appearance of secondary enamel knots in these regions, expressing FGF4. It has been suggested that it is a combination of repressed proliferation in the enamel knots and stimulated proliferation in the surrounding epithelium (stimulated by signals from the knots) that brings about the characteristic morphological changes associated with cusp formation.

As terminal differentiation of the different cell types of the tooth is completed, the differentiated cells secrete the characteristic extracellular proteins that give teeth their enormous strength. The mesenchyme first expresses alkaline phosphatase and tenascin, and later osteocin and Type I collagen which provide the structural consistency characteristic of the final teeth. The ameloblasts secrete the enamel.

Different types of teeth

During the bell stage, the characteristic morphologies of the different types of teeth begin to become apparent. For example, incisor and molar tooth buds can be distinguished by their pattern of cusps. The development of these features is likely to be achieved by a transcriptional code that regulates the fine details of morphogenesis and the pattern of deposition of the ECM components.

It has been suggested that this transcriptional code may be set up by a positional code established early in tooth development by different combinations of homeodomain transcription factors. A striking pattern of overlapping expression domains of a set of homeodomain proteins has been observed in the mesenchyme at the stage when tooth development is initiated (Fig. 11.26). The homeobox genes expressed in the region where the teeth will develop are not members of *Hox* clusters. The proteins encoded by these genes include Msx1 and 2; Dlx1, 2, 3, 5, 6, and 7; Barx1,

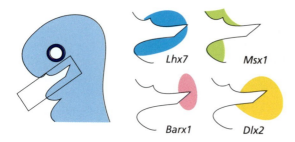

Fig. 11.26. An odontogenic homeobox code. Different home-obox genes are expressed at embryonic stage 10.5 in overlapping domains in the regions where teeth will develop. By analogy with the role of other members of this family in the establishment of differences along the anteroposterior axis of all embryos, it has been suggested that these genes provide a system of molecular information for the different classes of teeth. (After Tucker, A. and Sharpe, C. (1999) Molecular genetics of tooth morphogenesis and patterning: the right shape in the right place. *J. Dental Res.* **78**, 826–34.)

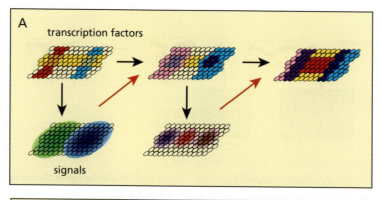

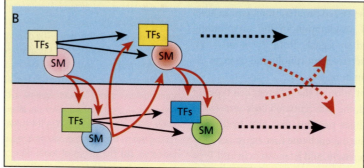

Fig. 11.27. Principles for patterning cellular assemblies in two dimensions. (A) A sheet of cells is endowed with a distribution of cell states that can be represented, for example, by individual cells or/and groups of cells expressing certain transcription factors (*top row*). Superimposed on this landscape there is an arrangement of signalling molecules (*bottom row*) that will interact with the transcription factors to generate new landscapes of transcription factors and signalling molecules. At each stage, the complexity of the pattern increases, until the cells differentiate. **(B)** Patterning of cellular ensembles in space results from sequential and mutually dependent interactions between suites of signalling molecules (SMs) and transcription factors (TFs). Thus, a particular stage has a spatial distribution of TFs and SMs and this leads to the establishment of a new suite of molecules (indicated by a different colour). If this process occurs in two layers of cells, for example as happens during development of the teeth, it can trigger a sequence of interactions that can persist for some time and pattern particular tissues.

Otlx2, and Lhx6 and 7. Barx1, Dlx1, and Dlx2, for example, are restricted to the presumptive molar region while Msx1 is only detected in the region where incisors will develop. Lhx6 and 7 are expressed throughout the oral region. Mice mutant for both *Dlx1* and *Dlx2* form no molars, supporting the involvement of these factors in specifying the molar tooth identity, but it has not been possible to demonstrate transformation of one tooth type into another as a result of mutations in any of these homeodomain factors. Nor is it clear how or when the proposed combinatorial code of homeodomain proteins is established.

General principles for the patterning of two-dimensional cellular assemblies

The emergence of pattern and shape during development is usually coupled to the appearance of different kinds of cells within a population whose growth is spatially constrained (e.g. in the form of an epithelial sheet). Within these constraints, the generation of cell diversity depends on the setting up of localized sources of long-range signalling molecules that create regions of competence and coarse-grained patterns of gene expression within the sheet. Within these patterns, new signalling centres emerge which initiate new rounds of long-range signalling or which mediate short-range interactions that refine the initial coarse-grained patterns of gene expression. Sucessive rounds of molecular interactions of this kind, in which a particular spatial pattern of signalling molecules and transcription factors serves as a pre-pattern for a new one, create complexity and pattern within the cell population (Fig. 11.27). Often, these regulatory interactions are coupled to patterns of cell proliferation that can serve to alter some parameters of long-range signalling and of cell interactions.

These general principles for the patterning of spatially constrained groups of cells can be extrapolated to two populations of cells (e.g. during epithelial–mesenchymal interactions). In this case, the pre-pattern of one population not only serves to pattern the sheet of cells in which it arises but also acts on the adjacent sheet and influences its development and patterning. This pattern will in turn feed back on the first one, creating an interdependence that can ensure coordinated patterning of the two populations (Fig. 11.27).

The large repertoire of transcription factors involved in the generation of patterns contrasts with the small number of different signalling molecules and explains how the same signalling event can produce different outcomes in different contexts.

SUMMARY

1. The simplest form of spatial organization of different cells in development is the formation of rows of cells within which different identities are defined. A variation of this basic theme is the generation of different stripes of cell identities within a sheet or an epithelium.

2. Patterns always emerge progressively through a sequence of long- and short-range inductive interactions. Usually, the process is initiated by the activity of long-range signalling molecules that create domains of competence where coarse-grained patterns of cell identities are specified. These patterns are then used to establish new sources of long-range signalling molecules or to mediate local cell interactions which refine the pattern induced by the initial signalling event.

3. The vulva of the nematode *Caenorhabditis elegans* provides an example in which three cells within a row of six cells acquire different identities through a sequence of long- and short-range interactions. This pattern of cell identities is then transformed into a three-dimensional structure, the vulva, when the individual cell identities manifest themselves during differentiation.

4. The larval epidermis of *Drosophila* provides an example of the generation and patterning of a large number of cell identities within a sheet of cells. This is achieved with reference to two axes, anteroposterior and dorsoventral, through the coupling of long-range signalling events to local cell interactions and proliferation.

5. The patterning of the *Drosophila* larval epidermis also shows how the same signalling molecules (in this case wingless and hedgehog) elicit different responses at different stages of development because the cells on which they act have different molecular constitutions.

6. In some situations, pattern results from interactions between two sheets of cells which regulate each other's patterning through sequential rounds of signalling events. During the development and patterning of the vertebrate teeth, for example, two tissues—one an epithelium and the other mesenchymal—pattern each other in a coordinated way.

7. A general principle of the patterning of these sheets of cells is that patterning relies on the activity of signalling centres that are fixed during a particular phase of development, but that change in location and activity over developmental time with a complexity that matches that of the pattern to which they contribute.

COMPLEMENTARY READING

Alexandre, C., Lecourtois, M., and Vincent, J. P. (1999) Wingless and Hedgehog pattern *Drosophila* denticle belts by regulating the production of short range signals. *Development* **126**, 5689–98.

Bejsovec, A. and Martinez Arias, A. (1991) Roles of *wingless* in patterning the larval epidermis of *Drosophila*. *Development* **113**, 471–85.

Greenwald, I. (1997) Development of the vulva. In *C. elegans II*. D. Riddle, T. Blumenthal, B. Meyer, and J. R Priess (Eds), pp.519–44. Cold Spring Harbor Laboratory Press.

Ingham, P. (1988) The molecular genetics of embryonic pattern formation in *Drosophila*. *Nature* **335**, 25–34.

Jernvall, J. and Thesleff, I. (2000) Reiterative signalling and patterning during mammalian tooth morphogenesis. *Mech. Dev.* **92**, 19–29.

Neubeüser, A., Peters, H., Balling, R., and Martin, G. (1997) Antagonistic interactions between FGF and BMP signalling pathways: a mechanism for positioning the sites of tooth formation. *Cell* **90**, 247–55.

Sanson, B., Alexandre, C., Fascetti, N., and Vincent, J. P. (1999) Engrailed and Hedgehog make the range of Wingless asymmetric in *Drosophila* embryos. *Cell* **98**, 207–16.

Chapter 12

Patterns in three dimensions

In this chapter, we continue with the progression in complexity that we began in the previous two chapters, where we discussed the generation of cell diversity (Chapter 10) and the patterning of diversified cell types within rows and sheets of cells (Chapter 11). Here, we discuss the development of patterns that arise from the organization of the cells in three 'cellular dimensions'.

All structures within developing embryos are three-dimensional but, in many cases, as in the neural tube, they arise from the ordered folding of an epithelium, and the developmental coordinates of a cell can be related to the two-dimensional plane of the epithelium from which it originated. In this chapter we shall discuss structures, such as limbs, where cells create a three-dimensional structure by computing information from two very different sources: from an epithelial plane, and also from a third dimension that allows them to grow and be patterned outwards from that plane. In this case, the three-dimensional structure does not arise from a simple folding of a sheet of cells but from information processing along three axes (Fig. 12.1). The situation is further complicated because in cases like the limb, the structures are composed of more than one layer of cells. Therefore, to ensure that the third dimension develops proportionately, exchanges of information between layers of cells also occur, similar to those described for the *Caenorhabditis elegans* vulva and the vertebrate teeth in Chapter 11. Because these interactions contribute to the final shape, they result in more than three developmental 'dimensions' to consider. A variation of these three-dimensional patterns involves a departure from a uniform solid geometry by the development of bifurcations of the basic pattern. Sequential bifurcations generate the sorts of complex, branching structures that characterize, for example, the development of the respiratory tracts of insects (tracheae) and vertebrates (lungs), or the vascular system of vertebrates.

The available evidence indicates that the generation and patterning of such complex structures relies not on a single signalling centre or on interactions between different signalling centres, but on a temporal succession of patterning centres that depend upon each other for their establishment and function. These signalling centres often make use of the same signalling molecules (FGF, Wnt, hh, EGF, and TGFβ/BMP) but they direct different outcomes at different stages of the patterning process, as a result of the different competence of the responding tissue. It is the information from these centres, coordinated and integrated in space and time, that generates pattern. Although considerable information has been accumulated about the role of these signalling networks in complex pattern formation, less is known about how the signalling information is interpreted by target cells.

Limbs and appendages

The wings of a bird or the the legs of a vertebrate develop as growth-driven external projections from the body wall of embryos (Fig. 12.1 and 12.2). This projection provides a third coordinate to the anteroposterior (AP) and dorsoventral (DV) ones that characterize the sheets of cells that form the body wall of the embryo. This third coordinate is called proximodistal (PD) with reference to the body wall.

The information processed by the cells in a developing limb or appendage affects both growth and patterning and ensures that the final structure is properly proportioned and adapted to its function. In vertebrates, but not in insects, this process of integration has an added layer of complexity in that limbs are the result of interactions between the ectoderm and the mesoderm; the exchange of information across these layers adds another variable that has to be considered in the analysis of limb development. The importance of the interactions between mesoderm and ectoderm in the final pattern can be demonstrated in simple experiments in which mesoderm and ectoderm from different regions of the embryo are exchanged and the result

Fig. 12.1. **From one dimension, to two dimensions, to three dimensions.** When cells grow out and away from a plane defined by the anterior (A), posterior (P), dorsal (D), and ventral (V) coordi- nates, they create structures along a new axis: the proximodistal axis. This requires the activity of molecules that promote, direct, and control growth and pattern along this axis.

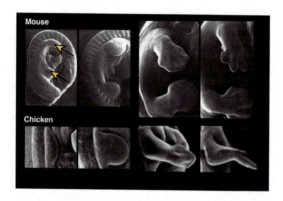

Fig. 12.2. **Examples of limbs and appendages.** Development (from left to right) of the limbs in a mouse (top) and a chicken (bottom) embryo. In the mouse, the fore and hind limbs appear first as bulges (arrows), at specific positions with respect to the de- veloping somites. The developing limbs then grow outward and become patterned. Note the emergence of the digits as the embryo and the limbs grow bigger. In the chicken only the forelimb is shown. This gives rise to the wing, which is a modification of the mouse forelimb. Note the similarity between the two limbs early in development. (Images courtesy of J. C. Izpisua Belmonte.)

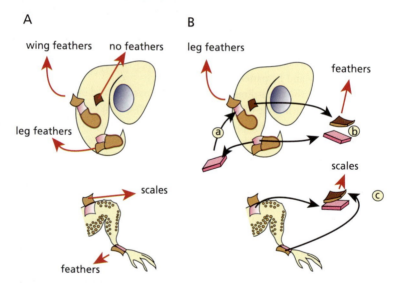

Fig. 12.3. **The mesoderm instructs the pattern of the limb ectoderm. (A)** The ectoderm of the wing and leg primordia of the chick embryo (*top*) gives rise to limb-specific patterns of feath- ers. There are also regions of the ectoderm that do not give rise to feath- ers at all. The ectoderm of the hindlimb, shown at a later stage of develoment (*bottom*), gives rise to scales or feathers depending on its position in the limb. **(B)** Recombinations between ectoderm and mesoderm indicate that the mesoderm is able to instruct the patterning of the ectoderm. (a) If the mesoderm from a leg bud is placed underneath the ectoderm of a wing bud, the ectoderm develops feathers characteristic of the leg rather than normal wing feathers. (b) Ectoderm that would normally not give rise to feathers, produces feathers when it is placed over mesoderm from a region that normally gives rise to feathers (c) Mesoderm from the region of a leg that normally produces scales, induces scales from ectoderm that would normally only produce feathers. (Adapted from Wessells, N. K. (1977) *Tissue interactions and development.* Benjamin.)

of the interaction observed (Fig. 12.3). For example, if chicken leg mesoderm is placed underneath the ectoderm that will give rise to the wing, it induces this tissue to produce leg feathers instead of the wing feathers that would normally be produced. Similar exchanges indicate that the mesoderm can instruct the ectoderm as to the structures that it will produce (Fig. 12.3).

Initiation of vertebrate limb development

In vertebrates, the development of limbs is initiated at specific AP and DV coordinates of the body wall. On the flank of the embryo, as somitogenesis proceeds, a domain comprising ectoderm and mesoderm becomes competent to develop limbs during stages 12 to 16 in the chicken and 9 and 10 in the mouse (Fig. 12.4). This area of competence can be revealed because it, and no other, can produce limbs if it is treated with an inducing agent, or if another part of the embryo is grafted to it (see also Chapter 1, Fig. 1.5). Within the competent region there is a fore- and a hindlimb domain which overlap, so that extra limbs that are artificially induced are either fore (anterior) or hind (posterior), depending on the position of the inducing agent within the area of competence (Fig. 12.4). These domains are examples of classical developmental fields; the cells within them have the prospective potency to become limb cells, while those in the regions where the limbs actually develop have limb identity as their prospective fate.

The domains from which the limbs develop are associated with the expression of three transcription factors: two T-box factors (Tbx5 for the fore- and Tbx4 for the hindlimb) and a PRD domain homeodomain protein (Pitx1) for the hindlimb (Fig. 12.4). Ectopic expression of these transcription factors can induce partial transformations of limbs. For example, forced expression of *Tbx5* in the hindlimb induces patterns of differentiation characteristic of a wing. However, these transformations are incomplete, suggesting that there is no simple code that specifies either fore- or hindlimb and that these transcription factors are operating in a more complex molecular environment about which we still know very little. Within the limb competence region, the expression of these transcription factors is followed by the appearance of bulges, the limb buds, which comprise ectoderm and mesoderm. At about the middle of the DV axis of the incipient bud a conspicuous thickening of the ectoderm emerges, the apical ectodermal ridge (AER) which, as it grows, defines the

distal-most point of the developing limb. The mesoderm and the ectoderm then proliferate along the PD axis, to generate the limb.

Our understanding of the early stages of limb development has come, primarily, from studies in chicken embryos. However, recent work with the mouse suggests that the basic process is the same in all vertebrates and that, as is always the case, the lessons from one embryo, for example the chick, can be applied (with a few strategic variations of gene deployment—see Chapter 10) to other vertebrate embryos. Fate-mapping studies and other experiments indicate that the limb is the outcome of a sequence of interactions between the mesoderm and the ectoderm (Fig. 12.5) that are reminiscent of the epithelial–mesenchymal interactions we have described for the development of the teeth (Fig. 11.24, 11.25). Transplantation experiments suggest that the initial signal arises from within the somites and results in a region of the intermediate mesoderm (the mesoderm nearest to the somite) and lateral plate mesoderm (between the intermediate mesoderm and the ectoderm of the body wall) becoming competent to define the overlying ectoderm as limb ectoderm.

The observation that certain tissues from developing embryos, for example an anteriorly located region within the head of the salamander, can induce the appearance of extra limbs if placed within the limb field (see Chapter 1, Fig. 1.5) suggests the existence of signalling molecules that can elicit the initial stages of limb development within a competent region. Beads soaked in FGF or forced expression of Wnt proteins will mimic this activity, indicating that these signalling molecules are active agents in the induction of limb development (Fig. 12.6). Studies on the expression of two *Wnt* genes (*Wnt2b* and *Wnt8c*) and of two *FGF* genes (*FGF8* and *FGF10*) during the initial stages of development, and regulatory interactions between them, have led to a model for the initiation of limb development (Fig. 12.6). In this model, an FGF8/FGF10 signalling relay that is initiated in the intermediate mesoderm and which is mediated and coordinated by Wnt signalling leads to the definition of the AER. FGF8, probably from the intermediate mesoderm, induces expression of *FGF10* in the lateral plate mesoderm and the activity of two Wnt proteins (Wnt2b and Wnt8c) restricts *FGF10* expression to the positions on the AP axis at which the induction of the fore and the hind limb bud occurs. FGF10 then induces the expression of *FGF8* in the overlying ectoderm. FGF10 probably acts through another Wnt protein, Wnt3a, which responds to FGF10 and is capable of activating the expression of *FGF8* in the ectoderm. The expression of *FGF8* in

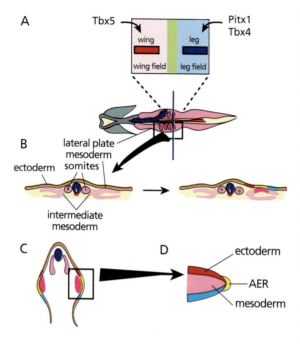

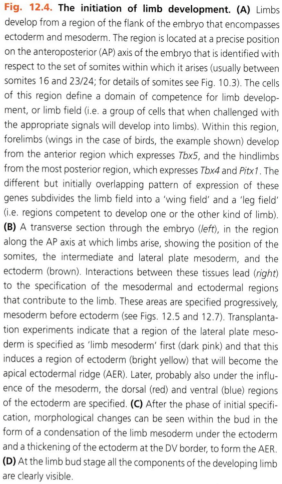

Fig. 12.4. The initiation of limb development. (A) Limbs develop from a region of the flank of the embryo that encompasses ectoderm and mesoderm. The region is located at a precise position on the anteroposterior (AP) axis of the embryo that is identified with respect to the set of somites within which it arises (usually between somites 16 and 23/24; for details of somites see Fig. 10.3). The cells of this region define a domain of competence for limb development, or limb field (i.e. a group of cells that when challenged with the appropriate signals will develop into limbs). Within this region, forelimbs (wings in the case of birds, the example shown) develop from the anterior region which expresses *Tbx5*, and the hindlimbs from the most posterior region, which expresses *Tbx4* and *Pitx1*. The different but initially overlapping pattern of expression of these genes subdivides the limb field into a 'wing field' and a 'leg field' (i.e. regions competent to develop one or the other kind of limb). **(B)** A transverse section through the embryo (*left*), in the region along the AP axis at which limbs arise, showing the position of the somites, the intermediate and lateral plate mesoderm, and the ectoderm (brown). Interactions between these tissues lead (*right*) to the specification of the mesodermal and ectodermal regions that contribute to the limb. These areas are specified progressively, mesoderm before ectoderm (see Figs. 12.5 and 12.7). Transplantation experiments indicate that a region of the lateral plate mesoderm is specified as 'limb mesoderm' first (dark pink) and that this induces a region of ectoderm (bright yellow) that will become the apical ectodermal ridge (AER). Later, probably also under the influence of the mesoderm, the dorsal (red) and ventral (blue) regions of the ectoderm are specified. **(C)** After the phase of initial specification, morphological changes can be seen within the bud in the form of a condensation of the limb mesoderm under the ectoderm and a thickening of the ectoderm at the DV border, to form the AER. **(D)** At the limb bud stage all the components of the developing limb are clearly visible.

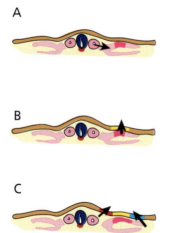

Fig. 12.5. Interactions between ectoderm and mesoderm during the initiation of limb development. The early stages of limb development are brought about by interactions between the mesoderm and the ectoderm. **(A)** The first step in the development of the limb is the definition of limb bud mesoderm. Transplantation experiments indicate that this relies on a signal (arrow) from the region of the somites and the intermediate mesoderm that selects some cells from the lateral plate mesoderm. **(B)** These cells then signal to the overlying ectoderm and induce it to become limb bud ectoderm. Fate-mapping studies indicate that this first induction specifies a subset of ectodermal cells that will become the AER (yellow). **(C)** Later, a second round of signalling from the region of the somites and intermediate mesoderm, and from the lateral plate mesoderm instructs the acquisition of dorsal (red) and ventral (blue) characteristics in both the mesoderm and the ectoderm. It is not clear if the ectoderm is patterned at the same time as the mesoderm or if it is patterned from the mesoderm once the mesoderm has acquired dorsal and ventral characteristics.

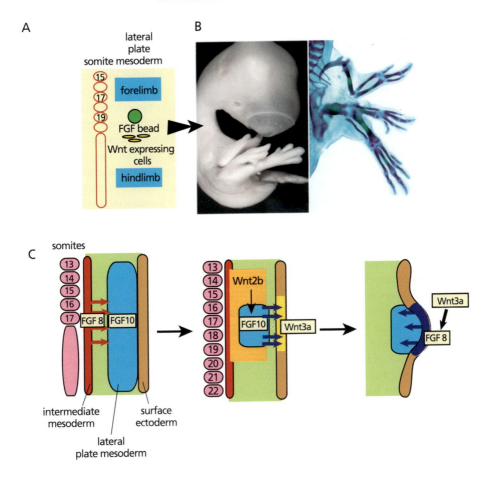

Fig. 12.6. The role of FGF and Wnt signalling in the initial stages of limb development. (A) Diagrammatic longitudinal section through a chick embryo in the limb-competent domain, showing an experiment in which either a bead soaked in FGF or cells expressing Wnt proteins are implanted between the positions where fore- and hindlimbs normally arise. **(B)** The result of this experiment is the appearance of an ectopic limb from the region where the bead was implanted. **(C)** Model for the action of FGFs in the induction of limb development. FGF (probably FGF8) signalling from the intermediate mesoderm, which is the mesoderm that lies between the somites and the lateral plate mesoderm, activates the expression of *FGF10* throughout a large region in the lateral plate mesoderm. *Wnt2b*, expressed by both the intermediate and lateral plate mesoderm, restricts the expression of *FGF10* to the region of the forelimb mesoderm (*Wnt8c* performs a similar role in the hindlimb region). Once restricted, FGF10 activates the expression of *FGF8* in the ectodermal region that will become the AER. FGF10 probably acts by inducing Wnt3a, which in turn induces the expression of *FGF8* in the AER. FGF8 then feeds back to the mesoderm to maintain the expression of *FGF10*, thus creating a feedback loop that maintains FGF signalling within and between these two tissues. (Adapted from Martin, G. (1998) The roles of FGFs in the early development of vertebrate limbs. *Genes Dev.* **12**, 1571–86; and Kawakami, Y. *et al.* (2001) WNT signals control FGF-dependent limb initiation and AER induction in the chick embryo. *Cell* **104**, 891–900.)

the ectoderm is restricted to the future AER and reinforces *FGF10* expression in the lateral plate mesoderm, thereby setting up a positive feedback loop (Fig. 12.6).

Establishment of AP and DV coordinates in the limb bud

Although we have a detailed description of the early stages of limb development and clear evidence that an interplay between FGFs and Wnts is important in this process, very little is known about what restricts the appearance of the limb primordia to their normal positions along the body axis. It is likely that the codes of *Hox* gene expression that define position along the AP axis (see Chapter 2, Fig. 2.8) contribute to specifying the position of the limb field as well as to the onset of expression of *Tbx4*, *Tbx5*, and *Pitx1*. Forelimbs always develop at the anterior border of expression of *Hoxc6* and *Hoxb8*, and hindlimbs near the anterior border of *Hoxc9* expression. Less is known about positioning along the DV axis. Some genes show localized expression patterns along this axis, suggesting that they might play a role in its definition. For example, *Wnt7a* is expressed in the dorsal ectoderm and *En1* is expressed in the ventral ectoderm. However, mouse embryos mutant for these genes initiate limb development normally, suggesting that, as is often the case, no single factor is responsible for instructing this complex developmental event. Often, functional redundancy between members of a gene family obscures the effects of loss of function of one member.

Because the limb bud emerges from the body wall, it is logical to think that it will inherit some of the body wall's AP and DV coordinates. Experimental manipulations show that, to a certain degree, this is the case, but they also show that the developing limb has an autonomous set of coordinates (Fig. 12.7). These experiments, in which a limb primordium is rotated with respect to both sets of coordinates at different times of development, show that the AP coordinates of the limb are established first and are probably derived from the main axis of the embryo. The DV axis is established later in a process that it is associated with the mesodermal–ectodermal interactions that define the limb bud (Fig. 12.7).

Initiation of appendage development in insects

The wings and legs of *Drosophila* have provided a very good model system for studying the molecular mechanisms that initiate limb development and endow the limb primordium with coordinates for patterning. In keeping with the general theme that we have encountered throughout this book, some of these molecular mechanisms are conserved in vertebrates, but with modifications to accommodate interactions between the mesoderm and ectoderm.

Like the limbs of vertebrates, the appendages of *Drosophila* also develop at defined AP and DV positions along the body axis. In this case, however, no mesodermal–ectodermal interactions are involved. Instead, cells from the ectodermal sheet are allocated to become appendages at coordinates selected by the intersection of two molecular systems within the ectoderm itself. Appendages develop from the three thoracic segments that are characterized by the combinatorial expression of the three *Hox* genes *Sex combs reduced*, *Antennapedia* and *Ultrabithorax* (see Chapter 2, Fig. 2.8). The precise AP and DV coordinates within every segment selected by the *Hox* positional system are determined by interactions between signalling from the BMP homologue dpp, the Wnt signalling molecule wingless, and the EGF receptor ligand spitz. Wingless creates a competence domain along the AP axis that becomes restricted to a particular region along the DV axis by antagonistic activities of dpp and spi (Fig. 12.8). Within this region the wing and the leg initially share a common primordium which then splits into two following changes in the patterns of expression and activity of dpp and spitz. This relative movement of the primordia is dependent on the levels of dpp and of spitz signalling (Fig. 12.8). The initial common primordium may be explained by the suggestion that *Drosophila* wings evolved from legs as auxiliary appendages that only recently became adapted for flying.

Cellular parameters of vertebrate limb development

Classical embryological experiments revealed that within the incipient limb bud there are specialized populations of cells that play different roles during the development and patterning of the limb. Two regions can be distinguished morphologically: the ectodermal thickening known as the AER and a group of mesodermal cells that lie underneath it, called the progress zone (PZ). The cells of the PZ proliferate and give rise to cartilage, whose patterning provides classic landmarks for experimental analysis of the patterning of the limb and muscle. A third specialized cell population, the zone of polarizing activity (ZPA), cannot be distinguished morphologically but was discovered through transplantation experiments (Fig. 12.9).

The functional importance of these populations of cells can be demonstrated experimentally. For example, removal of the AER abolishes limb development, and grafts of extra AERs, within the limb field on the flank of the embryo or within the bud itself, trigger the development of

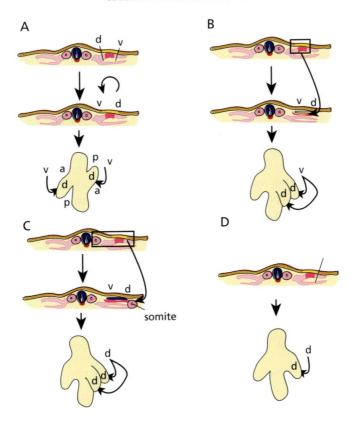

Fig. 12.7. Definition of coordinates in the developing limb of a chicken embryo. (A) 180° rotation of the presumptive wing mesoderm and ectoderm early in development (stage 13) results in the development of a wing with a normal DV polarity indicating that, by this stage, the DV axis has not been defined. In contrast, the AP polarity is inverted, indicating that the AP axis has been defined by this stage. The cartoon of the embryo shows a control limb (left) and an experimental one (right). **(B)** If presumptive wing ectoderm and mesoderm are excised from an embryo, rotated by 180° and placed between the mesoderm and ectoderm of a normal limb primordium, an ectopic limb arises that has the same DV polarity as the endogenous limb. This indicates that the DV polarity of the transplanted tissues has not been determined by the time of the transplant, and that they receive the same influences as the endogenous limb does. **(C)** If, in the same experiment as B, a piece of a somite is also transplanted, both the ectopic limb and the endogenous limb develop a mirror image polarity with two dorsal surfaces. Together with B, this experiment suggests that signals from the somite are important in establishing the DV polarity of the limb and that, in particular, they have the ability to determine the dorsal region first. **(D)** In this experiment, a filter is placed between a region of the lateral mesoderm and the prospective limb mesoderm. As a result, the limb develops with a double dorsal surface indicating that, like dorsal identity, ventral identity is also induced, in this case from the lateral mesoderm. (After Michaus, J., Lapointe, F., and Le Douarin, N. (1997). The dorsoventral polarity of the presumptive limb is determined by signals produced by the somites and the lateral somatopleura. *Development* **124**, 1453–63.)

extra limbs (Fig. 12.10). However, only mesoderm that is both within the limb field and underneath the AER is competent to respond; substitution of limb bud mesoderm by any other kind of mesoderm also abolishes limb development (Fig. 12.10). Thus an essential element in the development of the limb is the concerted activity of the AER on the PZ.

Transplantation experiments identified another region in the mesoderm that is important for the patterning of the limb: the ZPA. Transplantation of cells from the posterior region of the mesoderm to the anterior region of a developing bud induces a mirror image duplication of the pattern of digits (Fig. 12.10). The activity of the cells of the ZPA depends on two variables: their location in the bud and how long they reside in it (Fig. 12.10). The most pos-

terior digit always arises closest to the ZPA, with more anterior ones appearing progressively further away from it. The temporal factor is demonstrated by the observation that short exposures to the ZPA lead to the appearance of digits usually located further from the posterior region, while longer exposures induce progressively more posterior digits. These experiments led to the suggestion that the ZPA is the source of a substance that diffuses across the bud and specifies different digits in a concentration-dependent manner, with the highest concentration specifying the most posterior digits (see Chapter 9, Fig. 9.3 and 9.5). This hypothesis also explains the time effect since responses to higher concentrations would take a longer time to be elicited.

Altogether, these experiments led to the idea that the

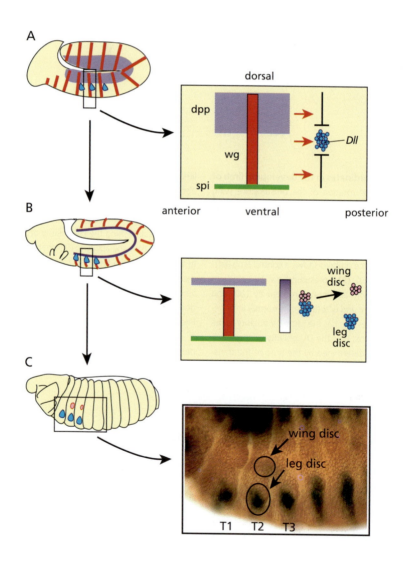

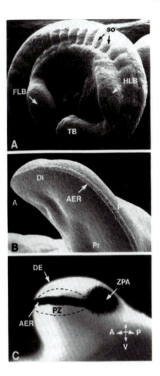

Fig. 12.9. Organization of the developing limb bud. **(A)** Mouse embryo showing the early signs of limb development in the bulging out of the fore- (FLB) and hindlimb (HLB) buds. The tail bud (TB) and the somites (SO) are indicated. **(B)** Early limb bud showing the apical ectodermal ridge (AER) as a thickening of the ectoderm. The basic coordinates are indicated: anterior (A), posterior (P), proximal (Pr), and distal (Di). **(C)** Early limb bud with the AER highlighted by the expression of FGF, the ZPA by the expression of Shh, and the PZ outlined. (Images courtesy of J. C. Izpisua Belmonte.)

Fig. 12.8. Initiation of appendage development in *Drosophila*. The appendages of *Drosophila*—six legs, two wings, and two balancing organs called halteres—develop from discrete sets of epidermal cells called imaginal discs that are set aside from the larval epidermis during embryogenesis. **(A)** The primordia of the ventral (legs) and dorsal (wing and haltere) appendages are first visible about two hours after gastrulation, as paired clusters of cells in the thoracic segments of the embryo (only one side shown). These clusters of cells (blue) express the homeodomain transcription factor Distalless (Dll), and develop under the influence of a cartesian system of molecular coordinates defined by the expression of dpp and wingless, as well as by the activity of the EGF receptor. At the time of appearance of the patches of Dll expression, the *wingless* (*wg*) gene is expressed in a narrow stripe (red) in every segment just anterior to the clusters of Dll-expressing cells, and dpp is expressed first in a broad band (purple) and then in a narrower stripe (see B) perpendicular to the stripes of *wg* expression. Experiments with mutants for elements of these different signalling pathways indicate that whereas wingless promotes the expression of Dll, dpp represses it. In this way, dpp defines a dorsal limit for the expression of Dll within the realm defined by wingless. The EGF receptor ligand spitz, diffusing from the ventral midline (see Chapter 9, Fig. 9.15 and Chapter 11, Fig. 11.15), produces a gradient, which creates an inhibitory field that defines the ventral limit of the cluster of Dll expressing cells. **(B)** After the initial expression of Dll, the patterns of wingless and dpp expression change. Associated with this transition, the cluster of Dll expression breaks into two populations, one of which (pink) migrates dorsally to give rise to the wing disc, and the other (blue) ventrally to give rise to the leg disc. At this stage, the levels of dpp and EGF receptor activity are sources of information for these changes: high levels of dpp and low EGF receptor activity promote the migration of the cells that will give rise to the wing and the haltere. Then, the cells invaginate and form sacs that are attached to the larval epidermis but distinct from it. **(C)** At the end of embryogenesis, as the larval epidermis differentiates, the primordia for the different appendages can be found as clusters of cells underneath the epidermis. During larval life they will remain in this position and will proliferate. In the photograph, the wing disc, which coincides with the position of the haltere disc, can be seen in T2 and T3. Leg discs are visible in T1, T2 and T3. The dark staining shows the expression of *wg* in these primordia. The primordia have some regional information that results in *wg* being expressed only in a subset of the cells of each leg disc. (After Goto, S. and Hayashi, S. (1997) Specification of the embryonic limb primordium by graded activity of Decapentaplegic. *Development* **124**, 125–32; Kubota, K. *et al.* (2000) EGF receptor attenuates dpp signalling and helps to distinguish the wing and the leg cell fates in *Drosophila*. *Development* **127**, 3769–76; and Bate, M. and Martinez Arias, A. (1991) The embryonic origin of the imaginal discs in *Drosophila*. *Development* **112**, 755–61).

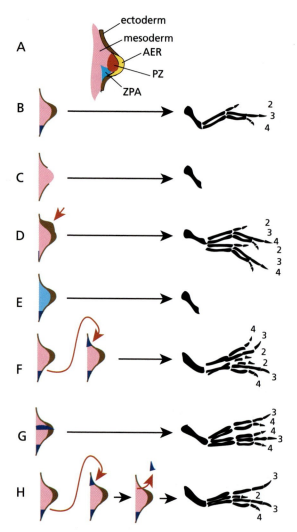

Fig. 12.10. Cellular parameters of limb development. (A) The chick limb develops from the limb bud through a series of interactions between the ectoderm and the mesoderm. Three populations of cells play an important role in this process: the apical ectodermal ridge (AER) in the ectoderm and two specialized populations of mesodermal cells: the progress zone (PZ) and the zone of polarizing activity (ZPA). The PZ, located underneath the AER, is an area of active proliferation and produces the cells that will form the muscles and the cartilage of the limb. The ZPA is located in the posterior region of the mesoderm and is important in the patterning of the mesoderm along the AP axis. **(B)** The limb bud grows and is progressively patterned; by the time the chick hatches, the limb shows the characteristic skeletal elements depicted on the right: from left to right, the humerus, radius and ulna, and the three digits (2, 3, and 4). **(C)** Removal of the AER aborts the growth and patterning of the limb at any stage of development. **(D)** Grafting of ectopic AERs produces extra limbs, indicating that the AER, once formed, has the ability to initiate and maintain the development of the limb. **(E)** The AER acts on the mesoderm, but only limb mesoderm. Substitution of the limb mesoderm by another kind of mesoderm (blue) does not allow the AER to elicit its effects and aborts the development of the limb. **(F)** A posterior domain of the mesoderm, the ZPA, has the property of organizing the patterning of the limb along the AP axis. If this region is transplanted to a more anterior position it elicits the development of ectopic digits with a reverse polarity. **(G)** The effects of the ZPA are dependent on its position within the limb bud. More posterior digits always appear closest to the ZPA, with more anterior digits appearing progressively further away from it (see also Fig. 9.5). **(H)** The effects of the ZPA are time-dependent; a ZPA graft elicits different digits depending on how long it is left in the limb bud. The shorter the time, the fewer ectopic digits it will elicit, and the more anterior they will tend to be.

development and patterning of the limb depends on spatially localized signalling centres: the AER controlling growth and patterning along the proximodistal axis, and the ZPA controlling patterning along the anteroposterior axis.

AER-driven initiation of PD patterning and growth

The requirement for the AER in the growth and patterning of the limb indicates that its appearance is a very important event in the development of the limb. The formation of the AER is a progressive process as revealed by the evolution of

the expression of *FGF8*, which is always associated with the AER. Initially *FGF8* is expressed over a broad domain that is refined as development progresses. The details of this refinement are beginning to emerge.

The AER does not emerge at a simple interface between cell populations that express different transcription factors or signalling molecules; rather, it develops within a landscape of gene expression that contains different boundaries (Fig. 12.11). Lineage and morphological studies in mouse embryos support the idea that the AER develops in a stepwise fashion, and reveal three transient lineage restriction boundaries that probably play a role in its formation (Fig. 12.11). Early in development there are two lineage restrictions along the DV axis: one at the dorsalmost border of the cells that express *FGF8*, and a second one at about the middle of this population. As develop-

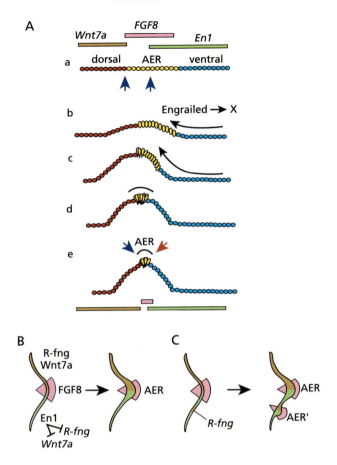

Fig. 12.11. The emergence of the AER. (A) The AER is derived from a population of ectodermal cells (yellow) that is induced by the underlying mesoderm, and is flanked by dorsal and ventral ectodermal cells. The expression of *FGF8* is closely associated with the development of the AER. The initial boundaries of the AER are probably loosely defined and are refined with time. When the AER is first visible its dorsal boundary is delimited by the expression of *Wnt7a*, while its dorsoventral midline is delineated by the expression of *En1*. (a) Lineage studies suggest that early in the development of the AER there are two boundaries of lineage restriction (blue arrows), that is, boundaries that are not crossed by clones of proliferating cells (see Chapter 8): one coincides with the dorsal-most domain of *En1* expression and the second lies at the dorsal boundary of the AER. At this stage no such restriction can be observed at the ventral boundary of the AER region. (b–d) As development proceeds, cells within the AER region alter their shapes and begin to bunch together at the tip of the emerging bud, where they aggregate and give rise to the thickening that characterizes the AER. (e) This compaction is accompanied by the disappearance of the lineage restriction at the dorsoventral midline and the appearance of a new lineage restriction (red arrow) at the ventral-most limit of the emerging AER. With the appearance of this boundary, the AER becomes a region of lineage restriction bounded both dorsally and ventrally. Little is known about what determines these events. It is likely that En1 is involved, either directly or indirectly (indicated by the arrow to substance 'X'), since the process of thickening is perturbed, but not abolished, in *En1* mutants. (Adapted from Loomis, C. *et al.* (2000) Analysis of the genetic pathway leading to formation of ectopic apical ectodermal ridges in *Engrailed-1* mutant limbs. *Development* **125**, 1137–48). **(B)** In chicken embryos, the boundary of expression of *Radical fringe* (*R-fng*) coincides with the domain at which the AER appears. The expression of *R-fng* is repressed by En1 ventrally. **(C)** Ectopic expression of *R-fng* in the ventral region of the limb elicits the appearance of new AERs (AER') at the interface of *En1* and *R-fng* expression. This suggests that interfaces of *R-fng* expressing and non expressing cells are required for the formation of the AER and that some of the effects of Engrailed in the process might be mediated thorugh its regulation of the expression of *R-fng*.

ment proceeds, the *FGF8*-expressing cells are drawn towards these borders, coinciding with the appearance of the AER. Later on, the central lineage restriction disappears and a new ventral one appears within the *FGF8*-expressing cells. These events coincide with the appearance of the mature AER which in this way becomes a cell population developmentally isolated from the dorsal and ventral ectoderm. The expression of *FGF8* becomes confined within these lineage restrictions, but the restrictions do not represent absolute limits for the expression of genes. For example, the expression of *En1* is limited not by the ventral border of the AER but by a boundary located towards the middle of this structure, where there is only a transient lineage restriction.

Little is known about what regulates the maturation of the AER. Because in *En1* mutants there are aberrations in the development of the AER, it is probable that En1 plays a role in the development of this structure. However, it is not clear what this role is or how it is carried out. Studies in chicken embryos have identified a candidate molecule for regulating the appearance of the AER: Radical fringe (R-fng), a glycosyl transferase that modulates Notch signalling (see Chapter 8) (Fig. 12.11). The *R-fng* gene is expressed in the dorsal ectoderm and is repressed ventrally by En1. Ectopic expression of *R-fng* in the ventral ectoderm triggers the appearance of new AERs at the border of *R-fng* expressing and non-expressing cells, suggesting that it is this border that sets up the AER in the normal animal (Fig. 12.11). The border depends on En1, which represses the expression of *R-fng* ventrally; this explanation can account for the effects of gain and loss of En1 function that have been observed in various experiments, because they will result in changes in the regulation of *R-fng* expression. However, although the boundaries of *R-fng* expression are clearly relevant for the emergence of the AER, implying a role for Notch signalling, very little is still known about how Notch signalling is activated or modulated during this process. The signal could come from the ectoderm or from the mesoderm, which is also subdivided into dorsal and ventral populations of cells.

Molecular mediators of growth and positional information in the growing bud

The experiments discussed in the last section indicate that there are sources of molecular information (the AER and the ZPA) that are strategically deployed within the limb primordium to direct and coordinate the development and

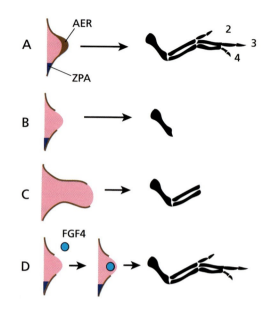

Fig. 12.12. The activity of the AER is mediated by FGF. (A) Development of the chick limb from a normal limb bud, showing the positions of the AER and the ZPA. **(B)** If the AER is removed, development of the limb is truncated. **(C)** If the AER is removed at a later stage, more elements of the limb develop but it is truncated at the stage at which the AER is removed. **(D)** A bead soaked in FGF4 can substitute for the AER and promote the growth and patterning of the limb.

patterning of the limb, and that there is a responding cell population, the PZ. Candidate signalling molecules that mediate the activity of the AER and the ZPA have been identified through experiments in which beads impregnated with candidate substances are implanted in developing chick limb buds and their effects analysed either as an alteration of the normal pattern or as the ability to substitute for a particular signalling region (Fig. 12.12 and 12.13).

As we have seen, when the AER forms, the expression of *FGF8* becomes restricted to this ectodermal derivative, which then also expresses *FGF2* and *FGF4*. The suggestion that these factors might play a role in the activity of the AER is confirmed by the observation that beads coated with any of the FGFs can completely substitute for the AER and rescue the growth and patterning defects created by its removal (Fig. 12.12). Consistent with this, FGF receptors are expressed in the limb mesoderm and are required for its growth and patterning.

The ZPA contains two different molecular activities, one associated with retinoic acid (RA) and the other with a member of the hedgehog family of signalling molecules,

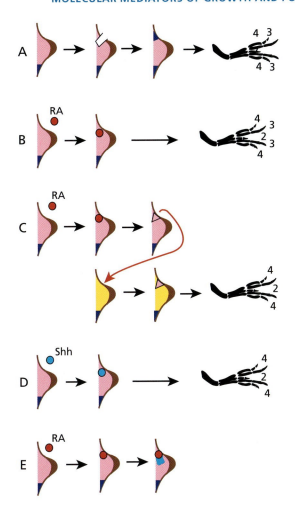

Fig. 12.13. Molecular activities of the ZPA. (A) The ZPA is defined experimentally as the region of posterior mesoderm that promotes the development of posterior digits when placed in the anterior region of a limb bud. **(B)** A bead soaked in retinoic acid (RA) can mimic the effects of the ZPA, and acts in a concentration-dependent manner (see Chapter 9, Fig. 9.8). **(C)** If RA were directly responsible for the patterning of the limb, one would expect an absolute requirement for RA during the patterning process. This was tested in an experiment in which a bead was soaked in RA and transplanted into a chicken limb bud. After a while, the bead was removed and the tissue was allowed to rest for a few hours to clear any RA that might have been left by the bead. The anterior region that had been exposed to RA was tested for ZPA activity by transplanting it into a limb bud from a quail embryo so that the host and donor could be distinguished. It was observed that, even though there was no RA, the anterior region of the chick limb bud behaved as a ZPA. This indicated that RA is not continuously required for the patterning process but that it plays a role in re-specifying the anterior mesoderm as ZPA which then can act to pattern the mesoderm. **(D)** The hh family member Sonic hedgehog (Shh) is expressed in the ZPA. Beads or cells containing Shh induce extra digits, like RA. **(E)** RA induces Shh (blue), but Shh does not induce RA (not shown), suggesting that normally it is Shh that directs the patterning of the digits. In the experiment in C, the RA induces Shh which then can stably influence the patterning of the digits.

Sonic hedgehog (Shh). Implanting beads coated with RA at different positions in the limb bud showed that RA acts in a concentration-dependent manner and can mimic the activity of the ZPA (Fig. 12.13 and see also Chapter 9). However, further experiments showed that, rather than mediating the activity of the ZPA directly, RA is required to induce the ZPA which, once established, can induce digits without the mediation of RA. In these experiments a RA-coated bead was implanted at the anterior end of a limb bud and after a while removed. Time was allowed for clearance of the RA and then the tissue in which the bead had been implanted was transplanted to another limb bud and shown to be capable of acting as a ZPA, even though it contained no RA (Fig. 12.13).

A more likely candidate to mediate the activity of the ZPA is Shh, which is expressed in the ZPA under the control of RA and can elicit similar effects to the ZPA and RA (Fig. 12.13). *Shh* mutant mice have severe limb deficiencies

that are consistent with its being a major requirement for limb development. However, although Shh does diffuse throughout the limb bud and can generate stable gradients, in some experiments its activity can not be related to this diffusion, and non-diffusible forms of hedgehog have been shown to elicit the same effects. One possible explanation for these observations is that there is another molecule, downstream of both Shh and RA, which mediates the effects of the ZPA. One candidate is BMP2, whose expression is under the control of Shh and which may play this role in the context of other signalling molecules. Alternatively, rather than there being a single molecule that diffuses from the ZPA and acts over a long range, a series of short-range influences may be superimposed on the activity of the long-range signalling molecules. Such an interplay could achieve, over time, a very refined pattern, just as interactions among transcription factors effect a refined pattern in the early *Drosophila* embryo (see Chapters 9 and 11). A third possibility is that patterning of the limb along the AP axis is achieved by a cooperation of RA, Shh, and other molecules acting sequentially but in an interdependent way; that is, although RA induces Shh, it might also be required for the patterning activities of Shh, and the same might be the case for Shh and some of its targets.

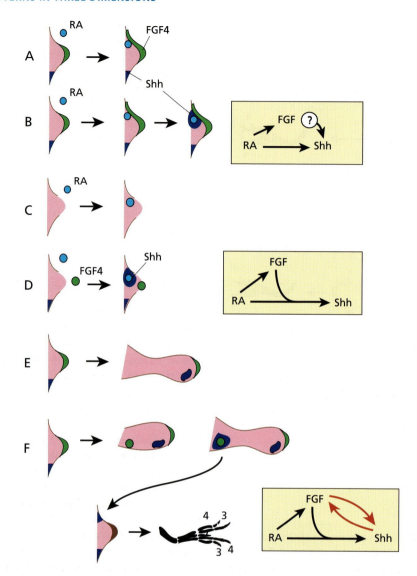

Fig. 12.14. Interactions between the ZPA and the AER stabilize the expression of FGF and Shh. (A) Implanting a bead soaked in RA extends the pattern of *FGF* expression in the AER in the direction of the implanted RA bead. **(B)** RA also induces the expression of *Shh* but the dynamics of this induction are different from those of its effects on *FGF*. The expression of *Shh* appears at least six hours later than that of *FGF*. This suggests that, perhaps, the inductive effects of RA on *Shh* require or are mediated by FGF (top box). **(C)** This can be tested by seeing whether RA can induce the expression of *Shh* in the absence of an AER. Implanting a bead soaked in RA cannot induce *Shh* expression in the absence of an AER. The loss of the AER also results in the loss of expression of *Shh* in the normal ZPA (purple triangle in A and B). **(D)** However, implanting one bead with RA and another with FGF into limbs that have lost the AER does result in the appearance of ectopic *Shh* expression and in the main-

tenance of the endogenous expression of *Shh*. Therefore, both RA and FGF are required for the onset of *Shh* expression (middle box). **(E)** As the limb grows out, *Shh* expression is restricted to a distal posterior domain. This, and the experiments in A–D, suggest that FGF may be involved in maintaining *Shh* expression throughout limb development. **(F)** This can be tested by implanting a bead impregnated with FGF into the proximal part of the growing limb. Under these circumstances, the expression of *Shh* is maintained in this proximal region. Furthermore, the cells that maintain expression of *Shh* have all the properties of the normal ZPA. For example, they can induce an ectopic mirror-image limb if transplanted to the anterior region of a limb bud. Thus, after being induced, Shh becomes involved in a positive feedback loop with FGF, maintaining its expression and linking the AER and the ZPA (bottom box).

Integration of information from the ZPA and the AER

When beads soaked in RA are applied to limb buds, the domain of expression of *FGF*, and the AER, extend towards the bead, suggesting that RA can influence the extent of the AER. This effect precedes the onset of *Shh* expression and therefore raises the possibility that the effects of RA on the expression of *Shh* are mediated by FGF from the AER. However, neither RA nor FGF alone is capable of inducing the expression of *Shh*, suggesting that a collaboration between them might be required for the expression of *Shh* in the ZPA (Fig. 12.14).

As the limb grows, the expression of *Shh* remains, like the activity of the ZPA with which it is associated, restricted to a small domain in the posterior distal mesoderm next to the AER (Fig. 12.14). This suggests that after their establishment, there is a close regulatory relationship between the AER and the ZPA. If a bead impregnated with FGF is implanted in the proximal growing part of the limb, *Shh* expression is maintained there, suggesting that FGF from the AER is likely to be responsible for maintaining expression of *Shh* in the ZPA (Fig. 12.14). This, together with the effects of RA on the AER, suggests that there is a regulatory loop between the ZPA and the AER that maintains the expression of *Shh* and *FGF*. This mechanism serves to stabilize the positional cues for the PZ (Fig. 12.14).

The role of the DV axis in limb patterning

Cells within a developing limb process information not only along the PD and AP axes, but also along the DV axis (Fig. 12.15). This is demonstrated by the observation that removal of the dorsal, but not the ventral, ectoderm of the limb bud abolishes limb development and the regulatory relationships that underlie it. Dorsal ectodermal cells express *Wnt7a*, and *Wnt7a* mutants display double-ventral limbs, consistent with the idea that Wnt7a plays a role in the patterning of the limb along the DV axis. Further evidence comes from experiments in which cells expressing *Wnt7a* can substitute for the dorsal ectoderm and promote limb development and patterning (Fig. 12.15). Wnt7a acts through its target, *Lmx1*, which encodes a LIM-family homeodomain transcription factor that is expressed in the underlying dorsal mesoderm.

The fact that a limb bud deprived of dorsal ectoderm shows defects in expression and regulation of the AP and PD patterning systems indicates that the cells within the PZ integrate signals from the three axes. Surprisingly, a limb bud deprived both of the AER and the dorsal ectoderm can be made to develop normally by addition of beads containing FGF and Wnt7a. These experiments suggest that Shh, Wnt7a, and FGF are necessary and sufficient to trigger the development of the limb and that most (if not all) of the subsequent molecular machinery resides in the developing mesoderm (Fig. 12.15).

Growth and patterning along the PD axis

The growth of the limb along the PD axis depends on the continuous activity of the AER and follows two basic principles. The first is that proximal pattern elements are specified before distal ones, and the second that the pattern element generated by a cell is dependent on the time it spends in the PZ. These principles have been inferred from experiments in which the distal region (including the AER) of a limb bud at a given developmental stage is transplanted on to a limb bud at a different stage that has had the AER and the PZ removed. Grafting of older tips on to younger limbs results in truncations whereas grafting of younger tips on to older buds results in extra pattern elements (Fig. 12.16).

Very little is known about how the different pattern elements are specified along the PD axis. One possibility is suggested by the involvement of certain transcription factors in the patterning of the *Drosophila* leg along this axis. The *Drosophila* leg develops as a simple epithelium that takes the shape of a cone which grows along its proximodistal axis (in other words, the cone elongates) as development proceeds. At a descriptive level, at least, the process is similar to the growth of the vertebrate limb. In the *Drosophila* leg, the specification of the proximal pattern elements depends on the translocation to the nucleus of a PBC-family homeodomain transcription factor, exd. Nuclear translocation of exd is controlled by another homeodomain protein, homothorax, which is only expressed in proximal cells of the leg. In these cells, homothorax binds to exd, in some way favouring its localization to the nucleus, where it interacts with specific *Hox* gene products to drive the transcription of specific targets that inhibit the development of distal structures (Fig. 12.16). Vertebrate homologues of exd (Pbx) and homothorax (Meis) are expressed in the developing limb in a similar pattern to their *Drosophila* counterparts, and ectopic expression experiments suggest that they might

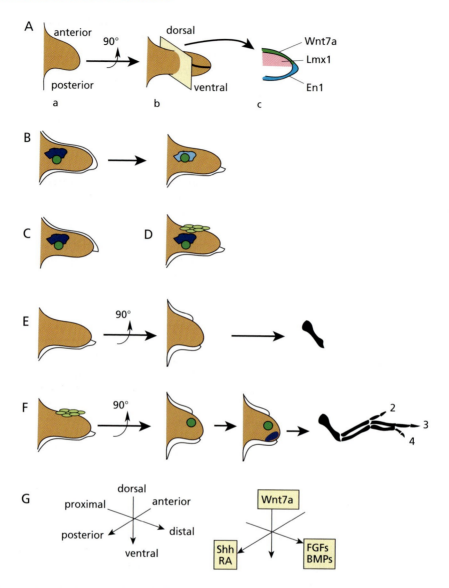

Fig. 12.15. Integration of information from the DV axis. (A) Rotating the familiar view of the limb bud through 90° reveals the DV axis, which provides a third molecular coordinate for the development and patterning of the limb. A section through the limb bud reveals both asymmetries in pattern about this axis, and asymmetric expression of certain genes in the ectoderm (*Wnt7a* dorsally and *En1* ventrally). The mesoderm is also divided into dorsal and ventral regions as indicated by the expression of *Lmx1* in a dorsal domain that corresponds to the domain of *Wnt7a* expression in the ectoderm. **(B)** As we have seen in Fig. 12.13, implanting a bead soaked in FGF (green bead) in the proximal region of the limb as it grows (*left*) results in a maintenance of the expression of Shh (blue patch surrounding the bead). If the dorsal ectoderm is removed (*right*), the limb does not develop normally and FGF fails to maintain Shh expression. **(C)** Removal of ventral ectoderm does not affect the

development of the limb or the maintenance of Shh expression by FGF. **(D)** A pellet of cells expressing *Wnt7a* can substitute for the dorsal ectoderm in limb development and in maintaining the expression of Shh by FGF. This suggests that *Wnt7a*, which is expressed by the dorsal ectoderm, plays an important role in the patterning of the limb. **(E)** Removal of both the dorsal ectoderm and the AER results in truncated limbs. (The 90° rotation is included in order to show both DV and AP views of the same limb.) **(F)** A combination of cells expressing *Wnt7a* and beads soaked in FGF can restore normal limb development and patterning to buds deprived of both AER and dorsal ectoderm. **(G)** The limb develops and patterns with regard to three molecular coordinates that have spatially localized sources: Wnt7a along the DV axis, Shh and RA along the AP axis, and FGFs and BMPs along the PD axis.

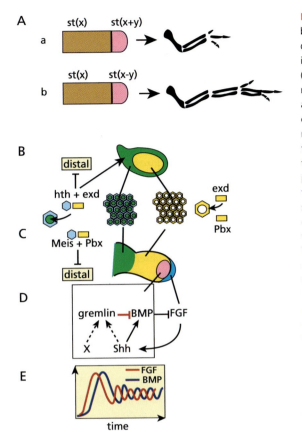

Fig. 12.16. The proximodistal (PD) axis. (A) Graft exchanges between developing limbs of different ages reveal that the mesodermal cells acquire information about their fates along the PD axis in sequence and in an irreversible manner. Grafting of an old limb (stage x+y) on to a young one (stage x) results in missing proximodistal structures, whereas grafting of a young PZ (stage x–y) onto an old one results in extra structures. This indicates that the time a cell spends in the PZ is a measure of its PD identity. **(B)** The development of the PD axis of the *Drosophila* leg has provided some clues to the organization of PD pattern in vertebrates. The leg is derived from the leg disc where, just as in a vertebrate limb, there is a proximal (green) and a distal (yellow) cell population. A key event in the definition of proximal versus distal is the activity of the transcription factor extradenticle (exd), which can be cytoplasmic or nuclear. Exd (see also Chapter 3, Fig. 3.8) is ubiquitously expressed, but whereas in the distal region it is cytoplasmic, proximally it is nuclear. The reason for the change of intracellular localization is the presence, only in the proximal cells, of the homothorax (hth) protein, which complexes with exd and takes it into the nucleus. The hth–exd complex represses the development of distal structures. Expression of hth distally perturbs the development of these structures. **(C)** There are homologues of exd and hth in vertebrates: Pbx for exd and Meis for hth. As in the case of the *Drosophila* proteins, although Meis is expressed throughout the limb, the expression of Pbx is restricted to the proximal region where the effects of ectopic expression suggest that it is also involved in the suppression of distal cell fates. **(D)** The observations summarized in A–C suggest that most of the patterning process is initiated in the distal region of the developing limb and more precisely in the PZ (pink). The proliferative activity in the PZ is dependent on the FGF supplied by the AER (blue). The control of FGF expression in the AER is a complex process that depends on a balance of positive and negative inputs mediated by Shh and some BMPs. This system provides an example of the crucial role that balances between signalling molecules play in pattern formation. Shh can maintain the expression of *FGF* in the ectoderm, but it also antagonizes it indirectly through BMP2 and BMP4, which are targets of Shh in the mesoderm, and which suppress the expression of *FGF*. The action of BMPs is kept in check by the activity of gremlin, an inhibitor of BMP activity like noggin and chordin (see Fig. 4.27), whose expression is under the control of Shh and other factors as indicated. **(E)** This basic set of regulatory relationships provides an interdependent system that can finely balance the signalling to the PZ. The initial activity of FGF and Shh would lead to the expression and activity of BMPs which would provide a brake on the expression and activity of FGF. The expression of gremlin introduces a control point which is not dependent on the feedback loop and whose regulation is likely to be important in the cessation of the activity of the AER when growth along the PD axis is complete.

play a similar role. For example, ectopic expression of Meis, which is usually restricted to the proximal region of the limb, induces truncations and proximalizations of the remaining structures. It is likely that the Hox cofactors for Pbx in the limb bud are encoded by members of the *Hoxa* and *Hoxd* clusters, which are known to be expressed in the bud, and that these transcription factor combinations are involved in creating the diversity of pattern elements that characterizes the proximodistal axis (see below).

The limb bud does not grow indefinitely and transplantation experiments show that the growth-inducing activity of the AER decays with time. Members of the BMP family of signalling molecules are key factors in this process and act by antagonizing the activity of the FGFs. BMP2 is expressed in the mesoderm under the control of Shh, and is a candidate for down-regulating the activity of the AER. However, there must be a way of modulating the activity of BMP2 so that it does not inhibit the activity of the AER too soon. This is provided by the BMP antagonist called gremlin, which is specifically expressed in the developing limb mesoderm, also under the control of Shh. Thus, the bal-

ance, over time, between the interactions of Shh and FGF on one hand, and of BMP2 and gremlin under the control of Shh on the other, creates a self-regulating network that controls the extent and probably also the rate of growth

(Fig. 12.16). Although we know the elements of the system, we still do not understand how they function.

It is possible that the cessation of growth correlates with an increase in the activity of BMPs. This could be achieved either by tinkering with the expression of any of the elements of the network, probably through the appearance of a new regulator in the limb program of gene expression that either up-regulates the expression of BMPs or down-regulates the expression of gremlin and related molecules.

Hox codes and the registration of the positional information

The *Hox* genes are expressed in an ordered pattern along the AP axis of the body and play a key role in regulating the development of different structures along this axis (see Chapter 2, Fig. 2.8). Some of these structures are the limbs, which develop at levels along the AP axis at which specific *Hox* genes are expressed.

There are four *Hox* clusters in vertebrates and two of these clusters, *Hoxa* and *Hoxd*, are expressed in ordered patterns in the developing limbs. The genes of the *Hoxd* cluster, for example, are expressed in the early limb bud in a nested pattern in the mesoderm (Fig. 12.17). The expression of these genes in the limb follows similar principles to those that regulate their expression along the AP axis: the genes are progressively activated, with genes located more 3′ in the complex being expressed more posteriorly, and once a gene has been activated at a particular AP level, it remains expressed posterior to that level (Fig. 12.17). During the early stages of development, the expression pattern of the different members of the *Hoxd* cluster correlates well with the pattern of digits that will emerge later, and responds to signalling from the ZPA in the same way as the resulting digits. This led initially to the suggestion that just

as there is a *Hox* code for the development of structures along the anteroposterior body axis, there is a related code for the specification of different digits. However, a simple correlation between *Hox* gene expression and particular digits breaks down later in development, when *Hoxd* genes are expressed dynamically in the mesoderm, ending up in a pattern that is the reverse of the initial one (Fig. 12.17). Genes from another cluster, *Hoxa*, also show complex changes in expression over time.

Genetic analysis of the *Hoxa* and *Hoxd* clusters indicates that they do not play simple roles in the specification of digits but probably function in the specification of proximodistal pattern elements of the cartilage. Mice mutant for genes of the *Hoxa* and *Hoxd* clusters show defects in the development of structures along the PD axis, but often a mutant phenotype is only observed if the animal is doubly mutant for corresponding genes from both clusters, suggesting a redundancy in their function (Fig. 12.17).

The *Hoxb* and *Hoxc* clusters are expressed in either the forelimb or the hindlimb, suggesting that these clusters are involved in instructing the differences between these limb types.

The patterning of the digits

The more information accumulates about the development and patterning of the limb, the more difficult it is to reconcile the observations with a simple model in which the digits are patterned by the concentration-dependent activity of a morphogen that emanates from the ZPA. It looks as if every molecule that comes to be expressed in the ZPA paves the way, molecularly speaking, for the expression of another ZPA factor and for the response of cells in the PZ. Thus, retinoic acid induces Shh, which in turn induces BMP2, gremlin, and others. These successive induc-

---

Fig. 12.17. Hox codes and the patterning in the limb. (A) The *Hox* genes are expressed in the developing limbs. The *Hoxa* and *Hoxd* clusters are expressed in both the fore- and the hindlimbs suggesting that they have basic functions in the development and patterning of both limb types. **(B)** In the initial stages of development (*left*), the *Hox* genes are expressed in nested patterns from anterior to posterior, with a progressive onset of expression from 5′ (anterior) to 3′ (posterior). However, this pattern changes over time and by the final stages of limb development (*right*) is the reverse of the initial pattern. **(C)** The phentoypes of mice mutant for *Hoxa* and *Hoxd* genes. Capitals indicate wild-type alleles, lower case mutants. The humerus (h), radius (r), and ulna (u) are indicated. Individual mutants for *Hoxa11* or *Hoxd11* display defects in the development of the radius and the ulna which become very severe in the double mutant, suggesting that the corresponding genes from different clusters have partially redundant roles in this process. **(D)** Analysis of mutants and of patterns of expression suggests a correlation between *Hox* gene function and the proximodistal pattern elements of the limb. (Images C and D from Davis, A. P. *et al.* (1995) Absence of radius and ulna in mice lacking *hoxa-11* and *hoxd-11*. *Nature* **375**, 791–5).

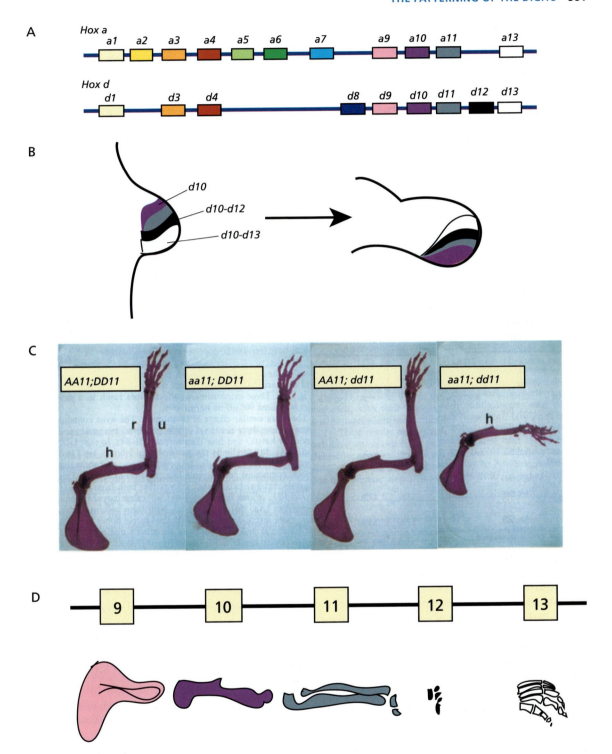

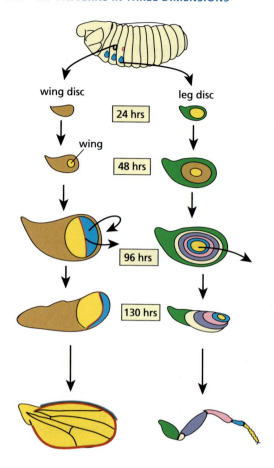

Fig. 12.18. The development of the wings and legs of Drosophila. Just before hatching, the *Drosophila* embryo contains small subepidermal sacs of about 30–40 cells, the imaginal discs, that will give rise to the dorsal (wing and haltere) and ventral (legs) appendages of the fly (see Fig. 12.8). Soon after hatching (at 24 hours of development), the discs begin to grow. Early in the development of the wing disc (*left*), a small population of cells (yellow) is set aside within the disc that will give rise to the wing proper. This population grows very fast during the larval period to generate a large circular sheet of cells with two well-defined subpopulations: a dorsal (yellow) and a ventral (blue) one. After proliferation, it undergoes a series of morphogenetic movements that will fold the dorsal and ventral halves back against each other and give the wing its final shape and organization. The line (red) that separates the dorsal and ventral regions is the line of folding and so becomes the wing margin, which will generate a row of densely packed sense organs. The wing surface (only the dorsal surface is shown) is criss-crossed by a series of veins with defined patterns and positions. The leg discs (*right*) develop in a different way. The initial stages define proximal (green) and distal (yellow) domains and the disc then grows by intercalation between these two cell populations. The intercalation gives rise to a series of concentric rings of cells that contribute to different regions of the leg in an ordered manner.

Development and patterning of the appendages of *Drosophila*

The appendages of *Drosophila* can be considered functional homologues of the vertebrate limbs in that they provide a basis for locomotion and are extensions of the body wall. As we have seen, it is possible to draw some parallels between the development of both types of structures, for example in the processes and molecules used to pattern the PD axis (Fig. 12.16). However, in other ways their modes of development and basic organization are very different. Most importantly, whereas epithelial–mesenchymal interactions are a very important element during vertebrate limb development, the patterning of the appendages of *Drosophila* involves a single layer of epidermal cells. Even within *Drosophila*, there are differences between legs and wings: both are derived from imaginal discs (Fig. 12.8) but whereas legs have an overt proximodistal pattern that is generated progressively, the PD pattern of the wing is not so obvious.

In the early stages of their development, leg and wing discs have a similar basic organization: a disc of cells with a centrally located population that will become the most

tions and the subsequent regulatory interactions, together with the dynamic expression patterns of members of the *Hoxa* and *Hoxd* clusters, suggest that the patterning of the digits is likely to be the result of a series of local interactions operating over time, rather than of a single diffusible molecule conveying positional information over space as previously thought.

Once the digits have formed, the fate of the intervening tissue varies in different species: in some (e.g. mice, chickens) this tissue disappears whereas in others (e.g. ducks) it does not. In those species where the webs or pads between digits are eliminated, this occurs by programmed cell death mediated by BMPs and their receptors (see Chapter 7, Fig. 7.25).

distal point of the appendage. However, soon after they begin to grow, their modes of development diverge. The leg grows by a variation of the vertebrate theme of the progress zone, by intercalation of sectors between the most proximal and the most distal positions of the disc. These sectors express different genes and each is associated with a particular structure in the final leg (Fig. 12.18). The wing disc, however, changes its geometry as it begins to grow. The most distal portion of the wing disc becomes the primordium of the wing blade and begins to grow quickly with reference to a line that separates dorsal and ventral cells and which defines the future margin of the wing (Fig. 12.18). As it grows, the wing blade becomes a conspicuous circle with a ventral and a dorsal cell population which, through a series of concerted morphogenetic movements, shape the cells into the final wing of the fly.

Because the strategies are so different for the two discs, the growth and patterning of the wing blade cannot be related easily to that of the leg discs or of the vertebrate limbs. Nevertheless, the development and patterning of the wing within the wing disc has been studied in great detail and has provided a number of useful insights into how patterning mechanisms can be coupled with growth over a large number of cells.

The initiation of wing development

The cells that will give rise to the wing proper are not specified in the embryo but early in the development of the larva, within the wing disc (Fig. 12.18 and 12.19). This group of cells is defined by the converging activity of three signalling pathways (wingless, dpp, and Notch) acting in the context of low levels of the transcription factor vestigial, which is expressed throughout the wing discs. The convergence of these signals at a specific point in the disc outlines a population of about 20 or 30 cells in which the levels of expression of *vestigial* are increased. A specific enhancer of *vestigial*, the 'boundary enhancer' (*vgBE*) integrates these early inputs. *Vestigial* is exclusively expressed in the wing disc and can be thought of as providing a molecular context for wing development just as *Tbx4* and *Tbx5* or *Pitx1* do in the early development of vertebrate limbs. That is, *vestigial* expression leads to the activation of wing specific genes and so channels the output of these signalling pathways towards wing development. Consis-

tent with this idea, activation of high levels of *vestigial* expression in other discs (for example, those for the legs) induces the development of wing tissue and, sometimes, full wings.

The very small primordium of cells that gives rise to the wing acquires two molecular references that will determine the patterning of the wing during its growth. One of these references is set up by Notch signalling and establishes a stripe of cells that express *wingless* at the boundary between dorsal and ventral cells. The second reference is a stripe of *dpp* expression that appears just anterior to the anteroposterior compartment boundary and that depends on hedgehog signalling (Fig. 12.19).

After being defined, and as it is endowed with AP and DV references, the wing begins to grow. The initial phases of growth are very dependent on events at the DV boundary and, like the initial stages in the development of vertebrate limbs, require an interface between cells expressing (dorsal) and not expressing (ventral) the glycosyltransferase fringe, a modulator of Notch signalling. Along the DV boundary, high levels of *vestigial* expression are maintained by Notch signalling, driven by its ligands Delta and Serrate, and acting on the *vgBE* enhancer. The growth of the wing cells in what will become the dorsal and ventral halves of the wing blade correlates with the activation of a second enhancer of the *vg* gene, the quadrant enhancer (*vgQE*), which directs the expression of *vg* in the growing cells of the wing (Fig. 12.20). This enhancer receives inputs from wingless and dpp and integrates these signals to promote the growth of the wing. Although it is clear that *vestigial* is absolutely required for the development of the wing, it alone cannot drive the growth or the patterning of the wing. The idea that it provides a molecular context for wing development suggests that during this phase, in addition to operating downstream of the wingless and dpp signalling pathways, it might also work in parallel with them, identifying wing specific target genes that will be modulated by these signals (Fig. 12.20).

The group of cells in which wing development is initiated displays many of the features characteristic of an 'organizer' in the sense in which this term is used in vertebrates (that is, a group of cells that has the ability to trigger and organize the development of particular structures). Indeed, multiple wings result if the spatial overlap of the three signalling pathways is artificially reiterated throughout the wing disc, just as transplantation of the organizer produces twinning of a vertebrate embryo (Fig. 12.21).

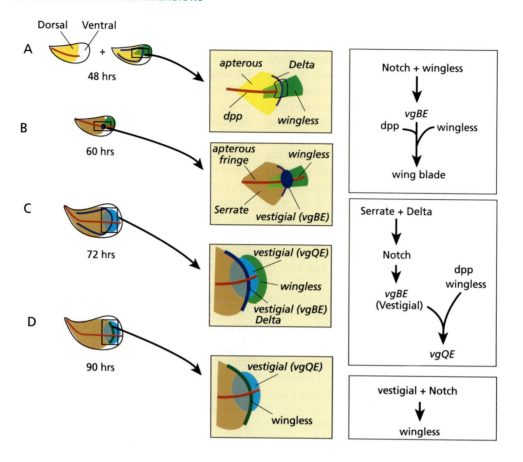

Fig. 12.19. The initiation of wing development. (A, B) The development of the wing proper, also called the wing blade, is initiated in the wing disc during larval life. At this stage, the wing disc is subdivided into a dorsal and a ventral cell population by the expression of the LIM homeodomain transcription factor apterous (yellow) in the dorsal cells. At about 50 hours of development, the expression of *vestigial*, which up until this stage is expressed at low levels throughout the wing disc, is up-regulated where the ligands for three signalling systems intersect: Wnt (wingless, green), BMP (dpp, red), and Notch (Delta, purple). Wingless and Delta increase the expression of *vestigial* through a specific enhancer, the *vgBE*, while dpp appears to act through a different part of the regulatory region. The cells that express high levels of vestigial will become the wing blade. **(B, C)** Shortly after the establishment of the wing primordium, high levels of vestigial expression are maintained by the combined activity of Delta and a second Notch ligand, Serrate, which is expressed throughout the dorsal cell population (brown) together with fringe, a modulator of Notch signalling activity. Vestigial, whose expression is driven by the *vgBE*, then cooperates with wingless and dpp signalling to activate its own 'quadrant enhancer' (*vgQE*) which identifies wing cells (light blue, grey) that begin to proliferate. Thus at this stage there are two populations of cells expressing vestigial under the influence of different signalling systyems. The *vgBE* maintains high levels of expression of vestigial at the dorsoventral boundary (purple), under the influence of Notch signalling driven by Delta and Serrate. The *vgQE* drives expression of vestigial in the cells (light blue, grey) that are growing to make the wing under the influence of dpp and wingless. **(D)** At the DV boundary, vestigial interacts with Notch signalling to activate the expression of *wingless* in a narrow stripe (dark green). This group of *wingless* expressing cells becomes a stable cell population and will give rise to the wing margin in the adult wing. At this stage the wing primordium expresses vestigial (blue, grey) and is crossed by two intersecting stripes of wingless and dpp (red) expression.

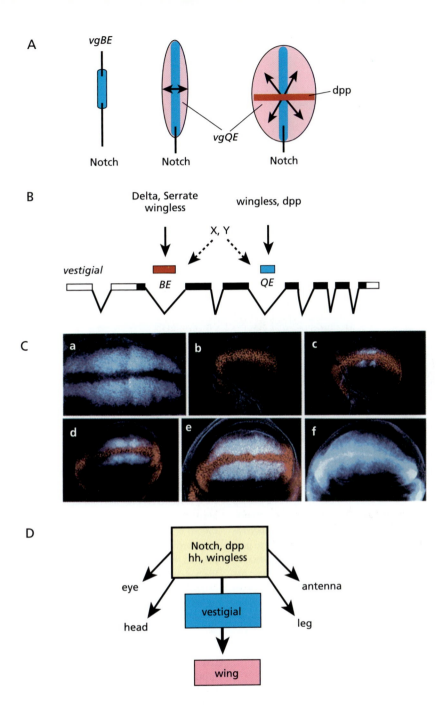

Fig. 12.20. Correlation of the development of the wing with the expression of *vestigial*. (A) Diagrammatic representation of how the wing grows away from its origin within the region (blue) that will become the wing margin. Arrows indicate expansion. **(B)** The *vestigial* (*vg*) gene contains two enhancers, located in introns, which drive expression in specific regions of the developing wing. Each of these enhancers responds to particular combinations of signals (X and Y represent additional, unknown signals). **(C)** Patterns of expression directed by the *vgQE* (blue) and the *vgBE* (red) revealed by the expression of a reporter gene throughout the wing disc. (a) The activity of the *vgQE* is excluded from the DV boundary. (b–e)

Wing discs at different stages: the activity of the *vgBE* is always restricted to the DV boundary and precedes that of the *vgQE*. The activity of the *vgQE* reflects the growth of the wing. (f) If both enhancers drive the same reporter, the normal pattern of expression of *vestigial* results. (From Kim, J. *et al.* (1996) Integration of positional signals and regulation of wing formation and identity by the *Drosophila vestigial* gene. *Nature* **382**, 133–8.) **(D)** The function of vestigial is to ensure that the activities of the different signalling pathways are targeted to wing-specific genes. As well as being downstream of these signalling events, vestigial probably also operates in parallel or in collaboration with them.

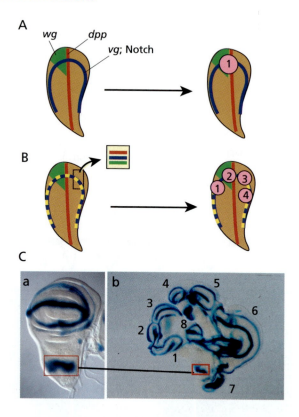

Fig. 12.21. Organizing the initial stages of wing development. (A) In the wild type, the wing emerges from a point within the wing disc where wingless, dpp, and Notch signalling overlap and collaborate with the activity of the transcription factor vestigial. This point of convergence is unique and defines a small cell population (1) that will grow into the wing blade. **(B)** It is possible to generate a line of artificial overlaps between the expression of *vestigial*, and wingless, Notch, and dpp signalling, for example by using the Gal4 expression system (see Fig. 10.2) to drive the expression of *wg* and *dpp* under the control of the *vgBE*. This would be predicted to result in the generation of several wing primordia (1–4). **(C)** Results of an experiment of the type described in B. (a) In a wild-type control, a reporter used to follow the expression of *wg* highlights the wing tissue (circle bisected by a stripe at the DV boundary) and a region that will give rise to part of the body wall (box). (b) Forced overlap of *wg*, *dpp*, *Notch* and *vestigial* expression along the DV boundary results int the development of seven independent wing pouches, each with its own margin. The expression of wingless in the body wall region is boxed for reference. (Image C from Klein, T. and Martinez Arias, A. (1999). The vestigial gene product provides a molecular context for the interpretation of signals during the development of the wing in *Drosophila*. *Development* **126**, 913–25.)

Patterning and growth of the wing

As the wing begins to grow, it is patterned with regard to the sources of signalling molecules located at the DV and AP boundaries: wingless and the Notch ligands Serrate and Delta at the DV boundary, and dpp and hedgehog at the AP boundary. Thus, the signalling centres become sources of gradients whose profiles change as the cells grow. However, patterning of the wing does not occur through a continuous and linear response of the cells to emerging graded inputs from the different signalling molecules, as would be required by the simple concept of positional information (see Chapter 9). Genetic analysis of wing development indicates that the patterning of the wing takes place progressively, through a sequence of landscapes of gene expression in which each pattern serves as a pre-pattern for the next one. Thus, the activity of the signalling molecules produced by the signalling centres is being changed by their outputs and, in addition, influenced by local interactions that create pattern independently of the long-range signalling molecules.

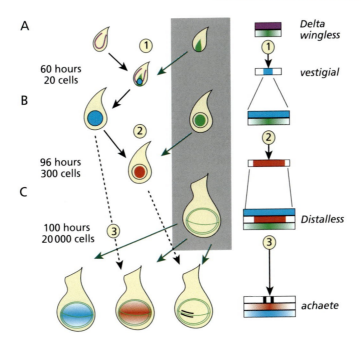

Fig. 12.22. Patterning of the wing along the DV axis. The expression pattern of each factor is shown in diagrams of the wing disc (*left*) and in a series of idealized cross sections through the dorsoventral boundary of the developing wing (*right*). **(A)** The initial stages of wing development, and in particular, interactions between Notch (purple) and wingless (green) signalling (1) lead to the definition of a small population of cells characterized by the expression of vestigial (blue) (for details, see Fig. 12.19). **(B)** Interactions between vestigial activity and wingless signalling lead to the activation of another transcription factor, Distalless (red) in a subset of the cells that express vestigial (2). **(C)** Regulatory events in the wing primordium lead to changes in the pattern of wingless expression. Two prominent features of the new pattern are a circle (double green lines) that outlines the wing tissue and a stripe that bisects this circle and subdivides it into dorsal (top) and ventral (bottom) cell populations. This pattern of wingless expression results in the modulation of vestigial and Distalless expression, and in the expression of genes of the *achaete–scute* complex (black) near the source of wingless along the DV boundary (3). The graded and nested expression of these genes suggests a correlation of their expression patterns with the diffusion of wingless from the stripe at the DV boundary. (Adapted from Martinez Arias, A. (2000). The informational content of gradients of Wnt proteins. *Science* STKE. (43):PE1.)

As we have seen, the DV patterning system plays a very important role in the initial stages of wing development. Later on, patterning along this axis is centred on a stripe of wingless expression that appears midway through development and whose major function is to modulate a succession of regulatory interactions that happen over time (Fig. 12.22). As a result, by the time the wing differentiates there is a set of nested patterns of gene expression, centred at the DV boundary, which contribute to the pattern along this axis. The second global coordinate of the wing disc, the AP axis (Fig. 12.23), not only contributes to the initial definition of the primordium through the deployment of dpp, but plays an important role in the patterning of the wing through the signalling activities of hh and dpp. Both molecules are involved in the generation of patterns of gene expression that over time serve as templates for the specification of the pattern elements that decorate the wing (Fig. 12.23). These effects are concentration-dependent and show many of the characteristics expected for morphogens (see Chapter 9). In addition to its effect on patterning, dpp has an effect on growth (see Chapter 7, Fig. 7.17).

The roles of hedgehog and dpp in the organization and patterning of the wing can be dramatically demonstrated by the generation of small clones of cells that are able to express hh or dpp ectopically within a normal developing wing. These clones act as new sources of hh and dpp that mimic the signalling from the AP boundary and change the overall pattern of the wing accordingly. These modifications can always be related back to the new sources of the signalling molecules (Fig. 12.24).

The joint activities of the boundary and quadrant enhancers of the *vestigial* gene indicate that, as they grow, the cells of the wing integrate inputs from both the AP and DV axes to determine their position, rate of proliferation, and the pattern elements that they differentiate (Fig. 12.25). However, very little is known about how this integration takes place at the level of wing-specific genes or how the networks that drive proliferation are regulated by vestigial and the signalling molecules.

The patterning of the wing: Veins

The adult wing is decorated by epidermal thickenings called veins, which form a series of lines parallel to the AP axis of the wing and provide landmarks of position within the blade. The development of these structures, which are elaborated during the last phases of development (the pupal stage in *Drosophila*) is presaged in the disc by the expression of several genes that outline the territories over which the veins will develop. These territories are under the control of local cell interaction systems that are triggered by the AP signalling system (Fig. 12.23).

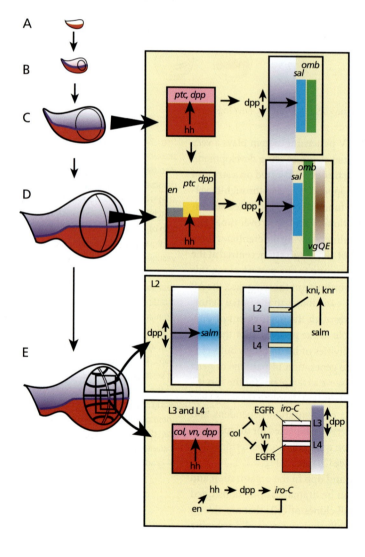

During the pupal stages the structures of the adult fly differentiate and the pattern of the veins is refined (Fig. 12.26). This process is accompanied by a redeployment of the expression and activity of many of the signalling systems and responders that figured in the large scale patterning processes during the main growth phase of the wing, namely dpp, Delta/Notch, and EGF receptor signalling. This can be most clearly seen in the case of *dpp*, which ceases to be expressed in a stripe along the AP axis and becomes integrated into local networks of gene activity and patterning associated with individual veins (Fig. 12.26). Interactions between these signalling systems refine the veins to their final size and shape and elicit their differentiation.

Strategies in the development and patterning of vertebrate limbs and the appendages of insects

Several analogies can be drawn between wing development in *Drosophila* and limb development in vertebrates. For example, members of the Wnt family of signalling proteins

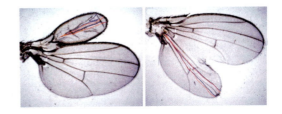

Fig. 12.24. Instructive effects of dpp in the patterning of the wing. Pattern reorganizations induced by generating clones of cells expressing *dpp* ectopically. The borders of the clones are indicated by coloured lines. The wing on the left shows a duplication arising from an anterior clone, and the one on the right a duplication arising from a posterior clone. In each case, the extra tissue reflects the local pattern and can be a perfect duplication, as in the case of the posterior clone. (From Zecca, M., Basler, K., and Struhl, G. (1995) Sequential organizing activities of engrailed, hedgehog and decapentaplegic in the *Drosophila* wing. *Development* **121**, 2265–78.)

Fig. 12.23. Patterning of the wing along the AP axis. (A) From its appearance in the embryo, the wing disc has a fixed reference along the anteroposterior axis, the AP boundary, defined by the interface between cells (blue) that express the transcription factor engrailed (en) and the signalling protein hedgehog (hh), and those that do not. **(B, C)** Hedgehog diffuses anteriorly and induces the expression of *dpp* in a narrow stripe of cells (red) adjacent to the *en/hh* expressing cells. Another consequence of the activity of hh is the modulation of the expression of its receptor patched (ptc) (see also Chapter 5, Fig. 5.23 and Chapter 9, Fig. 9.9, 9.23). Dpp diffuses from its spatially restricted source (pink shading) and induces the expression of two genes, *spalt* (*sal*) and *optomotor blind* (*omb*). **(D)** As development proceeds and the disc grows, the thresholds for hh and dpp signalling change and new targets (*en* and the *vgQE*) as well as new domains of pre-existing targets (*dpp*, *ptc*, *sal*, and *omb*) are defined. Towards the end of the larval period it is possible to observe three discrete thresholds for hh signalling in the expression of *dpp*, *ptc*, and *en*. The targets of dpp are expressed in graded domains over the wing tissue. **(E)** The mosaic of gene expression along the AP axis serves as a dynamic template for the patterning of structures such as the veins of the wing. Thus, late in the larval stages, discrete territories have been defined in the wing pouch which, together with the gradients of dpp and hh, define positions used as references to lay down the pattern of veins. These positions are

identified through the activation of transcription factors or of ligands for the EGF receptor pathway. Each vein arises where a specific suite of gene expression leads, through cell interactions, to the definition of narrow territories of transcription factor expression, unique for that vein. The second vein (L2) arises at the limit of the expression of the transcription factor spalt major (salm), whose expression is induced at a certain threshold of dpp. The development of the L2 vein requires two zinc-finger transcription factors, knirps (kni) and knirps-related (knr), whose expression is activated at a particular low concentration of salm. The third and fourth veins (L3 and L4) require the activity of the EGF receptor triggered by a ligand encoded by the *vein* (*vn*) gene, whose expression is activated by hedgehog late in development. The restriction of the activity of the protein encoded by *vein* to its borders of expression is achieved by the suppression of EGF receptor activity through the activity of a transcription factor, collier (col), which is expressed over the domain of *vein* expression, also under the control of hedgehog. The distinctive feature of L3 and L4 is the expression in L3 of a homeobox containing gene, *iro-C*, whose borders of expression are determined through repression by engrailed and activation by a particular level of dpp. (From de Celis, J. F. and Barrio, R. (2000) Function of the *spalt/spalt-related* gene complex in positioning the veins in the *Drosophila* wing. *Mech. Dev.* **91**, 31–41.)

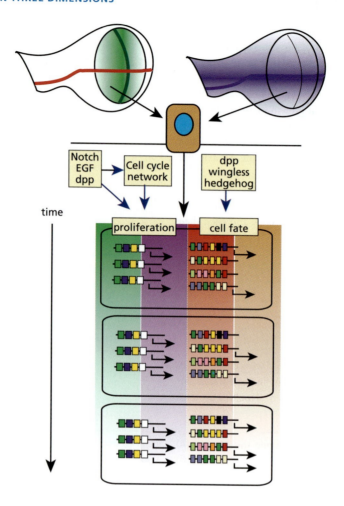

Fig. 12.25. Integration of AP and DV information by cells within the growing wing pouch. Any cell at a given point in the developing wing pouch integrates information from the two main reference signalling systems (AP and DV) with its intrinsic programs of gene expression to acquire position-specific fates. Over time, a cell will be changing and adjusting its repertoire of gene expression to the evolving patterns of signalling molecules and cell interactions. Although much is known about the control of position-specific gene expression through the integration of positional systems at enhancers like those of *vestigial*, much less is known about the control of proliferation. The diagram illustrates, diagrammatically, a cell adjusting its pattern of transcriptional activity as development proceeds. Potential inputs to the processes of proliferation and pattern formation are shown.

play essential roles in the initial stages of development of both structures. The development and patterning of the *Drosophila* wing with regard to the AP and DV boundaries can be compared with similar processes in the vertebrate limb which use the ZPA and the AER as references: the ZPA plays a role analogous to that of the AP boundary and the AER corresponds to the DV boundary. The similarities are reinforced by the observation that hedgehog and TGFβ/BMP family members operate at the ZPA and the AP boundary, while members of the *fringe* gene family are required at the DV border and the AER to drive the initial outgrowth of the limbs and the wings.

However, there are also important differences between the two systems. One is exemplified by the deployment of *engrailed* gene expression: posterior, regulating the expres-

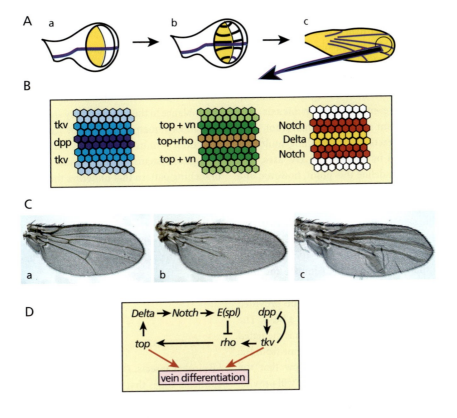

Fig. 12.26. The patterning of the veins during the pupal stage. (A) Pattern of *dpp* expression (red) in the wing disc, during the transition from the larva (a, b) to the pupa (c), and its association with the pattern of the developing veins (black in b). Yellow indicates the dorsal surface of the wing, which becomes folded onto the ventral surface in the pupa (see Fig. 12.18). **(B)** Patterns of expression of elements of three signalling pathways during the final stages of vein development. The vein territories outlined in the larva (see Fig. 12.23) become refined in the pupal stages by subdivision into a vein territory flanked by an intervein territory on either side. The expression of *dpp* (purple) becomes restricted to the vein region whereas that of its receptor *thickvein* (*tkv*) (blue) becomes associated with the intervein region. Elements of EGF receptor and Notch signalling pathways also show a differential deployment in vein and intervein regions: the EGF receptor encoded by *torpedo* (*top*) is expressed in veins and interveins, but its ligand encoded by *vein* (*vn*, green) is restricted to the intervein regions whereas the activator *rhomboid* (*rho*, brown) is expressed in the vein region. In the case of Notch signalling, Delta (yellow) is expressed at high levels by the veins and Notch (red) is activated in the intervein regions. **(C)** Effects of mutations in components of signalling pathways on the pattern of veins: (a) wild-type wing; (b) wing mutant for the EGF receptor ligand, vein, which causes a loss of veins; (c) wing mutant for Notch signalling, which causes a thickening of the veins due to a failure in the process of refinement. **(D)** Regulatory interactions between elements of the different signalling pathways during the pupal stages of vein development. Arrows indicate positive regulatory interactions, blunt ended lines negative ones. The regulatory relationships are inferred from genetic analysis of the process. (From de Celis, J. F. (1998) Positioning and differentiation of veins in the *Drosophila* wing. *Int. J. Dev. Biol.* **42**, 335–43.) (Images courtesy of J. F. de Celis.)

sion of *hh* in *Drosophila*, but ventrally, with an important but non-essential role in the vertebrate limb. Also, a Wnt family member, *Wnt7a*, is expressed dorsally in the vertebrate limb where it is important in coordinating the regulatory interactions between the different axes but this has no clear counterpart in the *Drosophila* wing, where the expression of *wingless* has no fixed spatial distribution but rather shows dynamic patterns of expression associated with the activity of several transcription factors.

These contrasting observations indicate that, attractive

as the analogies are, they might be more the result of chance than of design. Rather than revealing the existence of a conserved molecular 'cassette' involved in limb development, the parallels might reflect a convergence of molecular strategies with a limited number of molecular elements to solve one problem: the emergence of a proximodistal axis. The crucial event in setting up this axis is the definition of a distal point within a primordium. This point is defined by interactions of signalling molecules within the plane defined by the AP and DV axes. The nature of the signals that define this point, however, and even of those that will mediate its activity, will change from organism to organism.

In general, the signals are less important than the readiness of the cells to interpret the signals. FGF will only trigger limb development within the limb field (that is, in cells that are competent to initiate the limb program). Perhaps for this reason the actual source of FGF does not matter and experimentally it can be provided from another part of the embryo or even by a bead. The same is true for the activity of the ZPA, which is mediated by Shh. However, the Shh does not have to come from the ZPA itself: transplantation experiments show that the notochord or even the primordia of the teeth will provide the same activity. In *Drosophila*, wing development can be induced in the legs by expression of the vestigial protein, probably because vestigial allows cells to interpret signals common to other appendages in a wing specific manner.

Branched structures: The insect tracheal system and vertebrate lungs

A common feature of the circulatory, respiratory and nervous systems of most animals is that they are all branched cellular networks. All these networks form in a basically similar way, from an initial chain of cells with specialized branching points that serve as origins for new chains. Each new chain also has the ability to branch. By this means, densely branched structures can emerge that have a very large surface area. The function of these tree-like structures is to distribute their contents (blood in the vascular system, air in the respiratory system and information in the nervous system) over the widest possible field with a minimum use of cell numbers or space.

How these patterns are established is an interesting biological problem that has attracted the attention of many theoreticians. They have created models in which branching is the result of a basic algorithm iterated a number of times in space and time. However, the branching patterns that result from these models bear no more than a passing resemblance to the real ones: they are, for example, too regular and symmetric, and cannot account for the apparent structural whims imposed by functional demands on biological systems. In this section we deal with how the branching pattern that underlies the respiratory system of insects and land vertebrates is generated. Despite their phylogenetic separation, these two systems develop in quite similar ways and, not surprisingly by now, make use of exactly the same molecular processes and elements that we have discussed earlier in this chapter and in Chapters 10 and 11.

The tracheal system of *Drosophila*

Like other insects, *Drosophila* has an internal network of epithelial tubes that spans the whole animal and distributes gases throughout the body: the tracheal system. The cellular elements that will configure this system are laid down in the embryo, by a small number of cells that develop a stereotyped pattern of branches. This system is elaborated during larval life and modified in the adult as the animal grows. The essence of this network is a basic pattern of branching that is modified in a segment- and stage-specific manner. Here, we are concerned with the basic pattern that emerges during embryogenesis (Fig. 12.27).

The cells that will generate the tracheal system are derived from the sheet of ectodermal cells that, after gastrulation, will give also give rise to the nervous system and the epidermis. Like most structures in *Drosophila*, the tracheal system has a segmental structure, that is, every segment contributes a unit to it, and all of these units are brought together into a single network by coordinated fusion of the units from adjacent segments. The basic elements appear two hours after gastrulation as groups of about 80 ectodermal cells located half way along the DV axis and repeated on each side of every segment. These cells undergo concerted changes of shape with respect to their ectodermal neighbours (Fig. 12.27). As a result of these shape changes the cells invaginate and become arranged in epithelial sacs

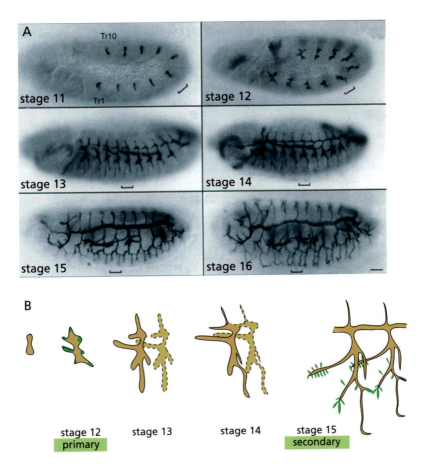

Fig. 12.27. The development of the tracheal system of *Drosophila.* **(A)** Invagination and progressive branching of the tracheal cells during embryogenesis, revealed by antibodies that highlight the lumen of the tubes. All embryos are shown on their side, with anterior to the left and dorsal at the top. The different stages span a period of about 10 hours. The square bracket highlights one of the tracheae (Tr5). **(B)** Development of Tr5 showing the patterns of primary and secondary branching and the way Tr5 joins to the posterior major branch. (From Samovlakis, C. *et al.* (1996) Development of the *Drosophila* tracheal system occurs by a series of morphologically distinct but genetically coupled branching events. *Development* **122**, 1395–1407.)

that remain associated with the outside through a small orifice: the tracheal pit. These cells will not divide again. At specific positions on the sacs the cells change shape and, without further proliferation, undergo a sequence of morphogenetic changes that generate rounds of primary and secondary branching (Fig. 12.27). The primary branching rounds are very stereotyped while the secondary branching is more variable, giving rise to a very large number of branches that eventually cover most of the surface of the animal. One of the branches from each segmental unit connects with the corresponding branch in each adjacent segment to provide a continuum along the length of the animal. The other branches navigate through the inside of the embryo following cues in the mesoderm and the nervous system. As a result, the inside of the embryo becomes covered by a network of tubes made up of specialized cells. Some of the tubes are formed by cells coming together, while others are formed by single cells rolling up upon themselves.

Cell-type specification in the tracheal system

The groups of cells in each segment that give rise to the tracheal system are called the tracheal placodes and are defined by the expression of two transcription factors—a

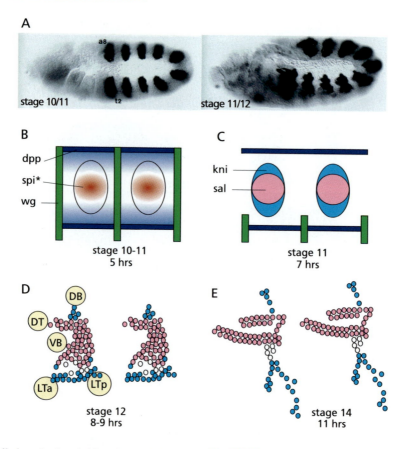

Fig. 12.28. Definition of cell populations in the tracheal pit.
(A) Expression of a reporter gene under the control of the promoter of the *tracheless* gene, which defines the cells that will become tracheal placodes at stage 10. After invagination, the cells from the placode begin to branch during stages 11 and 12. **(B)** At stage 10, the EGF receptor is activated within the tracheal placode in a graded manner (red) by a gradient of active ligand spitz (spi*), created by expression of the spitz regulator rhomboid in the centre of the placode. The placode is framed at this stage by longitudinal stripes of dpp expression (blue) dorsal and ventral to it and by stripes of wingless expression (green) anterior and posterior to it. These signalling molecules provide cells in the placode with a molecular landscape which, in the context of the activity of the tracheless and tango transcription factors, results in the regionalization of the placode. **(C)** The EGF receptor activates the expression of the transcription factor encoded by the *spalt* gene (sal), and dpp activity leads to the expression of *knirps* (kni). Knirps represses the expression of *spalt*, resulting in a refinemenet of the regionalization during stage 11, as shown. **(D)** After invagination, the first round of branching takes place, giving rise to five branches: the dorsal branch (DB), the dorsal trunk (DT), the visceral branch (VB), and the lateral trunk anterior (LTa), and posterior (LTp). Spalt (pink) and knirps (blue) are important in endowing cells with an identity that contributes to their behaviour. **(E)** During stage 14, the branches begin to outline the tracheal tree and to form the tubes that will make the tracheae.

bHLH protein called tracheless (trh) and a POU domain protein called drifter/ventral veins lacking (dfr/vvl)— about two hours after gastrulation. Tracheless forms heterodimers with another bHLH transcription factor encoded by the *tango* gene and promotes the cell shape changes that accompany the invagination of the primordium. The coordinates for the expression of trh and dfr/vvl are determined by the positional information laid down by the pair rule genes and the genes that they regulate; as we have seen in Chapter 11 (Fig. 11.16), these genes establish a dynamic mosaic of combinations of transcription factors and signalling molecules over the ectodermal surface. At the time of their appearance, the placodes are framed by stripes of wingless expression on either side along the AP axis and of dpp on either side along the DV axis (Fig. 12.28). These signalling molecules help define the AP and DV coordinates at which the placodes appear.

The combination of trh/tango and dfr/vvl activates two classes of genes: genes encoding other transcription factors that will develop cell diversity within the trachea and, acting directly or through a further tier of transcription factors, genes encoding molecules involved in triggering and bringing about the organized cell shape changes that accompany the invagination of the placode. Before invagination, different cells in the placode are characterized by the expression of genes encoding components of various signalling systems: the dpp receptor thickveins (tkv), the FGF receptor breathless (btl), and rhomboid (rho), which is an essential element for the activation of spitz, the ligand for the EGF receptor (see Chapter 9, Fig. 9.15). It is likely that the different cell fates within the placode emerge through an interaction between these transcription factors and the heterodimers between trh/tango and dfr/vvl.

By the time the placode invaginates it contains a number of different cells destined to define branches that will navigate in different directions inside the embryo. Each of these branches can be assigned an identity in terms of a suite of region-specific transcription factors. Three signal sources provide information to the cells of the placode: the dorsal and ventral stripes of dpp; a central source (promoted by rhomboid) of active EGF receptor ligand spitz; and the lateral stripes of wingless (Fig. 12.28). The cells make use of a mixture of these three signals to specify different regions within the placode: the zinc-finger-containing transcription factor knirps is expressed in the dorsal and ventral domains, and spalt in the middle domain. These transcription factors begin to provide identity to the groups of cells that will produce the branches. Thus, as in so many other examples we have discussed, the 80 cells of the placode are patterned by the major signalling pathways acting on combinations of transcription factors. The consequences of this mosaic of gene expression become visible during the process of invagination.

Branching morphogenesis in the tracheal system

Soon after invagination the placodes develop a pattern of six branches that is repeated in every segment (Fig. 12.28). The positions within the placode from which these branches arise are determined by the expression, in discrete clusters of cells surrounding the placode, of the gene branchless (bnl) which encodes an FGF molecule (Fig. 12.29). Since all the cells in the placode express the FGF receptor encoded by breathless (btl), the spatial specificity in this case is provided by the distribution of the ligand. Embryos that are mutant for either btl or bnl show no tracheal development beyond the invaginated placode stage, indicating that FGF signalling is necessary for the branching process.

As the cells in the placode invaginate they begin to elongate and move towards cells in the surrounding clusters which express FGF/branchless. This generates a first set of branches that migrate towards the sources of FGF, which acts as a chemoattractant. The leading cells of every branch are followed by further cells, which begin to organize themselves into the rudiments of a hollow tube (Fig. 12.29). Activation of the FGF receptor breathless in the leading cells of the branches activates the Ras signalling pathway in these cells; this has been inferred from the observation that mutants in elements of the Ras pathway also show impaired tracheal development. There are some Ras pathway components that function specifically in signalling by FGF and not EGF. One such transducer, the stumps/dof protein, is expressed by all cells in the placode and is necessary for tracheal development.

Thus, the first pattern of branches is determined by a combination of the spatial subdivision of the placode into different cell types, which probably distinguishes the cells that will lead the migration from those that will follow, and the pattern of FGF signalling, which creates a chemoattractive field and specifies the pattern of branching (Fig. 12.29).

Following the first round of branching, a second round begins (Fig. 12.29). Although triggered by FGF signalling, like the first round, it has a very different cellular basis. During primary branching, the cells at the tip of every primary branch are exposed to the highest concentrations of FGF. As a result one or two cells at these tips begin to express sprouty, which is an inhibitor of Ras signalling, and the transcription factor pointed.

Pointed is a *Drosophila* homologue of the Elk transcription factor, which is a target and effector of Ras signalling (see Chapter 5, Fig. 5.36), and is necessary for the development of a second branch in the cells at the leading edge of the primary branch. Sprouty, on the other hand, binds to some of the transducers of FGF signalling and creates a higher threshold for the FGF response in the cells just behind the tip. This ensures that only tracheal cells with high levels of FGF signalling—those near the tips of the primary branches—can respond to FGF and use the activity of pointed to initiate secondary branching (Fig. 12.29).

After the initial branching events, further branches are formed by individual cells rolling themselves into a tubular shape. The pattern of these branches is determined by new mesodermal and ectodermal patterns of breathless expression. These projections do not form in a rigidly defined pattern. Once again, terminal branching is under the control of the FGF signalling that is operative during the

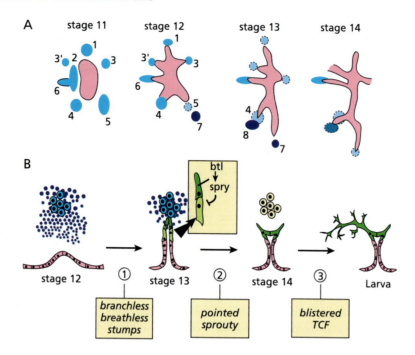

Fig. 12.29. Mechanisms for the generation of branches in the trachea. (A) All cells in the tracheal pit express the FGF receptor breathless (pink) and when they invaginate they spread over the basal surface of the epidermal and mesodermal cells following a pattern outlined by clusters of expression of branchless/FGF (light blue). The primary pattern of branches arises through a combination of the intrinsic patterns of gene expression of the cells of the placode, and the pattern of FGF. As development proceeds the pattern of branchless/FGF changes, as does the pattern of branching; some domains disappear (encircled by dashed lines) while new ones appear (dark blue). (After Sutherland, D. *et al.* (1996) branchless encodes a *Drosophila* FGF homologue that controls tracheal cell migration and the pattern of branching. *Cell* **87**, 1091–101.) **(B)** The process of branching follows a simple sequence of events that is reiterated at every branch point. Tracheal cells expressing breathless (btl) move towards the source of FGF which acts as a chemoatractant. This migration is led by some cells at the tip of the branch. When the cells reach the source they express the sprouty (spry) protein which modulates FGF signalling and inhibits branchless signalling in cells that are not at the tip. High levels of branchless induce the expression of pointed in the cells at the leading edge, which promotes secondary branching. Further branching, as occurs during the larval period, requires transcription factors of the TCF (ternary complex factor) family (see Chapter 5, Fig. 5.36).

previous round of branching, but its effects are mediated by yet another molecular mechanism, this time involving a MADS-family transcription factor encoded by the *blistered* gene (the *Drosophila* homologue of the gene encoding serum response sactor, SRF). TCF (ternary complex factor) is also required as part of the transcriptional complex through which blistered/SRF acts (see Chapter 5, Fig. 5.36). The *TCF* gene is turned on in the secondary branch cells, just before the final branching spree, under the control of FGF (Fig. 12.29).

The pattern of primary and secondary branching can be readily correlated with the pattern of expression of branchless. In *bnl* mutants there is no branching, while ectopic expression of *bnl* leads to extra branching. The transcriptional control of the expression pattern of the *bnl* gene is therefore a primary determinant of the shape and pattern of the tracheal tree. The pattern of *bnl* is dynamic and probably responds to a changing mosaic of gene activity resulting from interactions between cells in the ectoderm (see Chapter 11) and exchanges between the ectoderm and the mesoderm. This suggests that there is a complex control element upstream of the *bnl* promoter.

Branching depends on controlled interactions between signalling molecules and the cytoskeleton and therefore it is interesting that many of the molecules associated with a specific stage of branching are transcription factors. Finding the connection between these trancscription factors

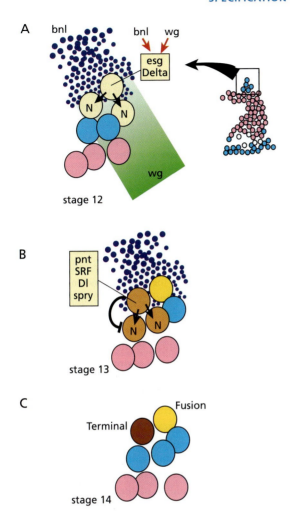

Fig. 12.30. Specification of cell types in the developing branches. (A) Within many branches there are, in addition to the bulk of the cells that make up the tube, two kinds of specialized cells: fusion cells, which promote fusion with branches from adjacent segments, and terminal cells, which send long bifurcated branches towards the target tissues. Individual cells within a given branch are selected to adopt either fate. The fusion cell is specified first and is selected from a small number of cells at the tip of the branch. In the absence of Notch signalling, more than one cell adopts this fate, indicating that it might arise through a mechanism similar to lateral inhibition. Cells at the tip express high levels of breathless (pale yellow) and in a way that is still not understood, one of these cells expresses the Notch ligand Delta and the zinc-finger transcription factor escargot (esg), and becomes the fusion cell. Wingless signalling is required for the stable adoption of this fate. As this cell becomes the fusion cell, Delta signals to Notch (N) in the surrounding cells, preventing them from adopting the same fate. **(B)** After the segregation of the fusion cell (bright yellow), the terminal cell is selected from among those that did not become fusion cells. In the absence of Notch signalling too many cells adopt the terminal fate (brown), suggesting that it is selected from among a small number of cells that have the potential to adopt this fate, through a second round of lateral inhibition. As in the case of the fusion cell, events that are not understood lead to one of the cells expressing Delta, which will signal to Notch in the surrounding cells and inhibit them from developing as terminal cells. The cell selected to be the terminal cell also expresses the transcription factors serum response factor (SRF) and pointed (pnt), which are involved in promoting the terminal fate. This cell also expresses sprouty (spry) which, in addition to promoting branching, reinforces the effects of lateral inhibition by making the surrounding cells less sensitive to the high levels of branchless. **(C)** After these two rounds of signalling, every branch contains a fusion cell (yellow) and a terminal cell (brown) in addition to the cells that will make up the tube.

and the cellular activities that drive the morphogenetic events associated with branching is an interesting challenge for the future.

Specification of cell types within the migrating branch

Within every branch there are three types of cells, each with a specific function in the construction of the tracheal tree: terminal cells which, as we have already seen, express SRF and give rise to long processes extending towards target tissues; fusion cells, characterized by the expression of a zinc-finger protein escargot, which form connections with branches from adjacent segments; and the cells that make up the rest of the tracheal tube (Fig. 12.30). The pattern of

different cells is built to a single-cell level of precision: for example, for the fusions that will generate a functional tree to occur correctly, only one cell should fuse with one other cell, and therefore only one cell should be specified as a fusion cell in every branch.

Observations and experiments with different mutants indicate that the specification of the terminal and fusion cell types is very similar to the specification of neural and muscle precursors (see Chapter 10). Combinations of basic signalling pathways acting on the constellation of transcription factors of particular cell populations generate 'equivalence groups', within which Notch signalling singles out particular cell fates (see also Chapters 8 and 10). In every branch, the first cell to be specified is the fusion cell, followed by the terminal cell (Fig. 12.30).

Notch signalling is an important element in the specification of each of these cells from a small cluster of equivalent cells. However, in this case, although all cells share the same potential, there is a spatial bias in Notch signalling and it is always one and the same cell from the equivalence group that develops as either the terminal or the fusion cell. The biasing factor may be the effects of signalling by branchless.

The development of the vertebrate lung

Just as insects have a tubular network to distribute and exchange gases throughout their bodies, so too have terrestrial vertebrates developed networks for this purpose. One of these networks is the circulatory system and the other the respiratory system. In land vertebrates, the key organ associated with the respiratory network is the lungs. The development of the lungs involves epithelial–mesenchymal interactions similar to those we have seen associated with the development of the teeth (Chapter 11) and the limb (this chapter), but with the added element of branching morphogenesis.

The lungs, which are connected to the exterior via the trachea or windpipe, consist of a tree-like system of bronchial branches whose narrowest elements, the bronchioli, terminate in finely lobed structures called alveoli. The alveoli are the sites of exchange of gases with the circulatory system, which flows through the mesenchymal tissue surrounding the bronchial network (Fig. 12.31).

The trachea and the lungs originate from the anterior foregut, which is at this stage a tubular epithelium of endoderm. An outpocketing from the foregut forms the tracheal rudiment, which separates from the foregut (which forms the oesophagus) and buds to form two bronchi. These buds result from interactions between the epithelial cells of the tracheal rudiment and the surrounding mesenchyme, in a manner reminiscent of the appearance of the limb bud. This apparent similarity goes further in that, just as in the limb, *FGF10* expression in the mesenchyme surrounding the tracheal rudiment is required for bronchial bud formation; *FGF10* mutant mice develop no lung buds even though the trachea develops normally. Again as in the limb, the location of lung initiation on the anterior–posterior axis is likely to be related to *Hox* gene expression and function. There is evidence that *FGF10* expression may be under the control of Shh signalling: some members of the Gli family of Shh effectors are expressed in the mesenchyme, and mice doubly mutant for two of these genes (*Gli2* and *Gli3*) fail completely to develop either lungs or trachea (Fig. 12.31).

The early development of the tubular epithelium that will generate the lungs is dependent on a number of transcription factors which probably lie downstream of Shh and FGF10. These include homeodomain transcription factors such as Nkx2.1, GATA family members, and HNFβ. In *Nkx2.1* mutants the trachea does not separate from the oesophagus, and the lungs, although they bud, do not develop very much. These genes create a regulatory network that stabilizes patterns of expression and maintains the epithelial–mesenchymal interactions that initiate branching morphogenesis.

The primary bronchi that arise by budding from the tracheal rudiment then undergo a specific pattern of secondary branching: in the mouse, branching in the right lobe produces three branches, whereas only one is produced on the left. As a result the left lung is smaller. The spatial coordinates of these processes are rather precise, leading to a defined pattern for the latter events and enabling efficient packaging of asymmetric organs like the heart. The secondary branches within the lobes grow and undergo several further rounds of branching to generate finer bronchioli, and eventually the alveoli which become interconnected with the capillaries. There are between six and eight generations of branches which, in humans, produce on the order of 17 million branches.

The similarities with the tracheal system of *Drosophila* are obvious. The budding of the primary bronchi and formation of the network of secondary branches within the lobes of the lung are very similar to the primary and secondary branching of the tracheal pit, and the finer branching of the bronchioli is comparable to the terminal branching of the tracheal system. As in *Drosophila*, the patterns of budding and secondary branching are stereotyped and reproducible. The precise mechanism that regulates these processes is not known, however there is evidence for an involvement of FGF signalling and of regulators of FGF signalling such as sprouty. FGF receptors are expressed in the respiratory epithelium and there are patches of *FGF10* expression throughout the mesenchyme near the distal epithelial tubules, a situation that is reminiscent of *breathless* and *branchless* expression in *Drosophila*. In tissue culture assays, beads soaked in FGF10 exert a chemoattractive effect on lung epithelium, and lungs fail to develop in mice mutant for *FGF10*. Therefore, the similarities at the cellular level extend to the molecular level.

There are, however, some differences between the pat-

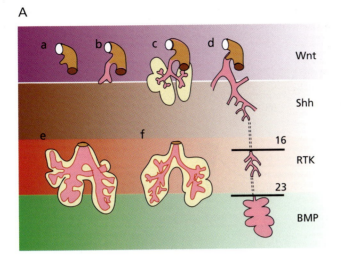

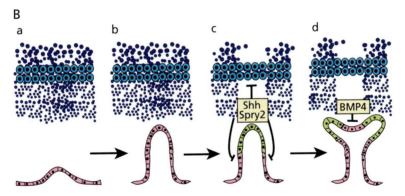

Fig. 12.31. Development and patterning of the vertebrate lung. (A) Stages in the development of the lung in mouse and human. All occur in the context of signalling by the the major signalling pathways, indicated by graded colours to reflect quantitative requirements. The development of the lungs is initiated in the anterior foregut, where an outpocketing of the endodermal epithelium forms the tracheal rudiment (a) and buds to form the two primary branches, or bronchi (b), which grow, surrounded by mesenchyme, and produce secondary branches that form the basic structural network underlying the lobes of the lung (c). Several further rounds of branching produce finer branches called bronchioli that terminate in sacs called alveoli, which function in the exchange of gases with the surrounding tissue (c, d). The first 16 rounds of branching are stereotyped, while the last seven (from 17 to 23) are not. In mice, the pattern of secondary branching of the bronchi results in lungs with four lobes (e), in contrast to the three lobes characteristic of humans (f). (Adapted from Warburton, D. *et al.* (2000) The molecular basis of lung morphogenesis. *Mech. Dev.* **92**, 55–81.) **(B)** Stages of branching morphogenesis during lung development. The process is very similar to tracheal branching in *Drosophila* and is also promoted by FGF. The major differences are that in this case, branching is mediated by interactions between an epithelium (the lungs) and a mesenchymal tissue (blue), which is the source of FGF, and that the process occurs in a highly proliferative tissue. There is also a feedback control by the cells at the leading edge of the lung epithelium on the expression of FGF in the mesenchyme. (a,b) FGF10 from the mesenchyme attracts cells from the lung epithelium and promotes the expression of *Sprouty2* (*Spry2*) and *Sonic hedgehog* (*Shh*) at the tip of the epithelium. Spry2 inhibits FGF signalling in the lateral cells of the branch and Shh suppresses the expression of *FGF10* in the mesenchyme. (c) This generates a new pattern of FGF10 expression and responsiveness which leads to branching. BMP4, whose expression is elicited at the tip by FGF10, also contributes to the branching pattern by suppressing proliferation of the epithelium (d). Reiterated rounds of these signalling events lead to the profuse branching pattern observed in the lungs.

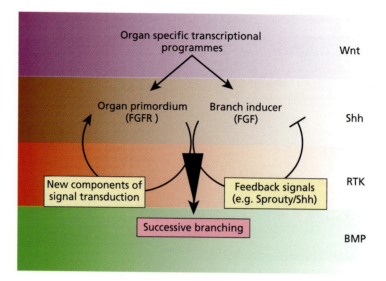

terning of the lungs and that of the tracheal system. One concerns the involvement of Shh in the branching process. Although, as in *Drosophila*, sprouty plays a role in the inhibition of FGF signalling in the epithelium, in lung development Shh acts on the mesenchyme to inhibit the expression of *FGF*, thereby contributing to the pattern of branching. Another important difference is that in the lung, branching morphogenesis takes place within a phase of growth, so branching must be coordinated with cell proliferation. The proliferation is driven by cytokines such as bombesin secreted by specialized cells which differentiate along the epithelium, and branching is associated with proliferative arrest mediated by BMP4 (Fig. 12.31).

Are there general principles for three-dimensional patterning?

The similarity of the mechanisms used in the patterning of the tracheal tree and the lungs suggests that they might represent a general mechanism of branching morphogenesis (Fig. 12.32). Such mechanisms might operate, for example, in the mammary gland, the vascular system or even the complex dendritic trees that determine neuronal function and the activity of the nervous system. It will be interesting to see whether successive cycles of spatially patterned FGF signalling also play a role in these systems.

An interesting and unresolved issue is how FGF signalling is connected to the cytoskeletal activity that is es-

Fig. 12.32. General principles of branching morphogenesis. Transcriptional programs associated with the development of branching structures define initial patterns of transcription of FGF receptor (FGFR) in the organ primordium and of a branch inducer, FGF, in the neighbourhood of the organ that will branch. At the same time, cells within this organ express different signalling molecules and transcription factors which make them competent to respond to the signal and which create specialized cells to pattern the particular structure. FGF is usually located near the organ primordium and acts as a chemoattractant for the cells that express the FGF receptor. The pattern of FGF defines a pattern of branching and induces two kinds of feedback: one to the organ itself, where new transcription factors and patterns of signalling molecules are created that will assist in further rounds of branching, and a second based on signals that prevent the response (like spry) or that suppress the expression of FGF and thus create new landscapes for branching. Reiterated cycles of these interactions generate successive bifurcations of the initial pattern and therefore branches. (Adapted from Metzger, R. and Krasnow, M. (1999) Genetic control of branching morphogenesis. *Science* **284**, 1635–39.)

sential for the execution of the branching pattern. Genetic analysis of tracheal development has identified a number of transcription factors and feedback modulators of Ras signalling as important elements of the response to FGF signalling. In Chapter 6 (Fig. 6.29), we saw how a transcription factor, snail, can regulate epithelial–mesenchymal transitions by regulating the expression of the cell adhesion molecule E-cadherin. It may be that related transcriptional control networks regulate the modulators of cytoskeletal activity that mediate the cell shape changes accompanying branching morphogenesis.

A comparison of the initial phases of branching morphogenesis and limb development suggests some features that might be common to both processes, in particular, the definition of a primordium for the structure in the plane of the epithelium and then the specification, within this primordium, of a region that will, by growth or morphogenesis, move out from out of the plane in which it is defined.

There is, however, an important difference. In the limb, the outgrowth is patterned from an internal set of axes that create a system of molecular coordinates to which the growing cells respond. In the trachea or the lungs, although there is an internal system of molecular coordinates that creates differences in the primordium, the patterning is driven by an external signalling system: FGF acting as a chemoattractant.

SUMMARY

1. The limbs of vertebrates and the legs and wings of insects result from the ordered outgrowth of cells from defined epithelial sheets. This process takes place with reference to a proximodistal (PD) axis that is perpendicular to the plane of the epithelium defined by its anteroposterior (AP) and dorsoventral (DV) axes.

2. The primordia for limbs and appendages arise within the plane of the body wall of the embryo at positions where the activity of several signalling pathways intersects, creating a grid of cellular differences. These signalling events are integrated over a domain of competence for limb or appendage development that is established by intrinsic gene expression programs of the cells.

3. Early in development, specific cell populations within the primordia are defined as signalling sources. In both insects and vertebrates, one of these signalling sources, usually located in the centre of the primordium, will become the most distal point of the limb or appendage, and will play a central role in organizing the outgrowth along the PD axis.

4. In vertebrates, the development and patterning of the limb is regulated by the concerted activities of three signalling centres: the apical ectodermal ridge (AER) and the dorsal cells in the ectoderm, and the zone of polarizing activity (ZPA) in the mesoderm. These centres promote the proliferation and patterning of a mesodermal population of competent cells: the progress zone (PZ).

5. The activity of the AER is mediated by FGF, that of the dorsal ectoderm by Wnt7a and that of the ZPA by Shh. Together, these three signalling molecules are sufficient to elicit the development and patterning of the limb mesoderm.

6. The wing of *Drosophila* is patterned with respect to two signalling centres located at the interface between dorsal and ventral cells (the DV boundary), and anterior and posterior cells (the AP boundary).

7. Dpp and hh mediate the patterning activities of the AP axis, and wingless and Notch those of the DV axis. These signalling molecules also play a role in the development of other structures, but in the wing they act through the transcription factor vestigial to implement wing specific programs of gene expression.

8. Analysis of the fine-grained patterning of the *Drosophila* wing indicates that this is a progressive process in which sequential and mutually dependent patterns of gene expression create complexity.

9. In both vertebrates and insects it is not understood how proliferation is coupled to patterning.

10. The tracheal system of *Drosophila* and the lungs of vertebrates are examples of a higher level of proximodistal patterning: branching morphogenesis. In this process, an initial outgrowth along the PD axis bifurcates to generate branches and this process is reiterated to generate a complex three-dimensional structure.

11. Branching morphogenesis relies on spatially controlled cycles of FGF signalling. In this process, FGF acts as a chemoattractant for an epithelial cell population and its pattern of expression determines the pattern of branching.

COMPLEMENTARY READING

Carroll, S. *et al.* (2001) *From DNA to diversity: Molecular genetics and the evolution of animal design.* Blackwell Science.

Hogan, B. (1999) Morphogenesis. *Cell* **96**, 225–33.

Johnson, R. and Tabin, C. (1997) Molecular models for vertebrate limb development. *Cell* **90**, 979–90.

Kim, J. *et al.* (1996) Integration of positional signals and regulation of wing formation and identity by the *Drosophila vestigial* gene. *Nature* **382**, 133–8.

Klein, T. and Martinez Arias, A. (1999) The vestigial gene product provides a molecular context for the interpretation of signals during the development of the wing in *Drosophila*. *Development* **126**, 913–25.

Metzger, R. and Krasnow, M. (1999) Genetic control of branching morphogenesis. *Science* **284**, 1635–9.

Index

Note: **bold** page numbers denote illustrations, where these are not covered by other text references